I0041833

PLANT ABIOTIC STRESS PHYSIOLOGY

VOLUME 1

Responses and Adaptations

Plant Abiotic Stress Physiology: Volume 1: Responses and Adaptations
Hardback ISBN: 978-1-77463-017-4

Plant Abiotic Stress Physiology: Volume 2: Molecular Advancements
Hardback ISBN: 978-1-77463-018-1

Plant Abiotic Stress Physiology, 2-Volume Set:
Hardback ISBN: 978-1-77463-016-7

PLANT ABIOTIC STRESS PHYSIOLOGY

VOLUME 1

Responses and Adaptations

Edited by
Tariq Aftab, PhD
Khalid Rehman Hakeem, PhD

AAP | APPLE
ACADEMIC
PRESS

First edition published 2022

Apple Academic Press Inc.
1265 Goldenrod Circle, NE,
Palm Bay, FL 32905 USA

4164 Lakeshore Road, Burlington,
ON, L7L 1A4 Canada

CRC Press
6000 Broken Sound Parkway NW,
Suite 300, Boca Raton, FL 33487-2742 USA

2 Park Square, Milton Park,
Abingdon, Oxon, OX14 4RN UK

© 2022 by Apple Academic Press, Inc.

Apple Academic Press exclusively co-publishes with CRC Press, an imprint of Taylor & Francis Group, LLC

Reasonable efforts have been made to publish reliable data and information, but the authors, editors, and publisher cannot assume responsibility for the validity of all materials or the consequences of their use. The authors, editors, and publishers have attempted to trace the copyright holders of all material reproduced in this publication and apologize to copyright holders if permission to publish in this form has not been obtained. If any copyright material has not been acknowledged, please write and let us know so we may rectify in any future reprint.

Except as permitted under U.S. Copyright Law, no part of this book may be reprinted, reproduced, transmitted, or utilized in any form by any electronic, mechanical, or other means, now known or hereafter invented, including photocopying, microfilming, and recording, or in any information storage or retrieval system, without written permission from the publishers.

For permission to photocopy or use material electronically from this work, access www.copyright.com or contact the Copyright Clearance Center, Inc. (CCC), 222 Rosewood Drive, Danvers, MA 01923, 978-750-8400. For works that are not available on CCC please contact mpkbookspermissions@tandf.co.uk

Trademark notice: Product or corporate names may be trademarks or registered trademarks and are used only for identification and explanation without intent to infringe.

Library and Archives Canada Cataloguing in Publication

Title: Plant abiotic stress physiology / edited by Tariq Aftab, PhD, Khalid Rehman Hakeem, PhD.
Names: Aftab, Tariq, editor. | Hakeem, Khalid Rehman, editor.
Description: First edition. | Includes bibliographical references and indexes. | Contents: Volume 1. Responses and adaptations -- Volume 2: Molecular advancements.
Identifiers: Canadiana (print) 20210335858 | Canadiana (ebook) 20210335955 | ISBN 9781774630167 (set) | ISBN 9781774630174 (v. 1 ; hardcover) | ISBN 9781774639443 (v. 1 ; softcover) | ISBN 9781774630181 (v. 2 ; hardcover) | ISBN 9781774639511 (v. 2 ; softcover) | ISBN 9781003180562 (v. 1 ; ebook) | ISBN 9781003180579 (v. 2 ; ebook)
Subjects: LCSH: Plants—Effect of stress on—Molecular aspects. | LCSH: Plant physiology. | LCSH: Plants—Adaptation.
Classification: LCC QK754.P53 2022 | DDC 572.8/2928—dc23

Library of Congress Cataloging-in-Publication Data

Names: Aftab, Tariq, editor. | Hakim, Khalid Rehman, editor.
Title: Plant abiotic stress physiology. Volume 1, Responses and adaptations / Tariq Aftab, Khalid Rehman Hakim.
Description: First edition. | Palm Bay, FL, USA : Apple Academic Press, [2022] | Includes bibliographical references and index. | Contents: 1. -- Innovations in crop production: an amalgamation of abiotic stress physiology and technology -- 2. -- Redox homeostasis managers in plants under environmental stresses -- 3. -- Reactive oxygen and nitrogen species: oxidative damage and antioxidative defense mechanism in plants under abiotic stress -- 4. -- Physiological mechanisms of plants involved in phosphorus nutrition and its deficiency management -- 5. -- Responses and adaptation of photosynthesis and respiration under challenging environment -- 6. -- Forward and reverse genetic approaches for improving abiotic stress tolerance in crop plants -- 7. -- Salinity-induced changes on different physiological and biochemical features of plants -- 8. -- Next-generation climate-resilient agricultural technology in traditional farming for food and nutritional safety in the modern era of climate change -- 9. -- CRISPR/Cas-mediated genome editing technologies in plants -- 10. -- Use of ornamental plants for the phytoremediation of metal-contaminated soils -- 11. -- Role of electromagnetic radiation in abiotic stress tolerance. | Summary: "Plant Abiotic Stress Physiology, 2-volume set highlights the various innovative and emerging techniques and molecular applications that are currently being used in plant abiotic stress physiology. Volume 1: Responses and Adaptations focuses on the responses and adaptations of plants to stress factors at the cellular and molecular levels and offers a variety of advanced management strategies and technologies. With contributions from specialists in the field, the volume discusses how plants have developed diverse physiological and molecular adjustments to safeguard themselves under challenging conditions and how emerging new technologies can utilize these plant adaptations to enhance plant resistance. Topics in this volume include redox homeostasis managers in plants, oxidative damage and antioxidative defense mechanism, photosynthesis and respiration under challenging environments, salinity-induced changes, genetics approaches for improving abiotic stress tolerance in crop plants, CRISPR/CAS-mediated genome editing technologies, and more. Agriculture today faces countless challenges to meet the rising need for sustainable food supplies and guarantees of high-quality nourishment for a quickly growing population. To assure sufficient food production, it is necessary to address the difficult environmental circumstances that are causing cellular oxidative stress in plants due to abiotic factors, which play a defining role in shaping yield of crop plants. This volume, in conjunction with Plant Abiotic Stress Physiology: Volume 2: Molecular Advancements, helps to meet these challenges by providing a rich source of information on plant abiotic stress physiology and effective management techniques"-- Provided by publisher.
Identifiers: LCCN 2021049565 (print) | LCCN 2021049566 (ebook) | ISBN 9781774630174 (v. 1 ; hardcover) | ISBN 9781774639443 (v. 1 ; paperback) | ISBN 9781003180562 (v. 1 ; ebook)
Subjects: LCSH: Plants--Effect of stress on.
Classification: LCC QK711.2 .P564 2022 (print) | LCC QK711.2 (ebook) | DDC 581.1--dc23/eng/20211013
LC record available at https://lccn.loc.gov/2021049565
LC ebook record available at https://lccn.loc.gov/2021049566

ISBN: 978-1-77463-017-4 (hbk)
ISBN: 978-1-77463-944-3 (pbk)
ISBN: 978-1-00318-056-2 (ebk)

Dedication

(1908 – 1999)

Hakim Abdul Hameed was a great philanthropist, thinker, visionary, an Indian physician of the traditional medicine system of Unani, founder-chancellor of Jamia Hamdard and a former chancellor of Aligarh Muslim University, India. He was honored by the Government of India in 1965 with the award of Padma Shri, the fourth highest Indian civilian award, and in 1992, the Government awarded him the third highest Indian honor of Padma Bhushan.

About the Editors

Tariq Aftab, PhD

Tariq Aftab, PhD, is currently an Assistant Professor in the Department of Botany at Aligarh Muslim University, India, where he earned his PhD. He is the recipient of a prestigious Leibniz-DAAD fellowship from Germany, a Raman Fellowship from the Government of India, and a Young Scientist Awards from the State Government of Uttar Pradesh (India) and Government of India. After completing his doctorate, he worked as Research Fellow at the National Bureau of Plant Genetic Resources, New Delhi, and as Postdoctorate Fellow at Jamia Hamdard, New Delhi, India. Dr. Aftab was also a Visiting Scientist at the Leibniz Institute of Plant Genetics and Crop Plant Research (IPK), Gatersleben, Germany, and in the Department of Plant Biology, Michigan State University, USA. He is a member of various scientific associations from India and abroad. He has edited seven books with with international publishers, including Elsevier Inc., Springer Nature and CRC Press (Taylor & Francis Group); co-authored several book chaptersa; and published over 60 research papers in peer-reviewed international journals. His research interests include physiological, proteomic, and molecular studies on medicinal and aromatic plants.

Khalid Rehman Hakeem, PhD

Khalid Rehman Hakeem, PhD, is Professor at King Abdulaziz University, Jeddah, Saudi Arabia. After completing his doctorate (Botany; specialization in Plant Eco-physiology and Molecular Biology) from Jamia Hamdard, New Delhi, India, in 2011, he worked as a Lecturer at the University of Kashmir, Srinagar, for a short period. Later, he joined Universiti Putra Malaysia, Selangor, Malaysia, and worked there as Postdoctorate Fellow in 2012 and Fellow Researcher (Associate Prof.) from 2013 to 2016. Dr. Hakeem has more than 10 years of teaching and research experience in plant eco-physiology, biotechnology and molecular biology, medicinal plant research, plant–microbe–soil interactions as well as in environmental studies. He is the recipient of several fellowships at both national and international levels; also, he has served as the visiting scientist at Jinan University, Guangzhou, China. Currently, he is involved with a number of international research projects with different government organizations.

To date, Dr. Hakeem has authored and edited more than 60 books with international publishers, including Springer Nature, Academic Press (Elsevier), and CRC Press. He also has to his credit more than 115 research publications in peer-reviewed international journals and 60 book chapters in edited volumes with international publishers.

At present, Dr. Hakeem serves as an editorial board member and reviewer of several high-impact international scientific journals from Elsevier, Springer Nature, Taylor & Francis, Cambridge, and John Wiley. He is included in the advisory board of Cambridge Scholars Publishing, UK. He is also a fellow of the Plantae group of the American Society of Plant Biologists, member of the World Academy of Sciences, member of the International Society for Development and Sustainability, Japan, and member of the Asian Federation of Biotechnology, Korea. Dr. Hakeem has been listed in Marquis Who's Who in the World, since 2014–2019. Currently, Dr. Hakeem is engaged in studying the plant processes at eco-physiological as well as molecular levels.

Contents

Contributors

Abdelghafar M. Abu-Elsaoud
Botany Department, Faculty of Science, Suez Canal University, Ismailia, Egypt

Awatif M. Abdulmajeed
Biology Department, Faculty of Science, University of Tabuk, Umluj 46429, Saudi Arabia

Muhammad Afzal
Plant Production Department, Food Science and Agricultural College, King Saud University, POX 2455-11451, Riyadh, Saudi Arabia

Adeel Ahmad
Institute of Soil and Environmental Sciences, University of Agriculture Faisalabad, Pakistan

Shakeel Ahmad
State Key Laboratory of Rice Biology, China National Rice Research Institute, Hangzhou 310006, China

Zahoor Ahmad
Department of field crops, Faculty of Agriculture, Cukurova University Adana, Turkey

Asgar Ahmed
Maize Breeding Division, Bangladesh Wheat and Maize Research Institute, Dinajpur 5200, Bangladesh

Muqadas Aleem
Department of Plant Breeding and Genetics, University of Agriculture, Faisalabad, 38040, Punjab, Pakistan
National Centre for Soybean Improvement, Key Laboratory of Biology and Genetics and Breeding for Soybean Ministry of Agriculture, State Key Laboratory of Crop Genetics and Germplasm Enhancement, Nanjing Agricultural University, Nanjing 210095, China

Haifa Abdulaziz S. Alhaithloul
Biology Department, College of Science, Jouf University, Sakaka 2014, Kingdom of Saudi Arabia

Muhammad Ashar Ayub
Institute of Soil and Environmental Sciences, University of Agriculture Faisalabad, Pakistan

Celaleddin Barutçular
Department of Field Crops, Faculty of Agriculture, Cukurova University Adana, Turkey

Tariq Ahmad Dar
Department of Botany, Government Degree College (Women), Pulwama, Kashmir, 190006, India

Sourav Garai
Department of Agronomy, Bidhan Chandra KrishiViswavidyalaya, Nadia, West Bengal, India

Abdelhalim I. Ghazy
Plant Production Department, Food Science and Agricultural College, King Saud University, POX 2455-11451, Riyadh, Saudi Arabia

Shreya Gupta
National Institute of Plant Genome Research, New Delhi, India

Khalid Rehman Hakeem
Department of Biological Sciences, King Abdulaziz University, Jeddah, Saudi Arabia

Farhana Hassan
Signal Transduction Lab, Department of Biotechnology, School of Biological Sciences,
University of Kashmir, Srinagar, J&K, 190006, India

Akbar Hossain
Bangladesh Wheat and Maize Research Institute, Dinajpur 5200, Bangladesh

Mst. Tanjina Islam
Department of Agronomy, Hajee Mohammad Danesh Science and Technology University,
Dinajpur 5200, Bangladesh

Muhammad Aamir Iqbal
Department of Agronomy, University of Poonch Rawalakot Azad Kashmir, Pakistan

Muhammad Munir Iqbal
Centre for Plant Genetics and Breeding, The University of Western Australia, 6009, Perth, WA, Australia

Shakra Jamil
Agricultural Biotechnology Research Institute, Ayub Agricultural Research Institute,
Faisalabad, 38000, Punjab, Pakistan

Riffat John
Department of Botany, University of Kashmir, Hazratbal Srinagar, Kashmir, 19006, India

Shamsa Kanwal
Agricultural Biotechnology Research Institute, Ayub Agricultural Research Institute,
Faisalabad, 38000, Punjab, Pakistan

Firdous A. Khanday
Signal Transduction Lab, Department of Biotechnology, School of Biological Sciences,
University of Kashmir, Srinagar, J&K, 190006, India

Sagar Maitra
Department of Agronomy, Centurion University of Technology and Management,
Paralakhemundi 761211, Odisha, India

Mousumi Mondal
Research Scholar, Department of Agronomy, Bidhan Chandra KrishiViswavidyalaya, Nadia,
West Bengal, India

Naveed Ul Mushtaq
Department of Bioresources, School of Biological Sciences, University of Kashmir,
Srinagar 190006, India

Khurram Naveed
Institute of Soil and Environmental Sciences, Faculty of Agriculture, University of Agriculture
Faisalabad 38000, Pakistan

Jagamohan Nayak
Research Scholar, Department of Agronomy, Bidhan Chandra KrishiViswavidyalaya, Nadia,
West Bengal, India

Amina Nisar
Department of Plant Breeding and Genetics, University of Agriculture, Faisalabad, 38040, Punjab, Pakistan

Deepu Pandita
Government Department of School Education, Jammu, Jammu and Kashmir, India

Sami ullah Qadir
Department of Environmental Sciences, Government Degree College, Kokernag, Kashmir, India

Vaseem Raja
Department of Botany, University of Kashmir, Hazratbal Srinagar, Kashmir, 19006, India

Aadil Rasool
Department of Bioresources, School of Biological Sciences, University of Kashmir, Srinagar 190006, India

Shabir A. Rather
School of Integrative Plant Sciences, Section of Plant Biology, Cornell University, Ithaca 14853, NY, USA

Muhammad Zia ur Rehman
Institute of Soil and Environmental Sciences, Faculty of Agriculture, University of Agriculture Faisalabad 38000, Pakistan

Reiaz ul Rehman
Department of Bioresources, School of Biological Sciences, University of Kashmir, Srinagar 190006, India

Umair Riaz
Institute of Soil and Environmental Sciences, Faculty of Agriculture, University of Agriculture Faisalabad 38000, Pakistan

Ayman El Sabagh
Department of Field Crops, Faculty of Agriculture, Siirt University, Turkey
Department of Agronomy, Faculty of Agriculture, Kafrelsheikh University, Kafr El-Sheikh, Egypt

Seerat Saleem
Department of Bioresources, School of Biological Sciences, University of Kashmir, Srinagar 190006, India

Talha Saleem
Institute of Soil and Environmental Sciences, Faculty of Agriculture, University of Agriculture Faisalabad 38000, Pakistan

Faran Salik
Department of Botany, University of Delhi, Delhi 110007, India

Wasifa Hafiz Shah
Department of Bioresources, School of Biological Sciences, University of Kashmir, Srinagar-190006, India

Rahil Shahzad
Agricultural Biotechnology Research Institute, Ayub Agricultural Research Institute, Faisalabad, 38000, Punjab, Pakistan

Karuna Sharma
Department of Botany, University of Delhi, Delhi 110007, India

Amrina Shafi
Education Department, Government of J&K, Srinagar, Kashmir, 190002, India

Aamar Shehzad
Maize Research Station, Ayub Agricultural Research Institute, Faisalabad 38000, Pakistan

Mona H. Soliman
Botany and Microbiology Department, Faculty of Science, Cairo University, Giza 12613, Egypt

Javaria Tabassum
State Key Laboratory of Rice Biology, China National Rice Research Institute, Hangzhou 310006, China

Rana Muhammad Sabir Tariq
Department of Agriculture & Agribusiness Management, University of Karachi, Karachi, Pakistan

Muhammad Umair
Institute of Soil and Environmental Sciences, Faculty of Agriculture, University of Agriculture
Faisalabad 38000, Pakistan

Umer Majeed Wani
Department of Biotechnology, University of Kashmir, Hazratbal, Srinagar, Kashmir, 19006, India

Ejaz Ahmad Waraich
Department of Agronomy, University of Agriculture, Faisalabad, Pakistan

Aisha Abdul Waris
Institute of Soil and Environmental Sciences, Faculty of Agriculture, University of Agriculture
Faisalabad 38000, Pakistan

Aqib Zeb
State Key Laboratory of Rice Biology, China National Rice Research Institute, Hangzhou 310006, China

Abbreviations

$\cdot OH$	hydroxyl radicals
$^1O_2$	singlet oxygen
2-DGE	two-dimensional gel electrophoresis
ABA	abscisic acid
ABF	ABA-binding factor
ABRE	ABA-responsive element
AOX	alternative oxidases
APX	ascorbate peroxidase
AQp	aquaporin
AsA	ascorbate
AsA	ascorbate
ASC	ascorbate anion
ASI	anthesis silking interval
BCCA	branched-chain amino acids
BEs	base editors
BR	brassinosteroid
BZR-1	Brassinazole Resistant 1
CAMs	calmodulins
cAPX	cytosolic APX
Cas9	CRISPR-associated protein 9
CAT	catalase
CBF	cold binding factor
CBLs	calcineurin B-like proteins
CDPK	calcium-dependent protein kinases
chlAPX	chloroplastic APX
CRISPR	clustered regularly interspaced short palindromic
CRT	C-repeat related
CSA	climate-smart agriculture
DEGs	differentially expressed genes
DHA	dehydroascorbate
DHAR	dehydroascorbate reductase
DREB	dehydration responsive element binding
DREB1	dehydration responsive element binding 1
DSBs	double-stranded breaks

EC	electrical conductivity
EDTA	ethylene diamine tetra-acetic acid
EGTA	ethylene glycol tetra-acetic acid
EMF	electromagnetic frequencies
EMR	electromagnetic radiation
ERF	ethylene response factor
ETC	electron transport chain
ETS	electron transport system
FTL	flowering locus T-like
GB	glycine betaine
GE	genome editing
GEBV	genomic estimated breeding value
GHGs	greenhouse gas
GM	genetically modified
GMOs	genetically modified organisms
GOI	gene of interest
GPX	glutathione peroxidase
GR	glutathione reductase
gRNA	guide RNA
GS	genomic selection
GS	glutathione synthetase
GSH	glutathione
GSNO	S-nitrosoglutathione
GSSG	glutathione dimers
GST	glutathione S-transferase
GWAS	genome-wide association studies
HDR	homology-directed repair
HEPN	higher eukaryotes and prokaryotes nucleotide
HKT	high-affinity K^+ transporter
HMs	heavy metals
IRT	infrared thermography
LEA	late embryogenesis abundance
LEDs	light-emitting diodes
LHC	light-harvesting complex
MABC	marker-based backcrossing
MAPK	mitogen-activated protein kinases
MARS	marker-assisted recurrent selection
MAS	marker-assisted selection
MDA	malondialdehyde

MDHA	monodihydroascorbate
MDHAR	monodehydroascorbate reductase
miRNAs	microRNAs
Mn-SOD	manganese SOD
MS	mass spectrometry
NGS	next-generation sequencing
NGSAT	next-generation sequencing agricultural technologies
NHEJ	nonhomologous end joining
NHX	sodium hydrogen exchanger
NHX	sodium/hydrogen (Na^+/H^+) exchanger gene
NiNOR	nitrite-nitric oxide reductase
NMR	nuclear magnetic resonance
NOS	nitric oxide synthase
NR	nitrate reductase
NSCCs	nonselective cation channels
OEC	oxygen-evolving complex
OH	hydroxyl radicals
OLEDs	organic-based LEDs
$ONOO^-$	peroxynitrite
PA	polyamine
PAM	protospacer adjacent motif
PCD	programmed cell death
PCR	polymerase chain reaction
PDH	pyruvate dehydrogenase
PFS	protospacer flanking sequences
PGPR	plant-growth-promoting rhizobacteria
PM	plasma membrane
PODs	peroxidases
PPDK	phosphopyruvate dikinase
PS	photosystem
PSI	photosystem I
PTM	post-translational modification
QTL	quantitative trait loci
QTLS	quantitative trait analysis
RAM	root apical meristem
RBOH	respiratory burst oxidase homolog
RGA	rapid generation advancement
RNp	ribonucleoprotein
RNS	reactive nitrogen species

ROS	reactive oxygen species
SAGE	serial analysis of gene expression
SAM	shoot apical meristem
SDS	sodium dodecyl sulfate
siRNAs	short interfering RNAs
SNP	single nucleotide polymorphism
SOD	superoxide dismutase
SOLiD	supported oligonucleotide ligation and detection
SOS	salt overly sensitive
SP	southern plateau
tAPX	thylakoidal
TCA	tricarboxylic acid cycle
TF	transcriptional factor
THS	terminal heat stress
TILLING	targeting induced local lesions IN genome
UAV	unmanned aerial vehicles
WUE	water use efficiency
YAG	yttrium aluminum garnet
ZFNs	zinc finger nucleases

Preface

Today, agriculture faces countless challenges in order to accomplish the rising requirement for a sustainable food supply and guarantee high-quality nourishment for a quickly mounting population. To assure sufficient food production, it is essential to escalate the yield per area of arable land. Plants are frequently imperiled by challenging environmental circumstances causing cellular oxidative stress due to abiotic factors that play a defining role in shaping yields of crop plants. This also determine the variance in scattering of the plant types across distinctives kinds of habitats. The capability of plants to respond and/or acclimatize to unfavorable environmental conditions is straightforwardly related to the tolerance capacity of plants to cope with abiotic stresses.

Ecophysiological methods have significantly enhanced our knowledge of mechanisms involved in basic plant processes. These include photosynthesis, respiration, plant-water relations, and plant responses to abiotic and biotic stresses, from immediate to evolutionary perspectives. These studies also postulate the foundation for ascending plant physiological developments in challenging environmental conditions. Evolving ecophysiological mechanisms and methodologies to improve plant responses to unfavorable environmental circumstances is important in creating significant tools for better understanding of plant adaptations to various abiotic stresses and sustaining the supply of goods and services as global climate change intensifies.

This volume, *Plant Abiotic Stress Physiology, Volume 1: Responses and Adaptations,* comprises 11 chapters contributed by specialists in the field, emphasizing a broad variety of themes, ranging from responses to adaptations in plants under abiotic stress factors.

Volume 2 of the 2-part *Plant Abiotic Stress Physiology* focuses on molecular advancements. It discusses how plants have developed diverse physiological and molecular adjustments to safeguard themselves under challenging conditions and how emerging new technologies can utilize these plant adaptations to enhance plant resistance. These include using plant-environment interactions in develop crop species that are resilient to climate change, taking genomics and phenomics approaches for the study of abiotic stress tolerance, employing methyl jasmonate and salicylic acid, harnessing the CRISPR/CAS system to strengthen plant stress resistance, and more.

—**Tariq Aftab**
Aligarh Muslim University, India

—**Khalid Rehman Hakeem**
King Abdulaziz University, Saudi Arabia

CHAPTER 1

Innovations in Crop Production: An Amalgamation of Abiotic Stress Physiology and Technology

SHREYA GUPTA[1] and SHABIR A. RATHER[2*]

[1]National Institute of Plant Genome Research, New Delhi, India

[2]School of Integrative Plant Sciences, Section of Plant Biology, Cornell University, Ithaca 14853, NY, USA

*Corresponding author. E-mail: rathershabir100@gmail.com.

ABSTRACT

Abiotic stresses are well known to alter plant growth and development globally. Their impacts are destructive and adversely affect crops at different stages of their life cycle. Abiotic stresses include drought, salinity, heat, and nutrients deficiency that are known to damage the agronomically essential crops. Thus, the need of the hour is to use advanced tools that understand the physiological, cellular, and molecular aspects including plant breeding and genomics approach to develop stress-tolerant crops efficiently. The holistic approach of understanding this coalescence of improving the crops under abiotic stress and the progress in the molecular technology includes application of biotechnological tools such as transgenics, marker-assisted breeding, targeted genome editing using clustered regularly interspaced short palindromic repeat (CRISPR)–CRISPR-associated protein 9, and bioinformatic tools. Thus, culminating research into the crosstalk signaling pathways, where plant growth hormones and their regulatory genes frame a multidimensional interactome under combined stress conditions. Besides the study of genome and transcriptome, the new era of proteome and metabolome has been utilized to understand their effect on photosynthetic pathways under abiotic stress. Water scarcity and nutrient deficiency are a major threat to food security worldwide and to cope up with drought stress, the transgenics approach seems promising. The new technologies have revolutionized the research and are extensively used to modify the target genes in crop plants that are important regulators during abiotic stress, which can further help in generating new varieties with novel stress-tolerant traits. The chapter will highlight the improvements in crop productivity by using advance ground-breaking techniques. Such advances are replacing the conventional techniques, not in terms of time and labor, but also in cost-effectiveness and thus promote agricultural sustainability and ensure food security globally.

1.1   INTRODUCTION

Abiotic stresses are well known to alter plant growth and development worldwide. Their impacts are destructive and adversely affect crops at different stages of their life cycle. The abiotic class of stresses encompasses salinity, cold, heat, light, drought, and nutrients deficiency, and heavy metal toxicity is known to damage the agronomically essential crops. These stresses damage the crops at different stages of plant growth and development from their germination to flowering and fruiting [1]. During primary metabolic processes such as photosynthesis and respiration, osmotic balance is majorly affected by these abiotic stresses that ultimately hamper physiochemical, molecular, and cellular pathways in the plant [2]. However, being sessile, plants are quite smart and have developed various mechanisms for coping with stress. The use of conventional practices for crop improvements under stress required time and labor and also the expensiveness of these methods have caused their shift toward the need for modern tools.

In addition, the complexity of these stresses and their responses are multidimensional and dynamics is more intensive that is not resolved by these previously used tools [3]. Thus, for a holistic approach of understanding this coalescence of improving the crops under abiotic stress and the progress in the molecular technology includes application of biotechnological tools such as transgenics, marker-assisted breeding, targeted genome editing using clustered regularly interspaced short palindromic repeat (CRISPR)–CRISPR-associated protein 9(Cas9), and bioinformatic tools. The modern technological advancements include genomics, transcriptomics, and proteomics approach combined with the systems biology that serves to have better potential to understand the effects of stress over crops more robustly and ubiquitously all over the plant parts in comparatively lesser time. Further, the study of crosstalk signaling pathways where plant growth hormones and their regulatory genes frame a multidimensional interactome under combined stress conditions seems a promising approach. For example, the genome sequencing projects of important crops have been successfully achieved.

In addition, many bioinformatics-based databases are available for many cereal crops. Most widely used tools such as CyVerse and galaxy are used for high-throughput analysis [4]. Even the molecular markers have shown to be an unusual approach for enhancing the crop yield, for example, for the screening quantitative trait loci (QTLs) for drought tolerance, scientists have used marker-assisted selection during drought stress. Therefore, instead of conventional techniques of breeding, the breeding approach utilizing markers seems to be a promising alternative [5]. The role of significant phytohormones including abscisic acid, ethylene, brassinosteroids (BRs), and their transcription factors, are being explored for their effect of improving the efficacy of abiotic stress tolerance in various species of crops. A dire threat to food security around the world is water scarcity and to cope up with drought stress, the transgenics approach seems promising. For example, one of the 13 *Flowering Locus T-Like (FTL)* genes, *OsFTL10*, is known to be induced by drought stress. It is shown that its overexpression in rice improves drought resistance [6]. Nutrient deficiency is a common problem in soil and to alleviate that we use chemical fertilizers, especially urea that degrades the environment. A novel transcription factor from rice, nitrogen-mediated tiller growth response 5, improves nitrogen use efficiency if overexpressed in wild-type rice [7]. The widely used genome editing tool CRISPR/Cas9 has revolutionized the research. This technology is used to modify the target genes in crop plants, which are important regulators during abiotic stress, for example, it is shown that CRISPR/Cas9-induced modifications in rice genes OsBADH2, OsMPK, and Os02g23823 that play an instrumental role in many pathways related to abiotic stress in rice [8]. This novel technique can further help to generate new varieties with novel stress-tolerant traits. This chapter will highlight the advancements in crop productivity due to modern ground-breaking techniques that aim to make cereal crops stress tolerant. Such advances could promote agricultural sustainability and ensure food security globally.

1.2 MECHANISMS IN PLANTS FOR COMBATING ABIOTIC STRESS

Abiotic stresses overall reduce the growth of plant and survival. The gross biomass due to stress conditions is also observed to be decreased [9].

Although, different stresses have different effects; however, the overall damage often has similar changes under different stress at physical, cellular, and molecular levels. For example, when a plant encounters stress, the first line of defense comes from the outermost barrier, which is a cell wall that shows resistance because of its higher resilience [10]. After the physical barrier, come the internal physiological, cellular, and molecular changes that are activated under stress, including changes in the osmoticum, ionic balance, ROS balance, various transporters, and channels are expressed. Various physiological processes such as photosynthesis, respiration, and propagation are also suppressed under stress. The reduced uptake of water and CO_2 availability under stress is also decreased. This eventually causes lower photosynthetic and transpiration efficiency thereby lowering the stomatal conductance. At the molecular level, various phytohormones and their signaling cascades begin under stress and their regulatory genes and targets are activated for combating stress [11]. Signaling molecules such as secondary messengers, accumulation in cytosolic calcium levels, and pH alteration also play a key role under abiotic stress tolerance [12].

1.3   CONVENTIONAL METHODS FOR IMPROVING ABIOTIC STRESS TOLERANCE

The traditional methods include the study of a wide range of genetic variability among sexually compatible crop plants. These techniques have thus led to the improvements of plants at only physical, biochemical, and cellular levels under abiotic stresses. In addition, the use of conventional breeding methods takes years to develop the desired cultivar with improved traits. These methods are thus time-consuming and expensive [13]. Many other limitations with these methods include a collection of germplasm, lesser genetic variations among germplasm collected, and the unwanted or undesired characters in the cultivar that gets accumulated unknowingly [14]. The current situation where multiple stresses affect plants and the complex regulated mechanisms of stress tolerance could not be achieved by such methods where crop breeding fails to understand the crosstalk between wide ranges of metabolic, molecular, and physiochemical pathways. Therefore, for a robust understanding and quick generation of desired crop plants to combat abiotic stress efficiently and effectively, advanced biotechnological tools have been developed, which aim to modify crops genetically.

1.4 ADVANCEMENTS IN TECHNOLOGY FOR ABIOTIC STRESS TOLERANCE

Several modern molecular approaches include the understanding of various plant hormones and their regulatory genes, the role of protein kinases and phosphatases that are activated under stress and are responsible for improving abiotic stress tolerance.

Table 1.1 shows regulatory genes that provide resistance against different abiotic stresses in the context of plants.

TABLE 1.1   Role of Regulatory Genes in Providing Tolerance to Abiotic Stress in Plants

Regulatory Genes	Abiotic Stress Response	References
ABF	Drought stress	[15]
bZIPs	Cold, drought, and salinity stress	[16]
BRL3 (BRI1-Like) BR receptor	Drought stress	[17]
CBFs	Drought, cold stress	[18]
Cold-regulated protein SICOR413IM1	Chilling	[19]
Cysteine protease gene	Salt stress	[20]
Catalase	Heat stress	[21]
Calcineurin B-like protein	Cold, drought, and salinity stress	[22]
DREB1 and DREB2	Drought, cold, and salinity	[16]
ERFs	Drought and cold stress	[16]
Glycosyltransferase	Abiotic stress	[23]
Heat-shock proteins	Heat and drought stress	[24]
Late embryogenesis abundant genes	Drought, cold, and salinity	[25]
Mitogen-activated protein kinase (MAPK)	Drought, cold, and salinity	[22]
NAC (OsNAC6, SNAC1, SNAC2) Transcription factors	Drought, cold, and salinity	[26]
Protein phosphataes 1a	Salinity stress	[27]
Proline dehydrogenase	Freezing and salinity stress	[28]
RUBISCO	Abiotic stress	[29]
Superoxide dismutase (SOD)	Oxidative stress	[30]
Trehalose	Drought and salinity	[31]
Ubiquitin Ligase E3	Drought stress	[32]
WRKY	Drought, temperature, salinity, and oxidative stress	[33]

1.4.1 PHENOMICS STUDY FOR EXPLORING TOLERANCE OF ABIOTIC STRESS

Phenotypic analysis of the plants due to abiotic stress is a pertinent parameter that leads to a detailed understanding later at the components hampering plant survival at the genetic and molecular level. Phenotyping is broad terms that characterize above as well as below-ground plant organs, assessing their development, adaptive responses, tolerance, and resistance during stress at spatiotemporal environmental dynamics [34]. These morphological studies are the basic foundation for a better breeding selection system. The effects of stress on the plants at the internal sites are highly complex, and the way plants adapt to these unfavorable conditions is a dynamic process. Different plants experience the extent of the stress that plants encounter in different ways [35]. In fact, within the same plant species, these effects vary depending upon the age and developmental stage of the plants. The earlier used plant phenotyping was time-consuming, destructive, and a bit difficult to manage the plant biomass. However, with the recent advancements in technology various modern plant phenotyping methods have been generated that include digital technologies that give appropriate information about the environmental pressures on the crop plants and help in providing knowledge of suitable plants that can tolerate abiotic stress with better resistance and morphology. The information includes measuring of complex adaptive traits about growth, yield, total plant biomass under abiotic stresses with higher precision and accuracy [36]. Some key innovations in this area include the use of software for image analysis, data collection, management, and annotation. This is also now using some robotics as efficient tools for phenotyping at a comparatively faster rate. Many previous reports have shown study of aerial plant parts, and have ignored the roots as they are less accessible. However, the latest studies have also examined the roots as they are in direct contact with the soil and any kind of nutrient or water stress is first encountered by roots [37].

Thus, the study of root system architecture is also now being investigated vigorously with high-throughput phenotypic analysis for a better understanding of stress adaptive responses. The latest advancements in phenotypic studies of plant parts without any damage includes techniques such as infrared thermography for shoots [38] and underground radars, and electrical resistivity imaging for shoots [37]. The satellite imaging is also an emerging tool, especially for precision agriculture that provides an overall observation of the crops by damage from any abiotic stress. The Scopus database contains bibliographic records of phenotypic research of plants, have been developed

[35]. To analyze the data at a large scale, many software have been developed, such as Crop Design and Plant Accelerator 1 [39]. Phenotyping is an essential initial step for analyzing crops and then improving their productivity and yield during abiotic stress conditions. It provides an authentic way to select beneficial traits for developing crops that can be suitable under specific environmental conditions by culminating the information from genetics, molecular biology, epigenetics, together with marker-assisted breeding and crop management practices.

1.4.2 USE OF MOLECULAR MAPPING

The molecular mapping and breeding involve approaches to identify particular loci in the genome, which confers to a specific desirable trait under particular abiotic stress that are pooled together for generating higher-yielding varieties. Many advances at the genomic level are developed for increasing stress tolerance in cereal crops. The molecular markers, QTL mapping, genome selection strategies, transcriptome profiling, and genome-wide association studies are considered valuable approaches for the attribution and recognition of the genes for a particular trait under a specific stress condition and to ameliorate tolerance of in plants by employing the technique of marker-assisted breeding [40]. Earlier used DNA molecular markers such as RAPD, RFLP, AFLP, and SSRs are now lagging, and the upcoming exciting way for studying variations at the loci can be identified by QTL mapping. These variations can occur at the expression levels, which can be selected and via breeding can be incorporated in the elite varieties of important cereal crops [41]. Several QTLs have been developed for different crops under different abiotic stresses such as nutrient deficiency, heat and cold stress, drought, flooding, and freezing tolerance in *Vitis vinifera* [42], phosphate deficiency tolerance in crops such as maize, common bean, rice, wheat [43], and the list is endless (Table 1.2). The precision and accuracy of QTL and its proficiency for providing stress tolerance not comparable with previously mentioned makers.

These examples will further allow new opportunities and combining them with other areas such as genetics and breeding, and the selection will help us in providing the key source to characterize genes for a trait that will eventually be a beneficial approach of providing the knowledge of genomics to plant breeding program in fields. Stress is a multigenic trait, and therefore many high-throughput methods are essential for understanding the mechanisms for better quality crops. Thus, for studying combined stress and its effects at the molecular level, several combined approaches are developed along with studying large populations and develop higher resolution linkage maps and methods like positional cloning for plant breeding to generate tolerance against abiotic stress in various species of crop plants.

TABLE 1.2 Role of Different QTLs Identified in Different Plant Species that Provide Abiotic Stress Tolerance

Cereal Crops	Abiotic Stress	QTL Identified	Trait Improvement	References
Barley	Hypoxia (waterlogging)	Membrane potential	Waterlogging and salt stress tolerance	[44]
Common bean (Phaseolus vulgaris)	Nutrient stress (low P and high Al)	Yd4.1^{BG}	Growth period, stable yield	[45]
Maize	Heat stress	Heat susceptibility index (HSI)	Leaf length, Plant height, Leaf greenness, Leaf growth rate	[46]
Rice	Salt stress	Shoot length	Salinity tolerance	[47]
Wheat	Drought stress	Drought sensitivity index	Grain filling and grain weight	[48]

1.4.3 STRATEGIES TO COMBAT STRESS BY STUDYING FUNCTIONAL AND REGULATORY GENES FROM A MOLECULAR PERSPECTIVE

Various regulatory genes, as well as functional genes, undergo activation when exposed to different types of abiotic stresses. Regulatory genes include signaling molecules such as calcium ions, transcription factors of different plant hormones, enzymes like kinases, and phosphatases, which together contribute to aiding stress tolerance by downstream activation of many genes [49]. One of the protein kinases known as SnRK2, whose overexpression has been shown to increase tolerance during higher salinity in the soil and is revealed in various crops such as rice, wheat, and maize [50]. The role of calcium-binding protein kinases has also been identified like calcineurin B-like protein-interacting protein kinase during cold stress tolerance in rice and cotton [51] and barley during drought stress [52]. The conserved protein kinases known as mitogen-activated protein kinases (MAPK) are also activated during stress conditions of the abiotic type. OsMAPK5 overexpression in rice is observed to be tolerant to stresses of the abiotic type like cold, drought, and high salinity. Another kinase known as *Arabidopsis thaliana* PR5 receptor-like kinase (AtPR5K) provides drought tolerance in ABA-dependent stress response [53]. PP2C (PP2C-A) phosphatases are core components in ABA signaling and are known to be stress responsive. In maize, overexpression of *ZmPP2C* leads to their higher survival rate during high NaCl levels as compared to wild-type maize [54]. It is further been shown that the chromatin modification via acetylation mediated by chromatin remodeler known as ATPase BRAHMA (BRM) of these PP2Cs have a role to play during salt stress tolerance [55]. The role of various transcription factors is also very important during stress conditions such as dehydration responsive element binding 1/cold binding factor (DREB1/CBF), DREB2, leucine zipper (bZIP), ethylene response factor (ERF)-like TF Sub1A, and ABA-responsive element-binding protein/ABA-binding factor (ABF), and are also known to provide tolerance against abiotic stress due to heat, cold, salt, and drought. It is reported that transgenic wheat with overexpression of *TaERF3* exhibits drought and salt stress tolerance [56].

In addition, two other DREB/CBF genes (*TaDREB3* and *TaCBF5L*) were overexpressed in barley and wheat that displayed improved drought and cold tolerance [57]. It is also shown that overexpression of the DREB gene from potato in cotton plants exhibited enhanced drought tolerance [58]. Similarly, overexpression of *WRKY1* and *WRKY33* in *Arabidopsis* improved drought and heat tolerance [59]. Even wheat overpressing *Arabidopsis WRKY30* also exhibited better tolerance during drought stress [58]. The NAC transcription factors in *Arabidopsis* such as *ANAC055* and *ANACO72* are known to provide drought and salinity. In sorghum, a number of NAC TFs are identified such as *SbNAC6*, *SbNAC17*, *SbNAC26*, *SbNAC46*, *SbNAC56* have known to improve tolerance against cold, heat, salinity, drought stress [60]. In wheat, *TaNAC47* is incited by drought, cold, and salt [61]. The overexpression of NAC from wheat *TaNAC29* is known to provide drought and salt tolerance in *Arabidopsis* [62].

In comparison to wild plants, transgenic rice plants exhibited improved tolerance toward drought due to the overexpression of *ONAC14* [63]. Over-expressing of *OsMYB6* in transgenic rice exhibits improved drought and salt tolerance [64]. MYB TF from maize *ZmMYBIF35,* when overexpressed in *Arabidopsis,* showed better survivability during cold stress in comparison to wild-type plants [65]. These findings indicate the wide range roles of important enzymes and regulatory genes communicating various signaling pathways during abiotic stress conditions, which can further be analyzed and characterized by genetic manipulations for enhancing efficient stress tolerance in crop plants. Such practices will eventually help in promoting farming in extreme environmental conditions and provide better crop stress management practices in the future.

1.4.4   HIGH-THROUGHPUT GENOMICS, TRANSCRIPTOMICS, PROTEOMICS, AND METABOLOMICS

The research concerning plant stress is getting major attention today and therefore, plant scientists are extensively exploiting various "omics" approaches comprising proteomics, genomics, metabolomics, and transcriptomics to generate vast and robust stress regulative mechanisms in plant species and to understand better the inter-relatedness of these stress responses at the genetic, molecular, physiological as well as cellular level [66]. These high-throughput technological approaches are beneficial in providing new insights and horizons in understanding multidimensional aspects about the abiotic stress tolerance mechanisms in plants to improve crops, which can withstand extremely harsh conditions without compromising yield and growth (Figure 1.1). These approaches aim to find out the essential regulatory genes involved in the signaling pathways that get activated under abiotic stress conditions and thereby enhancing the stress tolerance in the respected plant species in which they get expressed [67].

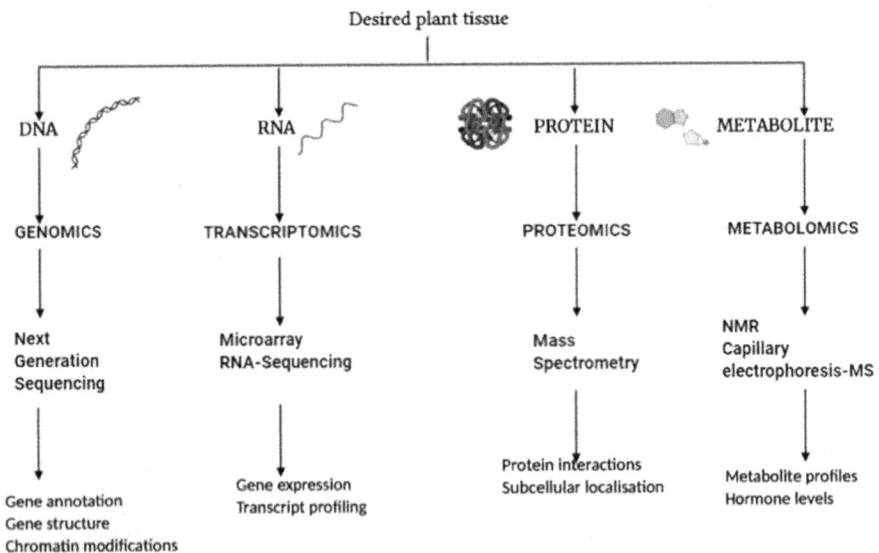

FIGURE 1.1   Flowchart showing various tools involved in proteomics, genomics, metabolomics, and transcriptomics that help in better understanding of abiotic stress mechanisms in plants.

Genomics involves the study of the genome, and therefore it gives us the useful biology of an organism by providing about the knowledge of various important gene sequences, gene structure, and annotations [68]. With the advancements in the sequencing the genomes, the study of genomics has become well authentic and is being exploited all over to understand the depth knowledge of abiotic stress. Various steps are carried out for genomics study, which includes isolation of DNA, its amplification, sequencing, assembly, quality assessments, and its annotation both structurally and functionally. The robust procedure provides useful information about the organism genome. The functional genomics approaches have helped recognize the crucial genes related to stress. The vast genomic data available online in the form of databases has to lead to the development of transcriptomics, proteomics, and metabolomics approach as well [69]. Further, the study of crop wild relatives would help in identifying novel key genes that can provide better tolerance where the genomic analysis of buckweed showed many genes related to drought and chilling stress [70].

Transcriptomics involves the study of RNA expression profile of an organism in space and time [71]. Dealing with RNA is a tricky thing as it is highly dynamic, unlike genome. These days many tools are being used to understand the transcriptome such as RNA sequencing, serial analysis of gene expression (SAGE), and microarrays [72, 73]. The availability of various online databases has led to the study of extensive transcriptome analysis especially for plants under abiotic stresses [69]. The transcriptome analysis of *Arabidopsis* indicated the altered transcripts levels of about 770 genes, which play an instrumental part during heat and drought stress together [74]. In chickpea, people have characterized drought and salt-stressed transcriptome using techniques such as next-generation sequencing (NGS) and SAGE [72]. Using the microarray technique, the transcriptome study of cotton seedlings under combined stress conditions is also conducted [75]. In maize, the effect of different stress such as heat, cold, salt, and drought showed about 2346, 1841, 1661, and 2019 genes are differentially regulated in these stresses, respectively, by using RNA sequencing approach [76]. The impact of osmotic stress in barley showed an increase in the genes that are involved in the biosynthesis of suberin using RNA sequencing, which further concluded that different genes are expressed differentially under different stress conditions. Thus, their role in signaling pathways could be exploited for better adaptability towards abiotic stress conditions [73].

Proteomics is the overall study of all the proteins in an organism expressed at a given time and space [77]. Like transcriptome, the proteome is also highly dynamic in nature, which is changed with age, developmental stage, and any stress condition in plants. The use of different types of mass spectrometry has helped in accomplished the proteome profiling, which measures mass and charge of protein fragments after digestion. Each mass spectrometric peak is like a unique fingerprint for that particular protein, which helps to identify the protein in a mixture [78, 79]. A technique such as two-dimensional gel electrophoresis (2-DGE) is also used in proteomics [79], which is utilized to compare the profiles of proteome due to stress conditions and identify all stress-tolerant proteins that are being expressed differentially and to exploit their roles in abiotic stress-induced signaling pathways [80]. The study of extracellular matrix proteome of rice plants due to drought stress revealed the proteins alterations concerning signaling, ROS scavenging, and carbohydrate metabolism [81]. Various studies have been carried out for proteome analysis under cadmium stress in different plants such as *Brassica juncea* [82], *Linum usitatissimum* [83], *A. thaliana* [84], and *Glycine max* [85]. The study of the nuclear proteome in chickpea under drought stress has also been studied [86]. The change in nuclear proteome during flooding in soybean has also been studied, which showed that H2, H3, and H4 were differentially regulated [87]. In tomato, the study of chloroplast proteome during drought stress revealed that chloroplast proteins interact with nuclear signaling proteins [88]. Recently, under waterlogging conditions, the proteome profile of barley is studied using 2-DGE and MS and identified vital players expressed during waterlogging condition [79].

Metabolomics is an emerging and an advance omics approach, which involves the study of metabolic profile at cellular or tissue level under any stress condition [89]. The metabolites are the molecules of small size having molecular masses, which are <2000 Da and are highly diverse in structure. The metabolites consist of aldehydes, hormones, amino acids, peptides, lipid molecules, ketones, steroids, and secondary metabolites. To scrutinize the metabolic profiling in plants due to stress conditions, various advancements in mass spectrometry have been developed such as nuclear magnetic resonance spectroscopy, liquid chromatography, and high-performance liquid chromatography or gas chromatography [91]. Plants under normal conditions have around 250,000 metabolites [91]. However, this number is significantly enhanced under stress conditions [92]. Therefore, these metabolites have a great potential and can be used as marker under different abiotic stresses [90]. Under drought stress, *Arabidopsis* metabolite profile showed an increase in several metabolites such as proline, gamma-aminobutyrate, and some metabolites involved in the citric acid cycle [93]. Under UV-B exposure, plants also showed an increase in metabolites involved in UV-B protection such as phenolics, flavonoids, and ascorbic acid [94]. In another study, it was seen that the branched-chain amino acids (BCAA) are enhanced under stress like drought and salinity in maize and wheat where they found that these BCAAs act as osmoprotectants [95]. In heat stress conditions, there is metabolite change in flavonoids, amino acids, polyamines, fatty acids, organic acids, and sugars. Several studies have shown changes in metabolites during salt stress as well such as changes in glycolysis metabolites, TCA cycle metabolites, and organic acids. Many studies have been carried in studying the effect of drought on metabolites and it is observed that polyols, sugar alcohols, sugars, phenols, amino acids are significantly higher in plants under drought stress [96–105].

New upcoming omics are also studied such as proteogenomics, miRNAomics, and lipidomics. All these omics approaches are connected via bioinformatics tools to analyze massive data accurately, efficiently [106]. This integration helps in understanding the cross-talks and biological networks that operate during abiotic stress conditions.

1.4.5 GENOME EDITING VIA CRISPR–CAS9

The conventional breeding practices have been widely used for crop improvement; however, they have shortcomings such as loss of genetic diversity and being time-consuming these practices could not meet up the demands of increasing human population globally. The use of modern genetic technology is considered to be the best and the fastest technology for developing varieties with new genotypes and phenotypes. Sophisticated and efficient technology has revolutionized the crop improvement for editing genomes [107]. The CRISPR–Cas9 consists of short repetitive sequences with invading spacers that target DNA, an operon of *Cas* genes, and a leader sequence rich in AT that encodes for Cas proteins [108]. The system recognizes DNA/RNA recognition rather than protein motifs and generates a double-stranded break. These programmed nucleases have revolutionized genetic engineering and opened up the new exciting and innovative ways of genome editing. The technique can not only induce mutations but can also either activate or repress a particular gene by CRISPR activation or CRISPR interference mechanisms, respectively [109]. The conventional genetic engineering methods have a significant drawback of random insertions of the transgene, which might affect other unwanted regions in the genome. This problem has been solved by this precise method of gene regulation as the designing of the guide RNA can prevent such off-target effects and provide stable transient expression. Although, the use of CRISPR has been well established and studied; however, very less information is available for plants [110]. Recently, the role of CRISPR/Cas9 in targeting multiple genes in the same organism has been optimized for various crops such as cotton, maize, rice, and wheat [111]. The use of CRISPR/Cas9-based editing system is also extended in genome-wide screening for trait improvement in crops [112]. Thus, this technique seems promising in generating novel allelic variations for a desirable trait. The method has great potential in developing crops that can withstand extreme stress conditions at a comparatively faster rate.

The plants response to different stresses is a very complex adaptive process, and therefore, it becomes more complicated to apply tools and methods for developing enhanced cultivar. Nevertheless, the gene-editing using CRISPR/Cas9 of maize AGROS8 has resulted in generating drought tolerance in maize [113].

In addition, to avoid the transgene insertions and unwanted off-targets, the new modified CRISPR/Cas9 with ribonucleoproteins is used in a proto-plasm fusion approach and has been successfully achieved for crops such as *Arabidopsis*, tobacco, and rice [114]. Lately, a bit more efficient method of transgene-free approach in wheat has been used, which is callus based editing tool with the transient expression of CRISPR/Cas9 DNA or RNA [115]. Being a transgene-free method, the generated crops will hopefully be acceptable in public as well. This remarkable advanced tool has open up the opportunities for crop improvement during abiotic stresses such as salt, cold, drought, and heat tolerance, especially when facing global warming. CRISPR has manipulated many other genes in order to explore their part in stress tolerance of the abiotic type. For example, brassinazole resistant 1 (BZR1) is involved in BR responses that is known to regulate thermotoler-ance via Feronia (FER) genes, which is confirmed by both CRISPR-bzr1 and BZR1 overexpression [116]. Table 1.3 shows the use of CRISPR in modifying various species of plants against abiotic stress.

Previous studies for exploring the utility of CRISPR mostly focus on the crop rice for abiotic stress tolerance. The mechanism of CRISPR/Cas9 is still emerging and developing. Very less information is available for plants, especially concerning abiotic stress improvement. However, this breakthrough gene-editing approach has great potential and has an excellent future for agricultural biotechnology.

1.4.6 USE OF SMALL RNAS IN PROVIDING RESISTANCE AGAINST ABIOTIC STRESS TOLERANCE

The complexity of abiotic stress has been explored scientifically, and many efforts have been made. The mechanisms include the alteration in gene

TABLE 1.3 Use of CRISPR/Cas9 Genome Editing in Different Plants Gene Sequences, which Led to The Improved Tolerance against Different Abiotic Stress Conditions

Plant Studied	Stress tolerance	Sequences Targeted	Role	References
Arabidopsis	Herbicide resistance	Bar gene	BAR gene confers glufosinate resistance	[117]
Linum usitatissimum	Glyphosate tolerance	50-Enolpyruvylshikimate-3-phosphate synthase (EPSPS)	EPSPS plays a role in the formation of aromatic amino acids	[118]
Oryza sativa	Salt stress tolerance	GT-1 component in the salt response of OsRAV2 (critical regulatory regions in its promoter)	RAV members involved in abiotic stress tolerance	[119]
Zea mays	Drought stress tolerance	ARGOS8	Plays a negative role in ethylene-mediated responses	[113]
Oryza sativa	Herbicide resistance	C827 and Acetolactate gene	Provide tolerance against herbicide	[120]
Solanum lycopersicum	Cold stress tolerance	C-repeat binding factor *SlCBF1*	Provide chilling tolerance	[121]
Triticum aestivum	Drought tolerance	TaDREB2	Provides dehydration tolerance	[122]
Oryza sativa	Drought tolerance	OsDREB	Provides dehydration tolerance	[123]

expression at both mRNA as well as protein levels. Various small RNAs have been investigated in exploring their roles in combating abiotic stress tolerance. He has been referred to as the molecules with tiny sizes and major roles [124]. The small RNAs include piwi interacting RNAs, short interfering RNAs (siRNAs), trans-acting siRNAs, and microRNAs (miRNAs) that play a role during drought, cold, heat, salinity as well as nutrient stresses in plants. During particular stress, these small RNAs are either upregulated or downregulated, which in turn downregulate the negative regulators or upregulate the positive regulators for providing tolerance [125]. High-throughput technologies as well as computational bioinformatics tools including NGS have played an immense role in identifying the downstream targets of these small RNAs.

In additiona, the latest RNA interference approach is another emerging way of developing the transgenics for better tolerance against abiotic stress in various species of crop plants. It has been shown that small RNAs target various regulatory sequences including the transcription factors involved in many hormone signaling pathways. It is also reported that miR167 targets auxin response factors ARF6 and ARF8 [126]. Various miRNAs including miR172, mirR171, miR168, miR167, miR159, and miR156 are upregulated during drought stress conditions [127]. Also, it has been reported that many of these small RNAs including miR827, miR529, miR408, miR403, miR399, miR398, miR397, miR396, miR395, miR393, miR390, and miR319 have been characterized and have been shown to regulate multiple stresses of the abiotic type including salt, heat, drought, cold, and high oxidative stress and nutrient deficiencies unusually low nitrogen in various plant species including *Arabidopsis, Medicago, Brachypodium, Glycine max, Oryza sativa, Triticum aestivum, and Zea mays* [128, 129]. Therefore, it is concluded that these small RNAs play an instrumental role in the regulation of abiotic stress in plants. Early identification of these small RNAs was based on genetic approaches, which were having many limitations including low efficiency, large consumption of time, and high expenses. With the entrance of new high-throughput computational technology, these limitations have been resolved. Various bioinformatics tools such as miRFinder, SplamiR, miRPlant, MiRCheck, findMiRNA, and miRDeep-P have been exclusively developed for the identification of these small RNAs and their target genes in plants [130–135].

Moreover, advance technologies such as degradome sequencing have been used to identify targets of these miRNA [136]. These technologies have not only identified these small RNAs but also have been annotated [137]. In addition, a database for miRNA sequences and their annotation is developed and is named as miRBase [138]. A recent study has shown the role of 22-nt plant siRNA in environmental stress adaptative responses [139]. Thus, these small RNAs have now become the potential candidates for enhanced crop productivity during abiotic stress tolerance for genetic engineering strategies, which will give a better understanding of transcriptional as well as post-transcriptional regulatory mechanisms in plants. Various methods have been employed for generating transgenics using these small miRNAs-based approaches such as target mimics, artificial miRNAs, and overexpression of these miRNAs [140]. Many transgenic plants have been developed by modifying these miRNA expressions such as transgenic *Arabidopsis* [141], *Oryza sativa* [142], *and Solanum lycopersicum* [143] by the overexpression of miR172, miR390, and miR399, respectively, for improving salt tolerance, heavy metal toxicity, and phosphorus deficiency, respectively. These small RNAs seem to be promising and useful for making the plants withstand extremes stresses of the environment.

1.4.7 *BIOINFORMATICS AND SYSTEMS BIOLOGY APPROACH*

The aim of this bioinformatics for stress tolerance in plants is to integrate with all the omics mentioned above approaches including proteomics, genomics, metabolomics, and transcriptomics. Bioinformatics has become fundamental in the way of data generation, storing the information in the form of databases, making it publically accessible, data integration as well as a compilation for the valuable source of information. The era of bioinformatics is revolutionizing and is providing a new perspective to the existing experiments in biology thus paving the way for improved scientific knowledge about the management of abiotic stress tolerance mechanism in various species of plants by adding new innovative functionalities with efficient software and tools [144].

The biological databases are the significant sources of all the information available from all the scientific labs globally [145]. The information contained in these databases is the collection and integration from different biological studies including genomics, transcriptomics, proteomics, metabolomics, and evolutionary history. Various databases are developed for studying plant stresses with the gene and protein sequences specifically involved during particular stress that eventually help the scientific community to understand plant stress biology better effectively. The list of these databases has been listed in Table 1.4 below.

Plant Abiotic Stress Physiology, Volume 1

TABLE 1.4 Different Databases That Have Been Generated Especially to Understand Stress Biology in Plants

Sr. No.	Databases	Description	Link	References
1	Plant Environmental Stress Transcript Database	It contains stress transcripts of the crops. The database allows the search regarding annotated transcripts, hypothetical conserved genes, and sequence alignments of ortholog sets based on different stress conditions.	(http://intranet.icrisat.org/gt1/tog/homepage.html)	[146]
2.	Plant Stress Gene Database	The database provides the orthologs and paralogs of different stress-responsive genes	(http://ccbb.jnu.ac.in/stressgenes/frontpage.html)	[147]
3.	PASmiR	It is the database for the molecular regulation of miRNA that is involved in plant abiotic stresses.	(http://pcsb.ahau.edu.cn:8080/PASmiR/)	[148]
4.	TAIR	The Arabidopsis Information Resource provides information about all the 30,000 genes of *Arabidopsis*.	(http://arabidopsis.org)	[149]
5.	STIFDB (Stress-Responsive Transcription Factor Database)	The database shows the predicted transcription factors and their binding sites, which are involved during abiotic stresses in *Arabidopsis, Oryza* species.	(http://caps.ncbs.res.in/stifdb/)	[150]
6.	DroughtDB (Drought Stress Gene Database)	This database gives information about all the drought-responsive genes in crops such as maize and barley.	(http://pgsb.helmholtz-muenchen.de/droughtdb/)	[151]
7.	PlantPReS (Plant Proteome Response to Stress)	It is a protein-based database that gives the information of proteins involved in stress.	(http://www.proteome.ir/)	[69]
8.	PSPDB (Plant Stress Protein Database)	Protein-based database that gives functional information about the protein involved during stress.	(http://bioclues.org/pspdb/)	[152]
9.	Rice SRTFDB (Rice Stress-Responsive Transcription Factor Database)	It gives a detailed knowledge of the TF in rice during high salt and drought conditions.	(http://www.nipgr.res.in/RiceSRTFDB.html)	[153]
10.	QlicRice	The database gives information about the abiotic stress-responsive QTL in rice.	(http://nabg.iasri.res.in:8080/qlic-rice/)	[154]

Not only these databases have been developed, many additional latest bioinformatics tools such as sequence alignments, which are based on computational algorithms like dynamic programming [155, 156]; efficient, other algorithms like BLAST and FASTA have also been developed. These tools have been considered important for solving many puzzles in biological systems. In addition, molecular docking and dynamics for studying the protein-ligand binding have been extensively used by various computational tools such as AutoDock, SANJEEVINI, and AutoDock Vina. In fact, the interaction of DREB proteins with dehydration-responsive element/C repeat has been observed using molecular docking method [157]. The structure prediction of the proteins is another important aspect that plays a crucial role during stress conditions as different types of proteins are involved in various metabolic processes. Under particular stress, protein structure, as well as function, is altered, which in turn cause harmful effects. Thus, bioinformatics tools have played an immense role in predicting the 3D structure of proteins using different methods like fold recognition, homology-based modeling, template-based prediction, ab initio protein structure prediction [158]. Some of the web-based tools for protein structure prediction are MODELLER, RaptorX, I-TASSER, PEP-FOLD, and many others are being developed [159]. Thus, bioinformatics tools have been extensively used in this era where the goal of generating high elite varieties, which can tolerate various stresses of the environment. Therefore, in silico approaches plays an important role during plant stress at genome, transcripts, and protein levels. Although a lot many efforts have been made and many new advance methods are being developed, there still exist many challenges including the plant stress tolerance mechanisms during combined stress to generate a whole interactome with all the information available for all the crops species to increase the knowledge for crop improvement during adversities of the environment, which no doubt are on the rise.

1.5 CONCLUSION

Abiotic stresses have detrimental effects on crops worldwide. Their impact is destructive and adversely affects crops at different stages of their life cycle. This chapter highlighted the advancements in the crop productivity by advance techniques, which develop elite crop varieties that can withstand extremes of the environment. These advances have additional benefits over the previously used conventional techniques in terms of their cost-effectiveness, less time and labor, and thus ensure food security globally.

The genome-based approaches are now extensively used to investigate the mechanisms of abiotic stress tolerance. The use of molecular mapping, omics approaches, bioinformatics tools, genome editing with CRISPR/Cas9 is breakthrough innovations that have played an important role in making crops abiotic stress-tolerant. All these techniques have different ways of understanding the abiotic stress mechanisms in plants, and ultimately identifying the essential marker genes responsible for improving stress tolerance. This further can be transgenically engineered in desired crops for withstanding stress, thus providing abiotic stress tolerance. Not only these, but other reverse genetics approaches could also be combined with these modern techniques for a robust understanding of stress biology in plants. Further, the unexplored area in plant stress biology is the study of combined stress conditions. Thus, the construction of a multidimensional interactome between different hormone networks, their signaling pathways, and their associated regulatory and target genes under combined stress conditions could be useful to explore the depths of abiotic stress tolerance mechanisms that operate in plants.

KEYWORDS

- abiotic stress tolerance
- molecular mapping
- OMICS approach
- small RNAs
- CRISPR–Cas9
- bioinformatics tools
- crop improvement

REFERENCES

1. Rodriguez-Uribe, L.; Higbie, S. M.; Stewart, J. M.; Wilkins, T.; Lindemann, W.; Sengupta-Gopalan, C.; Zhang, J. Identification of salt responsive genes using comparative micro-array analysis in Upland cotton (*Gossypium hirsutum* L.). *Plant Science*, **2011**, *180* (3), 461–469.
2. Chaves, M. M.; Flexas, J.; Pinheiro, C. Photosynthesis under drought and salt stress: regulation mechanisms from whole plant to cell, *Annals of Botany*, **2009**, *103* (4), 551–560.

3. Zhang, X.; Jin, S.; Liu, D.; Lin, Z.; Zeng, F.; Zhu, L.; Guo, X. Cotton biotechnology: challenge the future for cotton improvement; *Advances in Plant Biotechnology* (Rao, GP, Zhao, Y.; Radchuk, VV and Bhatnagar, SK, eds); **2008**; pp. 241–301.

4. Peri, S.; Roberts, S.; Kreko, I. R.; McHan, L. B.; Naron, A.; Ram, A.; Nelson, A. D. Read mapping and transcript assembly: a scalable and high-throughput workflow for the processing and analysis of ribonucleic acid sequencing data. *Frontiers in Genetics*, **2020**, *10*, 1361.

5. Sahebi, M.; Hanafi, M. M.; Rafii, M. Y.; Mahmud, T. M. M.; Azizi, P.; Osman, M.; Miah, G. Improvement of drought tolerance in rice (*Oryza sativa* L.): Genetics, genomic tools, and the WRKY gene family. *BioMed Research International*, **2018**, *2018*, 3158474.

6. Fang, M.; Zhou, Z.; Zhou, X.; Yang, H.; Li, M.; & Li, H. Overexpression of OsFTL10 induces early flowering and improves drought tolerance in Oryza sativa L. *PeerJ*, **2019**, *7*, 6422.

7. Wu, K.; Wang, S.; Song, W.; Zhang, J.; Wang, Y.; Liu, Q.; Zhao, Y. Enhanced sustainable green revolution yield via nitrogen-responsive chromatin modulation in rice. *Science*, **2020**, *367*, 6478.

8. Miao, J.; Chi, Y.; Lin, D.; Tyler, B. M.; Liu, X. Mutations in ORP1 conferring oxathi-apiprolin resistance confirmed by genome editing using CRISPR/Cas9 in *Phytophthora capsici* and *P. sojae*. *Phytopathology*, **2018**, *108* (12), 1412–1419.

9. Negrão, S.; Schmöckel, S. M.; & Tester, M. Evaluating physiological responses of plants to salinity stress. *Annals of botany*, **2017**, *119* (1), 1–11.

10. Le Gall, H.; Philippe, F.; Domon, J. M.; Gillet, F.; Pelloux, J.; Rayon, C. Cell wall metabolism in response to abiotic stress. *Plants*, **2015**, *4* (1), 112–166.

11. Govind, G.; Harshavardhan, V. T.; Hong, C. Y. Phytohormone signaling in response to drought; In: *Salt and Drought Stress Tolerance in Plants*: Springer, Cham.; **2020**; pp. 315–335.

12. Rajasheker, G.; Jawahar, G.; Jalaja, N.; Kumar, S. A.; Kumari, P. H.; Punita, D. L.; Kishor, P. B. K. Role and regulation of osmolytes and ABA interaction in salt and drought stress tolerance; In: *Plant Signaling Molecules*: Woodhead Publishing; **2019**; pp. 417–436.

13. Zafar, S. A.; Zaidi, S. S. E. A.; Gaba, Y.; Singla-Pareek, S. L.; Dhankher, O. P.; Li, X.; Pareek, A. Engineering abiotic stress tolerance via CRISPR/Cas-mediated genome editing. *Journal of Experimental Botany*, **2020**, *71* (2), 470–479.

14. Ashraf, M. Inducing drought tolerance in plants: recent advances. *Biotechnology Advances*, **2010**, *28* (1), 169–183.

15. Fujita Y.; Fujita M.; Satoh R.; Maruyama K.; Parvez MM.; Seki M.; Hiratsu K.; Ohme-Takagi M.; Shinozaki K.; Yamaguchi-Shinozaki K. AREB1 is a transcription activator of novel ABRE-dependent ABA signaling that enhances drought stress tolerance in Arabidopsis. *Plant Cell*, **2005**, *17*, 3470–3488.

16. Chen H.; Liu L.; Wang L.; Wang S.; Cheng X. VrDREB2A, a DREB-binding transcription factor from Vigna radiata, increased drought and high-salt tolerance in transgenic *Arabidopsis thaliana*. *Journal* of *Plant Research*, **2016**, *129*, 263–273.

17. Martignago, D.; Rico-Medina, A.; Blasco-Escaméz, D.; Fontanet-Manzaneque, J. B.; Caño-Delgado, A. I. Drought resistance by engineering plant tissue-specific responses. *Frontiers in Plant Science*, **2019**, *10*, 1676.

18. Ito Y.; Katsura K.; Maruyama K.; Taji T.; Kobayashi M.; Seki M.; Shinozaki K.; Yamaguchi-Shinozaki K. Functional analysis of rice DREB1/CBF-type transcription

factors involved in cold-responsive gene expression in transgenic rice. *Plant Cell Physiol*, **2006**, *47*, 141–153.

19. Ma X.; Chen C.; Yang M.; Dong X.; Lv W.; Meng Q. Cold-regulated protein (SlCOR413IM1) confers chilling stress tolerance in tomato plants. *Plant Physiology and Biochemistry*, **2018**, 124, 29–39.

20. Zheng L.; Chen S.; Xie L.; Lu Z.; Liu M.; Han X.; Qiao G.; Jiang J.; Zhuo R.; Qiu W. Overexpression of cysteine protease gene from Salix matsudana enhances salt tolerance in transgenic Arabidopsis. *Environmental and Experimental Botany*, **2018**, *147*, 53–62.

21. Chiang CM.; Chen SP.; Chen LFO.; Chiang MC.; Chien HL.; Lin KH. Expression of the broccoli catalase gene (BoCAT) enhances heat tolerance in transgenic Arabidopsis. *Journal of Plant Biochemistry and Biotechnology*, **2014**, *23*, 266–277.

22. Boudsocq M.; Laurière C. Osmotic signaling in plants. Multiple pathways mediated by emerging kinase families. *Plant Physiology*, **2005**, *138*, 1185–1194

23. Li Q.; Yu HM.; Meng XF.; Lin JS.; Li YJ.; Hou BK. Ectopic expression of glycosyltransferase UGT76E11 increases flavonoid accumulation and enhances abiotic stress tolerance in Arabidopsis. *Plant Biology*, **2018**, *20*, 10–19.

24. Jacob P.; Hirt H.; Bendahmane A. The heat-shock protein/chaperone network and multiple stress resistance. *Plant Biotechnology Journal*, **2017**, *15*, 405–414.

25. Magwanga RO.; Lu P.; Kirungu JN.; Lu H.; Wang X.; Cai X.; Zhou Z.; Zhang Z.; Salih H.; Wang K. Characterization of the late embryogenesis abundant (LEA) proteins family and their role in drought stress tolerance in upland cotton. *BMC Genetics*, 2018, *19*, 6.

26. Wang JY.; Wang JP.; Yang HF. Identification and functional characterization of the NAC gene promoter from *Populus euphratica*. *Planta*, **2016**, *244*, 417–427.

27. Liao YD.; Lin KH.; Chen CC.; Chiang CM. Oryza sativa protein phosphatase 1a (OsPP1a) involved in salt stress tolerance in transgenic rice. *Molecular Breeding*, 2016, *36*, 22.

28. Cabassa-Hourton C.; Schertl P.; Bordenave-Jacquemin M.; Saadallah K.; Guivarc'h A.; Lebreton S.; Planchais S.; Klodmann J.; Eubel H.; Crilat E. Proteomic and functional analysis of proline dehydrogenase 1 link proline catabolism to mitochondrial electron transport in *Arabidopsis thaliana*. *The Biochemical Journal*, **2016**, *473*, 2623–2634.

29. Lin HH.; Lin KH.; Syu JY.; Tang SY.; Lo HF. Physiological and proteomic analysis in two wild tomato lines under waterlogging and high temperature stress. *Journal of Plant Biochemistry and Biotechnology*, **2016**, *25*, 87–96.

30. Shafi A.; Pal AK.; Sharma V.; Kalia S.; Kumar S.; Ahuja PS.; Singh AK. Transgenic potato plants overexpressing SOD and APX exhibit enhanced lignification and starch biosynthesis with improved salt stress tolerance. *Plant Molecular Biology Reporter*, **2017**, *35*, 504–518.

31. Ge LF.; Chao DY.; Shi M.; Zhu MZ.; Gao JP.; Lin HX. Overexpression of the trehalose-6-phosphate phosphatase gene OsTPP1 confers stress tolerance in rice and results in the activation of stress responsive genes. *Planta*, **2008**, *228*, 191–201.

32. Yang L.; Wu L.; Chang W.; Li Z.; Miao M.; Li Y.; Yang J.; Liu Z.; Tan J. Overexpression of the maize E3 ubiquitin ligase gene ZmAIRP4 enhances drought stress tolerance in Arabidopsis. *Plant Physiology and Biochemistry*, **2018**, *123*, 34–42.

33. Wu X.; Shiroto Y.; Kishitani S.; Ito Y.; Toriyama K. Enhanced heat and drought tolerance in transgenic rice seedlings overexpressing OsWRKY11 under the control of HSP101 promoter. *Plant Cell Reports*, **2009**, *28*, 21–30.

34. Pieruschka, R.; Schurr, U. Plant phenotyping: past, present, and future. *Plant Phenomics*, **2019**, *2019*, 7507131.

35. Costa, C.; Schurr, U.; Loreto, F.; Menesatti, P.; Carpentier, S. Plant phenotyping research trends, a science mapping approach. *Frontiers in Plant Science*, **2019**, *9*, 1933.

36. Fiorani, F.; Schurr, U. Future scenarios for plant phenotyping. *Annual Review of Plant Biology*, **2013**, *64*, 267–291.

37. Zhu, J.; Ingram, P. A.; Benfey, P. N.; Elich, T. From lab to field, new approaches to phenotyping root system architecture. *Current Opinion in Plant Biology*, **2011**, *14* (3), 310–317.

38. Munns, R. Plant adaptations to salt and water stress: differences and commonalities; In: *Advances in Botanical Research*. Academic Press; **2011**; *57*; pp. 1–32.

39. Mir, R. R.; Reynolds, M.; Pinto, F.; Khan, M. A.; Bhat, M. A. High-throughput phenotyping for crop improvement in the genomics era. *Plant Science*, **2019**, *282*, 60–72.

40. Deshmukh, R.; Sonah, H.; Patil, G.; Chen, W.; Prince, S.; Mutava, R.; Nguyen, H. T. Integrating omic approaches for abiotic stress tolerance in soybean. *Frontiers in Plant Science*, **2014**, *5*, 244.

41. Langridge, P.; Paltridge, N.; Fincher, G. Functional genomics of abiotic stress tolerance in cereals. *Briefings in Functional Genomics*, **2006**, *4* (4), 343–354.

42. Reynolds, A. G. Grapevine breeding in France—A historical perspective; In: *Grapevine breeding programs for the wine industry*. Woodhead Publishing; **2015**; pp. 65–76.

43. Maharajan, T.; Ceasar, S. A.; Ajeesh krishna, T. P.; Ramakrishnan, M.; Duraipandiyan, V.; Naif Abdulla, A. D.; Ignacimuthu, S. Utilization of molecular markers for improving the phosphorus efficiency in crop plants. *Plant Breeding*, **2018**, *137* (1), 10–26.

44. Gill, M. B.; Zeng, F.; Shabala, L.; Zhang, G.; Fan, Y.; Shabala, S.; Zhou, M. Cell-based phenotyping reveals QTL for membrane potential maintenance associated with hypoxia and salinity stress tolerance in barley. *Frontiers in Plant Science*, **2017**, *8*, 1941.

45. Diaz, L. M.; Ricaurte, J.; Tovar, E.; Cajiao, C.; Terán, H.; Grajales, M.; Raatz, B. QTL analyses for tolerance to abiotic stresses in a common bean (*Phaseolus vulgaris* L.) population. *PLoS One*, **2018**, *13* (8), 0202342.

46. Van Inghelandt, D.; Frey, F. P.; Ries, D.; Stich, B. QTL mapping and genome-wide prediction of heat tolerance in multiple connected populations of temperate maize. *Scientific Reports*, **2019**, *9* (1), 1–16.

47. Jahan, N.; Zhang, Y.; Lv, Y.; Song, M.; Zhao, C.; Hu, H.; Hu, J. QTL analysis for rice salinity tolerance and fine mapping of a candidate locus qSL7 for shoot length under salt stress. *Plant Growth Regulation*, **2020**, *90* (2), 307–319.

48. Gahlaut, V.; Jaiswal, V.; Tyagi, B. S.; Singh, G.; Sareen, S.; Balyan, H. S.; Gupta, P. K. QTL mapping for nine drought-responsive agronomic traits in bread wheat under irrigated and rain-fed environments. *PLoS one*, **2017**, *12* (8), 0182857.

49. Barajas-Lopez, J. D. D.; Moreno, J. R.; Gamez-Arjona, F. M.; Pardo, J. M.; Punkkinen, M.; Zhu, J. K.; Fujii, H. Upstream kinases of plant Sn RK s are involved in salt stress tolerance. *The Plant Journal*, **2018**, *93* (1), 107–118.

50. Nguyen, H. C.; Lin, K. H.; Ho, S. L.; Chiang, C. M.; Yang, C. M. Enhancing the abiotic stress tolerance of plants: from chemical treatment to biotechnological approaches. *Physiologia Plantarum*, **2018**, *164* (4), 452–466.

51. Tang, W.; Thompson, W. A. Role of the Arabidopsis calcineurin B-like protein-interacting protein kinase CIPK21 in plant cold stress tolerance. *Plant Biotechnology Reports*, **2020**, 1–17.

52. Fedorowicz-Strońska, O.; Koczyk, G.; Kaczmarek, M.; Krajewski, P.; Sadowski, J. Genome-wide identification, characterisation and expression profiles of calcium-dependent

protein kinase genes in barley (*Hordeum vulgare* L.). *Journal of Applied Genetics*, **2017**, *58* (1), 11–22.

53. Baek, W.; Lim, C. W.; Luan, S.; Lee, S. C. The RING finger E3 ligases PIR1 and PIR2 mediate PP2CA degradation to enhance abscisic acid response in Arabidopsis. *The Plant Journal*, **2019**, *100* (3), 473–486.

54. He, Z.; Wu, J.; Sun, X.; & Dai, M. The maize clade A PP2C phosphatases play critical roles in multiple abiotic stress responses. *International Journal of Molecular Sciences*, **2019**, *20* (14), 3573.

55. Nguyen, N. H.; Jung, C.; Cheong, J. J. Chromatin remodeling for the transcription of type 2C protein phosphatase genes in response to salt stress. *Plant Physiology and Biochemistry*, **2019**, *141*, 325–331.

56. Rong, W.; Qi, L.; Wang, A.; Ye, X.; Du, L.; Liang, H.; Zhang, Z. The ERF transcription factor Ta ERF 3 promotes tolerance to salt and drought stresses in wheat. *Plant Biotechnology Journal*, **2014**, *12* (4), 468–479.

57. Yang, Y.; Al-Baidhani, H. H. J.; Harris, J.; Riboni, M.; Li, Y.; Mazonka, I.; Haefele, S. DREB/CBF expression in wheat and barley using the stress-inducible promoters of HD-Zip I genes: impact on plant development, stress tolerance and yield. *Plant Biotechnology Journal*, **2020**, *18* (3), 829–844.

58. El-Esawi, M. A.; Alayafi, A. A. Overexpression of StDREB2 transcription factor enhances drought stress tolerance in cotton (*Gossypium barbadense* L.). *Genes*, **2019**, *10* (2), 142.

59. He, G. H.; Xu, J. Y.; Wang, Y. X.; Liu, J. M.; Li, P. S.; Chen, M.; Xu, Z. S. drought-responsive WRKY transcription factor genes TaWRKY1 and TaWRKY33 from wheat confer drought and/or heat resistance in Arabidopsis. *BMC Plant Biology*, **2016**, *16* (1), 116.

60. Kadier, Y.; Zu, Y. Y.; Dai, Q. M.; Song, G.; Lin, S. W.; Sun, Q. P.; Lu, M. Genome-wide identification, classification and expression analysis of NAC family of genes in sorghum [*Sorghum bicolor (L.) Moench*]. *Plant Growth Regulation*, **2017**, *83* (2), 301–312.

61. Zhang, L.; Zhang, L.; Xia, C.; Zhao, G.; Jia, J.; Kong, X. The novel wheat transcription factor TaNAC47 enhances multiple abiotic stress tolerances in transgenic plants. *Frontiers in Plant Science*, **2016**, *6*, 1174.

62. Huang, Q.; Wang, Y.; Li, B.; Chang, J.; Chen, M.; Li, K.; He, G. TaNAC29, a NAC transcription factor from wheat, enhances salt and drought tolerance in transgenic Arabidopsis. *BMC Plant Biology*, **2015**, *15* (1), 268.

63. Shim, J. S.; Oh, N.; Chung, P. J.; Kim, Y. S.; Choi, Y. D.; Kim, J. K. Overexpression of OsNAC14 improves drought tolerance in rice. *Frontiers in Plant Science*, **2018**, *9*, 310.

64. Tang, Y.; Bao, X.; Zhi, Y.; Wu, Q.; Guo, Y.; Yin, X.; Liu, W. Overexpression of a MYB family gene, OsMYB6, increases drought and salinity stress tolerance in transgenic rice. *Frontiers in Plant Science*, **2019**, *10*, 168.

65. Baillo, E. H.; Kimotho, R. N.; Zhang, Z.; Xu, P. Transcription factors associated with abiotic and biotic stress tolerance and their potential for crops improvement. *Genes*, **2019**, *10* (10), 771.

66. Gupta N.; Singh A.; Zahra S.; Kumar S. *PtRFdb: a database for plant transfer RNA-derived fragments. Database*, Oxford, **2018.**

67. Mehta, S.; James, D.; Reddy, M. K. Omics technologies for abiotic stress tolerance in plants: current status and prospects. In: *Recent Approaches in Omics for Plant Resilience to Climate Change*, Springer, Cham.; **2019**; pp. 1–34.

68. Gilliham M.; Able JA.; Roy SJ. Translating knowledge about abiotic stress tolerance to breeding programmes. *The Plant Journal*, **2017**, *90* (5), 898–917.

69. Mousavi SA.; Pouya FM.; Ghaffari MR.; Mirzaei M.; Ghaffari A.; Alikhani M.; Ghareyazie M.; Komatsu S.; Haynes PA.; Salekdeh GH. PlantPReS: a database for plant proteome response to stress. *Journal of Proteomics*, **2016**, *143*, 69–72.

70. Zhang L.; Li X.; Ma B.; Gao Q.; Du H.; Han Y.; Li Y.; Cao Y.; Qi M.; Zhu Y.; Lu H. The tartary buckwheat genome provides insights into rutin biosynthesis and abiotic stress tolerance. *Molecular Plant*, **2017**, *10* (9), 1224–1237.

71. Shen W.; Li H.; Teng R.; Wang Y.; Wang W.; Zhuang J. Genomic and transcriptomic analyses of HD-Zip family transcription factors and their responses to abiotic stress in tea plant (*Camellia sinensis*). *Genomics*, **2018**.

72. Molina C.; Zaman-Allah M.; Khan F.; Fatnassi N.; Horres R. The salt-responsive transcriptome of chickpea roots and nodules via deepSuperSAGE. *BMC Plant Biology*, **2011**, *11*, 31.

73. Kreszies T.; Shellakkutti N.; Osthoff A.; Yu P.; Baldauf JA.; Zeisler-Diehl VV.; Ranathunge K.; Hochholdinger F.; Schreiber L. Osmotic stress enhances suberization of apoplastic barriers in barley seminal roots: analysis of chemical, transcriptomic and physiological responses. *New Phytologist*, **2018**, *221* (1), 180–194.

74. Rizhsky L.; Liang H.; Shuman J.; Shulaev V.; Davletova S.; Mittler R. When defense pathways collide. The response of Arabidopsis to a combination of drought and heat stress. *Plant Physiology*, **2004**, *134*, 1683–1696.

75. Zhu YN.; Shi DQ.; Ruan MB.; Zhang LL.; Meng ZH. Transcriptome analysis reveals crosstalk of responsive genes to multiple abiotic stresses in cotton (*Gossypium hirsutum* L.*). PLoS One*, **2013**, *8* (11), 80218.

76. Le DT.; Nishiyama R.; Watanabe Y. Differential gene expression in soybean leaf tissues at late developmental stages under drought stress revealed by genome-wide transcriptome analysis. *PLoS One*, **2012**, *7*, 49522.

77. Tyers M, Mann M. From genomics to proteomics. *Nature*, **2003,** *422*, 193–197.

78. Nakagami H.; Sugiyama N.; Ishihama Y.; Shirasu K. Shotguns in the front line: phosphoproteomics in plants. *Plant and Cell Physiology*, **2012**, *53*, 118–124.

79. Luan H.; Shen H.; Pan Y.; Guo B.; Lv C.; Xu R. Elucidating the hypoxic stress response in barley (*Hordeum vulgare* L.) during waterlogging: a proteomics approach. *Scientific Reports*, **2018**, *8* (1), 9655.

80. Liu B.; Zhang N.; Zhao S. Proteomic changes during tuber dormancy release process revealed by iTRAQ quantitative proteomics in potato. *Plant Physiology and Biochemistry*, **2015**, *86*, 181–190.

81. Pandey A.; Rajamani U.; Verma J.; Subba P.; Chakraborty N.; Datta A. Identification of extracellular matrix proteins of rice (*Oryza sativa* L.) involved in dehydration-responsive network: a proteomic approach. *Journal of Proteome Research*, **2010**, *9*, 3443–3464.

82. Alvarez S.; Berla BM.; Sheffield J.; Cahoon RE.; Jez JM. Comprehensive analysis of the *Brassica juncea* root proteome in response to cadmium exposure by complementary proteomic approaches. *Proteomics*, **2009**, *9* (9), 2419–2431.

83. Hradilova J.; Rehulka P.; Rehulkova H.; Vrbova M.; Griga M. Comparative analysis of proteomic changes in contrasting flax cultivars upon cadmium exposure. *Electrophoresis*, **2010**, *31* (2), 421–431.

84. Semane B.; Dupae J.; Cuypers A.; Noben JP.; Tuomainen M. Leaf proteome responses of Arabidopsis thaliana exposed to mild cadmium stress. *The Journal of Plant Physiology*, **2010**, *167* (4), 247–254.

Plant Abiotic Stress Physiology, Volume 1

85. Ahsan N.; Nakamura T.; Komatsu S. Differential responses of microsomal proteins and metabolites in two contrasting cadmium (Cd) accumulating soybean cultivars under Cd stress. *Amino Acids,* **2012**, *42* (1), 317–327.

86. Subba P.; Kumar R.; Gayali S.; Shekhar S.; Parveen S. Characterization of the nuclear proteome of a dehydration-sensitive cultivar of chickpea and comparative proteomic analysis with a tolerant cultivar. *Proteomics,* **2013**, *13* (12–13), 1973–1992.

87. Yin X.; Komatsu S. Quantitative proteomics of nuclear phosphoproteins in the root tip of soybean during the initial stages of flooding stress. *Journal of Proteomics,* **2015**, *119*, 183–195.

88. Tamburino R.; Vitale M.; Ruggiero A.; Sassi M.; Sannino L.; Arena S. Chloroplast proteome response to drought stress and recovery in tomato (*Solanum lycopersicum* L.*)*. *BMC Plant Biology,* **2017**, *17*, 40.

89. Kumar M.; Kuzhiumparambil U.; Pernice M.; Jiang Z.; Ralph Peter J. Metabolomics: an emerging frontier of systems biology in marine macrophytes. *Algal Research,* **2016**, *16*, 76–92.

90. Parida AK.; Panda A.; Rangani J. Metabolomics-guided elucidation of abiotic stress tolerance mechanisms in plants. In: *Plant metabolites and regulation under environmental stress*: Academic, San Diego, CA; **2018**; pp. 89–131.

91. Kim HK.; Choi YH.; Verpoorte R. NMR-based metabolomic analysis of plants. *Nature Protocols,* **2010**, *5*, 536–549.

92. Muthuramalingam P.; Krishnan SR.; Pandian S.; Mareeswaran N.; Aruni W.; Pandian SK.; Ramesh M. Global analysis of threonine metabolism genes unravel key players in rice to improve the abiotic stress tolerance. *Scientific Reports,* **2018**, *8* (1), 9270.

93. Urano K.; Maruyama K.; Ogata Y.; Morishita Y.; Takeda M. Characterization of the ABA-regulated global responses to dehydration in Arabidopsis by metabolomics. *The Plant Journal,* **2009**, *57* (6), 1065–1078.

94. Kusano M.; Tohge T.; Fukushima A.; Kobayashi M.; Hayashi N. Metabolomics reveals comprehensive reprogramming involving two independent metabolic responses of Arabidopsis to UV-B light. *The Plant Journal,* **2011**, *67* (2), 354–369.

95. Witt S.; Galicia L.; Lisec J.; Cairns J.; Tiessen A. Metabolic and phenotypic responses of greenhouse-grown maize hybrids to experimentally controlled drought stress. *Molecular Plant,* **2012**, *5* (2), 401–417.

96. Luengwilai K.; Saltveit M.; Beckles DM. Metabolite content of harvested Micro-Tom tomato (Solanum lycopersicum L.) fruit is altered by chilling and protective heat-shock treatments as shown by GC–MS metabolic profiling. *Postharvest Biology and Technology,* **2012**, *63*, 116–122.

97. Shin H.; Oh S.; Kim K.; Kim D. Proline accumulates in response to higher temperatures during dehardening in peach shoot tissues. *Journal of Horticulture,* **2016**, *85* (1), 37–45.

98. Rivero RM.; Mestre TC.; Mittler R.; Rubio F.; Garcia-Sanchez F.; Martinez V. The combined effect of salinity and heat reveals a specific physiological, biochemical and molecular response in tomato plants. *Plant, Cell & Environment,* **2014**, *37*, 1059–1073.

99. Joshi V.; Joung JG.; Fei Z.; Jander G. Interdependence of threonine, methionine and isoleucine metabolism in plants: accumulation and transcriptional regulation under abiotic stress. *Amino Acids,* **2010**, *39* (4), 933–947.

100. Wang H.; Tang X.; Wang H.; Shao H-B. Proline accumulation and metabolism-related genes expression profiles in Kosteletzkya virginica seedlings under salt stress. *Frontiers in Plant Sciencei,* **2015**, *6*, 792.

101. Pang Q.; Zhang A.; Zang W.; Wei L.; Yan X. Integrated proteomics and metabolomics for dissecting the mechanism of global responses to salt and alkali stress in Suaeda corniculata. *Plant Soil*, **2016**, *402*, 379–394.

102. Wenzel A.; Frank T.; Reichenberger G.; Herz M.; Engel KH. Impact of induced drought stress on the metabolite profiles of barley grain. *Metabolomics*, **2015**, *11*, 454–467.

103. Alcázar R.; Bitrián M.; Bartels D.; Koncz C.; Altabella T.; Tiburcio AF. Polyamine metabolic canalization in response to drought stress in Arabidopsis and the resurrection plant *Craterostigma plantagineum. Plant Signaling & Behavior*, **2014**, *6*, 243–250.

104. Khan N.; Bano A.; Rahman MA.; Rathinasabapathi B.; Babar MA. UPLC-HRMS-based untargeted metabolic profiling reveals changes in chickpea (*Cicer arietinum*) metabolome following long-term drought stress. *Plant, Cell & Environment*, **2018**, *42* (1), 115–132.

105. Griesser M.; Weingart G.; Schoedl-Hummel K.; Neumann N.; Becker M.; Varmuza K. Severe drought stress is affecting selected primary metabolites, polyphenols, and volatile metabolites in grapevine leaves (*Vitis vinifera* cv. Pinot noir). *Plant Physiology and Biochemistry*, **2015**, *88*, 17–26.

106. Franz M.; Lopes CT.; Huck G.; Dong Y.; Sumer O.; Bader GD. Cytoscape.js: a graph theory library for visualisation and analysis. *Bioinformatics*, **2016**, *32* (2), 309–311.

107. Shan, Q.; Wang, Y.; Li, J.; Zhang, Y.; Chen, K.; Liang, Z.; Gao, C. Targeted genome modification of crop plants using a CRISPR-Cas system. *Nature Biotechnology*, **2013**, *31* (8), 686–688.

108. Hille, F.; Charpentier, E. CRISPR-Cas: biology, mechanisms and relevance. *Philosophical Transactions of the Royal Society B: Biological Sciences*, **2016**, *371* (1707), 20150496.

109. Bortesi, L.; Fischer, R. The CRISPR/Cas9 system for plant genome editing and beyond. *Biotechnology Advances*, **2015**, *33* (1), 41–52.

110. Piatek, A.; Ali, Z.; Baazim, H.; Li, L.; Abulfaraj, A.; Al-Shareef, S.; Mahfouz, M. M. RNA-guided transcriptional regulation in planta via synthetic dC as9-based transcription factors. *Plant Biotechnology Journal*, **2015**, *13* (4), 578–589.

111. Zafar, S. A.; Zaidi, S. S. E. A.; Gaba, Y.; Singla-Pareek, S. L.; Dhankher, O. P.; Li, X.; Pareek, A. Engineering abiotic stress tolerance via CRISPR/Cas-mediated genome editing. *Journal of Experimental Botany*, **2020**, *71* (2), 470–479.

112. Mahas, A.; Mahfouz, M. Engineering virus resistance via CRISPR–Cas systems. *Current Opinion in Virology*, **2018**, *32*, 1–8.

113. Shi, J.; Gao, H.; Wang, H.; Lafitte, H. R.; Archibald, R. L.; Yang, M.; Habben, J. E. ARGOS 8 variants generated by CRISPR-Cas9 improve maize grain yield under field drought stress conditions. *Plant Biotechnology Journal*, **2017**, *15* (2), 207–216.

114. Woo, J. W.; Kim, J.; Kwon, S. I.; Corvalán, C.; Cho, S. W.; Kim, H.; Kim, J. S. DNA-free genome editing in plants with preassembled CRISPR-Cas9 ribonucleoproteins. *Nature Biotechnology*, **2015**, *33* (11), 1162–1164.

115. Zhang, Y.; Liang, Z.; Zong, Y.; Wang, Y.; Liu, J.; Chen, K.; Gao, C. Efficient and transgene-free genome editing in wheat through transient expression of CRISPR/Cas9 DNA or RNA. *Nature Communications*, **2016**, *7* (1), 1–8.

116. Yin, Y.; Qin, K.; Song, X.; Zhang, Q.; Zhou, Y.; Xia, X.; Yu, J. BZR1 transcription factor regulates heat stress tolerance through FERONIA receptor-like kinase-mediated reactive oxygen species signaling in tomato. *Plant and Cell Physiology*, **2018**, *59* (11), 2239–2254.

117. Hahn, F.; Mantegazza, O.; Greiner, A.; Hegemann, P.; Eisenhut, M.; Weber, A. P. An efficient visual screen for CRISPR/Cas9 activity in *Arabidopsis thaliana. Frontiers in Plant Science*, **2017**, *8*, 39.

118. Sauer, N. J.; Narváez-Vásquez, J.; Mozoruk, J.; Miller, R. B.; Warburg, Z. J.; Woodward, M. J.; Walker, K. A. Oligonucleotide-mediated genome editing provides precision and function to engineered nucleases and antibiotics in plants. *Plant Physiology*, **2016**, *170* (4), 1917–1928.

119. Duan, Y. B.; Li, J.; Qin, R. Y.; Xu, R. F.; Li, H.; Yang, Y. C.; Yang, J. B. (2016). Identification of a regulatory element responsible for salt induction of rice OsRAV2 through ex situ and in situ promoter analysis. *Plant Molecular Biology*, **2016**, *90* (1–2), 49–62.

120. Shimatani, Z.; Kashojiya, S.; Takayama, M.; Terada, R.; Arazoe, T.; Ishii, H.; Ezura, H. Targeted base editing in rice and tomato using a CRISPR-Cas9 cytidine deaminase fusion. *Nature Biotechnology*, **2017**, *35* (5), 441–443.

121. Li, R.; Li, R.; Li, X.; Fu, D.; Zhu, B.; Tian, H.; Zhu, H. Multiplexed CRISPR/Cas9-mediated metabolic engineering of γ-aminobutyric acid levels in *Solanum lycopersicum*. *Plant Biotechnology Journal*, **2018**, *16* (2), 415–427.

122. Kim, D.; Alptekin, B.; Budak, H. CRISPR/Cas9 genome editing in wheat. *Functional & Integrative Genomics*, **2018**, *18* (1), 31–41.

123. Yang, Y.; Sornaraj, P.; Borisjuk, N.; Kovalchuk, N.; Haefele, S. M. Transcriptional network involved in drought response and adaptation in cereals; In: *Abiotic and biotic stress in plants–recent advances and future perspectives*, InTech; **2016**; pp. 3–29.

124. Shriram, V.; Kumar, V.; Devarumath, R. M.; Khare, T. S.; Wani, S. H. MicroRNAs as potential targets for abiotic stress tolerance in plants. *Frontiers in Plant Science*, **2016**, *7*, 817.

125. Banerjee, S.; Sirohi, A.; Ansari, A. A.; & Gill, S. S. (2017). Role of small RNAs in abiotic stress responses in plants. *Plant Gene*, *11*, 180–189.

126. Wu Y.; Wei B.; Liu H.; Li T.; Rayner S. MiRPara: a SVM-based software tool for prediction of most probable microRNA coding regions in genome scale sequences. *BMC Bioinformatics*, **2011**, *12*, 107.

127. Liu, H.H.; Tian, X.; Li, Y.J.; Wu, C.A.; Zheng, C.C. Microarray-based analysis of stressregulated microRNAs in Arabidopsis thaliana. *RNA*, **2008**, *14*, 836–843.

128. Gentile, A.; Dias, L.I.; Mattos, R.S.; Ferreira, T.H.; Menossi, M. MicroRNAs and drought responses in sugarcane. *Frontiers in Plant Science.*, **2015**, *6*, 58.

129. Kantar, M.; Unver, T.; Budak, H.; Regulation of barley miRNAs upon dehydration stress correlated with target gene expression. *Functional & Integrative Genomics*, **2010**, *10*, 493–507.

130. Huang, T.H.; Fan, B.; Rothschild, M.F.; Hu, Z. L.; Li, K.; Zhao, S.H. MiRFinder: an improved approach and software implementation for genome-wide fast microRNA precursor scans. *BMC Bioinformatics*, **2007**, *8*, 341.

131. Thieme, C.J.; Gramzow, L.; Lobbes, D.; Theissen, G. SplamiR–prediction of spliced miRNAs in plants. *Bioinformatics*, **2011**, *27*, 1215–1223.

132. An, J.; Lai, J.; Sajjanhar, A.; Lehman, M.L.; Nelson, C.C. miRPlant: an integrated tool for identification of plant miRNA from RNA sequencing data. *BMC Bioinformatics*, **2014**, *15*, 275.

133. Jones-Rhoades, M.W.; Bartel, D.P. Computational identification of plant microRNAs and their targets, including a stress-induced miRNA. *Molecular Cell*, **2004**, *14*, 787–799.

134. Mathelier, A.; Carbone, A. MIReNA: Finding microRNAs with high accuracy and no learning at genome scale and from deep sequencing data. *Bioinformatics*, **2010**, *26*, 2226–2234.

135. Yang, X.; Li, L. miRDeep-P: A computational tool for analyzing the microRNA transcriptome in plants. *Bioinformatics*, **2011**, *27*, 2614–2615.

136. Yang, J. H.; Li, J. H.; Shao, P.; Zhou, H.; Chen, Y. Q.; Qu, L. H. starBase: a database for exploring microRNA–mRNA interaction maps from Argonaute CLIP-Seq and Degradome-Seq data. *Nucleic Acids Research*, **2011**, *39* (suppl_1), D202-D209.

137. Rosewick, N.; Momont, M.; Durkin, K.; Takeda, H.; Caiment, F.; Cleuter, Y.; Georges, M. Deep sequencing reveals abundant noncanonical retroviral microRNAs in B-cell leukemia/lymphoma. *Proceedings of the National Academy of Sciences*, **2013**, *110* (6), 2306–2311.

138. Tripathi, A.; Goswami, K.; & Sanan-Mishra, N. Role of bioinformatics in establishing microRNAs as modulators of abiotic stress responses: the new revolution. *Frontiers in Physiology*, **2015**, *6*, 286.

139. Wu, H.; Li, B.; Iwakawa, H. O.; Pan, Y.; Tang, X.; Ling-hu, Q.; Zhang, X. Plant 22-nt siRNAs mediate translational repression and stress adaptation. *Nature*, **2020**, *581* (7806), 89–93.

140. Zhou, M.; & Luo, H. MicroRNA-mediated gene regulation: potential applications for plant genetic engineering. *Plant Molecular Biology*, **2013**, *83* (1–2), 59–75.

141. Li, Z.; Wang, S.; Cheng, J.; Su, C.; Zhong, S.; Liu, Q.; Zheng, B. Intron lariat RNA inhibits microRNA biogenesis by sequestering the dicing complex in Arabidopsis. *PLoS Genetics*, **2016**, *12* (11), 1006422.

142. Ding, Y.; Ye Y.; Jiang, Z.; Wang, Y.; Zhu C. MicroRNA390 is involved in cadmium tolerance and accumulation in rice. *Frontiers in Plant Science*, **2016**, *7*, 235.

143. Gao, H. J.; Yang, H. Y.; Bai, J. P.; Liang, X. Y.; Lou, Y.; Zhang, J. L.; Chen, Y. Ultrastructural and physiological responses of potato (*Solanum tuberosum* L.) plantlets to gradient saline stress. *Frontiers in Plant Science*, **2015**, *5*, 787.

144. Esposito, A.; Colantuono, C.; Ruggieri, V.; Chiusano, M. L. Bioinformatics for agriculture in the Next-Generation sequencing era. *Chemical and Biological Technologies in Agriculture*, **2016**, *3* (1), 9.

145. Attwood, T. K.; Gisel, A.; Eriksson, N. E.; Bongcam-Rudloff, E. Concepts, historical milestones and the central place of bioinformatics in modern biology: a European perspective. *Bioinformatics-Trends and Methodologies*, **2011**, *1*.

146. Balaji, J.; Crouch, J. H.; Petite, P. V.; Hoisington, D. A. A database of annotated tentative orthologs from crop abiotic stress transcripts. *Bioinformation*, **2006**, *1* (6), 225.

147. Prabha, R.; Ghosh, I.; Singh, D. P. Plant Stress Gene Database: a collection of plant genes responding to stress condition. *ARPN Journal of Science and Technology's*, **2011**, *1*, 28–31.

148. Zhang, S.; Yue, Y.; Sheng, L.; Wu, Y.; Fan, G.; Li, A.; Wei, C. PASmiR: a literature-curated database for miRNA molecular regulation in plant response to abiotic stress. *BMC Plant Biology*, **2013**, *13* (1), 33.

149. Berardini, T. Z.; Reiser, L.; Li, D.; Mezheritsky, Y.; Muller, R.; Strait, E.; Huala, E. The Arabidopsis information resource: making and mining the "gold standard" annotated reference plant genome. *Genesis*, **2015**, *53* (8), 474–485.

150. Naika, M.; Shameer, K.; Mathew, O. K.; Gowda, R.; Sowdhamini, R. STIFDB2: an updated version of plant stress-responsive transcription factor database with additional stress signals, stress-responsive transcription factor binding sites and stress-responsive genes in Arabidopsis and rice. *Plant and Cell Physiology*, **2013**, *54* (2), e8-e8.

151. Alter, S.; Bader, K. C.; Spannagl, M.; Wang, Y.; Bauer, E.; Schön, C. C.; Mayer, K. F. DroughtDB: an expert-curated compilation of plant drought stress genes and their homologs in nine species. *Database*, **2015**.

152. Kumar, S. A.; Kumari, P. H.; Sundararajan, V. S.; Suravajhala, P.; Kanagasabai, R.; Kishor, P. K. PSPDB: plant stress protein database. *Plant Molecular Biology Reporter*, **2014**, *32* (4), 940–942.
153. Priya, P.; Jain, M. RiceSRTFDB: a database of rice transcription factors containing comprehensive expression, cis-regulatory element and mutant information to facilitate gene function analysis. *Database*, **2013**.
154. Smita S.; Lenka S.K.; Katiyar A.; Jaiswal P.; Preece J.; Bansal K.C. QlicRice: a web interface for abiotic stress responsive QTL and loci interaction channels in rice. Database (Oxford) **2011**: bar037.
155. Needleman, S. B.; Wunsch`, C. D. A general method applicable to the search for similarities in the amino acid sequence of two proteins. *Journal of Molecular Biology*, **1970**, *48* (3), 443–453.
156. Smith, T. F.; Waterman, M. S. Identification of common molecular subsequences. *Journal of Molecular Biology*, 1981, *147* (1), 195–197.
157. Nawaz, M.; Iqbal, N.; Idrees, S.; Ullah, I. DREB1A from Oryza sativa var. IR6: homology modelling and molecular docking. *Turkish Journal of Botany*, **2014**, *38* (6), 1095–1102.
158. Wang, S.; Sun, S.; Li, Z.; Zhang, R.; Xu, J. Accurate de novo prediction of protein contact map by ultra-deep learning model. *PLoS Computational Biology*, **2017**, *13* (1), e1005324.
159. Kumar, S.; Shanker, A. Bioinformatics Resources for the stress biology of plants; In: *Biotic and Abiotic Stress Tolerance in Plants*, Springer, Singapore; **2018**; pp. 367–386.

CHAPTER 2

Redox Homeostasis Managers in Plants under Environmental Stresses

VASEEM RAJA[1,2*], SAMI ULLAH QADIR[3], UMER MAJEED WANI[4],
TARIQ AHMAD DAR[2], SABEEHA BASHIR[1], and RIFFAT JOHN[1*]

[1]*Department of Botany, University of Kashmir, Hazratbal Srinagar,
Kashmir, 19006, India*

[2]*Department of Botany, Government Degree College (Women), Pulwama,
Kashmir, 190006, India*

[3]*Department of Environmental Sciences, Government Degree College,
Kokernag, Kashmir, India*

[4]*Department of Biotechnology, University of Kashmir, Hazratbal,
Srinagar, Kashmir, 19006. India*

*Corresponding author. E-mail: wrajamp2009@gmail.com;
riffat_iit@yahoo.com.*

ABSTRACT

Reactive oxygen species (ROS) are reflected as the unwanted byproducts of aerobic metabolism. Most dominating ROS include singlet oxygen, hydroxyl radical, superoxide radical ($O_2^{\bullet-}$), and hydrogen peroxide (H_2O_2). Because of their higher oxidizing metabolic activities and fast rates of electron flow chloroplast, mitochondria, and peroxisomes are the cellular organelles releasing ROS. In nature, there always exists a delicate balance between ROS production and scavenging; however, environmental stresses alter this delicate balance leading to oxidative stress. During the course of evolution, plants have adjusted themselves to the promising physiological, biochemical, and molecular states to avoid lethal effects of ROS as well as to display them as signaling molecules. A steady state between ROS production in different organelles and their scavenging through different metabolic pathways exists in plants to use them as signaling molecules. This steady state in various cellular organelles is achieved through the ROS gene network. Each cellular compartment in plants is equipped with its own ROS homeostasis control and ROS signaling is altered depending upon the cell type, with evolution, plants have adjusted to the changing environments and have learnt ways to scavenge the lethal effects of ROS through various enzymatic and nonenzymatic antioxidants, which operate in different cellular organelles to scavenge the ROS.

2.1   INTRODUCTION

Reactive oxygen species (ROS) are reflected as the unwelcomed companions of aerobic metabolism that have accompanied the aerobic life about 2.2–2.7 billion years ago [1]. Most abundant and devastating ROS comprise of hydroxyl radical, singlet oxygen ($^1O_2$), hydrogen peroxide (H_2O_2), superoxide radical (O_2^{*-}), etc. [2]. Any disparity between ROS production and scavenging leads to oxidative damage to proteins, lipids, DNA, and RNA molecules and even lead to cell death [3, 4]. During the course of evolution, plants have adopted mechanism to avoid the lethal effects of ROS and also uses them as signaling molecules [5–7].

As their byproduct ROS are continuously being produced from several metabolic pathways functioning in various organelles. Because of their higher oxidizing metabolic activities and fast rates of electron flow Chloroplast, mitochondria, and peroxisomes are the cellular organelles producing ROS [8, 9]. It is necessary that a steady state should exist between ROS production in different organelles and their scavenging [10].

Each cellular compartment in plant cell is equipped with its own ROS homeostasis control. ROS signaling in plants is altered depending upon the cell type, level of stress, and stage of development. ROS production under diverse abiotic stress situations in cells is predicted by several ROS antennas to produce specific stress signals that lead to acclimation responses. In cells, various redox reactions help in decoding ROS signals that can alter protein function and structure that regulates transcription factor binding to DNA affecting transcription [6, 11]. As such, in cells, ROS-driven metabolic regulation and signal transduction are tightly controlled by this redox system [6, 11, 12]. Abiotic stress conditions disrupt several metabolic activities, which triggers ROS production; however, ROS as signaling molecules are also produced in abiotic stress signal transduction network. Plants should maintain the redox pool to counter the effects of stress. It has been validated that during metabolic disruption, ROS released alters the redox status of numerous enzymes and controls metabolic fluxes in the cellular environment as a consequence of altered redox status changes in metabolic reactions occur to counter the effects of stress [13]. Metabolic ROS-driven modifications also lead to alterations in proteins regulating key events of transcription and translation [12]. Upon the perception of stress NADPH oxidases activated by calcium and phosphorylation (respiratory burst oxidase homolog, RBOH) at plasma membrane produce signaling ROS [14, 15]. Acclimation responses are activated as a result of alterations in the redox state of regulatory proteins, transcription, and translation thereby mitigating effects of stress on metabolism and also reducing metabolic ROS levels.

2.2   ROS PRODUCTION IN VARIOUS ORGANELLES

During the normal life cycle plants are exposed to numerous abiotic stress conditions, exposure to these environmental cues leads to ROS production. Chloroplast, mitochondria peroxisomes, and apoplast are the major sources of ROS production under such conditions [11, 15–17]. In almost all plant species including algae and other higher plants, chloroplasts are the main seats of photosynthesis and one of the major sources of ROS [18, 19]. Equipped with a complex network of thylakoids, the chloroplasts are capable of harvesting light energy through light-capturing apparatus. Oxygen generated in the chloroplasts during the process of photosynthesis accepts electrons that travel through several photosystems results in the formation of superoxide radicals. Therefore, among the ROS centers present in photosystems (PSI and PSII), electron transport chain (ETC) and triplet chlorophyll mark chloroplasts the major ROS source [20]. In the course of normal photosynthesis electron flow from excited photosystems is directed toward $NADP^+$ that generates a reducing power in the form of NADPH, which helps in reduction of final electron acceptor through Calvin cycle; however, low light conditions at PSII favor $^1O_2$ formation, considered to be the natural byproduct of photosynthesis. It has been deliberated through numerous finding that ROS production in chloroplast is associated with that of hypersensitive response [21]. It has also been demonstrated that chloroplast-generated ROS are able to transmit the wound-induced programmed cell death (PCD) through maize tissues [22].

Mitochondria are the main source of ROS as well as the primary producers of toxic metabolic byproducts mainly in the form of H_2O_2 [16, 23]. Mitochondrial ETC is able to harbor electrons with adequate amount of available free energy that leads to direct reduction of O_2 molecules, which is the inevitable source of mitochondrial ROS associated with aerobic respiration [11]. Prominent among the ROS-producing sites in mitochondria are Complex I and III of ETC [24, 25]. From the studies, it has been shown that 15% of the O_2 utilized in mitochondria leads to the production of H_2O_2. Upon reaction with Fe^{2+} and Cu^+ [1], H_2O_2 produces highly lethal OH that in turn can pass through mitochondrial membrane and cause membrane peroxidation resulting in the formation of several cytotoxic compounds that may lead to cellular damage upon reaction with protein, lipids, and nucleic acids [26, 27]. Two sites known to produce O_2 in peroxisomes are oxidation of xanthine and hypoxanthine to uric acid via xanthine oxidase in the organelle matrix and the second, peroxisome membranes where O_2 is released by the peroxisome's ETC composed of a flavoprotein NADH and cytochrome b [28]. Photorespiratory glycolate oxidase reactions, disproportionation of O_2 radicals, β-oxidation of fatty acids, enzymatic reactions of flavinoxidases are among the important metabolic processes responsible for H_2O_2 generation in peroxisomes [15, 17, 29].

2.3 ANTIOXIDANT DEFENSE MECHANISM: REDOX HOMEOSTASIS IN PLANTS

2.3.1 ENZYMATIC ANTIOXIDANTS

During the course of organic evolution, plants have adopted a vast number of distinct biochemical mechanisms to tackle with the damaging ROS from different cellular compartments [30–32]. Glutathione (GSH)–ascorbate (AsA) cycle (Halliwell–Asada cycle), operates in cytosol, mitochondria, plastids, and peroxisomes helping in quenching of highly toxic ROS. During the process, the first enzyme of the cycle APX (ascorbate peroxidase) detoxifies H_2O_2 by oxidizing AsA to monohydroascorbate radical (MDHA), which in turn is either directly reduced back to AsA by monodehydroascorbate reductase (MDHAR) or breaks down into AsA and dehydroascorbate (DHA). This DHA is reduced by dehydroascorbate reductase (DHAR) to AsA by oxidizing GSH, which is regenerated from oxidized glutathione dimers (GSSG) by NADPH-dependent glutathione reductase (GR) [1, 30, 65] (Figure 2.1). GSH and ascorbic acid along with enzymes such as APX, DHAR, and GR modulate the oxidation state of the cell [34]. In the AsA–GSH cycle, H_2O_2 generated from ROS metabolism is converted to H_2O at the expense of NADPH.

In addition to the AsA–GSH cycle, other well-studied antioxidant enzymes reported to play a crucial role in removing ROS is superoxide dismutase (SOD) that convert O_2^- into H_2O_2 and catalases (CATs) and PODs that further metabolize H_2O_2 to H_2O [35, 36]. While CAT is primarily localized in peroxisomes, isoforms of peroxidase, and SOD are distributed throughout the cell and can be found in cytosolic, mitochondrial, and chloroplastic compartments. Glutathione *S*-transferase (GST) is an important enzyme involved in detoxifying toxic products of lipid peroxidation and other electrophilic xenobiotics in both plants and animals by catalyzing the conjugation of electrophilic toxic molecules with reduced GSH, which are then transported out of the cytosol by GSH pumps [37].

The important ROS scavenging enzymes include SOD, found in almost all cellular compartments, the water–water cycle in chloroplasts, the AsA–GSH cycle in chloroplasts, cytosol, mitochondria, apoplast and peroxisomes, glutathione peroxidase (GPX), and CAT in peroxisomes [1]. The extensively studied enzyme SOD dismutates superoxide (O_2^{*-}) to H_2O_2, which in turn is detoxified by the activity CAT and a diversity of PODs disintegrating H_2O_2 into water and molecular oxygen [38].

2.3.2 NONENZYMATIC ANTIOXIDANTS

In plants, a major role in ROS signaling [41] has been played by nonenzymatic complex molecules, which comprise of major cellular redox buffers glutathione (γ-glutamyl-cysteinyl-glycine, GSH), AsA, carotenoids, phenolic compounds, and tocopherol [39, 40]. Apart from acting as enzymes cofactors, these antioxidant molecules interact with various cellular components thereby modulating various processes such as cell elongation, mitosis, cell death, and senescence that directly affect plant growth and development [42]. Decreased content of nonenzymatic antioxidant in mutants has shown to be oversensitive to stress [43, 44].

2.3.2.1   AsA

It is one of the most abundant low molecular weight antioxidant that plays a major defensive role against oxidative stress caused by elevated levels of ROS [45–47]. The common name of organic molecule L-*threo*-hexenon-1,4-lactone is vitamin C or L-ascorbic acid (L-AA). The ascorbic acid is one of the most dominant weak acid present in plant cell, at physiological pH

(5–7) it dissociates into its predominant form that is ascorbate anion (ASC). It plays a major role in mitigating the harmful effects of oxidative stress caused by increased ROS levels. Because of its tendency to donate electrons, it acts as a powerful antioxidant in both nonenzymatic and enzymatic reactions. Smrinoff–Wheeler pathway acts as a major contributor of AsA in plants. In Smrinoff–Wheeler pathway, galacturonic acid reductase reduces D-galacturonic acid to L-galactonic acid, which is further converted to L-galactono-1,4-lactone. The L-galactono-1,4-lactone is subsuently oxidized L-galactono-1,4-lactone dehydrogenase enzyme to AsA. After its synthesis in mitochondria by L-galactono-γ-lactone dehydrogenase it is transported to other cell organelles through facilated diffusion or by proton-electrochemical gradient. In addition of being present in abundance in photosynthetic tissue [49], it is also found in major plat cell types, organelles, and apoplast in plants [48]. About 90% of AsA is localized in cytoplasm, a considerable amount is transported to the apoplast unlike different antioxidants, where it is present in millimolar concentration. Apoplastic AsA provides the front-line defense against possibly damaging external oxidants [50]. AsA prevents the oxidative damage of sensitive macromolecules. AsA is mostly present in chloroplast in reduced form under normal physiological conditions, furthermore, it also takes part as a violaxanthin de-epoxidase cofactor, thus, sustaining dissipation of excess excitation energy [51]. Reaction of AsA with $O_2 \cdot -$, H_2O_2 protects the membrane from oxidative damage regeneration of α-tocopherol from tocopheroxyl radical also protects the membrane and preserves the enzyme activities that contain prosthetic transition metal ions [52]. AsA plays a major role in the removal of H_2O_2 through AsA–GSH cycle [30, 31].

AsA oxidation occurs in two successive steps, first generating MDHA and later DHA. In AsA–GSH cycle, APX utilizes two AsA molecules to reduce H_2O_2 to water with subsequent generation of MDHA. MDHA has a short life time and therefore it quickly dismutates into AsA and DHA or is reduced to by NADP (H) dependent enzyme MDHAR TO AsA [30, 32, 53]. DHA being extremely unstable is rapidly decomposed to tartarate and oxalate if pH values exceeds more than 6.0 [52]. In order to preclude this dissociation, AsA is rapidly formed by DHA reduction with the help of enzyme DHAR utilizing the reducing equivalents from GSH [54]. AsA level has been reported to alter in response to various stresses [55, 56]. The stability between the rates of AsA synthesis and turnover in connection with antioxidant demand, determines the levels of AsA under the environmental stress conditions [57]. In plants, abiotic stress tolerance is provided by increasing the expression of enzymes involved in synthesis of AsA. An important step present in the Smirnoff–Wheeler pathway for AsA biosynthesis in higher plants causes the synthesis of GDP-L-galactose from GDP-D-mannose with the help of enzyme GDP-Mannose 3_, 5-epimerase (GME Studies have shown that increase in the expression of two members of the GME gene family results in the enhanced accumulation of AsA and thereby increased tolerance in tomato plants under various abiotic stresses [58]. Increased AsA content has been shown to confer oxidative stress tolerance in *Arabidopsis* [60]. Overexpressing strawberry D-galacturonic acid reductase that converts D-galacturonate to L-galacturonate, leading to the production of AsA and thereby results in improved tolerance against abiotic stress in case of potato plants [59].

2.3.2.2 GSH

GSH that acts as a chief source of nonprotein (reduced sulfur) is present in the cells of majority of prokaryotes and eukaryotes [61–63]. GSH reduced form having the formulae c-glu-cys-gly is a tripeptide thiol that interchanges with the oxidized form, GSSG continuously [5, 30, 63]. In certain plants, GSH tripeptide homologs are present, in which other amino acid replace carboxy-terminal gly. GSH a nonprotein thiol having low molecular weight prevents the plant cells from the oxidative damages induced by ROS. In almost all the compartments of cell-like chloroplasts, mitochondria, cytosol, endoplasmic reticulum, and vacuoles it has been detected [5, 63]. Specific isoforms of γ-glutamyl-cysteinyl synthetase (γ-ECS) and glutathione synthetase (GS) in specific compartments of the plant cell synthesize GSH in chloroplast and cytosol. The redox state of cell is achieved by maintaining the balance between glutathione disulfide (GSSG) and GSH. An important role has been played by GSH in various biological processes due to its reducing power, involving signal transduction, growth/division, conjugation of metabolites, regulation of sulfate transport, synthesis of proteins and nucleic acids, enzymatic regulation, detoxification of xenobiotics, synthesis of phytochelatins for metal chelation, and stress-responsive gene expression [64]. GSH acts in many ways as an antioxidant. It directly functions as a scavenger of free radical by reacting chemically with $\cdot OH$, $O_2 \cdot -$, and H_2O_2. GSH protects the macromolecules (i.e., lipids, proteins, DNA) either by the acting as proton donor in the presence of ROS or by formation of organic free radicals directly with reactive electrophiles (glutathiolation), yielding GSSG. It is also involved in the production of another possible antioxidant AsA, through the AsA–GSH cycle. GSH by utilizing enzyme DHAR recycles AsA from its oxidized to reduced form [30, 65]. GSH, shown to act as a gene expression regulator, is the precursor for the synthesis of phytochelatins, which binds to the heavy metals and also acts as a substrate for the GSH S-transferases, which catalyzes the conjugation of GSH with potentially dangerous xenobiotics such as herbicides. Overexpressing GSH synthetase enzyme that is involved in the biosynthesis of GSH shows no effect on GSH level and was incompetent to increase the ozone tolerance and photoinhibition resistance [64], in hybrid poplar (*Populus tremula* × *P. alba*). However, overexpressing γ-ECS shows less sensitivity in Indian mustard toward cadmium stress and increases tolerance against chloroacetanilide herbicides in poplar plants [66, 67]. Eltayeb [67] observed that higher levels reduced GSH presents larger defense against oxidative in transgenic potato caused by various environmental stresses.

2.3.2.3 TOCOPHEROLS

Tocopherols are involved in scavenging of oxygen free radicals, $^1O_2$ and lipid peroxy radicles and they belong to the group of lipophilic antioxidants [68–70]. Tocopherols are only synthesized by photosynthetic species and only found in green areas of plants. The biosynthetic pathway of tocopherol employs only two compounds as precursors homogentisic acid and phytyl diphosphate. At least 5 enzymes 2-methyl-6-phytylbenzoquinol methyltransferase (VTE3), 4-hydroxyphenylpyruvate dioxygenase, γ-tocopherol methyltransferase (VTE4) homogentisate phytyl transferases (VTE2), tocopherol cyclase (VTE1), are convoluted in the synthesis of tocopherols, exclusive of the bypass pathway of phytyl-tail synthesis and utilization [71]. Tocopherols protects the PSII function and structure by quenching and chemically reacting with O_2 in chloroplast and thereby safeguarding the lipids and other cell membrane components. Tocopherols act as an efficient source of free radical trap by terminating the propagation of chain in lipid oxidation. Completely reduced phytyl chain of tocopherol and entirely substituted benzoquinone ring act as antioxidants in redox reaction with $^1O_2$ [72]. Quenching of O_2 by tocopherols is highly effective, and nearly about 220 molecules of O_2 are neutralized by single molecule of α-tocopherol *before* being degraded in vitro [73]. AsA, GSH [73], or coenzyme Q [74] helps in regeneration of reduced tocopherol back from its oxidized state. Tolerance to water deficit, salinity, and chilling in different plant species is seen to be enhanced by increasing the buildup of α-tocopherol [75, 76]. From studies, it is seen that GSH pools and endogenous AsA pathways in leaves are affected by metabolically engineering the biosynthesis of tocopherol. Studies have proposed that gene expression of APX, DHAR, and MDHAR that encode enzymes for Halliwell–Asada cycle is enhanced [31, 71].

2.3.2.4   CAROTENOIDS

Carotenoids capable of detoxifying various forms of ROS are lipophelic antioxidants [68–70], found in both plants and microorganisms. Carotenoids, as an antioxidant scavenges the $^1O_2$ and quenches the excited chlorophyll (Chl*) and triplet sensitizer (3Chl*) to prevent $^1O_2$ formation and thereby protects the photosynthetic apparatus from oxidative damage. They also influence plant development and biotic/abiotic stress responses by acting as substrates for signaling molecules [77]. The carotenoids ability to scavenge, inhibit, or reduce triplet chlorophyll production can be accounted for in terms of their chemical specificity. Carotenoids allow easy uptake of energy from the exited molecules and release of excess energy as heat due presence of isoprene chain residues containing enormous conjugated double bonds [39]. Under saline conditions, higher content of carotenoids helps in adapting sugarcane plant [78].

2.3.2.5   PHENOLIC COMPOUNDS

Phenolics comprise of diverse secondary metabolites (lignin, tannins, flavonoids, and hydroxycinnamate esters) having antioxidant function. They are present lavishly in tissues of plant [79, 80]. The aromatic ring of polyphenols together with the substituents enhances its biological property also involving antioxidant activity. Because of their robust capability to donate hydrogen atoms and electrons, phenols have been shown to overtake well-known antioxidants such as α-tocopherol and AsA during in vitro antioxidant assays. Polyphenols can directly impede the peroxidation of lipids by lipid alkoxyl radical trapping, can chelate transition metal ions and can also directly scavenge molecular species of active oxygen. They also reduce the fluidity of membrane by modifying the packing order of lipid [81]. Furthermore, it is shown that PODs oxidize phenylpropanoids and flavonoids and act in scavenging of H_2O_2 phenolic/AsA/POD system. Studies have shown that phenolic metabolism is induced in plants in response to variety of environmental stresses [82]. ROS serves as a signal to phenolic accumulation in response to the Cu^{2+} stress and also causes increase in total phenolic concentration in dark-grown roots of lentil [83].

2.3.3   ENZYMATIC ANTIOXIDANTS

The enzymatic line of defense comprises of a number of antioxidant enzymes such as SOD, CAT, GPX, enzymes of AsA–GSH cycle APX, monodehydro-ascorbate reductase (MDHAR), DHAR, and GR [30, 65]. These enzymes operate in different subcellular compartments and respond in concert when cells are exposed to oxidative stress. Table 2.1 shows various antioxidant enzymes that play important role in scavenging stress-induced ROS generated in plants.

TABLE 2.1   Major Antioxidative Enzymes

Enzyme	Abbreviation	Enzyme Commission Number
GR	Glutathione reductase	1.6.4.2
SOD	Superoxide dismutase	1.15.1.1
APX	Ascorbate peroxidase	1.11.1.11
CAT	Catalase	1.11.1.6
DHAR	Dehydroascorbate reductase	1.8.5.1
MDHAR	Monodehydroascorbate reductase	1.6.5.4
GST	Glutathione S-transferases	2.5.1.18
GPX	Glutathione peroxidase	1.11.1.9

2.3.3.1   SOD

SOD (1.15.1.1) is the key enzyme playing main role in defense against oxidative stress [30, 65]). The enzyme belongs to a group of metalloenzymes that dismutates $O_2^{\cdot-}$ to H_2O_2 and O_2 [65]. It is present in most of the subcellular compartments that generate activated oxygen Three isozymes of SOD copper/zinc SOD (Cu/Zn–SOD), manganese SOD (Mn-SOD), and iron SOD (Fe-SOD) are reported in plants [85, 85]. All forms of SOD are nuclear encoded and targeted to their respective subcellular compartments by an amino-terminal targeting sequence [86]. MnSOD is localized in mitochondria, whereas Fe-SOD is localized in chloroplasts. Cu/Zn-SOD is present in three isoforms, which are found in the cytosol, chloroplast, and peroxisome and mitochondria [86–88]. Eukaryotic Cu/Zn-SOD is cyanide sensitive and presents as dimer, whereas the other two (Mn-SOD and Fe-SOD) are cyanide insensitive and may be dimer or tetramers [88]. SOD activity has been reported to increase in plants exposed to various environmental stresses, including drought and metal toxicity [55]. Increased activity of SOD is often correlated with increased tolerance of the plant against environmental stresses. It was suggested that SOD can be used as an indirect selection criterion for screening drought-resistant plant materials [89]. Overproduction of SOD has been reported to result in enhanced oxidative stress tolerance in plants [30]. H_2O_2 produced is removed by a number of different peroxidase enzymes such as those in Halliwell–Asada cycle that uses AsA, reduced GSH, and NADPH as electron donors [30, 65, 86]. Three SOD enzymes are distinguished by their locations and covalently linked catalytic metal ions. Little difference exists among three types of SOD enzymes regarding their enzymatic properties and the amino acid sequence [90]. Transcription of SOD encoding genes is under the regulation of oxidative stress and it has been established that transcription of mitochondrial MnSOD gets activated under the high ROS levels in mitochondria [91]. The induction of SOD in plant cells in response to different stressful environments reflects its important role in the defense mechanism of plants. Water-deficit conditions generally increase the activities of SOD to scavenge O_2 radicals [92]. Overexpression of SOD in chloroplast also enhanced the activities of FeSOD, APX, DHAR), and MDHAR and the concentrations of GSH and AsA. Increased MDHAR, DHAR, and GR activities have also been observed in transgenic maize overexpressing MnSOD [93]. The overactivity of MnSOD was found to protect the membranes but not the PSII reaction center against MV-induced oxidative damage [94, 95]. MnSOD activity can also protect the plant from visible injury caused by ozone [96, 97]. Altered expression of SOD may in turn lead to the expression of other enzymes associated with the stress resistance [98]. The transgenic *Arabidopsis* plants showed higher tolerance to salt as compared to the wild plants. Further analyzes revealed that despite the enhanced activities of Mn–SOD, the activities of other antioxidative enzymes such as Cu/Zn–SOD, Fe–SOD, CAT, and APX of transgenic plants treated with salt were markedly higher than those of wild-type plants, indicating the enhanced ability of the antioxidant metabolites to scavenge/detoxify ROS in transgenic plants [91, 99–101].

2.3.3.2 CAT

Among antioxidant enzymes, CAT (1.11.1.6) was the first enzyme to be discovered and characterized. It is a ubiquitous tetrameric heme-containing enzyme that catalyzes the dismutation of two molecules of H_2O_2 into water and oxygen [102, 103]. It has high specificity for H_2O_2, but weak activity against organic peroxides. Plants contain several types of H_2O_2-degrading enzymes; however, CATs are unique as they do not require cellular reducing equivalent. CATs have a very fast turnover rate, but a much lower affinity for H_2O_2 than APX. The peroxisomes are major sites of H_2O_2 production. CAT scavenges H_2O_2 generated in this organelle during photorespiratory oxidation, β-oxidation of fatty acids, and other enzyme systems such as XOD coupled to SOD [104, 105]. In plants, there are three main CAT isoforms: CAT1, CAT2, and CAT3. In total leaf extracts, CAT2 is much less abundant than CAT1, which represents ~80% of leaf CAT activity [106]. Though there are frequent reports of CAT being present in cytosol, chloroplast, and mitochondria, the presence of significant CAT activity in these is less well established [107]. Environmental stresses cause either enhancement or depletion of CAT activity, depending on the intensity, duration, and type of the stress [108, 109]. In transgenic tobacco plants, having 10% wild-type, CAT activity showed accumulation of GSSG and a fourfold decrease in AsA, indicating that CAT is critical for maintaining the redox balance during the oxidative stress [110]. Overexpression of a CAT gene from *Brassica juncea* introduced into tobacco, enhanced its tolerance to Cd-induced oxidative stress [111]. Increased catalytic activity in tobacco transgenic with sense cDNA of CAT from cotton has been found to have reduced photorespiratory loss, while the antisense constructs decreased the specific activity of the CAT, which in turn led to the corresponding increase in the CO_2 compensation point [112]. Efficiency of *KatE* to protect plants against abiotic stress was also confirmed by Mohamed et al. [113] during *Agrobacterium tumefaciens* mediated transformation in tomato. Approximately threefold higher CAT activity (61.9 μm mg^{-1} protein) and lesser ion leakage in transgenics as compared to wild-type plants resulted in not only resistance to paraquat mediated oxidative stress but also to high light illumination (800 μmol m^{-2} s^{-1}), drought and chilling stress. Transgenic *Arabidopsis* plants overexpressing hot pepper *CaCat1* were shown to have increased paraquat resistance, but not to wounding, as defined by H_2O_2 levels. The results suggest that wounding may downregulate *CaCat1* expression at the posttranscriptional level in transgenic *Arabidopsis* in order to maintain the wound-signaling process in transgenic lines [114]. Al-Taweel et al. [115] examined the influence of salt stress on the repair of PSII and the synthesis of D1 in wild-type tobacco (*Nicotiana tabacum* "Xanthi") and in transformed plants expressing the CAT gene *katE* from *Escherichia coli*. The high tolerance of katE transgenic tobacco plants was suggested to be due to increased ability of the chloroplast's translational machinery of these plants to scavenge H_2O_2 under salt stress.

2.3.3.3 GPX

GPX a widely accepted stress enzyme and a protein with heme group that mostly prefers guaiacol and pyragallol aromatic electron-donating groups, GPX with four sealed disulfide bonds coupled structurally with a pair of Ca^{2+} ions [116, 117]. Associated with many pivotal biosynthetic processes including cell wall lignification, IAA degradation, ethylene biosynthesis, and wound healing many of the isoenzymes of this system occurs in plants within specialized tissues such as vacuoles, cell walls, and the cytosol and protects plants against various biotic and abiotic stressful conditions [118, 119]. Under stressful conditions, GPXs effectively quenches various forms of reactive oxygen and peroxy radicals and have been reported to increase under stressful environmental conditions comprising heavy metal pollution, herbicide stress, ozone, and stress from various polyacyclic aromatic hydrocarbons [108, 120, 121]. Tayefi-Nasrabadi et al. [121] also established that in salt-tolerant safflower plants the greater protection to salt-induced oxidative damage is provided through increase in activity of GPX, efficient catalytic activity, and introduction of some definite isoenzymes in comparison to salt-sensitive cultivars.

2.3.3.4 ENZYMES OF ASA–GSH CYCLE

For cell to respond quickly and to sense oxidative stress caused by a particular stress, the change in AsA/DHA and GSH/GSSG ratio is one of the pivotal indication. This AsA–GSH cycle commonly called as Halliwell–Asada recycling pathway for AsA and GSH regeneration, which also helps in the detoxification of H_2O_2. This successive oxidation–reduction cycle involves AsA, GSH together with NADPH, the catalysis of which is carried out by enzymes like APX, MDHAR, DHAR, and GR as represented in (Figure 2.1). The important subcellular organelles like cytosol, Chloroplasts, peroxisomes, and mitochondria are the four differently important sites in which AsA–GSH cycle exists [30, 31, 65]. This cycle has a prominent role in alleviating oxidative stress caused by different environmental factors [30, 65, 108].

2.3.3.5 APX

Among one of the principal components of AsA–GSH enzyme cycle, APX plays a pivotal role in regulation of different ROS levels within different cell compartments. This enzyme uses a pair of AsA molecules to reduce H_2O_2 to H_2O with a simultaneous generation of two MDHA molecules. APX belongs to super first-class family of heme PODs, synchronized by redox signaling and water [30, 65, 122]. In higher plants at different subcellular locations, chemically and enzymatically five distinct isoenzymes [65, 122] including (cytosolic, stromal, thylakoidal (tAPX), mitochondrial, and peroxisomal) [30, 65, 123]. of APX have been reported based on the arrangement of amino acids. APX found within different cell organelles helps in scavenging of H_2O_2 within these cell sites. Cytosolic APX (cAPX) helps in the elimination of H_2O_2 produced within the cytosol complex of the cells, apoplast, or that disseminated from organelles [30, 65]. On the basis of subcellular location, three types of APX have been identified (i.e., chloroplastic, cytosolic, and glyoxysomal). cAPX is likely involved in pathogen response, whereas glyoxysomal APX is involved in detoxification of H_2O_2 produced during fatty acid β-oxidation and photorespiration [70, 124, 125]. Chloroplastic APX (chlAPX) comprises tAPX and stromal (APX) isoforms and scavenge the H_2O_2 produced during photosynthesis [126]. Teixeira et al. [127] identified about eight isoforms of APX in rice genome and reflect the correlation between complexity of antioxidant system and large number of APX encoding genes in rice.

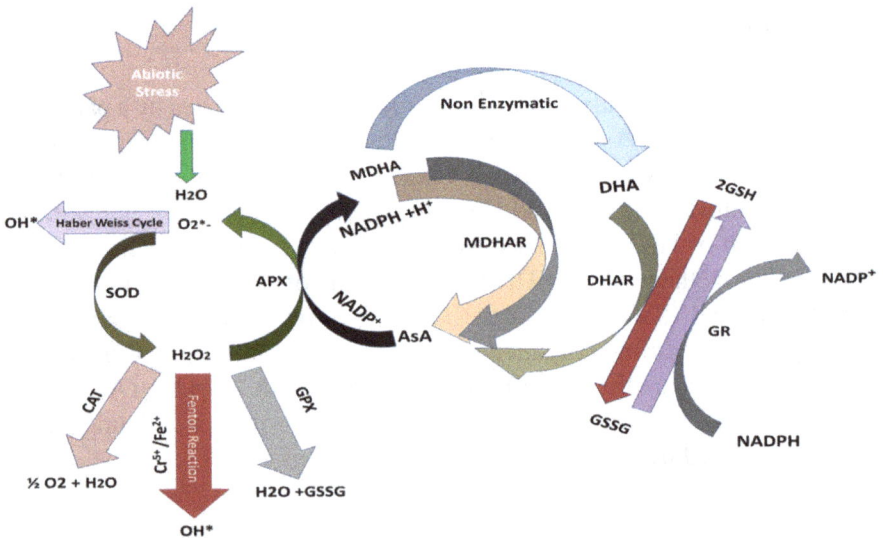

FIGURE 2.1   Ascorbate–glutathione pathway.

In plant cells, the APX is viewed to be one of the most widely spread antioxidant enzymes and isoforms of APX have much-advanced attraction for H_2O_2 than CAT, thus under stressful conditions APXs acts as competent scavengers of H_2O_2 [128]. The reports of increasing APX activity in response to different abiotic environmental stresses, for example drought, salinity, metal toxicity, chilling, and UV exposure [55, 108, 129]. cAPX-gene overexpression in transgenic plants of tomato (*Lycopersicon esculentum* L.) derived from pea (*Pisum sativum* L.) was reported to ameliorated oxidative damage caused by chilling and salt stress [130]. Similarly, overexpression of the caAPX gene increased tolerance to oxidative stress [131]. The detoxification of H_2O_2 by APX is followed by a set of reactions catalyzed by MDHAR, DHAR, and GR. All these four enzymes including APX are part of an AsA–GSH cycle, an important antioxidant mechanism in plants for the detoxification of ROS produced during respiratory chain reactions, operating in different cellular compartments including chloroplasts, peroxisomes, and cytosol [127, 132], and has important role in overcoming photosynthetic stress and senescence [123, 133].

2.3.3.6   MDHAR

MDHAR (1.6.5.4) generated in APX catalyzed reaction MDHA radical has a very short lifespan, if not quickly reduced, it disparate to AsA and DHA [30, 31, 65]. MDHAR belongs to FAD enzyme group, which help in the generation of AsA from MDHA radical by catalysis by the use of the regeneration of AsA from the MDHA radical using NAD(P)H as the electron donor [134]. It is the only reported enzyme that uses the MDA organic radical (MDA) as a substrate and has the ability of reducing phenoxyl radicals liberated by horseradish peroxidase coupled with H_2O_2 [135]. MDHAR isoenzymes reported to be present in various cellular organelles including chloroplasts [65, 30, 136] cytosol, mitochondria, and peroxisomes [65, 123]. within chloroplasts, MDHAR plays two important physiological functions: firstly, MDHAR helps in the maintenance of the regeneration cycle of AsA from MDHA and secondly, it aid in the mediation of the photoreduction of dioxygen to $O2^{\cdot-}$ when MDHA as substrate is not present [137]. The increase in the activity of MDHAR has been witnessed in several plant studies under varied conditions of stressful environmental conditions [55, 108]. Enhanced tolerance to salt and polyethylene glycol stresses in tobacco was successfully reported due to the overexpression of *Arabidopsis* MDHAR gene [138]. Tomato chloroplastic isoforms of MDHAR enhanced tolerance for temperature and methyl viologen when it was overexpressed and examined in transgenic *Arabidopsis* plants [68, 71].

2.3.3.7   DHAR

A monomeric thiol enzyme DHAR (EC 1.8.5.1) is commonly reported and abundantly present in dry seeds, roots, and green shoots, which uses GSH as a substrate and helps in catalyzing the reduction of DHA to AsA [31, 32, 139] there by helps in performing the important function of maintaining AsA in its reduced form. DHAR overexpression in plants such as tobacco, maize, and potato is reported to escalate the levels of AsA content, which help us in determining that DHAR plays an significant roles in shaping the AsA pool size [140, 141]. The increased activity of DHAR in different plants under different stressful environmental conditions notably among them was drought, metal toxicity, and chilling stress [55, 108, 129]. Constant upregulation of gene programming cytosolic DHAR was reported from *L. japonicas*, reported to be more tolerant to salt stress than other kinds of legumes. This DHAR upregulation was interrelated to its role in AsA recycling in the apoplast [142]. Overexpression of *Arabidopsis* cytosolic AtDHAR1 in transgenic potato reported to show higher tolerance to herbicide, drought, and salt stresses [138].

2.3.3.8   GR

GR (GR, EC 1.6.4.2), a NAD(P)H-dependent enzyme that brings about the catalytic reduction of GSSG to GSH there by maintains good cellular GSH/GSSG ratio. GR is a member of flavoenzymes group, containing an indispensable disulfide group. The catalysis mechanism is a two-step process; in first step NADPH reduces the moiety of Flavin, when the Flavin gets oxidized, the reduction of redox-active disulfide bridge leads to the production of thiolate anion and a cysteine. In the second involves the reduction of GSSG through the interchange reactions of thiol sulfide group [143]. GSH as an antioxidant, it participates in enzymatic and nonenzymatic oxidation–reduction processes, GSH gets oxidized to GSSG. In this AsA–GSH cycle, in this reaction process GSH gets oxidized the catalysis of which is brought about by DHAR. Although, its presence in cell organelles (chloroplasts, cytosol, mitochondria, peroxisomes) about 80% of GR activity is by chloroplastic isofroms in photosynthetic tissues of plants [144]. In chloroplast of plants H_2O_2 produced by Mehler reaction, GSH and GR are the two enzymes involved in its detoxification. The increased activity of GR has been documented by several authors under different stressful environmental conditions [55, 108, 129, 145]. The increased susceptibility to chilling stress was reported to be associated with increased activity of GR in relation to antisense-mediated depletion of tomato chloroplast [146].

The role of GR and GSH in the H_2O_2 scavenging in plant cells has been well established in the Halliwell–Asada pathway [147]. GR is involved in the recycling of reduced GSH, providing a constant intracellular level of GSH. High GSH levels, in particular high GSH/GSSG ratios, are associated with low H_2O_2 concentrations. GSH plays a significant role in redox balance of the cell [148] and detoxification of xenobiotics and heavy metals [149]. Overexpression of GR in *N. tabacum* and Populus plants leads to higher foliar AsA contents and improved tolerance to oxidative stress [64, 150]. Due to the complexity of ROS detoxification system, overexpressing one component of antioxidative defense system may or may not change the capacity of the pathway as a whole [151]). Transformation of a poplar hybrid, *Populus tremula × Populus alba*, with the bacterial genes for either *GR* (*gor*) or GS (*gshll*) resulted in the overexpression of GR in the chloroplast, which led to increase in the antioxidant capacity of the leaves and improvement in the capacity to withstand oxidative stress [64]. An important pathway by which plants detoxify heavy metals is through sequestration with heavy-metal-binding peptides called phytochelatins or their precursor, GSH. In order to develop transgenic plant varieties with increased bioaccumulation or tolerance competency to different metals and metalloids. In addition, instantaneous expression of several antioxidant enzymes, mostly those of Cu/Zn-SOD, APX, and DHAR, in chloroplasts are reported to be more operational than those expressed singly or in combination of pairs for developing transgenic plant varieties with improved tolerance to combined environmental stresses [26, 64, 147]. Therefore, in order to achieve multiple environmental stresses tolerance, more importance is placed to develop those transgenic plants with multiple overexpressing of different antioxidants.

2.4 ROS AS SIGNALING MOLECULES UNDER STRESS

Plants are equipped with a web of stress signals consisting of redox homeostasis, antioxidant signaling, and continuous production and scavenging ROS at cellular level [30, 152]. The ability of the cell to maintain hemostasis in order to overcome plethora of stress responses, several ROS are involved in signaling process thereby initiating expression of a network of genes actively participating in signal related transduction pathways, which advocates that ROS plays an active role in signaling through activation and control of various stress responses. It has also been validated that ROS are not only involved in stress responses but they are the active participants of many vital processes as energy metabolism and regulation of cell function [18]. The process by which plants achieve a state of hemostasis under stress conditions is activated by retrograde signaling and through regulation of systematic signaling pathways [153, 154].

ROS on the one hand are reported to be potentially damaging but on the other hand are actively involved in various signal transduction processes during abiotic and biotic stress. In addition, they play a crucial role during PCD and variety of other developmental processes [1, 155]. Oxidative burst, i.e., a high level of ROS production, has been demonstrated to be essential for many of these processes. RBOH genes that code NADPH oxidases associated with plasma membranes are the key players associated with the ROS-related signal transduction [14, 152, 156]. H_2O_2 produced by cytosolic membrane-bound NADPH oxidase increases abruptly during both abiotic and biotic stress conditions, is considered as a signaling molecule prevailing during stressful conditions [157]. Metabolic imbalance in the plants during stressful conditions leads to the generation of ROS, channeled to function as a signaling molecule to trigger defense mechanism and to achieve acclimation, which in turn leads to alleviate stressful conditions causing oxidative damage [10, 158]. ROS produced in different cell compartments bring about alterations at transcriptional level involving nuclear transcriptome, but the mechanism behind this signal transduction is poorly understood [159]. Over-production of H_2O_2 and O_2 in peroxisomes promotes oxidative damage but moderate levels act as signaling molecules mediating pathogen induce PCD in plants [160, 161].

Significant progress has been attained in understanding the part played by ROS in signaling and it is increasingly becoming clear now that these are major signaling molecules actively involved in several important processes in plants [162]. ROS acts as a biological signal by influencing the expression of several genes thereby regulating stress [157, 163]. It has been demonstrated that at an elevated level of ROS triggers changes in gene expression, such changes happen to be occur through the oxidation process involving different modules of signaling pathways that in turn switch on the transcriptional factors, possibly those being redox sensitive [157]. A comparative analysis of transcriptome in response to different stress conditions suggests that out of 286 O_2-responsive transcripts [164] 180 overlapped with H_2O_2 responsive transcripts [165]. It is, therefore, easy to understand why O_2 and H_2O_2 signaling function in the same cascade because most of the O_2 generated in the cell gets dismutated to H_2O_2 either spontaneously or through the catalytic conversion through SOD.

Adachi [166] has identified a WRKY TF that is phosphorylated by MAPK and a W-box in the promoter region of *Nicotiana tabacum* RBOH, inter connecting the phosphorylation events of MAPK in response to pathogen recognition with the accumulation of RBOH protein. The different members of the NAC family of TFs [167–169], APETALA2/ethylene response TF redox responsive transcription factor 1 that are regulated by different WRKYs [170], and different zinc-finger proteins such oxidative stress 2 [171] like ROS-response regulatory proteins were identified besides TFs, calcium (Ca) waves comprise important components of systemic signaling in plants [172–175]. Several findings summarized that RBOH acts as a central hub in the cellular ROS-signaling network [152, 156] and this RBOH functions along with MAPK pathways to integrate ROS signals and modulate cell-to-cell signal propagation in local and systemic signaling [15, 175]. These MAPK pathways take part in retrograde signaling from the chloroplast to the nucleus [18, 176]. Thus, MAPK pathways are of fundamental as well as far-reaching importance in converting ROS signals in to protein phosphorylation.

In plants, chloroplast redox state is well recognized in understanding redox-regulated gene expression [12] and any alteration in redox states of chloroplast triggers changes in chloroplast proteins, AsA, GSH, plastoquinone, and ROS along with ferridoxin system suggesting their key role as signaling components of the chloroplast. Production of $^1O_2$ and O_2^{*-} in plants is associated with PSI and PSII. In comparison with other ROS molecules, $^1O_2$ exhibits some equivocal characteristics during signaling process related to PCD [162, 177]. ROS generation in plants is believed to function as a second messenger as that of highly regulated Ca^{2+}.

2.5 MANIPULATION OF GSH–ASA PATHWAY

Large number of genes are activated in a plant under stress, resulting in increased levels of several enzymes, proteins, and metabolites responsible for stress tolerance (Table 2.2). Plants possess various enzymatic (SOD, APX, POD, and CAT) and nonenzymatic antioxidants (AsA, GSH) to keep the balance between the production and removal of ROS. *Arabidopsis thaliana* plants responded to salt stress by upregulation of genes involved in defense against oxidative stress [38] suggesting that antioxidants play a crucial role in protecting cells from oxidative damage under salt stress in *P. patens*, while studying the role of various antioxidants in salt tolerance in rice genotypes, reported that salt-sensitive genotype showed an increased level of $O2^-$, H_2O_2 and MDA and a decreased level of AsA, GSH, and thiol with lower antioxidant enzyme activity compared to salt-tolerant genotypes. Higher level of antioxidants (AsA and GSH) with higher antioxidant enzyme activities (SOD, APX, GR, GPX, and CAT) are the main determinants for depicting tolerance in salt-tolerant rice genotypes [178].

Salt-tolerant *Plantago maritima* showed an increased activity of SOD, CAT, GR, and APX with a lower level of MDA and a better protection mechanism against oxidative stress than the salt-sensitive *P. media* under salt stress [64]. PODs and NADP-dehydrogenases are the key antioxidative enzymes in olive plant under salt stress [179].

Chilling-induced oxidative stress leads to an alteration in antioxidant enzyme activities. Cucumber and maize under chilling stress showed an increased activity of antioxidant enzymes (SOD, APX, GR, MDHAR, and DHAR) increased during chilling [180–182]. However, if the duration of chilling stress is too long, the defense system may not remove overproduced ROS effectively, which may result in severe damage or even death [183].

Various enzymatic and nonenzymatic components are involved in defense mechanism against various metal stresses [55, 184, 185]. The increased activities of enzymatic (SOD, GPX, APX, MDHAR, DHAR, and GR) as well as nonenzymatic antioxidants in metal-treated plants suggests the involvement antioxidant defense mechanism in adaptive response to metal ions [55, 56, 184, 185].

After exposure to arsenite [As(III)] and arsenate [As(V)], the comparative antioxidant profiling of tolerant (TPM-1) and sensitive (TM-4) variety of *Brassica juncea* L. showed the higher level of GSH and antioxidant enzymes in TPM-1 than in TM-4, thus allowing TPM-1 to tolerate higher concentrations than TM-4 [56]. In *Picea asperata* seedlings, UV-B induced overproduction of ROS resulting in oxidative stress that enhanced the efficiency of antioxidant defense system consisting of antioxidant enzymes SOD, CAT, APX, and GPX, carotenoids, and UV-B absorbing compounds [186]. In *Arabidopsis*, peroxidase-related enzymes are induced preferentially by UV-B exposure. Gao and Zhang [187] observed that AsA-deficient mutant *vtc1* was found to be more sensitive to UV-B treatment than the wild-type plants suggesting that AsA is an important antioxidant for UV-B radiation. *Vicia faba* leaves infected with yellow mosaic virus showed enhanced antioxidant enzyme activities suggesting an important role of ROS scavenging system in managing ROS generated in response to biotic stress [188].

The AsA–GSH cycle operates in four different subcellular locations, including the cytosol, mitochondria, chloroplast, and peroxisomes and plays an important role in combating oxidative stress induced by environmental stress [30, 32, 65] MDHAR showed increased in many plants under environmental stress [30, 32, 65]. Overexpression of *Arabidopsis* MDHAR gene in tobacco increased its tolerance to polyethylene glycol and salt stresses [67]. Overexpression of tomato chloroplastic MDHAR arabidopsis increased its tolerance to temperature and methyl viologen-mediated oxidative stress [71].

DHAR is a monomeric thiol enzyme found abundantly in green shoots, roots, dry seeds, and etiolated shoots. Overexpression of DHAR in maize, tobacco, and potato leaves increased the AsA content suggesting an important role of DHAR in determining AsA pool size [141]. Various abiotic stresses (drought, chilling, and metal toxicity) increases the DHAR activity in plants [32, 189]. *L. japonicas* under salt stress, with consistent upregulation of cytosolic DHAR gene was found to be more tolerant than other legumes. This upregulation of DHAR was correlated to its role in recycling of AsA in apoplast [142]. Overexpressing arabidopsis cytosolic AtDHAR1 in transgenic potato increases its tolerance to salt, drought, and herbicide stresses [138].

Overexpression of GR in Populus and *N. tabacum* leads to higher foliar AsA content and hence improves their tolerance to oxidative stress [190]. Due to the complexity of ROS scavenging system, overexpression of one component of this system may not change the capacity of the whole pathway [151]. Overexpression of combination of these antioxidant enzymes in transgenic plants has synergistic effect on stress tolerance [191–193].

Overexpression of SOD in chloroplast increased the activities of FeSOD, DHAR, APX, and MDHAR and hence increased the concentration of AsA and GSH. Overexpression of MnSOD in transgenic maize increased GR, DHAR, and MDHAR activities [93]. Overexpression of SOD leads to enhancement of the tolerance to MV-dependent oxidative stress only if other antioxidant enzymes (APX, MDHAR, DHAR, AsA, and GSH) are also present at an increased level [94] suggesting that the overexpression of SOD would increase the oxidative stress tolerance provided the other antioxidant enzymes did not limit the oxygen-radical-scavenging capacity [94, 95, 181]. Overexpression of Mn–SOD in transgenic *Arabidopsis* increased the activity of Mn–SOD above twofold than that of as wild type and showed higher tolerance to salt stress compared to the wild type [99, 181]. Enhanced activity of Mn–SOD was supplemented with the increased activities of other antioxidant enzymes (Cu/Zn–SOD, Fe–SOD, CAT, and APX) in transgenic plants indicating the enhanced ability of the antioxidant metabolites to scavenge/detoxify ROS in transgenic *Arabidopsis* than those of wild type under salt stress [181, 194]. In another study, it was found that the increased APX expression observed in CuZn-SOD overexpressing transgenics to be unstable and APX activities returned to normal levels within two or three sexual generations. The ratio of specific enzyme activity of SOD and APX is more important than the total enzyme activity of each enzyme in abiotic stress protection [195].

APX localized in cytosol, chloroplasts, mitochondria, and peroxisomes is a key enzyme that scavenges the potentially harmful H_2O_2 by utilizing AsA as a specific electron donor [31, 32, 65]. APX scavenges H_2O_2 through a series of reactions catalyzed by MDHAR, GR, and DHAR. All these four enzymes including APX are part of AsA–GSH cycle, an important antioxidant mechanism for the detoxification of ROS in plants [31, 32]. Tobacco Overexpressing peroxisomal *APX* gene of *Arabidopsis* showed an increased protection against aminotriazole induced oxidative stress, which otherwise inhibits CAT activity in peroxisomes and glyoyxsomes, and leads to H_2O_2 accumulation in these organelles [33, 196]. However, these plants did not show increased protection against paraquat-induced oxidative damage which leads to the ROS production in chloroplasts, suggesting the organelle-specific role of APX in protection against oxidative stress. Tobacco plants overexpressing *Capsicum annuum* APX-like 1 gene (*CAPOA1*) showed a twofold increase in peroxidase activity with an increased tolerance to MV-mediated oxidative stress and resistance to the pathogen, *Phytophthora nicotianae* along with no morphological abnormalities [197]. Overexpression and suppression of spinach tAPX in transgenic tobacco showed three times increase in the level of tAPX in transgenics with increased tolerance to chilling and photooxidative stress and no change in morphology compared to the wild type. However, the transgenic plants with suppressed levels of tAPX did not reach to maturity [198]. Overexpression of tAPX in the chloroplasts of transgenic tobacco and *Arabidopsis* plants enhanced their tolerance to paraquat and high light-induced photooxidative stress [199], while characterizing single gene mutants of *Arabidopsis* lacking either sAPX or tAPX gene, mutant plant under high light intensity stress accumulated the oxidized proteins and H_2O_2 at higher levels than their wild types, suggesting that both chlAPXs, particularly tAPX are important for gene regulation and photoprotection of *Arabidopsis* leaves under photooxidative stress [195, 200, 201]. Despite the inactivation of APX in the chloroplast, the transformation of tobacco chloroplast with *E. coli* CAT gene (with higher affinity for H_2O_2 than plant CAT) under tomato *rbcS3C* promoter enhanced plant resistance to drought stress-induced photooxidation [202] with no chlorophyll destruction [203]. The overexpression of same gene under CaMV35S promoter in Japonica rice (*Oryza sativa*) increased the plants tolerance to the salt stress with increased CAT activity.

GR maintains the GSH pool in cell by utilizing NADPH for reducing the disulfide bonds of GSSG [204, 205]. Halliwell–Asada pathway reveals the role of GR and GSH in scavenging H_2O_2 in plant cells [195, 206]. GR reduces GSH, by providing a constant intracellular level of GSH [12, 207]). GR acts as a reductant during the formation of AsA from DHA catalyzed by DHAR [30, 65].

TABLE 2.2 Antioxidant Enzymes Activated in Response to Oxidative Stress Induced by Various Environmental Stresses

Plant Species	Stress Type	Antioxidant Enzyme	Reference
Oryza sativa	Drought	SOD, GPX, APX, MDHAR, DHAR, GR	[108]
Beta vulgaris		SOD, CAT, GPX	[208]
Triticum aestivum		SOD, APX and GR	[209]
Oryza sativa	Salinity	SOD, CAT, GPX, APX, GR	[178]
Oryza sativa		GPX	[210]
Olea europaea		SOD, CAT, and GR	[211]
Zea mays	Chilling	APX, MDHAR, DHAR, GR, SOD	[212]
Oryza sativa	Metals Al	SOD, GPX, APX	[213]
Oryza sativa	Ni	SOD, GPX, APX	[129]
Oryza sativa	As	SOD, GPX, APX	[55]
Oryza sativa	Mn	SOD, GPX, APX, GR	[56]
Linum usitatissimum	Fungal	GPX, CAT	[214]
Vicia faba	Viral	POD, CAT, APX, SOD	[188]

2.6 TRANSCRIPTIONAL FACTORS: A KEY TO STRESS RESPONSIVE GENE EXPRESSION

Increased ROS levels trigger specific targets to facilitate gene expression. Oxidation of signaling pathways occurs upon changes in transcriptional activity that in turn activates transcriptional factors or act directly by modification of redox-sensitive transcriptional factors. Mitogen-activated protein kinases are affected by ROS molecules to facilitate indirect activation of transcriptional factors [30]. Recently, a link between ROS-response ZAT12 a zinc finger protein and iron regulation were deciphered, which elucidated that a very delicate equilibrium existing between iron and ROS are crucial for plant growth and development [215, 216]. From the study, it has been demonstrated that ZAT 12 in response to overaccumulation of ROS is upregulated, which results in the suppression of FER-like iron deficiency-induced transcription factor thereby preventing risk of ROH formation. Moreover, in recent years, different members of the NAC family of TFs [217–221], APETALA2/ethylene response Tf redox responsive transcription factor 1 that are regulated by different WRKYs [222], and different zinc-finger proteins such oxidative stress 2 [192] like ROS-response regulatory proteins were identified. Transcriptional factors in response to abiotic stress have been extensively studied by many researchers during the last century [218–223]. Several differentially expressing stress-inducible genes have also been identified, during the course of study, the predictions have also been made about common cis-acting promoter elements controlling the similar expression pattern [224]. Furthermore, it is also argued that these cis-acting promoter elements are recognized by specific transcriptional factors to trigger gene expression in response to stress perception as demonstrated by (CBF/DREB1) regulon. Super transcriptional factor family AP2/ERF is unique to plant species. CBF/DREB1 belong to this small group. DRE/CRT a cis-acting element localized in promoters of various drought and cold-inducible genes is recognized by this small group of proteins [225, 226].

Under stress conditions, these transcriptionally stress-inducible genes transactivate downstream gene expression. Near about 40 downstream target proteins have been identified that are regulated through a single CBF3/DREB3 protein. All these downstream genes share a commonality that they are cold inducible, possesses c-repeat DRE/CRT sequence in their promoters, and are upregulated upon ectopic expression of CBF3/DREB1A [218, 219, 221]. CBF/DREB1 transcriptional factors are the key players of cold and freezing tolerance in plants, attributed to this regulon through widespread transcriptional reconfiguration and reprogramming. During low-temperature responses CBF/DREB1 TFs are considered to be master regulators; however, during other set of adverse environmental conditions the additional function attributed to this regulon is the weakly induction of DREB1F/DDF1, CBF4/DREB1D, and DREB1E/DDF2 transcriptional factors [227, 228]. The regulatory function of these genes was confirmed from the comperative genomic studies of their orthologs [225, 228]. Bioengineering of CBF/DREB1in plants seems to promising in achieving stress tolerance because of its utmost importance in regulating plant responses to low temperature and to understand the mechanism of expression of this regulon [229, 230, 221]. CBF/DREB1 genes are induced upon perception of cold in this regard two conserved sequences have been identified which cause the induction ICEr1 and ICEr2 of CBF expression region [230].

2.3 CONCLUSION AND FUTURE PERSPECTIVES

In plants, unfavorable environmental conditions lead to increased production and accumulation of ROS. To protect themselves from the detrimental effects of these lethal ROS molecules plants have developed an intricate antioxidant defense mechanism. Furthermore, to overcome several environmental insults an information cascade that starts with efficient signal perception is also required as an adaptive response. AsA–GSH pathway plays a central role in cellular metabolism and signaling. ROS have pleiotropic effects in plants. Under normal growth conditions, ROS released from specific compartments play a very decisive roles in plant metabolism and signaling. However, ROS overproduction results in oxidative stress, which can cause cell damage and even can lead to cell death. The antioxidant defense system must keep ROS under control in order to prevent oxidative stress and allow productive functions of ROS to continue. Recently, profound understandings of the strategic roles of enzymes such as APX, SOD, and CAT in antioxidant defense have come to lame light through various technological advancements. Individually, these enzymes play a fundamental role in antioxidant defense. From our view point it seems to be of potential interest and importance to explore the role of enzymes involved in the synthesis and metabolism of AsA and GSH. Several recent studies have highlighted the vital biological roles played by AsA–GSH pathway in plants; however, molecular strategies hold a key to future developmental processes. Manipulation of AsA–GSH pathway in plants through overexpression of its individual genes or pyramiding all the genes of the pathway could play a significant role in plants to combat multiple environmental stresses. Molecular technologies are making an important contribution in understanding the novel ways that will allow exploitation of ROS and antioxidants molecules to manipulate redox status of cells and recognize characters that could be utilized for sustainability of agricultural production for breeding programs.

KEYWORDS

- antioxidant enzymes
- environmental stresses
- glutathione-ascorbate cycle
- reactive oxygen species (ROS)

REFERENCES

1. Mittler, R. (2017). ROS are good. *Trends in Plant Science, 22*(1), 11–19.
2. Chaudhari, P., Ye, Z., & Jang, Y. Y. (2014). Roles of reactive oxygen species in the fate of stem cells. *Antioxidants & Redox Signaling, 20*(12), 1881–1890.
3. Barna, B., Fodor, J., Harrach, B. D., Pogány, M., & Király, Z. (2012). The Janus face of reactive oxygen species in resistance and susceptibility of plants to necrotrophic and biotrophic pathogens. *Plant Physiology and Biochemistry, 59*, 37–43.
4. Nath, M., Bhatt, D., Prasad, R., Gill, S. S., Anjum, N. A., & Tuteja, N. (2016). Reactive oxygen species generation-scavenging and signaling during plant-arbuscular mycorrhizal and *Piriformospora indica* interaction under stress condition. *Frontiers in Plant Science, 7*.
5. Foyer, C. H. & Noctor, G. (2003). Redox sensing and signaling associated with reactive oxygen in chloroplasts, peroxisomes and mitochondria, *Physiologia Plantarum*, 119 (3), 355–364.
6. Dietz, K. J. (2015). Efficient high light acclimation involves rapid processes at multiple mechanistic levels. *Journal of experimental botany, 66*(9), 2401–2414.
7. Mignolet-Spruyt, L., Xu, E., Idänheimo, N., Hoeberichts, F. A., Mühlenbock, P., Brosché, M., & Kangasjärvi, J. (2016). Spreading the news: subcellular and organellar reactive oxygen species production and signalling. *Journal of Experimental Botany, 67*(13), 3831–3844.
8. Sandalio, L. M., Rodríguez-Serrano, M., Romero-Puertas, M. C., & Luis, A. (2013). Role of peroxisomes as a source of reactive oxygen species (ROS) signaling molecules. In *Peroxisomes and Their Key Role in Cellular Signaling and Metabolism* (pp. 231–255). Springer Netherlands.
9. Demidchik, V. (2015). Mechanisms of oxidative stress in plants: from classical chemistry to cell biology. *Environmental and Experimental Botany, 109*, 212–228.
10. Mittler, R., Vanderauwera, S., Gollery, M., & Van Breusegem, F. (2004). Reactive oxygen gene network of plants. *Trends in Plant Science, 9*(10), 490–498.
11. Dietz, K.J. (2016). Thiol-based peroxidases and ascorbate peroxidases: why plants rely on multiple peroxidase systems in the photosynthesizing chloroplast? *Molecular Cell* 39, 20–25.
12. Foyer, C. H., & Noctor, G. (2016). Stress-triggered redox signalling: what's in pROSpect? *Plant, Cell & Environment, 39*(5), 951–964.
13. Miller, G. A. D., Suzuki, N., Ciftci-Yilmaz, S., & Mittler, R. (2010). Reactive oxygen species homeostasis and signalling during drought and salinity stresses. *Plant, Cell & Environment, 33*(4), 453–467.
14. Suzuki, N., Miller, G., Morales, J., Shulaev, V., Torres, M. A., & Mittler, R. (2011). Respiratory burst oxidases: the engines of ROS signaling. *Current Opinion in Plant Biology, 14*(6), 691–699.
15. Gilroy, S., Białasek, M., Suzuki, N., Górecka, M., Devireddy, A. R., Karpiński, S., & Mittler, R. (2016). ROS, calcium, and electric signals: key mediators of rapid systemic signaling in plants. *Plant Physiology, 171*(3), 1606–1615.
16. Huang, S., Van Aken, O., Schwarzländer, M., Belt, K., & Millar, A. H. (2016). The roles of mitochondrial reactive oxygen species in cellular signaling and stress response in plants. *Plant Physiology, 171*(3), 1551–1559.
17. Kerchev, P. I., Waszczak, C., Lewandowska, A., Willems, P., Shapiguzov, A., Li, Z., & Van Der Kelen, K. (2016). Lack of glycolate oxidase 1, but not glycolate oxidase

2, attenuates the photorespiratory phenotype of catalase 2-deficient *Arabidopsis*. *Plant Physiology, 171*, 1704–1719.

18. Dietz, K.J. (2016). Thiol-based peroxidases and ascorbate peroxidases: why plants rely on multiple peroxidase systems in the photosynthesizing chloroplast? *Molecules and Cells, 39*, 20–25.

19. Takagi, D., Takumi, S., Hashiguchi, M., Sejima, T., & Miyake, C. (2016). Superoxide and singlet oxygen produced within the thylakoid membranes both cause photosystem I photoinhibition. *Plant Physiology, 171*(3), 1626–1634.

20. Suzuki, N., Koussevitzky, S., Mittler, R. O. N., & Miller, G. A. D. (2012). ROS and redox signalling in the response of plants to abiotic stress. *Plant, Cell & Environment, 35*(2), 259–270.

21. Mur, L. A., Kenton, P., Lloyd, A. J., Ougham, H., & Prats, E. (2008). The hypersensitive response; the centenary is upon us but how much do we know? *Journal of Experimental Botany, 59*(3), 501–520.

22. Gray, J., Janick-Buckner, D., Buckner, B., Close, P. S., & Johal, G. S. (2002). Light-dependent death of maize lls1 cells is mediated by mature chloroplasts. *Plant Physiology, 130*(4), 1894–1907.

23. Kalogeris, T., Bao, Y., & Korthuis, R. J. (2014). Mitochondrial reactive oxygen species: a double edged sword in ischemia/reperfusion vs preconditioning. *Redox Biology, 2*, 702–714.

24. Marchi, S., Giorgi, C., Suski, J. M., Agnoletto, C., Bononi, A., Bonora, M., & Rimessi, A. (2012). Mitochondria-ros crosstalk in the control of cell death and aging. *Journal of Signal Transduction, 2012*.

25. Steffens, B. (2014). The role of ethylene and ROS in salinity, heavy metal, and flooding responses in rice. *Frontiers in Plant Science, 5*:685.

26. Grene, R., (2002) Oxidative stress and acclimation mechanisms in plants. in: C.R. Somerville, E.M. Myerowitz (Eds.), The *Arabidopsis* Book. American Society of Plant Biologists, Rockville, MD.

27. Rhoads, D. M., Umbach, A. L., Subbaiah, C. C., & Siedow, J. N. (2006). Mitochondrial reactive oxygen species. Contribution to oxidative stress and interorganellar signaling. *Plant Physiology, 141*(2), 357–366.

28. Reumann, S., Chowdhary, G., & Lingner, T. (2016). Characterization, prediction and evolution of plant peroxisomal targeting signals type 1 (PTS1s). *Biochimica et Biophysica Acta (BBA)—Molecular Cell Research, 1863*(5), 790–803.

29. Sandalio, L. M., & Romero-Puertas, M. C. (2015). Peroxisomes sense and respond to environmental cues by regulating ROS and RNS signalling networks. *Annals of Botany, 116*(4), 475–485.

30. Raja, V., Majeed, U., Kang, H., Andrabi, K. I., & John, R. (2017). Abiotic stress: Interplay between ROS, hormones and MAPKs. *Environmental and Experimental Botany, 137*, 142–157.

31. Qadir, S. U., Raja, V., Siddiqui, W. A., Abd_Allah, E. F., Hashem, A., Alam, P., & Ahmad, P. (2019). Fly-ash pollution modulates growth, biochemical attributes, anti-oxidant activity and gene expression in pithecellobium dulce (Roxb) benth. *Plants, 8*(12), 528.

32. Raja, V., Qadir, S. U., Alyemeni, M. N., & Ahmad, P. (2020). Impact of drought and heat stress individually and in combination on physio-biochemical parameters, antioxidant responses, and gene expression in Solanum lycopersicum. *3 Biotech, 10*, 1–18.

33. Cao, S., Du, X. H., Li, L. H., Liu, Y. D., Zhang, L., Pan, X., ... & Lu, H. (2017). Overexpression of Populus tomentosa cytosolic ascorbate peroxidase enhances abiotic stress tolerance in tobacco plants. *Russian Journal Of Plant Physiology, 64*(2), 224–234.
34. Hausladen, A., & Alschesr, R. G. (1993). Glutathione. In: Alscher, R.; Hess, J. eds. *Antioxidants in Higher Plants.* Boca Radon: CRC Press, 1–30.
35. Kangasjarvi, J., Talvinen, J., Utriainen, M., & Karjalainen, R. (1994). Plant defense systems induced by ozone. *Plant, Cell & Environment, 17*(7), 783–794.
36. Inze, D., & Van Montagu, M. (1995). Oxidative stress in plants. *Current Opinion in Biotechnology, 6*(2), 153–158.
37. Hayes, J. D., & Pulford, D. J. (1995). The glutathione S-transferase supergene family: regulation of GST and the contribution of the Isoenzymes to cancer chemoprotection and drug resistance part II. *Critical Reviews in Biochemistry and Molecular Biology, 30*(6), 521–600.
38. Wang, F., Liu, J., Zhou, L., Pan, G., Li, Z., & Cheng, F. (2016). Senescence-specific change in ROS scavenging enzyme activities and regulation of various SOD isozymes to ROS levels in psf mutant rice leaves. *Plant Physiology and Biochemistry, 109*, 248–261.
39. Mittler, R. (2002). Oxidative stress, antioxidants and stress tolerance. *Trends in Plant Science, 7*(9), 405–410.
40. Johnson, S. M., Doherty, S. J., & Croy, R. R. D. (2003). Biphasic superoxide generation in potato tubers. A self-amplifying response to stress. *Plant Physiology, 131*(3), 1440–1449.
41. Vranova, E., Inzé, D., & Van Breusegem, F. (2002). Signal transduction during oxidative stress. *Journal of Experimental Botany, 53*(372), 1227–1236.
42. De Pinto, M. C., & De Gara, L. (2004). Changes in the ascorbate metabolism of apoplastic and symplastic spaces are associated with cell differentiation. *Journal of Experimental Botany, 55*(408), 2559–2569.
43. Gao, Q., & Zhang, L. (2008). Ultraviolet-B-induced oxidative stress and antioxidant defense system responses in ascorbate-deficient vtc1 mutants of *Arabidopsis thaliana. Journal of Plant Physiology, 165*(2), 138–148.
44. Semchuk, N. M., Lushchak, V., Falk, J., Krupinska, K., & Lushchak, V. I. (2009). Inactivation of genes, encoding tocopherol biosynthetic pathway enzymes, results in oxidative stress in outdoor grown *Arabidopsis* thaliana. *Plant Physiology and Biochemistry, 47*(5), 384–390.
45. Farooq, A., Bukhari, S. A., Akram, N. A., Ashraf, M., Wijaya, L., Alyemeni, M. N., & Ahmad, P. (2020). Exogenously applied ascorbic acid-mediated changes in osmoprotection and oxidative defense system enhanced water stress tolerance in different cultivars of safflower (*Carthamus tinctorious* L.). *Plants, 9*(1), 104.
46. El-Beltagi, H. S., Mohamed, H. I., & Sofy, M. R. (2020). Role of ascorbic acid, glutathione and proline applied as singly or in sequence combination in improving chickpea plant through physiological change and antioxidant defense under different levels of irrigation intervals. *Molecules, 25*(7), 1702.
47. Hussain, I., Siddique, A., Ashraf, M. A., Rasheed, R., Ibrahim, M., Iqbal, M., ... & Imran, M. (2017). Does exogenous application of ascorbic acid modulate growth, photosynthetic pigments and oxidative defense in okra (Abelmoschus esculentus (L.) Moench) under lead stress?. *Acta Physiologiae Plantarum, 39*(6), 144.
48. Shao, H. B., Chu, L. Y., Lu, Z. H., & Kang, C. M. (2008). Primary antioxidant free radical scavenging and redox signaling pathways in higher plant cells. *International Journal of Biological Sciences, 4*(1), 8.

49. Smirnoff, N., Running, J. A., & Gatzek, S. (2004). Ascorbate biosynthesis: a diversity of pathways. In *Vitamin C: its Functions and Biochemistry in Animals and Plants* (pp. 7–29), Taylor & Francis.
50. Barnes, J., Zheng, Y., & Lyons, T. (2002). Plant resistance to ozone: the role of ascorbate. In *Air Pollution and Plant Biotechnology* (pp. 235–252). Springer, Tokyo.
51. Smirnoff, N. (2000). Ascorbic acid: metabolism and functions of a multi-facetted molecule. *Current Opinion in Plant Biology, 3*(3), 229–235.
52. Noctor, G., & Foyer, C. H. (1998). Ascorbate and glutathione: keeping active oxygen under control. *Annual Review of plant biology, 49*(1), 249–279.
53. Miyake, C., & Asada, K. (1994). Ferredoxin-dependent photoreduction of the monode-hydroascorbate radical in spinach thylakoids. *Plant and Cell Physiology, 35*(4), 539–549.
54. Asada, K. (1996). Radical production and scavenging in the chloroplasts. In *Photosynthesis and the Environment* (pp. 123–150). Springer Netherlands.
55. Mishra, S., Jha, A. B., & Dubey, R. S. (2011). Arsenite treatment induces oxidative stress, upregulates antioxidant system, and causes phytochelatin synthesis in rice seedlings. *Protoplasma, 248*(3), 565–577.
56. Srivastava, S., & Dubey, R. S. (2011). Manganese-excess induces oxidative stress, lowers the pool of antioxidants and elevates activities of key antioxidative enzymes in rice seedlings. *Plant Growth Regulation, 64*(1), 1–16.
57. Chaves, M. M., Pereira, J. S., Maroco, J., Rodrigues, M. L., Ricardo, C. P. P., Osório, M. L., Carvalho, I., Faria, T. & Pinheiro, C. (2002). How plants cope with water stress in the field? Photosynthesis and growth. *Annals of Botany, 89*(7), 907–916.
58. Zhang, C., Liu, J., Zhang, Y., Cai, X., Gong, P., Zhang, J., & Ye, Z. (2011). Overexpression of SlGMEs leads to ascorbate accumulation with enhanced oxidative stress, cold, and salt tolerance in tomato. *Plant Cell Reports, 30*(3), 389–398.
59. Hemavathi., Upadhyaya, C. P., & Young K. E. (2009). Overexpression of strawberry d-galacturonic acid reductase in potato leads to accumulation of vitamin C with enhanced abiotic stress tolerance, *Plant Science, 177*(6): 659–667.
60. Wang, Z., Xiao, Y., Chen, W., Tang, K., & Zhang, L. (2010). Increased vitamin C content accompanied by an enhanced recycling pathway confers oxidative stress tolerance in *Arabidopsis. Journal of Integrative Plant Biology, 52*(4), 400–409.
61. Gaucher, C., Boudier, A., Bonetti, J., Clarot, I., Leroy, P., & Parent, M. (2018). Glutathione: antioxidant properties dedicated to nanotechnologies. *Antioxidants, 7*(5), 62.
62. Kurylenko, O. O., Dmytruk, K. V., & Sibirny, A. (2019). Glutathione metabolism in yeasts and construction of the advanced producers of this tripeptide. In *Non-conventional Yeasts: from Basic Research to Application* (pp. 153–196). Springer, Cham.
63. Armeni, T., & Principato, G. (2020). Glutathione, an over one billion years ancient molecule, is still actively involved in cell regulatory pathways. In *The First Outstanding 50 Years of "Università Politecnica delle Marche"* (pp. 417–429). Springer, Cham.
64. Foyer, C. H., Souriau, N., Perret, S., Lelandais, M., Kunert, K. J., Pruvost, C., & Jouanin, L. (1995). Overexpression of glutathione reductase but not glutathione synthetase leads to increases in antioxidant capacity and resistance to photoinhibition in poplar trees. *Plant Physiology, 109*(3), 1047–1057.
65. Pandey, P., Singh, J., Achary, V., & Reddy, M. K. (2015). Redox homeostasis via gene families of ascorbate-glutathione pathway. *Frontiers in Environmental Science, 3*, 25.
66. Gullner, G., Kömives, T., & Rennenberg, H. (2001). Enhanced tolerance of transgenic poplar plants overexpressing γ-glutamylcysteine synthetase towards chloroacetanilide herbicides. *Journal of Experimental Botany, 52*(358), 971–979.

67. Eltayeb, A. E., Yamamoto, S., Habora, M. E. E., Matsukubo, Y., Aono, M., Tsujimoto, H., & Tanaka, K. (2010). Greater protection against oxidative damages imposed by various environmental stresses in transgenic potato with higher level of reduced glutathione. *Breeding Science, 60*(2), 101–109.

68. Ali, S. S., Ahsan, H., Zia, M. K., Siddiqui, T., & Khan, F. H. (2020). Understanding oxidants and antioxidants: Classical team with new players. *Journal of Food Biochemistry, 44*(3), e13145.

69. Santos-Sánchez, N. F., Salas-Coronado, R., Villanueva-Cañongo, C., & Hernández-Carlos, B. (2019). Antioxidant compounds and their antioxidant mechanism. In *Antioxidants*. IntechOpen.

70. Janků, M., Luhová, L., & Petřivalský, M. (2019). On the origin and fate of reactive oxygen species in plant cell compartments. *Antioxidants, 8*(4), 105.

71. Li, F., Wu, Q. Y., Sun, Y. L., Wang, L. Y., Yang, X. H., & Meng, Q. W. (2010). Overexpression of chloroplastic monodehydroascorbate reductase enhanced tolerance to temperature and methyl viologen-mediated oxidative stresses. *Physiologia Plantarum, 139*(4), 421–434.

72. Fryer, M. J. (1992). The antioxidant effects of thylakoid vitamin E (α-tocopherol). *Plant, Cell & Environment, 15*(4), 381–392.

73. Fukuzawa, K., Tokumura, A., Ouchi, S., & Tsukatani, H. (1982). Antioxidant activities of tocopherols on Fe^{2+}-ascorbate-induced lipid peroxidation in lecithin liposomes. *Lipids, 17*(7), 511–513.

74. Kagan, V. E., Fabisiak, J. P., & Quinn, P. J. (2000). Coenzyme Q and vitamin E need each other as antioxidants. *Protoplasma, 214*(1–2), 11–18.

75. Guo, Z. F., Ou, W. Z., Lu, S. Y., & Zhong, Q. (2006). Differential responses of antioxidative system to chilling and drought in four rice cultivars differing in sensitivity. *Plant Physiology and Biochemistry, 44*(11–12), 828–836.

76. Bafeel, S. O. and Ibrahim, M. M. (2008). Antioxidants and accumulation of α-tocopherol induce chilling tolerance in *Medicago sativa, International Journal of Agriculture and Biology, 10*(6), 593–598.

77. Li, F., Vallabhaneni, R., Yu, J., Rocheford, T., & Wurtzel, E. T. (2008). The maize phytoene synthase gene family: overlapping roles for carotenogenesis in endosperm, photomorphogenesis, and thermal stress tolerance. *Plant Physiology, 147*(3), 1334–1346.

78. Gomathi, R., & Rakkiyapan, P. (2011). Comparative lipid peroxidation, leaf membrane thermostability, and antioxidant system in four sugarcane genotypes differing in salt tolerance. *International Journal of Plant Physiology and Biochemistry, 3*(4), 67–74.

79. Grace, S. C., & Logan, B. A. (2000). Energy dissipation and radical scavenging by the plant phenylpropanoid pathway. *Philosophical Transactions of the Royal Society of London B: Biological Sciences, 355*(1402), 1499–1510.

80. Abbas, M., Saeed, F., Anjum, F. M., Afzaal, M., Tufail, T., Bashir, M. S., ... & Suleria, H. A. R. (2017). Natural polyphenols: An overview. *International Journal of Food Properties, 20*(8), 1689–1699.

81. Arora, A., Byrem, T. M., Nair, M. G., & Strasburg, G. M. (2000). Modulation of liposomal membrane fluidity by flavonoids and isoflavonoids. *Archives of Biochemistry and Biophysics, 373*(1), 102–109.

82. Michalak, A. (2006). Phenolic compounds and their antioxidant activity in plants growing under heavy metal stress. *Polish Journal of Environmental Studies, 15*(4).

83. Janas, K. M., Amarowicz, R., Zielińska-Tomaszewska, J., Kosińska, A., & Posmyk, M. M. (2009). Induction of phenolic compounds in two dark-grown lentil cultivars with different tolerance to copper ions. *Acta Physiologiae Plantarum, 31*(3), 587–595.

84. Fridovich, I. (1989). Superoxide dismutases. An adaptation to a paramagnetic gas. *The Journal of Biological Chemistry, 264*(14), 7761–7764.

85. Racchi, M. L., Bagnoli, F., Balla, I. and Danti, S. (2001). Differential activity of catalase and superoxide dismutase in seedlings and *in vitro* micropropagated oak (*Quercus robur* L.), *Plant Cell Reports,* 20(2): 169–174.

86. Bowler, C., Montagu, M. V., & Inzé, D. (1992). Superoxide dismutase and stress tolerance. *Annual Review of Plant Biology, 43*(1), 83–116.

87. Bueno, P., Varela, J., Gimenez-Gallego, G., & del Rio, L. A. (1995). Peroxisomal copper, zinc superoxide dismutase (characterization of the isoenzyme from watermelon cotyledons). *Plant Physiology, 108*(3), 1151–1160.

88. Del Río, L. A., Pastori, G. M., Palma, J. M., Sandalio, L. M., Sevilla, F., Corpas, F. J., & Hernández, J. A. (1998). The activated oxygen role of peroxisomes in senescence. *Plant Physiology, 116*(4), 1195–1200.

89. Zaefyzadeh, M., Quliyev, R. A., Babayeva, S. M. and Abbasov, M. A. (2009). The effect of the interaction between genotypes and drought stress on the superoxide dismutase and chlorophyll content in durum wheat landraces, *Turkish Journal of Biology, 33*(1): 1–7.

90. Gupta, A. S., Heinen, J. L., Holaday, A. S., Burke, J. J., & Allen, R. D. (1993). Increased resistance to oxidative stress in transgenic plants that overexpress chloroplastic Cu/Zn superoxide dismutase. *Proceedings of the National Academy of Sciences, 90*(4), 1629–1633.

91. Tang, Y., Bao, X., Zhi, Y., Wu, Q., Guo, Y., Yin, X., ... & Liu, W. (2019). Overexpression of a MYB family gene, OsMYB6, increases drought and salinity stress tolerance in transgenic rice. *Frontiers in Plant Science, 10,* 168.

92. Vanisri, S., Sreedhar, M., Jeevan, L., Pavani, A., Chaturvedi, A., Aparna, M., ... & Raju, C. S. (2017). Evaluation of rice genotypes for chlorophyll content and scavenging enzyme activity under the influence of mannitol stress towards drought tolerance. *International Journal of Current Microbiology and Applied Sciences, 6*(12), 2907–2917.

93. Smith, A.H. and Foyer, C.H., (2000). Bundle sheath proteins are more sensitive to oxidative damage than those of the mesophyll in maize leaves exposed to paraquat or low temperatures. *Journal of Experimental Botany, 51*(342), pp.123–130.

94. Slooten, L., Capiau, K., Van Camp, W., Montagu, M.V., Sybesma, C., Inze,' D. (1995). Factors affecting the enhancement of oxidative stress tolerance in transgenic tobacco overexpressing manganese superoxide dismutase in the chloroplasts. *Plant Physiology.* 107:737–775

95. Breusegem Van, F., Slooten, L., Stassart, J. M., Botterman, J., Moens, T., Van Montagu, M., & Inze, D. (1999). Effects of overproduction of tobacco MnSOD in maize chloroplasts on foliar tolerance to cold and oxidative stress. *Journal of Experimental Botany, 50*(330), 71–78.

96. Mishra, P., & Sharma, P. (2019). Superoxide Dismutases (SODs) and their role in regulating abiotic stress induced oxidative stress in plants. *Reactive Oxygen, Nitrogen and Sulfur Species in Plants: Production, Metabolism, Signaling and Defense Mechanisms,* 53–88.

97. Lee, J. K., Woo, S. Y., Kwak, M. J., Park, S. H., Kim, H. D., Lim, Y. J., ... & Lee, K. A. (2020). Effects of Elevated Temperature and Ozone in Brassica juncea L.: Growth, Physiology, and ROS Accumulation. *Forests*, *11*(1), 68.

98. McKersie, B. D., Murnaghan, J., Jones, K. S., & Bowley, S. R. (2000). Iron-superoxide dismutase expression in transgenic alfalfa increases winter survival without a detectable increase in photosynthetic oxidative stress tolerance. *Plant Physiology*, *122*(4), 1427–1438.

99. Shafi, A., Pal, A. K., Sharma, V., Kalia, S., Kumar, S., Ahuja, P. S., & Singh, A. K. (2017). Transgenic potato plants overexpressing SOD and APX exhibit enhanced lignification and starch biosynthesis with improved salt stress tolerance. *Plant Molecular Biology Reporter*, *35*(5), 504–518.

100. Guan, Q., Liao, X., He, M., Li, X., Wang, Z., Ma, H., ... & Liu, S. (2017). Tolerance analysis of chloroplast OsCu/Zn-SOD overexpressing rice under NaCl and $NaHCO_3$ stress. *PLoS One*, *12*(10), e0186052.

101. Che, Y., Zhang, N., Zhu, X., Li, S., Wang, S., & Si, H. (2020). Enhanced tolerance of the transgenic potato plants overexpressing Cu/Zn superoxide dismutase to low temperature. *Scientia Horticulturae*, *261*, 108949.

102. Vighi, I. L., Benitez, L. C., Amaral, M. N., Moraes, G. P., Auler, P. A., Rodrigues, G. S., ... & Braga, E. J. B. (2017). Functional characterization of the antioxidant enzymes in rice plants exposed to salinity stress. *Biologia Plantarum*, *61*(3), 540–550.

103. Ighodaro, O. M., & Akinloye, O. A. (2018). First line defense antioxidants-superoxide dismutase (SOD), catalase (CAT) and glutathione peroxidase (GPX): Their fundamental role in the entire antioxidant defense grid. *Alexandria Journal of Medicine*, *54*(4), 287–293.

104. Devi, S. S., Saha, B., & Panda, S. K. (2020). Differential Loss of ROS homeostasis and activation of anti oxidative defense response in tea cultivar due to aluminum toxicity in acidic soil. *Current Trends in Biotechnology & Pharmacy*, *14*(1).

105. Corpas, F. J., Barroso, J. B., Palma, J. M., & Rodriguez-Ruiz, M. (2017). Plant peroxisomes: a nitro-oxidative cocktail. *Redox Biology*, *11*, 535–542.

106. Mancini, A., Buschini, A., Restivo, F. M., Rossi, C., & Poli, P. (2006). Oxidative stress as DNA damage in different transgenic tobacco plants. *Plant Science*, *170*(4), 845–852.

107. Mhamdi, A., Queval, G., Chaouch, S., Vanderauwera, S., Van Breusegem, F. and Noctor, G. (2010). Catalase function in plants: a focus on *Arabidopsis* mutants as stress-mimic models, *Journal of Experimental Botany*, 61(15): 4197–4220.

108. Sharma, P., & Dubey, R. S. (2005). Drought induces oxidative stress and enhances the activities of antioxidant enzymes in growing rice seedlings. *Plant Growth Regulation*, *46*(3), 209–221.

109. Han, Q. H., Huang, B., Ding, C. B., Zhang, Z. W., Chen, Y. E., Hu, C., ... & Yuan, M. (2017). Effects of melatonin on anti-oxidative systems and photosystem II in cold-stressed rice seedlings. *Frontiers in Plant Science*, *8*, 785.

110. Willekens, H., Chamnongpol, S. and Davey, M. (1997). Catalase is a sink for H_2O_2 and is indispensable for stress defense in C-3 plants, *EMBO Journal*, 16(16): 4806–4816.

111. Guan, Z., Chai, T., Zhang, Y., Xu, J., & Wei, W. (2009). Enhancement of Cd tolerance in transgenic tobacco plants overexpressing a Cd-induced catalase cDNA. *Chemosphere*, *76*(5), 623–630.

112. Brisson, L. F., Zelitch, I., & Havir, E. A. (1998). Manipulation of catalase levels produces altered photosynthesis in transgenic tobacco plants. *Plant Physiology*, *116*(1), 259–269.

113. Mohamed, E. A., Iwaki, T., Munir, I., Tamoi, M., Shigeoka, S., & Wadano, A. (2003). Overexpression of bacterial catalase in tomato leaf chloroplasts enhances photo-oxidative stress tolerance. *Plant, Cell & Environment*, *26*(12), 2037–2046.

114. Lee, S. H., & An, C. S. (2006). Ectopic overexpression of the hot pepper Cat1 in *Arabidopsis* enhances resistance to paraquat, but not to wounding. *Journal of Plant Biology*, *49*(6), 421–426.

115. Al-Taweel, K., Iwaki, T., Yabuta, Y., Shigeoka, S., Murata, N., & Wadano, A. (2007). A bacterial transgene for catalase protects translation of D1 protein during exposure of salt-stressed tobacco leaves to strong light. *Plant Physiology*, *145*(1), 258–265.

116. Schuller, D. J., Ban, N., Van Huystee, R. B., McPherson, A. and Poulos, T. L. (1996). The crystal structure of peanut peroxidase, *Structure*, *4*(3): 311–321.

117. Rodrigo, R., & Libuy, M. (2014). Modulation of plant endogenous antioxidant systems by polyphenols. In *Polyphenols in Plants* (pp. 65–85). Academic Press.

118. Asada, K. (1992). Ascorbate peroxidase–a hydrogen peroxide-scavenging enzyme in plants. *Physiologia Plantarum*, *85*(2), 235–241.

119. Kobayashi, K., Kumazawa, Y., Miwa, K., & Yamanaka, S. (1996). ε-(γ-Glutamyl) lysine cross-links of spore coat proteins and transglutaminase activity in *Bacillus subtilis*. *FEMS Microbiology Letters*, *144*(2–3), 157–160.

120. Shah, K., Kumar, R. G., Verma, S. and Dubey, R. S. (2001). Effect of cadmium on lipid peroxidation, superoxide anion generation and activities of antioxidant enzymes in growing rice seedlings, *Plant Science*, 161(6): 1135–1144.

121. Tayefi-Nasrabadi, H., Dehghan, G., Daeihassani, B., Movafegi, A. and Samadi, A. (2011). Some biochemical properties of guaiacol peroxidases as modified by salt stress in leaves of salt-tolerant and salt-sensitive safflower (*Carthamus tinctorius* L.cv.) cultivars, *African Journal of Biotechnology*, 10(5): 751–763.

122. Patterson, W. R. and Poulos, T. L. (1995). Crystal structure of recombinant pea cytosolic ascorbate peroxidase, *Biochemistry*, 34(13): 4331–434.

123. Jimenez, A., Hernandez, J. A., del Río, L. A., & Sevilla, F. (1997). Evidence for the presence of the ascorbate-glutathione cycle in mitochondria and peroxisomes of pea leaves. *Plant Physiology*, *114*(1), 275–284.

124. Gill, S. S., & Tuteja, N. (2010). Reactive oxygen species and antioxidant machinery in abiotic stress tolerance in crop plants. *Plant Physiology and Biochemistry*, *48*(12), 909–930.

125. Hasanuzzaman, M., Hossain, M. A., da Silva, J. A. T., & Fujita, M. (2012). Plant response and tolerance to abiotic oxidative stress: antioxidant defense is a key factor. In *Crop stress and its management: Perspectives and Strategies* (pp. 261–315). Springer, Dordrecht.

126. Nakano, Y., & Asada, K. (1981). Hydrogen peroxide is scavenged by ascorbate-specific peroxidase in spinach chloroplasts. *Plant and Cell Physiology*, *22*(5), 867–880.

127. Teixeira, M. C., Monteiro, P., Jain, P., Tenreiro, S., Fernandes, A. R., Mira, N. P. & Sá-Correia, I. (2006). The YEASTRACT database: a tool for the analysis of transcription regulatory associations in Saccharomyces cerevisiae. *Nucleic Acids Research*, *34* (suppl_1), D446–D451.

128. Wang, J., Zhang, H. and Allen, R. D. (1999). Overexpression of an *Arabidopsis* peroxisomal ascorbate peroxidase gene in tobacco increases protection against oxidative stress, *Plant and Cell Physiology*, 40(7): 725–732.

129. Maheshwari, R., & Dubey, R. S. (2009). Nickel-induced oxidative stress and the role of antioxidant defense in rice seedlings. *Plant Growth Regulation*, *59*(1), 37–49.

130. Wang, Y., Wisniewski, M., Meilan, R., Cui, M., Webb, R. and Fuchigami, L. (2005). Overexpression of cytosolic ascorbate peroxidase in tomato confers tolerance to chilling and salt stress, *Journal of the American Society for Horticultural Science*, *130*(2): 167–173.

131. Wang, J., Wu, B., Yin, H., Fan, Z., Li, X., Ni, S., ... & Li, J. (2017). Overexpression of CaAPX induces orchestrated reactive oxygen scavenging and enhances cold and heat tolerances in tobacco. *BioMed Research International*, *2017*, *4049534*.

132. Smirnoff, N., Running, J. A., & Gatzek, S. (2004). Ascorbate biosynthesis: a diversity of pathways. *Vitamin C: its Functions and Biochemistry in Animals and Plants*, 7–29.

133. Mittova, V., Tal, M., Volokita, M. & Guy, M. (2004). Up-regulation of the leaf mitochondrial and peroxisomal antioxidative systems in response to salt-induced oxidative stress in the wild salt tolerant tomato species *Lycopersicon pennellii*. *Plant, Cell & Environment*, *26*, 845–856.

134. Hossain, M. A., & Asada, K. (1985). Monodehydroascorbate reductase from cucumber is a flavin adenine dinucleotide enzyme. *Journal of Biological Chemistry*, *260*(24), 12920–12926.

135. Sakihama, Y., Mano, J., Sano, S., Asada, K. and Yamasaki, H. (2000). Reduction of phenoxyl radicals mediated by monodehydroascorbate reductase, *Biochemical and Biophysical Research Communications*, 279(3): 949–954

136. Hossain, M. A., Burritt, D. J., & Fujita, M. (2016). Cross-stress tolerance in plants: molecular mechanisms and possible involvement of reactive oxygen species and methylglyoxal detoxification systems. *Abiotic Stress Response in Plants*, Wiley, 323–375.

137. Miyake, C., & Asada, K. (1994). Ferredoxin-dependent photoreduction of the monode-hydroascorbate radical in spinach thylakoids. *Plant and Cell Physiology*, *35*(4), 539–549.

138. Eltayeb, A. E., Yamamoto, S., Habora, M. E.E., Yin, L., Tsujimoto, H. and Tanaka, K. (2011). Transgenic potato overexpressing *Arabidopsis* cytosolic *AtDHAR1* showed higher tolerance to herbicide, drought and salt stresses, *Breeding Science*, 61(1): 3–10.

139. Ushimaru, T., Maki, Y., Sano, S., Koshiba, K., Asada, K. and Tsuji, H. (1997). Induction of enzymes involved in the ascorbate-dependent antioxidative system, namely, ascorbate peroxidase, monodehydroascorbate reductase and dehydroascorbate reductase, after exposure to air of rice (*Oryza sativa*) seedlings germinated under water, *Plant and Cell Physiology*, *38*(5): 541–549.

140. Chen, Z., Young, T. E., Ling, J., Chang, S. C., & Gallie, D. R. (2003). Increasing vitamin C content of plants through enhanced ascorbate recycling. *Proceedings of the National Academy of Sciences*, *100*(6), 3525–3530.

141. Qin, A., Shi, Q. and Yu, X. (2011). Ascorbic acid contents in transgenic potato plants overexpressing two dehydroascorbate reductase genes, *Molecular Biology Reports*, *38*(3): 1557–1566.

142. Rubio, M. C., Bustos-Sanmamed, P., Clemente, M. R. and Becana, M. (2009). Effects of salt stress on the expression of antioxidant genes and proteins in the model legume Lotus japonicus, *New Phytologist*, *181*(4): 851–859

143. Ghisla, S., & Massey, V. (1989). Mechanisms of flavoprotein-catalyzed reactions. In *EJB Reviews 1989* (pp. 29–45). Springer, Berlin, Heidelberg.

144. Edwards, E. A., Rawsthorne, S., & Mullineaux, P. M. (1990). Subcellular distribution of multiple forms of glutathione reductase in leaves of pea (*Pisum sativum* L.). *Planta*, *180*(2), 278–284.

145. Hernandez, J. A., Jimenez, A., Mullineaux, P., & Sevilia, F. (2000). Tolerance of pea (*Pisum sativum* L.) to long-term salt stress is associated with induction of antioxidant defenses. *Plant, Cell & Environment, 23*(8), 853–862.

146. Shu, D. F., Wang, L. Y., Duan, M., Deng, Y. S. and Meng, Q. W. (2011). Antisense-mediated depletion of tomato chloroplast glutathione reductase enhances susceptibility to chilling stress, *Plant Physiology and Biochemistry, 49*(10): 1228–1237.

147. Bray, E. A. (2000). Response to abiotic stress. *Biochemistry and molecular biology of plants*, 1158–1203.

148. Chen, S., Vaghchhipawala, Z., Li, W., Asard, H., & Dickman, M. B. (2004). Tomato phospholipid hydroperoxide glutathione peroxidase inhibits cell death induced by Bax and oxidative stresses in yeast and plants. *Plant Physiology, 135*(3), 1630–1641.

149. Alscher, R. & Hess, J., eds. (1993). *Antioxidants in Higher Plants.* Boca Raton: CRC Press.

150. Aono, M., Kubo, A., Saji, H., Tanaka, K., & Kondo, N. (1993). Enhanced tolerance to photooxidative stress of transgenic *Nicotiana tabacum* with high chloroplastic glutathione reductase activity. *Plant and Cell Physiology, 34*(1), 129–135.

151. Lee, S. C., Kwon, S. Y., & Kim, S. R. (2009). Ectopic expression of a cold-responsive CuZn superoxide dismutase gene, SodCc1, in transgenic rice (*Oryza sativa* L.). *Journal of Plant Biology, 52*(2), 154–160.

152. Baxter, A., Mittler, R., Suzuki, N., (2014). ROS as key players in plant stress signalling. *Journal of Experimental Botany, 655*, 1229–1240. doi:http://dx.doi.org/10.1093/jxb/ert375.

153. Møller, I. M., & Sweetlove, L. J. (2010). ROS signalling–specificity is required. *Trends in Plant Science, 15*(7), 370–374.

154. Jaspers, P., & Kangasjärvi, J. (2010). Reactive oxygen species in abiotic stress signaling. *Physiologia Plantarum, 138*(4), 405–413.

155. Torres, M.A., Dangl, J.L., (2005). Functions of the respiratory burst oxidase in biotic interactions, abiotic stress and development. *Current Opinion in Plant Biology.* 84, 397–403. doi:http://dx.doi.org/10.1016/j.pbi.2005.05.014.

156. Willems, P., Mhamdi, A., Stael, S., Storme, V., Kerchev, P., Noctor, G., et al., (2016). The ROS wheel: refining ROS transcriptional foot prints in *Arabidopsis. Plant Physiology* 171, 1720–1733. doi:http://dx.doi.org/10.1104/pp.16.00420.

157. Laloi, C., Apel, K., Danon, A., 2004. Reactive oxygen signalling: the latest news. *Current Opinion in Plant Biology* 73, 323–328. doi:http://dx.doi.org/10.1016/j.pbi.2004.03.005.

158. Davletova, S., Schlauch, K., Coutu, J., Mittler, R., (2005). The zinc-finger protein Zat12 plays a central role in reactive oxygen and abiotic stress signaling in *Arabidopsis. Plant Physiology,* 1392, 847–856. doi:http://dx.doi.org/10.1104/ pp.105.068254.

159. Apel, K., Hirt, H., 2004. Reactive oxygen species: metabolism, oxidative stress, and signal transduction. *Annual Review of Plant Biology, 55*, 373–399. doi:http://dx.doi.org/ 10.1146/annurev.arplant.55.031903.141701.

160. McDowell, J.M., Dangl, J.L., (2000). Signal transduction in the plant immune response. *Trends in Biochemical Science, 252*, 79–82. doi:http://dx.doi.org/10.1016/S0968-00049901532-7.

161. Grant, J.J., Loake, G.J., (2000). Role of reactive oxygen intermediates and cognate redox signaling in disease resistance. *Plant Physiology 124*(1), 21–30. doi:http://dx.doi. org/ 10.1104/pp.124.1.21.

162. Pitzschke, A., Forzani, C., Hirt, H., (2006). Reactive oxygen species signaling in plants. *Antioxidants & Redox Signaling, 8*(9–10), 1757–1764. doi:http://dx.doi.org/10.1089/ ars.2006.8.1757.

163. Neill, S.J., Desikan, R., Clarke, A., Hurst, R.D., Hancock, J.T., (2002). Hydrogen peroxide and nitric oxide as signalling molecules in plants. *The Journal of Experimental Botany, 53*(372), 1237–1247. doi:http://dx.doi.org/10.1093/jexbot/53.372.1237.

164. Kindgren, P., Eriksson, M.J., Benedict, C., Mohapatra, A., Gough, S.P., Hansson, M., Strand, Å., (2011). A novel proteomic approach reveals a role for Mg-protoporphyrin IX in response to oxidative stress. *Plant Physiology, 141*(4), 310–320. doi:http://dx.doi.org/10.1111/j.1399–3054.2010.01440.x.

165. Woodson, J.D., Chory, J., (2008). Coordination of gene expression between organellar and nuclear genomes. *Nature Reviews Genetics, 9*, 383–395.

166. Adachi, H., Nakano, T., Miyagawa, N., Ishihama, N., Yoshioka, M., Katou, Y., (2015). WRKY transcription factors phosphorylated by MAPK regulate a plant immune NADPH oxidase in Nicotiana benthamiana. *Plant Cell, 27*, 2645–2663. doi:http://dx.doi.org/10.1105/tpc.15.00213.

167. Fang, Y., Liao, K., Du, H., Xu, Y., Song, H., Li, X., (2015). A stress responsive NAC transcription factor SNAC3 confers heat and drought tolerance through modulation of reactive oxygen species in rice. *The Journal of Experimental Botany, 66*, 6803–6817. doi: http://dx.doi.org/10.1093/jxb/erv386.

168. Chen, Y. E., Cui, J. M., Li, G. X., Yuan, M., Zhang, Z. W., Yuan, S., & Zhang, H. Y. (2016). Effect of salicylic acid on the antioxidant system and photosystem II in wheat seedlings. *Biologia Plantarum, 60*(1), 139–147.

169. Zhu, J.K. (2016). Abiotic stress signaling and responses in plants. *Cell*, 167(2), 313–324

170. Matsuo, M., Johnson, J.M., Hieno, A., Tokizawa, M., Nomoto, M., Tada, Y., et al., (2015). High REDOX RESPONSIVE TRANSCRIPTION FACTOR1 levels result in accumulation of reactive oxygen species in *Arabidopsis* thaliana shoots and roots. *Molecular Plant, 8*, 1253–1273. doi:http://dx.doi.org/10.1016/j. molp.2015.03.011.

171. He, L., Ma, X., Li, Z., Jiao, Z., Li, Y., Ow, D.W., (2016). Maize OXIDATIVE STRESS 2 homologs enhance cadmium tolerance in *Arabidopsis* through activation of a putative SAM dependent methyl transferase gene. *Plant Physiology, 171*, 1675– 1685. doi:http://dx.doi.org/10.1104/pp.16.00220.

172. Suzuki, N., Miller, G., Salazar, C., Mondal, H.A., Shulaev, E., Cortes, D.F., Shuman, J.L., Luo, X., Shah, J., Schlauch, K., Shulaev, V., Mittler, R., (2013). Temporal-spatial interaction between reactive oxygen species and abscisic acid regulates rapid systemic acclimation in plants. *Plant Cell, 25*, 3553–3569. doi:http://dx.doi.org/ 10.1105/tpc.113.114595.

173. Carmody, M., Crisp, P.A., d'Alessandro, S., Ganguly, D., Gordon, M., Havaux, M., (2016). Uncoupling high light responses from singlet oxygen retrograde signaling and spatial-temporal systemic acquired acclimation in *Arabidopsis*. *Plant Physiology, 171*, 1734–1749. doi:http://dx.doi.org/10.1104/pp.16.00404.

174. Choi, W.G., Toyota, M., Kim, S.H., Hilleary, R., Gilroy, S., (2014). Salt stress-induced Ca^{2+} waves are associated with rapid, long-distance root to-shoot signaling in plants. *The Proceedings of the National Academy of Sciences, 111*(17), 6497–6502. doi:http://dx.doi.org/10.1073/ pnas.1319955111.

175. Evans, M.J., Choi, W.G., Gilroy, S., Morris, R., (2016). A ROS-assisted calcium wave dependent on AtRBOHD and TPC1 propagates the systemic response to salt stress in *Arabidopsis* roots. *Plant Physiology, 171*, 1771–1784. doi:http://dx.doi.org/ 10.1104/ pp.16.00215.

176. Vogel, M.O., Moore, M., König, K., Pecher, P., Alsharafa, K., Lee, J., et al., (2014). Fast retrograde signaling in response to high light involves metabolite export,

MITOGEN-ACTIVATED PROTEIN KINASE6, and AP2/ERF transcription factors in *Arabidopsis. Plant Cell, 26,* 1151–1165. doi:http://dx.doi.org/10.1105/ tpc.113.121061.

177. Triantaphylidès, C., Havaux, M., (2009). Singlet oxygen in plants: production, detoxification and signaling. *Trends in Plant Science, 144,* 219–228. doi:http://dx.doi. org/10.1016/j. tplants.2009.01.008.

178. Mishra, P., Bhoomika, K., & Dubey, R. S. (2013). Differential responses of antioxidative defense system to prolonged salinity stress in salt-tolerant and salt-sensitive Indica rice (*Oryza sativa* L.) seedlings. *Protoplasma, 250*(1), 3–19.

179. Kharbech, O., Houmani, H., Chaoui, A., & Corpas, F. J. (2017). Alleviation of Cr (VI)-induced oxidative stress in maize (Zea mays L.) seedlings by NO and H2S donors through differential organ-dependent regulation of ROS and NADPH-recycling metabolisms. *Journal of Plant Physiology, 219,* 71–80.

180. Anwar, A., Yan, Y., Liu, Y., Li, Y., & Yu, X. (2018). 5-aminolevulinic acid improves nutrient uptake and endogenous hormone accumulation, enhancing low-temperature stress tolerance in cucumbers. *International Journal of Molecular Sciences, 19*(11), 3379.

181. Li, Z., Xu, J., Gao, Y., Wang, C., Guo, G., Luo, Y., ... & Hu, J. (2017). The synergistic priming effect of exogenous salicylic acid and H_2O_2 on chilling tolerance enhancement during maize (*Zea mays* L.) seed germination. *Frontiers in Plant Science, 8,* 1153.

182. Sun, Y., He, Y., Irfan, A. R., Liu, X., Yu, Q., Zhang, Q., & Yang, D. (2020). Exogenous brassinolide enhances the growth and cold resistance of maize (*Zea mays* L.) seedlings under chilling stress. *Agronomy, 10*(4), 488.

183. Ali, M. A., Fahad, S., Haider, I., Ahmed, N., Ahmad, S., Hussain, S., & Arshad, M. (2019). Oxidative stress and antioxidant defense in plants exposed to metal/metalloid toxicity. In: *Reactive Oxygen, Nitrogen and Sulfur Species in Plants: Production, Metabolism, Signaling and Defense Mechanisms* (pp. 353–370), Wiley.

184. Santovito, G., Trentin, E., Gobbi, I., Bisaccia, P., Tallandini, L., & Irato, P. (2020). Non-enzymatic antioxidant responses of mytilus galloprovincialis under cadmium-induced oxidative stress risk. Preprints 2020, 2020070235 (doi: 10.20944/preprints202007.0235. v1).

185. Han, C., Liu, Q., & Yang, Y. (2009). Short-term effects of experimental warming and enhanced ultraviolet-B radiation on photosynthesis and antioxidant defense of Picea asperata seedlings. *Plant Growth Regulation, 58*(2), 153–162.

186. Gao, Q., & Zhang, L. (2008). Ultraviolet-B-induced oxidative stress and antioxidant defense system responses in ascorbate-deficient vtc1 mutants of *Arabidopsis thaliana*. *Journal of Plant Physiology, 165*(2), 138–148.

187. Radwan, D. E. M., Fayez, K. A., Mahmoud, S. Y., & Lu, G. (2010). Modifications of antioxidant activity and protein composition of bean leaf due to Bean yellow mosaic virus infection and salicylic acid treatments. *Acta Physiologiae Plantarum, 32*(5), 891–904.

188. Xie, X., He, Z., Chen, N., Tang, Z., Wang, Q., & Cai, Y. (2019). The roles of environmental factors in regulation of oxidative stress in plant. *BioMed Research International, 2019,* 9732325.

189. Rubio, M. C., Bustos-Sanmamed, P., Clemente, M. R. and Becana, M. (2009). Effects of salt stress on the expression of antioxidant genes and proteins in the model legume Lotus japonicus, *New Phytologist, 181*(4): 851–859

190. Wu, D., Sun, Y., Wang, H., Shi, H., Su, M., Shan, H., ... & Li, Q. (2018). The SlNAC8 gene of the halophyte Suaeda liaotungensis enhances drought and salt stress tolerance in transgenic *Arabidopsis* thaliana. *Gene, 662,* 10–20.

191. He, R., Zhuang, Y., Cai, Y., Agüero, C. B., Liu, S., Wu, J., ... & Zhang, Y. (2018). Overexpression of 9-cis-epoxycarotenoid dioxygenase cisgene in grapevine increases drought tolerance and results in pleiotropic effects. *Frontiers in Plant Science, 9*, 970.

192. Li, Q., Jin, C., Wang, G., Ji, J., Guan, C., & Li, X. (2020). Enhancement of endogenous SA accumulation improves poor-nutrition stress tolerance in transgenic tobacco plants overexpressing a SA-binding protein gene. *Plant Science, 292*, 110384.

193. Yan, H., Li, Q., Park, S. C., Wang, X., Liu, Y. J., Zhang, Y. G., ... & Ma, D. F. (2016). Overexpression of CuZnSOD and APX enhance salt stress tolerance in sweet potato. *Plant Physiology and Biochemistry, 109*, 20–27.

194. Pandey, S., Fartyal, D., Agarwal, A., Shukla, T., James, D., Kaul, T., ... & Reddy, M. K. (2017). Abiotic stress tolerance in plants: myriad roles of ascorbate peroxidase. *Frontiers in Plant Science, 8*, 581.

195. Liu, J. X., Feng, K., Duan, A. Q., Li, H., Yang, Q. Q., Xu, Z. S., & Xiong, A. S. (2019). Isolation, purification and characterization of an ascorbate peroxidase from celery and overexpression of the AgAPX1 gene enhanced ascorbate content and drought tolerance in *Arabidopsis*. *BMC Plant Biology, 19*(1), 1–13.

196. Sarowar, S., Kim, E. N., Kim, Y. J., Ok, S. H., Kim, K. D., Hwang, B. K., & Shin, J. S. (2005). Overexpression of a pepper ascorbate peroxidase-like 1 gene in tobacco plants enhances tolerance to oxidative stress and pathogens. *Plant Science, 169*(1), 55–63.

197. Duan, M., Feng, H. L., Wang, L. Y., Li, D., & Meng, Q. W. (2012). Overexpression of thylakoidal ascorbate peroxidase shows enhanced resistance to chilling stress in tomato. *Journal of Plant Physiology, 169*(9), 867–877.

198. Ishikawa, T., & Shigeoka, S. (2008). Recent advances in ascorbate biosynthesis and the physiological significance of ascorbate peroxidase in photosynthesizing organisms. *Bioscience, Biotechnology, and Biochemistry, 72*(5), 1143–1154.

199. Giacomelli, L., Masi, A., Ripoll, D. R., Lee, M. J., & van Wijk, K. J. (2007). *Arabidopsis thaliana* deficient in two chloroplast ascorbate peroxidases shows accelerated light-induced necrosis when levels of cellular ascorbate are low. *Plant Molecular Biology, 65*(5), 627–644.

200. Maruta, T., Noshi, M., Tanouchi, A., Tamoi, M., Yabuta, Y., Yoshimura, K., ... & Shigeoka, S. (2012). H_2O_2-triggered retrograde signaling from chloroplasts to nucleus plays specific role in response to stress. *Journal of Biological Chemistry, 287*(15), 11717–11729.

201. Ahmad, P., Jaleel, C. A., Salem, M. A., Nabi, G., & Sharma, S. (2010). Roles of enzymatic and nonenzymatic antioxidants in plants during abiotic stress. *Critical Reviews in Biotechnology, 30*(3), 161–175.

202. Miyagawa, Y., Tamoi, M., & Shigeoka, S. (2000). Evaluation of the defense system in chloroplasts to photooxidative stress caused by paraquat using transgenic tobacco plants expressing catalase from *Escherichia coli*. *Plant and Cell Physiology, 41*(3), 311–320.

203. Gill, S. S., Anjum, N. A., Hasanuzzaman, M., Gill, R., Trivedi, D. K., Ahmad, I., ... & Tuteja, N. (2013). Glutathione and glutathione reductase: a boon in disguise for plant abiotic stress defense operations. *Plant Physiology and Biochemistry, 70*, 204–212.

204. Morgan, B., Ezeriņa, D., Amoako, T. N., Riemer, J., Seedorf, M., & Dick, T. P. (2013). Multiple glutathione disulfide removal pathways mediate cytosolic redox homeostasis. *Nature Chemical Biology, 9*(2), 119–125.

205. Avashthi, H., Pathak, R. K., Pandey, N., Arora, S., Mishra, A. K., Gupta, V. K., ... & Kumar, A. (2018). Transcriptome-wide identification of genes involved in Ascorbate–Glutathione

cycle (Halliwell–Asada pathway) and related pathway for elucidating its role in antioxidative potential in finger millet (Eleusine coracana (L.)). *3 Biotech, 8*(12), 499.

206. Sakhno, L.O., Yemets, A.I. and Blume, Y.B., (2019). The role of ascorbate-glutathione pathway in reactive oxygen species balance under abiotic stresses. In *Reactive Oxygen, Nitrogen and Sulfur Species in Plants: Production, Metabolism, Signaling and Defense Mechanisms*, Wiley. pp.89–111.

207. Sayfzadeh, S., Habibi, D., Taleghani, D. F., Kashani, A., Vazan, S., Qaen, S. H. S. & Rashidi, M. (2011). Response of antioxidant enzyme activities and root yield in sugar beet to drought stress. *International Journal of Agriculture and Biology, 13*(3), 357–362.

208. Sgherri, C., Milone, M. T. A., Clijsters, H., & Navari-Izzo, F. (2001). Antioxidative enzymes in two wheat cultivars, differently sensitive to drought and subjected to subsymptomatic copper doses. *Journal of Plant Physiology, 158*(11), 1439–1447.

209. Mittal, R., & Dubey, R. S. (1991). Influence of salinity on ribonuclease activity and status of nucleic acids in rice seedlings differing in salt tolerance. *Plant Physiology and Biochemistry-New Delhi-, 18*, 57–57.

210. Valderrama, R., Corpas, F. J., Carreras, A., Gómez-rodríguez, M.V., Chaki, M., Pedrajas, J. R., & Barroso, J. B. (2006). The dehydrogenase-mediated recycling of NADPH is a key antioxidant system against salt-induced oxidative stress in olive plants. *Plant, Cell & Environment, 29*(7), 1449–1459.

211. Fryer, M. J., Andrews, J. R., Oxborough, K., Blowers, D. A., & Baker, N. R. (1998). Relationship between CO_2 assimilation, photosynthetic electron transport, and active O_2 metabolism in leaves of maize in the field during periods of low temperature. *Plant Physiology, 116*(2), 571–580.

212. Sharma, P., & Dubey, R. S. (2007). Involvement of oxidative stress and role of antioxidative defense system in growing rice seedlings exposed to toxic concentrations of aluminum. *Plant Cell Reports, 26*(11), 2027–2038.

213. Ashry, N. A., & Mohamed, H. I. (2012). Impact of secondary metabolites and related enzymes in flax resistance and/or susceptibility to powdery mildew. *African Journal of Biotechnology, 11*(5), 1073–1077.

214. Singh, R., Singh, S., Parihar, P., Mishra, R. K., Tripathi, D. K., Singh, V. P., ... & Prasad, S. M. (2016). Reactive oxygen species (ROS): beneficial companions of plants' developmental processes. *Frontiers in Plant Science, 7*, 1299.

215. Le, C. T. T., Brumbarova, T., & Bauer, P. (2019). The interplay of ROS and iron signaling in plants. In *Redox Homeostasis in Plants* (pp. 43–66). Springer, Cham.

216. Saidi, M. N., Mergby, D., & Brini, F. (2017). Identification and expression analysis of the NAC transcription factor family in durum wheat (Triticum turgidum L. ssp. durum). *Plant Physiology and Biochemistry, 112*, 117–128

217. Shiriga, K., Sharma, R., Kumar, K., Yadav, S. K., Hossain, F., & Thirunavukkarasu, N. (2014). Genome-wide identification and expression pattern of drought-responsive members of the NAC family in maize. *Meta Gene, 2*, 407–417.

218. Borrill, P., Harrington, S. A., & Uauy, C. (2017). Genome-wide sequence and expression analysis of the NAC transcription factor family in polyploid wheat. *G3: Genes, Genomes, Genetics, 7*(9), 3019–3029.

219. Zhang, Y., Li, D., Wang, Y., Zhou, R., Wang, L., Zhang, Y., ... & Zhang, X. (2018). Genome-wide identification and comprehensive analysis of the NAC transcription factor family in Sesamum indicum. *PLoS One, 13*(6), e0199262.

220. Filichkin, S. A., Ansariola, M., Fraser, V. N., & Megraw, M. (2018). Identification of transcription factors from NF-Y, NAC, and SPL families responding to osmotic stress in multiple tomato varieties. *Plant Science, 274*, 441–450.

221. Matsuo, M., Johnson, J. M., Hieno, A., Tokizawa, M., Nomoto, M., Tada, Y., ... & Böhmer, F. D. (2015). High REDOX RESPONSIVE TRANSCRIPTION FACTOR1 levels result in accumulation of reactive oxygen species in *Arabidopsis* thaliana shoots and roots. *Molecular Plant, 8*(8), 1253–1273.

222. Nakashima, K., Takasaki, H., Mizoi, J., Shinozaki, K., & Yamaguchi-Shinozaki, K. (2012). NAC transcription factors in plant abiotic stress responses. *Biochimica et Biophysica Acta (BBA)-Gene Regulatory Mechanisms, 1819*(2), 97–103.

223. Hernandez-Garcia, C. M., & Finer, J. J. (2014). Identification and validation of promoters and cis-acting regulatory elements. *Plant Science, 217*, 109–119.

224. Wang, C. T., Yang, Q., & Yang, Y. M. (2011). Characterization of the ZmDBP4 gene encoding a CRT/DRE-binding protein responsive to drought and cold stress in maize. *Acta Physiologiae Plantarum, 33*(2), 575–583.

225. Msanne, J., Lin, J., Stone, J. M., & Awada, T. (2011). Characterization of abiotic stress-responsive *Arabidopsis* thaliana RD29A and RD29B genes and evaluation of transgenes. *Planta, 234*(1), 97–107.

226. Fujita, Y., Fujita, M., Shinozaki, K., & Yamaguchi-Shinozaki, K. (2011). ABA-mediated transcriptional regulation in response to osmotic stress in plants. *Journal of Plant Research, 124*(4), 509–525.

227. Lee, S. C., Lim, M. H., Yu, J. G., Park, B. S., & Yang, T. J. (2012). Genome-wide characterization of the CBF/DREB1 gene family in Brassica rapa. *Plant Physiology and Biochemistry, 61*, 142–152.

228. Hussain, S. S., Kayani, M. A., & Amjad, M. (2011). Transcription factors as tools to engineer enhanced drought stress tolerance in plants. *Biotechnology Progress, 27*(2), 297–306.

229. Shi, Y., Ding, Y., & Yang, S. (2018). Molecular regulation of CBF signaling in cold acclimation. *Trends in Plant Science, 23*(7), 623–637.

230. Hiraki, H., Uemura, M., & Kawamura, Y. (2019). Calcium signaling-linked CBF/DREB1 gene expression was induced depending on the temperature fluctuation in the field: views from the natural condition of cold acclimation. *Plant and Cell Physiology, 60*(2), 303–317.

CHAPTER 3

Reactive Oxygen and Nitrogen Species: Oxidative Damage and Antioxidative Defense Mechanism in Plants under Abiotic Stress

AMRINA SHAFI[1,2*], FARHANA HASSAN[2], and FIRDOUS A. KHANDAY[2*]

[1]*Education Department, Government of J&K, Srinagar, Kashmir, 190002, India*

[2]*Signal Transduction Lab, Department of Biotechnology, School of Biological Sciences, University of Kashmir, Srinagar, J&K, 190006, India*

Corresponding author. E-mail: amrinashafi7@gmail.com; khandayf@kashmiruniversity.ac.in.

ABSTRACT

Plants are versatile in terms of tolerance and adaptive mechanisms toward severe abiotic stresses. The most deleterious effects of abiotic stresses are the production of highly toxic species known as reactive oxygen species (ROS) and reactive nitrogen species (RNS). These species are produced in different cellular compartments within the plant body as a by-product of metabolic processes in the form of radical and nonradical species; however, their production is amplified when plants encounter abiotic stresses. These ROS and RNS molecules at low concentration act as signaling molecules but at higher concentration enforce harsh oxidative stress damaging cell protein, lipids, and nucleic acid. Thus, homeostasis of ROS and RNS is required to maintain at optimal levels to prevent damage. Plants have antioxidant system that comprises two components, i.e., enzymatic and nonenzymatic antioxidant defense system, which maintains ROS and RNS homeostasis inside cells. Thus, there is a delicate equilibrium between ROS and RNS production, and their scavenging system inside the cell, which helps the plant to survive from the toxic effects of these species. In this chapter, we will be discussing ROS and RNS generation, sites of production, and their vital role in acting as messenger molecules as well as inducers of oxidative damage under various abiotic stresses. Further, this chapter will focus on the mechanism of ROS production, oxidative stress, and antioxidant defense mechanisms under several stressful environmental conditions.

3.1   INTRODUCTION

Higher plants that dwell on land have to face naturally harsh environmental conditions (different physical, chemical, natural calamities) regularly. The sole reason that makes plants to tolerate and adapt in such harsh conditions is because of their sessile or sedentary lifestyle [1]. Thus, anything beyond their ability can be stressful to them such as temperature variations, water paucities or submerging, heavy metal stress, high salinity, UV radiation, extremes of soil pH, mechanical stresses like wind, wounding, and gaseous pollutants. These abiotic stresses affect the growth, physiology, and yield of plants [1], and the most deleterious effects of abiotic stresses are the production of enormous amounts of reactive oxygen and nitrogen species (ROS and RNS). Even though plants naturally produce these reactive species (ROS and RNS) as metabolic by-products; however, their production is overwhelmed in plants due to imbalance in cell redox homeostasis and enzymatic destruction during abiotic stresses. Reactive oxygen species (ROS) are known to be a partially reduced forms of oxygen generated constantly in various cell compartments with its own ROS signature. This signature can change depending on the developmental stage and the stress level, which plants encounter [2]. Reactive nitrogen species (RNS) comprises nitric oxide (NO) and its derived molecules, which are synthesized throughout stress conditions resulting in nitrosative stress similar to oxidative stress. Apart from this, ROS and/or RNS is synthesized via diverse reactions in various cell compartments, such as during photorespiration and cellular respiration in mitochondria and peroxisomes, photosynthesis in chloroplasts, and during redox reactions in the plant cell cytosol.

Besides acting as deleterious molecules at high concentrations, ROS and RNS are considered to be important signaling molecules in virtually all developmental, physiological, and stress acclimation responses of plants but at optimal or low concentrations [3]. These reactive species act as a double-edged sword [4], either acting as secondary messengers in various key physiological and metabolic processes or induce oxidative and nitrosative damages under abiotic stress conditions. To ensure survival, plants are equipped with defense machinery such as enzymatic and nonenzymatic scavenging compounds, which keep these reactive species under check [4]. These two components work hand in hand to scavenge ROS/RNS and also maintain a minimal amount of these molecules for plants to carry its physiological and metabolic processes. However, during stress, the regulatory mechanisms fail to accomplish this task and the delicate balance between ROS/RNS generation and elimination is perturbed. As a result, the concentration of these reactive molecules increases inside cells leading to degradation of vital biomolecules, which eventually leads to cellular death [4].

Unlike ROS, RNS in plants remain unresolved and have caught specific attention due to their interactions with ROS molecules [3]. It is a known fact that cell signaling requires the involvement of a mammoth number of molecules, whose cross-talk results in proper communication or transduction of information. Similarly, there is a cascade of events involved in ROS and RNS production, signaling, and transduction. Both ROS and RNS have diverse roles to play in plant system acting as both signaling molecules and key regulators of various processes such as metabolism, growth and development, solute transport, and response to abiotic stresses [3]. In this chapter, we will provide comprehensive coverage of ROS and RNS, their mechanism of generation, site of production, scavenging mechanisms, and their diverse roles in plants. Further, we will also discuss the signaling cross-talk of ROS/RNS with different components such as MAPKs, Ca^{2+}, and Rboh, and the genes and transcription factors (TFs) involved in ROS/RNS signaling.

3.2 ABIOTIC STRESSES AND THEIR CONSEQUENCES

Plants are confronted with several different abiotic stresses in the field conditions, which reduce their fitness capabilities and in turn the crop production (Figure 3.1). Among these stresses, salt stress is a foremost ecological impediment, which limits agricultural development and yield across the globe, as it magnifies the salinization of arable land globally. Approximately 19.5% of agricultural land is saline [5], and this rate is ever increasing due to drastic climate change over the globe. This persistent salinity stress puts an ionic threat to the plants by accumulating excess Na^+ and Cl^- ions, which disturb redox homeostasis. It is also associated with limited water availability, less nutrient uptake, changed metabolic, and photosynthetic activity [6]. Thus leading to an ionic and nutritional imbalance in plants, water leakage from cells through osmosis a process called desiccation. Even most plants cannot survive at concentrations exceeding 200 mM of NaCl [7]. Among various consequences of salinity stress, ROS production is measured as the main restraints, which diminish plant growth and yield by prompting a major disruption in plant ionic homeostasis, thereby distressing cell metabolism [8]. Further, salt stress decreases stomatal conductance and lowers transpiration, which leads to lower intracellular CO_2 concentration and availability of CO_2 for the Calvin cycle is reduced. As a result, oxidized $NADP^+$ gets depleted and electrons are transferred to O_2 to generate O_2^- (Mehler reaction). This superoxide is spontaneously dismutated by the action of SOD (superoxide dismutase) to a more stable ROS, that is, H_2O_2. Thus, salt stress impedes with carbon metabolism and leads to ROS generation [9]. Also, salinity stress is reported to be associated with the activation of Rboh [10], the main ROS-producing enzyme at the apoplast. Salt stress leads to NO synthesis through the involvement of polyamines (PAs) [11].

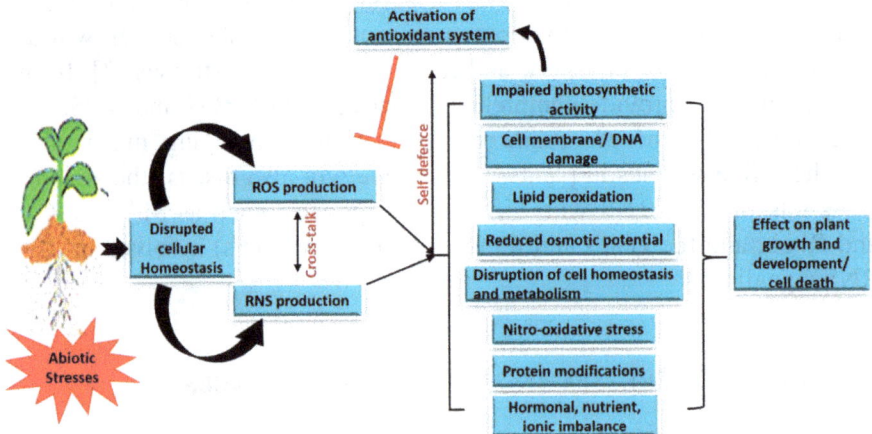

FIGURE 3.1 Schematic model of cross-talk between reactive oxygen species (ROS) and reactive nitrogen species (RNS) in plant responses to abiotic stress. The dual role of ROS and RNS, at high concentrations, can produce oxidative cell injuries leading to altered growth and development or even cell death. However, plants have an extensive array of antioxidant systems where ROS acts as a signaling molecule in diverse metabolic and biological processes.

Drought (water stress) occurs when humidity is low and air temperature is high, which results in an imbalance between transpiration flux and soil water intake [12]. It occurs due to many reasons, which include low rainfall, temperature fluctuations, and high light intensity, salinity [13]. Drought stress results in the reduced biomass production as it affects the plant growth and development [14], reduces seed germination, leaf size, stem extension, and root proliferation, and disturbs plant water relations [15]. It leads to enhanced metabolite flux through the photorespiratory pathway, which increases the oxidative load on tissues and generates ROS [14]. A drought stress response is similar to that of the salinity response. It reduces water potential, closes stomata and reduces CO_2 assimilation, damages chloroplast, and lead to ROS production [13].

Temperature stresses can also cause havoc on plants. As we know like any other organism, plant too has an optimal temperature where plant can show its best productivity. At temperature >30 °C both developmental and growth parameters are affected including seed germination and plant productivity [16]. Most sensitive to temperature stresses is photosynthesis, the temperature slightly higher than that of optimal, inhibit photosynthesis [17]. High temperatures destroy photochemical reactions and carbon metabolism is disturbed [18]. Heat stress leads to stomatal closure, reduction in intracellular CO_2 concentration, and water content is reduced in leaves leaving similar physiological effects as that of salinity and drought. Also, high temperature is thought to affect PSII, Rubisco, and ATP synthase [19], besides directly inactivating oxygen-evolving complex. Similarly, the temperature in the range of 0–15 °C is considered stressful for many plant species, which limit their productivity and distribution [20]. Chilling stress (too low temperature) can damage membranes including thylakoids due to disruption of lipid structure, protein denaturation, and leakage of solutes and free electrons as a result of ROS generation [21]. Maeda et al. reported, that low temperatures enhance ROS production, declines antioxidant machinery, leading to lipid peroxidation [22]. Chilling results in the formation of intra- and extra-cellular ice crystal, which leads to cellular impairment, desiccation, and disturbance of the plasma membrane (PM) [23], resulting in decreased rate of water and nutrients uptake, which eventually leads to cell starvation and death. Plants subjected to low temperature also generate NO through the activity of NOS (nitric oxide synthase) and nitrate reductase (NR) activity [24, 25].

Exposure of plants to irradiance far above the light saturation point of photosynthesis is known to be high-light stress. The process of photosynthesis in plants is carried by light energy, which is the vital prerequisite for plant growth. Thus, light stress has a huge influence on the plant's life cycle as most affected cell organelles are chloroplasts. Under high light intensity, PSII undergoes photoinhibition, putting the reaction center of PSII at risk. Thus decreases the rate of photoelectron transport chain (ETC), induces ROS production that further inactivates PSII [26]. This light-induced inactivation of PSII and its repair occurs simultaneously. Photoinhibition becomes apparent when the rate of photodamage to PSII exceeds the rate of its repair [27].

Rapid industrial development has contaminated natural resources for a few decades. Increased levels of heavy metals affect ecosystems and cause stress and toxicity in plants, animals, and microorganisms [28]. Heavy metals negatively impact growth, photosynthesis, cell division, reproduction, antioxidant activity, and diversity [28]. Aluminum (Al) being third most abundant metal in soil poses a great threat to the plants growing in acidic soil, wherein at pH below 5, Al^{3+} enters the root cells and interfere with important processes like cell division, cell elongation in root tips, leading to root damage and reduced water, and nutrient uptake [29]. Besides, Al promotes an imbalance in free radical and antioxidant levels in the cell, leading to oxidative toxicity [30]. Al is not a redox element but has the ability in facilitating oxidative stress via electrostatic bonding with oxygen donor ligands (e.g., carboxylate or phosphate groups) accountable for the cellular toxicity. Al^{3+} damages cellular elements such as lipids, proteins, and nucleic acids, leading to cell death. Cadmium (Cd) is additional heavy metal, which is known to be most toxic to plants, as plant roots uptake it easily and then accumulates in the higher part of plants, affecting crop yield and threatening global food. Cd can also enhance ROS generation via NADPH oxidase (Rboh) [31], leading to impaired cellular redox homeostasis and oxidative damage to biomolecules such as carbohydrates, proteins, DNA, and membrane lipids [32]. These biological fluctuations resulting from Cd stress affect plant-required minerals and nutrients uptake [33]. Environmental conditions like low pH or high reducing conditions lead to the formation of NO through the reduction of nitrite, which has been reported in the apoplast of barley aleurone layer [34]. Ozone and UVB radiation have been shown to induce NOS activity in some plants [24].

3.3 ROS AND RNS GENERATION AND SITE OF PRODUCTION UNDER DIFFERENT ABIOTIC STRESSES

ROS and RNS are synthesized via several reactions in diverse cellular compartments (chloroplasts, mitochondria, peroxisomes, and during various redox reactions) [35]. ROS generation is an inevitable process of aerobic organisms and the main source of ROS generation in plant cells is the chloroplast. The normal electron flow from photosystem (PS) centers to $NADP^+$ leads to the generation of NADPH, which is utilized in the Calvin cycle for giving electrons to the final electron acceptor the CO_2. During stress, the ETC gets overloaded and supply of $NADP^+$ is decreased. So electrons instead of reducing $NADP^+$, are leaked out from ferredoxin to O_2 to generate O_2^- through Mehler reaction ($2O_2 + 2Fd_{red} \rightarrow 2O_2^- + 2Fd_{ox}$). It is also thought that electrons from iron–sulfur centers of PSI are leaked out during drought, salinity, heat stress, to generate superoxide [36]. Under strong light conditions, the half-life of singlet Chl˙ is increased, which gets converted into triplet Chl˙ which upon reaction with $3O_2$ (triplet oxygen), produce $^1O_2$ (singlet oxygen) [37] especially at PS II [38]. Mitochondria serves as an effective site for ROS production in aerobic organisms. Under normal conditions, ETC and ATP synthesis are under tight control, but stress either inhibits or modifies the ETC components, leading to over reduction of electron carriers and in turn formation of ROS. During stress NAD^+-linked substrates are limited, electrons instead of flowing from complex I to complex II, and flow in the reverse direction that increases the production of ROS at complex I. Similarly, at complex III during stress the fully reduced ubiquinone after donating an electron to cytochrome C1 gets converted into ubisemiquinone radical, which is highly unstable reduced electron carrier in complex IIII capable of reducing O_2 to O_2^- [36].

Peroxisomes serve as an additional site for ROS production a part of normal metabolism like chloroplasts and mitochondria. During high-light stress in C3 plants, photorespiration occurs and it increases the production of ROS through glycolate oxidase reaction that accounts for the majority of H_2O_2 production in peroxisomes. β-oxidation of fatty acids also leads to ROS generation. Xanthine oxidase catalyzes the oxidation of xanthine and hypoxanthine to uric acid and is alternative site for O_2^- production inside peroxisomes [39]. Under stress conditions, misfolded proteins get accumulated that lead to ER stress. During ER stress, ROS accumulation occurs due to dysregulated disulfide bond formation and breakage, which further leads to mitochondrial dysfunction and enhanced ROS production [40]. Another site of ROS production in ER is the NAD(P)H-dependent electron transport involving Cyt p_{450} produces O_2^- [36]. Rboh or NADPH Oxidase serves as a unique oxidoreductase at PM involved in superoxide production. Many studies have reported the enzyme to be activated in response to stress, which leads to superoxide production.

RNS another set of free radicals are produced under stress conditions, which lead to nitrosative stress. Major RNS in plants includes free radicals such as NO, nitrogen dioxide (NO_2), and nonradicals S-nitrosothiols, nitrous acid (HNO_2), dinitrogen trioxide, peroxynitrite ($ONOO^-$), nitroxyl anion, nitrate (NO_3^-), nitrosonium cation, dinitrogen tetroxide, and nitryl chloride

[41]. RNS and their derivatives are generated at peroxisomes, chloroplast, and mitochondria [42], besides at apoplast, cytoplasm, PM [43]. Oxidative and reductive pathways have been proposed for the production of RNS. The oxidative pathway includes NOS-like activity, PAs, and hydroxylamine-based production. While as, reductive pathway include NR with nitrite-nitric oxide reductase (NiNOR) complex, NR with NO-forming nitrite reductase complex, mitochondrial and plastid ETC, peroxisomal NOS-like activity, and S-nitrosoglutathione (GSNO) reductase activity [34]. NO is the main nitrogen reactive species and other species are derived from it through many reactions [44]. In mammals, NOS catalyzes the generation of NO but such enzyme has not been found in plant genome except some NOS-like enzymes have been reported in marine green alga *Ostreoccus tauri* and *Ostreoccus lucimarinus* [45]. Further, NOS-like activity was also reported in peroxisomes of vascular plants [25] and was characterized in *Arabidopsis* mitochondria [46].

NO is mainly formed from the activity of cytoplasmic NR, which is a potent enzyme for ·NO synthesis during various developmental processes and abiotic stress [34]. NO is also formed as an intermediate product due to the activity of PM-bound NR and NiNOR, which synthesize it from NO_3^- during photosynthesis [47]. Peroxidases are considered the essential sites responsible for NO production under salinity stress in *Arabidopsis* [48]. Lower levels of $OONO^-$ are continuously produced in chloroplasts but in response to stress, the level goes on increasing [48]. ETC at chloroplast and mitochondria is thought to induce NO production. Involvement of mitochondrial ETC in ·NO production was investigated in several plant species [49]. Mitochondria are also reported to produce NO under anoxia [49]. Complex III and IV of mitochondria are considered the sites for NO production, Complex III can produce NO via electron leakage to nitrite. It is said that in the absence of oxygen, Fe^{2+} donates the electron from nitrite reduction to NO [50]. NO produced then contribute with other RNS and is the only known biological molecule that reacts faster with superoxide and is synthesized at higher concentrations than SOD level. After reacting with superoxide, it generates $ONOO^-$ [48]. The main site of superoxide production is chloroplast, mitochondria, PM, are also the sites of production of RNS. Besides this, $OONO^-$ is also produced by NR in the presence of oxygen and NAD(P)H. From $ONOO^-$ other reactive species are generated such as NO_3^-, OH^-, and NO_2. Even at physiological pH, $OONO^-$ rapidly combines with proton and generates highly reactive hydroxyl radical (OH^-). $OONO^- + H^+ \leftrightarrow HOONO \rightarrow HO· + NO_2^-$. This reaction is thought to produce hydroxyl radical far more effectively than Fenton/Haber-Weiss reaction [51, 52].

3.4 SCAVENGING MECHANISM OF ROS AND RNS IN PLANTS UNDER NORMAL AND STRESS CONDITIONS

Plants have developed immense ways of protecting themselves from harsh environmental conditions and one of the sophisticated protective mechanism is a thin hydrophobic waxy layer outside the aerial parts of the plant known as cuticle [53]. As an interface between plants and environment, it guards plant against various abiotic stresses, prevents evaporation of water from the epidermis, limiting water loss [54]. Plants being poikilothermic organisms underwent through membrane dysfunction by extreme high and low temperatures. Accordingly, nature protects them by increasing the degree of unsaturation in fatty acid layers, to balance membrane fluidity against heat stress and rigidity against cold stress. Unsaturation of fatty acids has been linked with the protection of PSII against cold-induced photoinhibition [54]. Heat-shock proteins also called molecular chaperons guard cell against the havoc of misregulated proteins, assist in protein folding, protein assembly, its transport and degradation. Thus this machinery help in restoring cellular homeostasis [54, 55]. Besides these general defense mechanisms, plants are equipped with sophisticated nonenzymatic and enzymatic antioxidant components, which protect them against harsh environmental conditions. Some of the nonenzymatic and enzymatic antioxidants include:

3.4.1 NONENZYMATIC ANTIOXIDANTS OR COMPATIBLE SOLUTES

These are highly soluble organic compounds, which get accumulated to high concentrations inside plant cells with little disturbances. It includes some sugars, amino acids, proline, glycine betaine, mannitol, raffinose, and trehalose. The characteristic properties of these solutes is that they are ROS scavengers and osmoprotectants, stabilizing proteins, and PMs [54]. Some of these osmolytes are as follows:

- Proline is most studied compatible solute in plants, which get accumulated in plants during stress. As an osmolyte, proline acts as a powerful antioxidant and helps in osmotic adjustment during abiotic stresses like salinity, drought, and extreme temperatures [56]. It usually gets accumulated in cytoplasm and chloroplast stroma, wherein it reduces the cellular water potential and enables water to move into the cell's interior, protects plants against dehydration and toxic effects of otherwise increasing ionic concentration [57]. Proline synthesis in chloroplast provides NADP$^+$ for accepting electrons and avoid the then generation of dangerous ROS from being formed if oxygen acts as electron acceptor [58] and reduces the chances of photoinhibition at chloroplast. Thus act as an electron sink and redox buffer by maintaining NADPH/NADP$^+$ ratio during its metabolism [57]. Proline is also thought to scavenge dangerous hydroxyl radicals (OH) [59] besides act as a precursor for antioxidant glutathione (GSH). Proline also acts as a signaling molecule and regulate the expression of many stress-responsive genes [58].

- GSH a potent radical scavenger, comprises glycine, cysteine, and glutamic acid [60], and is present in cytosol, chloroplast [61], vacuoles, ER, and mitochondria [60]. It has very high reductive potential due to the presence of nucleophilic cysteine residue and acts as a cellular buffer by maintaining the reduced state inside the cell during stressful conditions [60, 62], also plays a crucial role in replenishing ascorbic acid in ascorbate (AsA)–GSH cycle.
- Ascorbic acid (vitamin C) a potent soluble antioxidant having low molecular weight, is synthesized in mitochondria and transported to other organelles [63]. It neutralizes ROS molecules by reacting with them. It prevents membranes from oxidative damage by regenerating tocopherol molecules from tocopherol radical [4] and also acts as a substrate for ascorbate peroxidase (APX) to detoxify H_2O_2 into H_2O [64].
- Carotenoids is a lipophilic antioxidant of plastids known to protect photosynthetic machinery by acting as chain breakers of LPO cycle, scavenging the singlet oxygen, and produces heat, and also dissipate the excess energy via xanthophyll cycle [4]. It also protects against high-light damage by quenching the excited chlorophyll of photosynthetic machinery [60, 61].
- Flavonoids are diverse phenolic compounds exclusively synthesized by plants as secondary metabolites. Besides, they directly interact with ROS ($^1O_2$ and H_2O_2) and act as substrates for peroxidases [60]. They are also reported to reduce $^1O_2$ damage caused to the outer envelope of the chloroplastic membrane [4].
- α-Tocopherol is a lipophilic antioxidant known to protect lipids and other membrane constituents of chloroplasts by quenching excess energy of O_2^-. Also, it is considered as a free radical trap by stopping the LPO cycle from propagating. These are specific to the thylakoid membranes to preserve their membrane integrity, scavenge $^1O_2$, OH$^-$, and some lipid radicals thus prevent lipid peroxidation [60]. Tocopherols have

been shown to play a crucial role during low-temperature acclimation in plants [22, 64].

- Dehydrins belong to the late embryogenesis abundant class of proteins, which are induced by abiotic stresses like drought and salt temperature. They are thought to stabilize membranes, sequester ions, act as chaperones and scavenge free radicals [65].

3.4.2 ANTIOXIDANT ENZYME SYSTEM

These include a class of sophisticated enzymes, which help in converting toxic ROS molecules into less toxic or nontoxic molecules. There are several groups of enzymes that work together in a coordinated manner and assist nonantioxidant system in detoxifying ROS and RNS molecules. Some of these enzymes are as follows:

- SODs are membrane-bound metalloenzymes, which are considered the immediate defense line against superoxide damage during stressful conditions [56, 66]. The enzyme is responsible for rapid dismutation of O_2^- to H_2O_2 and O_2 [60]. This reaction limits the possibility of formation of more dangerous hydroxyl radicals (OH^-) from Haber–Weiss reaction.
- Catalase (CAT) enzyme is mainly responsible for detoxifying H_2O_2 into water and oxygen [67]. It occupies a central place in detoxifying H_2O_2 particularly because it does it without any reducing equivalent, unlike other antioxidant enzymes. It has the highest turn-over rates among antioxidant enzymes, wherein one CAT molecule reduces 6 million H_2O_2 molecules per minute [61]. These enzymes are commonly found in peroxisomes, with some activity reported in mitochondria, chloroplast, cytosol [68].
- Glutathione reductase (GR) enzyme is found in chloroplasts with small amounts being present in cytosol, mitochondria, and peroxisomes. Besides preventing thiol groups from getting oxidized it is an efficient ROS detoxifying members. It is involved in AsA–GSH cycle wherein it catalyzes the reduction of GSSH to GSH [69].
- Peroxidases are a diverse group of enzymes involved in detoxification of H_2O_2 using different compounds as substrates [60]. APX is an important enzyme of AsA–GSH cycle. APX detoxifies H_2O_2 in the cytosol through AsA–GSH cycle and in chloroplast through the water-water cycle [70]. APX utilizes ascorbic acid to reduce H_2O_2 to

H_2O and in this reaction, AsA is converted into monodihydroascorbate (MDHA), which can be recycled back to AsA by either getting reduced by MDHA reductase that uses NADPH as reducing equivalent, or by spontaneous conversion of MDHA to DHA (dehydro-AsA). This, in turn, is reduced to ascorbic acid by using enzyme DHA reductase. Thus regenerates cellular pool of ascorbic acid. This enzyme brings reducing power from GSH and oxidizes it into GSSG. Finally, this oxidized GSH is reduced back to reduced form by GR, which use NADPH as reductant [71, 64]. This is called AsA–GSH cycle or Foyer–Halliwell–Asada pathway. APX of the thylakoid membrane is involved in detoxification of H_2O_2 into H_2O in a chain of reactions called the water-water cycle [71]. During high-light stress, electrons are leaked out from water at PSII and are transferred to O_2 by PSI, result in the formation of O_2^- which is subsequently dismutated to H_2O_2 by SOD. Here comes the role of APX into play and it reduces this H_2O_2 back to H_2O using reducing power from AsA. Then here oxidized AsA is reduced by Ferredoxin-GSH and NADPH-dependent pathways. Guaiacol peroxidase (GPX) is a heme-containing enzyme utilizing H_2O_2 produced during stress in the process of lignin biosynthesis [4].

3.5   EMERGING ROLES OF ROS AND RNS UNDER NORMAL AND ABIOTIC STRESSES

ROS plays a vital role in the growth and development of plant system (Figure 3.2) such as in seed dormancy and germination [72]. ROS especially H_2O_2 control several processes at molecular level such as stability of mRNA, the process of translation, also by mediating DNA binding and facilitating the nucleo-cytoplasmic traffic, it controls the activities of several TFs [8]. At low concentrations ROS act as a secondary messenger (Figure 3.2) in several plant responses such as ABA-mediated stomatal closure, auxin-mediated root gravitropism, jasmonic acid-mediated lignin biosynthesis, salicylic acid-mediated hypersensitive response plus osmotic stress, and GA-mediated programmed cell death [36]. During germination, ROS are involved in endosperm mobilization and programmed cell death [72]. Hydroxyl radicals are involved in the loosening of the cell wall polysaccharide chains and mediate the breakdown of chitosan during germination process [73]. H_2O_2 triggers the breaking of dormancy by decreasing the levels of ABA (abscisic acid) and increasing GA (gibberellic acid) [74]. ROS also activates calcium (Ca^{2+})

channels and MAPKs during radicle enlargement [75]. Phytohormones are involved in promoting cell expansion in growing roots via induction of ROS production. During the development of the lateral root, auxin is involved in cell expansion process and keeping pericycle initials separate and ABA balances the equilibrium between cellular proliferation and cell differentiation in both the meristem and lateral primordia of the root [76]. Peroxidases also determine lateral root emergence through ROS signaling that promotes the transition from cellular proliferation to cell differentiation [72]. At the tip of pollen tubes, ROS is important for the efficient growth of female gametophyte [77] and is involved in initiating and degradation of cellular contents for recycling purpose during the senescence process [78]. ROS also play a role in senescence through the expression of WRKY53 gene, regulates seed germination through GA/ABA metabolism through the involvement of ZEP36 in coregulation of ABA-inhibited seed germination and crown root development in rice via activation of WOX11, a WUSCHEL-related Homeobox TF [79].

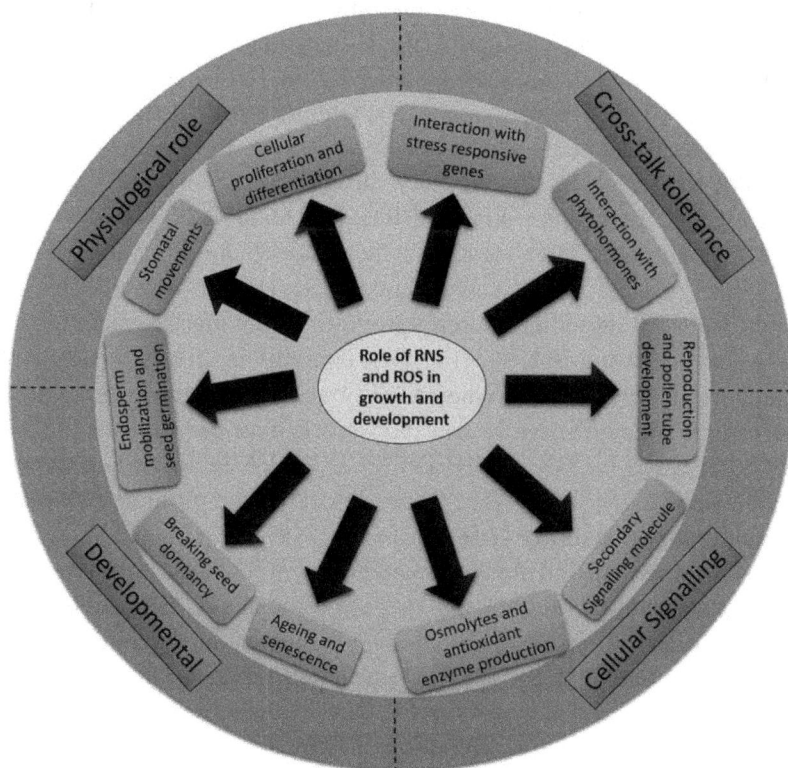

FIGURE 3.2   Versatile roles of ROS and RNS in plant growth and developmental processes.

In case of meristem development, maintenance of shoot apical meristem (SAM) and root apical meristem (RAM) relies on the exchange of information, SAM and RAM activity is affected by ROS, redox components, and phytohormones [80]. The upregulation of WUS gene expression and differentiation in the peripheral zone of SAM is associated with O_2^- and H_2O_2, respectively [72]. During the development of terminal plant organ such as a leaf, a complex agreement between cell multiplication and cell enlargement is required [81]. O_2^- mediated loosening of the cell wall is responsible for this cell enlargement. Whereas the stiffness is enhanced by the production of H_2O_2, which increases cross-linking. ROS along with proper pH and Ca^{2+} ions maintain polar cell growth of root hair cells [82]. This calcium release, during root hair development, is enhanced by the Rboh-mediated ROS production and this ROS burst is also essential for pollen tube rupture and sperm release [83]. The position of transition zone in the vegetative apical meristem is determined by the ROS accumulation in the microenvironment. The meristematic zone requires O_2^- for cell division and elongation zone needs H_2O_2 for cell differentiation, the ratio of O_2^- to H_2O_2 need to reach a certain level for cells to elongate [79]. This balance in the ratio of ROS in the transition zone is maintained by a TF UPB1 (UPBEAT1) and has been shown to affect UPB1 expression itself, which forms a feedback loop and plays a role in ROS homeostasis and root growth. ROS is responsible for controlling root tip stem cell activity through brassinosteroids (BRs) and binding of BR to receptor kinase BRI1 (BRASSINOSTEROID INSENSITIVE 1) increases H_2O_2 levels, which induce modification of BZR1 (BRASSINAZOLE-RESISTANT1) and BES1 (BRI-EMSSUPPSSOR 1), the key TFs in BR signaling thereby promoting root meristem development [79]. O_2^- activates the WUSCHEL gene activity in *Arabidopsis* SAM to maintain stem cell activities, whereas H_2O_2 promotes cell differentiation in the peripheral zone of SAM. Lower ROS levels have been shown to activate TF TCP (TEUSINTE BRANCHED/CYCLOIDEA/PCF), which expresses cell cycle-related genes CYCA2; 3 and CYCB1;1 and promote cell division and maintain SAM stability [79].

Apart from growth and developmental roles, ROS also acts as a secondary messenger and induce specific signals during the acquisition of cross-stress tolerance (Figure 3.1), which is usual thing to happen during long and intense exposure to oxidative stress [3]. ROS at higher concentrations is a potent toxic molecule, which leads to oxidative stress in plants and can threaten cell viability [33, 84]. Increased production of H_2O_2 and lipid peroxidation leads to plant cell death via SA signaling [85]. OH are considered most dangerous among all ROS, OH, and O_2^- (superoxide) capture an H atom from chains of polyunsaturated fatty acids and initiate a chain of reactions, which lead to lipid peroxidation [86], thus generating more stable molecules such as malondialdehyde (MDA), alkanes, lipid epoxides and alcohols [36]. Studies have shown that ROS interacts with diverse and vital phytohormones such as GA, brassinolide, and ABA to improve tolerance for salinity conditions [56, 66, 87]. Oxidative stress modify the amino acids in protein structure, particularly $^1O_2$ and OH attack the thiol groups and S-containing amino acids, such as Met and Cys [36] and an irreversible modification of protein side chains through carbonylation that negatively affects the structure and function of proteins [88]. Superoxide radical irreversibly inactivate enzymes containing Fe-S centers leading to their inactivation [88]. OH being capable of reacting with purines and pyrimidines as well as to the sugar residues, cause the majority of damage to the DNA molecules [89], whereas $^1O_2$ can react only with guanine bases, leading to genomic instability and abnormalities in protein synthesis [60].

RNS have gained much surveillance now a days owing to their crucial role in several plant growth and developmental processes (Figure 3.2) under normal as well as in stress conditions [25]. RNS such as NO, $ONOO^-$, NO_2, and HNO_2 are potent players in the process of seed dormancy, germination, stomatal movements, reproduction and pollen tube growth, ageing, and senescence, and in several stress response processes in plants [25]. NO is involved in plant physiological processes, such as germination, leaf expansion, hormone regulation, and response to stress conditions [85]. NO being multifunctional molecule in plants, activate ROS scavenging enzymes under abiotic stress conditions such as drought, salinity, Al toxicity and regulate plant physiology, disease resistance, and stress tolerance [33]. During drought tolerance, it enhances proline synthesis and osmolyte metabolism [33], closes stomata in an ABA-mediated pathway [90] and mediates responses of several phytohormones such as SA, JA, and ethylene in the cell's response to certain stresses in a very complex signaling network [91]. NO also acts as a modulator of the relative water content in Zea mays leaves through the activation of the vacuolar H^+-ATPase [92] and alleviated the negative effect of ROS, via reducing lipid free radicals, superoxide anions (O_2^-), hydrogen peroxide (H_2O_2), and through the expression of antioxidant enzymes [51, 93]. NO protects against salt stress by expressing the PM H^+-ATPase to balance K^+: Na^+ ion ratio [94]. Also, PAs are said to protect against salinity stress through the involvement of NO [46]. In Zea mays, NO is thought to reduce salt stress by releasing nanoparticle known as chitosan nanoparticles (CS NPs) [33]. NO has been reported to alleviate Cd stress and improve plant growth and biomass yield in Typha angustifolia [95]. Arsenic (As^{3+}) toxicity in rice, mung bean [96], and wheat [97] is reported to alleviate by exogenous application of NO, which reduce ROS and As^{3+}-induced MDA levels [98].

Most interesting roles RNS plays via post-translational modification (PTMs) such as S-nitrosylation, where in a strong nonenzymatic antioxidant GSH is modified into GSNO, which serve as an essential element in abiotic stress responses and in plant immunity [3, 99]. S-nitrosylation inhibits the activity of CAT and glycolate oxidase in peroxisomes, which may be regulating the levels of H_2O_2 in the cell [41, 100]. *S*-nitrosylation and tyrosine nitration are two PTMs mediated by NO, which occur in both plants and animals. *S*-nitrosylation is a rapid, reversible PTM wherein NO is covalently bound to thiol groups of cysteine residues giving rise to formation of S-nitrosothiols (SNOs). In a similar way, the NO group attached to a tyrosine residue, known as tyrosine nitration has been found in case of both plants and animals [101]. SNOs are considered as a key signaling molecule involved in plant stress responses and the most important SNOs is GSNO, which is a major NO reservoir and a phloem mobile long-distance signaling molecule. GSNO is also able to regulate protein function and gene expression during basic plant developmental processes and in various responses toward environmental stresses [99]. S-nitrosylated proteins have been detected in response to stresses such as drought, cold, salinity stress, and during pathogen attack [102]. Similarly, tyrosine nitration can transduce NO signal during hypersensitive defense response in plants [33]. Tyrosine nitration also negatively modulate cytosolic NADP-malic enzyme activity during cold stress in plants [8]. In plants, this nitrotyrosine is used as a marker of nitrosative stress in abiotic stresses similar to lipid peroxidation, which is considered as a mark for the oxidative stress. Other RNS molecule, the $ONOO^-$, is a powerful oxidant, able to damage many biological molecules [103] by destroying the magnetic coupling between QA (plastoquinone) and Fe^{2+} in PSII of chloroplast by either oxidization or nitration of the iron coordinated histidine residues and induces the release of iron [104]. $ONOO^-$ trigger lipid peroxidation and results in the oxidation of unsaturated fatty acids of membranes. $ONOO^-$ can also rapidly oxidize tocopherols to α-tocopheryl quinone, a form that is not easily repaired by cellular reductants [51]. $ONOO^-$ can also react with DNA by modifying nucleobases, sugar-phosphate backbone extracting a hydrogen atom from the deoxyribose sugar, resulting in the generation of breaks in DNA [105].

3.6 SIGNALING MECHANISMS OF ROS AND RNS UNDER ABIOTIC STRESS CONDITIONS

ROS and RNS are notorious for being vital signaling molecules in plants and key regulators of a variety of cellular processes (Figure 3.3) including metabolism, plant growth and development, response to various abiotic and biotic stresses, solute transport, and apoptosis [3]. Sewelam et al. [106] have discussed the central role of ROS in cellular signaling battery in plants under stressed conditions and the probable cross-talk with other signaling components (Figure 3.3), such as TFs, phytohormones, calcium, redox homeostasis, G-proteins, and kinases [107]. ROS is either recognized by specific TFs-like heat-shock factors NPR1 or by inhibition of phosphatase, which depends on the type of stress plants encounters [108]. Amongst ROS, H_2O_2 is frequently used as a signaling molecule due to its higher stability and ability to diffuse the biological membranes easily [106]. ROS perception switches on the signal transduction pathway within the cytoplasm of the cell and activates MAPK (mitogen-activated protein kinases) cascades [108]. MAPK in turn phosphorylates and activates numerous downstream molecules such as TFs, other kinases, phosphatases, and cytoskeleton-associated proteins [108]. In *Arabidopsis* genome, nearly 20 MAPK, 10 MAPKK, and 60 MAPKKK encoding genes were identified and MAPK8 was reported to connect protein phosphorylation, Ca^{2+}, and ROS in signaling pathways [106]. H_2O_2 has been shown to activate MAPKKK, ANP1 that phosphorylates AtMPK3 and AtMPK6 [106]. In *Arabidopsis*, a MAPKKK, MEKK1 is activated during abiotic stresses like a salt, cold, wound, and drought, and then MEKK1 activates another module of the cascade such as MKK2-MPK4/6 [108]. However, H_2O_2 gets activated by upstream kinase MKK6 and then it activates MPK3 and MPK6 in rice [108]. FeSOD antioxidant enzyme is encoded by FSD2 and FSD3 genes in *Arabidopsis*, and these genes are activated by MEKK1 via MKK5–MPK6-coupled signaling [35, 109].

MAPKs also lead to ROS production via activation of NADPH oxidase or Rboh [110]. MAPK has been shown in cross-talk with Rboh through the involvement of WRKY TF [111]. During root tip growth elongation ROOT HAIR DEFECTIVE 2(RHD/2) gene is responsible for triggering Rboh-mediated ROS production via MAPK cascade [72]. Rbohs are central players in the generation of ROS for signaling purposes in the way that they are deliberate producers of ROS, present in immediate contact with cell's environment integrating several signaling components such as Ca^{2+}, protein phosphorylations, Ca^{2+}-dependent protein kinases (CDPKs), Racs, MAPKs with the ROS production (Figure 3.3). Rboh genes have been reported to get induced in response to various biotic and abiotic stresses and most of the Rboh effects are mediated via ROS signaling pathways [106]. ROS produced during ozone-mediated activation of Rboh involve G-proteins to receive signals, which then directly activate RbohD and F [112]. Through the involvement of RbohD and H_2O_2, cell-to-cell signal is self-propagated in the form of ROS wave [113]. Wherein initial ROS induces Ca^{2+} release, which activates NADPH oxidase (Rboh), which in turn produces ROS and leads to activation of calcium channels to allow further Ca^{2+} influx in the cytosol (Figure 3.3) thus forms a forward loop and propagate signal in the form of Ca^{2+} and ROS wave to the adjacent cells [114].

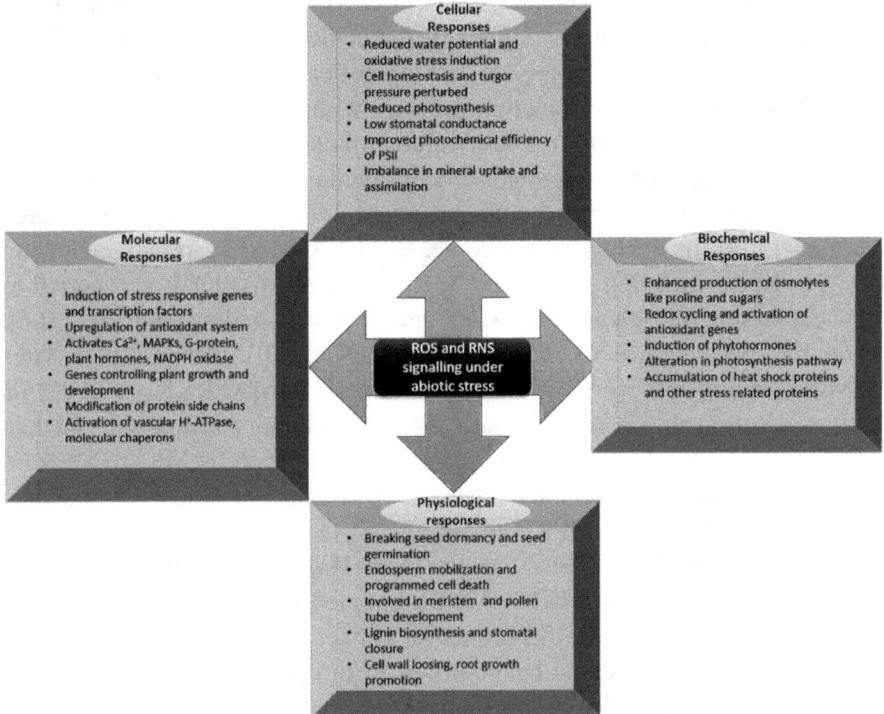

FIGURE 3.3 Various physiological, molecular, biochemical, and cellular responses under abiotic stresses mediated by ROS and RNS signaling.

Calcium is a versatile secondary messenger involved in all ROS signaling pathways and pathways leading to abiotic stress responses (Figure 3.3). Upon stimulation, Ca^{2+} is mobilized from its reservoirs and opening of Ca^{2+}

channels like CNGCs (Cyclic nucleotide-gated channels) and tonoplast two pore channel 1 (TPC1). Ca^{2+} is sensed by various sensors including calmodulins (CAMs), calcineurin B-like proteins (CBLs), CDPKs [54]. Further, CAMs interact with kinases, phosphatases, and TFs and CBLs interact with CBL-interacting protein kinases and propagate the signal in the cell [54]. In some cases, Ca^{2+} elevations have been reported to be upstream of ROS production, in others, it has been reported to be downstream of ROS production [115]. Jayakannan et al. [116] showed that under salt stress, ROS and RNS have a vital role in regulating membrane transporters, thereby stabilizing the ionic effects of salinity. ROS and RNS have been reported to be involved in many mechanisms underlying high-temperature responses and tolerance. As Cheng et al. [117] showed that 2-Cys peroxidases (PRX) are involved in high-temperature stress responses by regulating autophagosome formation and AsA–GSH metabolism in chloroplasts. Moreover, Qiao et al. [118] showed that OsANN1 and annexin protein with calcium-binding and ATPase activities, regulates H_2O_2 levels in rice under heat stress [35].

3.7   GENES AND TFS INVOLVED IN ROS/RNS SIGNALING PATHWAYS UNDER STRESS

Plants have evolved and developed complex signaling pathways in response to abiotic stress, which helps them to regulate sets of stress-responsive genes (Figure 3.3) encoding protein kinases, phosphatases, transcriptional factors (TFs), proteins involved in ROS scavenging or detoxification, phytohormone pathway, and calcium signaling [107, 119]. ROS can interact with NO in various stresses to induce a defense response, Sewelam et al. [106] and Chan et al. [107] have discussed in details about the cross-talk between NO and ROS molecules, regulating ABA biosynthesis that modulates stomatal shutting. Wang et al. [120] have reported H_2O_2 mediated activation of MAPK6 to synthesize NO in *Arabidopsis*. High throughput sequencing of pepper have revealed that RNS regulate 1945 genes in leaves and roots and during fruit ripening roughly 2987 genes are modulated, 498 of which were modulated by NO alone, 5 members of the WRKY gene family and four members of MYB TF family are reported to be involved in abiotic stress responses [121]. NO and H_2O_2 were found to overlap in signaling pathways in acclimation to salinity [122]. Also, ABA-induced H_2O_2 production was reported to generate NO in maize leaves, which then activate MAPK and increase the expression of antioxidant enzymes [119]. NO regulates the function of various essential enzymes during salt stress via covalent binding to thiol groups of Cysteine amino acid [2]. NO besides performing its signaling role via nitrosylation, also act as Ca^{2+}-mobilizing messenger in cells [106]. S-nitrosylation has many effects such as inducing some TFs, which affect their binding to DNA, inactivate Rbohs [123] and can also act as a negative regulator of MYB TFs that are vital regulators of abiotic stress responses, which can lead to metabolic reprogramming to keep homeostasis under stress conditions [2]. De Abreu Neto and Frei [124] had shown a meta-analysis of microarray experiments in rice, where they have reported overexpression of ROS-related genes among the differentially expressed genes (DEGs). They have reported about 990 and 1727 DEGs under oxidative stress (ozone and H_2O_2), respectively, and among these DEGs 33 were ROS-related genes. Wei et al. [125] conducted a similar analysis in cassava (*Manihot esculenta*) under abiotic stresses and has identified nearly 85 MeWRKY genes. Among these TFs, 78 MeWRKY genes were differentially expressed in response to drought stress and 9 MeWRKY genes were modulated after NaCl, mannitol, cold, H_2O_2, and ABA treatments demonstrating role of MeWRKY genes in stress response and redox signaling pathways. Redox-sensitive genes associated with ROS/RNS signaling include GPX, CAT, SOD, quinone reductase, GR, thioredoxin reductase, thioredoxin, metallothionein, cyclooxygenase-2, and γ-glutamyl cysteine synthase (required for GSH synthesis) [121].

TFs are vital guiding proteins, which are directly involved in abiotic stress responses and are important downstream elements of various signaling cascades that change the expression of stress-responsive genes and enhance tolerance to environmental stresses (Figure 3.3). There are many stress-responsive TFs such as ABA-responsive element (ABRE)-binding factors (AREBs/ABFs), DELLAs, NACs (NAM, ATAF, and CUC), WRKYs, APETALA2/Ethylene response factor (AP2/ERF) superfamily, Zinc finger bZIP (basic leucine zipper) families [54]. Several H_2O_2-responsive TFs have been reported to be involved in stress like drought, heat, and high light [121]. Genes for zinc finger domains scuh as ZAT7, ZAT10, and ZAT12 are upregulated by oxidative stress in *Arabidopsis* thaliana [119]. C2H2-type Zinc finger TF has been shown to negatively regulate stomatal closure by regulating ROS homeostasis genes and ROS is involved in antioxidant gene induction through C2H2-type ZFP, ZFP 182 in ABA-induced antioxidant defense response. Then ZFP36 was also found to be necessary for ABA-induced antioxidant defense and also play a role in ROS homeostasis during stress resistance in rice [119]. ZFP179 encodes a salt responsive Zinc finger protein related to tolerance against salt and oxidative stress, OSTZF1 a CCCH-Tendom Zinc finger protein confers tolerance to oxidative stress in rice by increasing the expression of ROS-scavenging enzymes, SERF1 and ERF TF regulates H_2O_2-mediated signaling during initial salinity responses in rice [119]. SUB1A another ERF TF increases expression of ROS-scavenging enzymes genes, enhance tolerance to oxidative stress, GmNAC 29 TF change the ROS production under abiotic stress by activating the expression of ROS encoding enzymes and NAC, GmNAC2 are negative regulators and participate in ROS signaling pathways through modulation of ROS-scavenging genes [119]. Histidine kinase and DREB2A TFs also regulate ROS via osmotic stress signaling [42]. ROS is associated with heat-shock TFs and heat-shock proteins in heat stress response [126]. ROS involved in cold stress acclimation involves DREB1A, DREB2A, KIN1, KIN2, COR 15A, COR 47, C-repeat related (CRT), CRT/DRE-binding factor1 (CBF1) have been shown to enhance cold tolerance in tomato plants [126]. Mostly, TFs that are involved in stress physiology were found to be related to ROS and RNS metabolism. Zhang et al. [127] and Fang et al. [128] demonstrated the roles of stress-related transcription for the regulation of ROS metabolism. Zhang et al. reported that ABRE-binding factor (PtABF) overexpression resulted in enhanced dehydration tolerance in Poncirus trifoliate, where PtABF interaction with ICE1 (Inducer of CBF Expression 1) affected stomatal development [127]. Besides, antioxidant enzymes and PA biosynthesis were induced in PtABF overexpressing plants, causing less cellular damage under dehydration. Similarly in another study, Fang et al. demonstrated that overexpression of SNAC3 TF in rice resulted in high temperature, drought, and oxidative stress tolerance [128]. Furthermore, direct binding of SNAC3 TF to the promoters of CATA, APX8, and RbohF genes that are vital constituents of ROS defense and signaling was shown [35].

3.8 CONCLUSION

Various modern trends and advancements have been made for understanding the physiology of abiotic stresses (Figure 3.1). ROS and RNS signaling are mostly widely studied and have been found to be associated with almost every aspect of plant metabolism and cellular functions. Latest advanced techniques such as transcriptomics, proteomics, and novel microscopy have made immense evolution, especially for finding new TFs, protein–protein interactions and in vivo optical manipulation of intracellular structures. This information will be a future challenge to disentangle the complete genome network providing abiotic stress tolerance. Novel roles and involvement of ROS and RNS are being suggested, which has enhanced our under-standing of the underlying signaling mechanisms, yet there is still a lack of vital information on role of these molecules under abiotic stresses. Thus, we need spatial data about ROS and RNS production in the cell and new understandings into their specific signaling. As we know that there is huge lacuna about the information involved in initiation, signaling, the sensing and response mechanisms of ROS and RNS, and how a cell maintains a homeostasis between ROS and RNS production and scavenging. Further, we also need to know more about the cross-talk between the pathways (cellular redox changes, calcium signaling, phytohormones), which is mediated by RNS, ROS, and other messenger molecules. The comprehensive information of the mechanism and regulation of the cross-talk by ROS, RNS, and additional signaling molecules in plant cells under abiotic stress conditions will significantly expand our understanding of ROS and RNS involvement in plant stress physiology.

KEYWORDS

- **reactive oxygen species**
- **nitrogen species**
- **antioxidant enzymes**
- **abiotic stresses**
- **signaling**
- **homeostasis**
- **scavenging**
- **nonenzymatic defense system**
- **secondary messengers**
- **transcription factors**

REFERENCES

1. Cramer, GR.; Urano, K.; Delrot, S.; Pezzotti, M.; Shinozaki, K.; Effects of abiotic stress on plants: a systems biology perspective. *BMC Plant Biol.* **2011,** *11(1),* 1–4.
2. Choudhury, F.K.; Rivero, R.M.; Blumwald, E.; Mittler, R.; Reactive oxygen species, abiotic stress and stress combination. *Plant J.* **2017,** *90(5),* 856–67.
3. Turkan, I.; ROS and RNS: key signalling molecules in plants. *J. Exp. Bot.* **2018,** *69(14),* 3313.
4. Das, K.; Roychoudhury, A.; Reactive oxygen species (ROS) and response of antioxidants as ROS-scavengers during environmental stress in plants. *Front. Environ. Sci.* **2014,** *2,* 53.

5. Agarwal, P.; Agarwal, P.K.; Gohil, D.; *Transcription Factor-Based Genetic Engineering for Salinity Tolerance in Crops. In Salinity Responses and Tolerance in Plants*, Springer, Cham, **2018,** pp. 185–211.

6. Liang, W.; Ma, X.; Wan, P.; Liu, L.; Plant salt-tolerance mechanism: A review. *Biochem. Bioph. Res. Co.* **2018,** *495(1),* 86–91.

7. Zhou, J.C.; Fu, T.T.; Sui, N.; Guo, J.R.; Feng, G.; et al. The role of salinity in seed maturation of the euhalophyte Suaeda salsa. *Plant Biosyst.* **2016,** *150(1),* 83–90.

8. Chaki, M.; Begara-Morales, J.C.; Barroso, J.B.; Oxidative stress in plants. *Antioxidants.* **2020,** *9,* 481.

9. Hossain, M.S.; Dietz, K.J.; Tuning of redox regulatory mechanisms, reactive oxygen species and redox homeostasis under salinity stress. *Front. Plant Sci.* **2016,** *7,* 548.

10. Rejeb, K.B.; Benzarti, M.; Debez, A.; Bailly, C.; Savouré, A.; et al. NADPH oxidase-dependent H_2O_2 production is required for salt-induced antioxidant defense in *Arabidopsis* thaliana. *J. Plant Physiol.* **2015,** *174,* 5–15.

11. Saha, J.; Brauer, E.K.; Sengupta, A.; Popescu, S.C.; Gupta, K.; et al. Polyamines as redox homeostasis regulators during salt stress in plants. *Front. Environ. Sci.* **2015,** *3,* 21.

12. Lipiec, J.; Doussan, C.; Nosalewicz, A.; Kondracka, K.; Effect of drought and heat stresses on plant growth and yield: a review. *Int. Agrophys.* **2013,** *27(4)*.

13. Salehi-Lisar, S.Y.; Bakhshayeshan-Agdam, H.; Drought stress in plants: causes, consequences, and tolerance. In *Drought Stress Tolerance in Plants*, Springer, Cham, **2016;** Vol. 1; p 1–16.

14. Farooq, M.; Wahid, A.; Kobayashi, N.; Fujita, D.B.; Basra, S.M.; Plant drought stress: effects, mechanisms and management. In *Sustainable Agriculture*, Springer, Dordrecht, **2009,** p 153–188.

15. Nadeem, M.; Li, J.; Yahya, M.; Sher, A.; Ma, C.; et al. Research progress and perspective on drought stress in legumes: A review. *Int. J. Mol. Sci.* **2019,** *20(10),* 2541.

16. Hasanuzzaman, M.; Nahar, K.; Alam, M.; Roychowdhury, R.; Fujita, M.; Physiological, biochemical, and molecular mechanisms of heat stress tolerance in plants. *Int. J. Mol. Sci.* **2013,** *14(5),* 9643–84.

17. Yamamoto, Y.; Aminaka, R.; Yoshioka, M.; Khatoon, M.; Komayama, K.; et al. Quality control of photosystem II: impact of light and heat stresses. *Photosynth. Res.* **2008,** *98(1–3),* 589–608.

18. Wang, J.Z.; Cui, L.J.; Wang, Y.; Li, J.L.; Growth, lipid peroxidation and photosynthesis in two tall fescue cultivars differing in heat tolerance. *Biol. Plant.* **2009,** *53(2),* 237–42.

19. Asthir, B.; Protective mechanisms of heat tolerance in crop plants. *J. Plant Interact.* **2015,** *10(1),* 202–10.

20. Theocharis, A.; Clément, C.; Barka, E.A.; Physiological and molecular changes in plants grown at low temperatures. *Planta.* **2012,** *235(6),* 1091–105.

21. Khan, T.A.; Fariduddin, Q.; Yusuf, M.; Lycopersicon esculentum under low temperature stress: an approach toward enhanced antioxidants and yield. *Environ. Sci.* **2015,** *22(18),* 14178–88.

22. Maeda, H.; Sakuragi, Y.; Bryant, D.A.; DellaPenna, D.; Tocopherols protect Synechocystis sp. strain PCC 6803 from lipid peroxidation. *Plant Physiol.* **2005,** *138(3),* 1422–35.

23. Yamazakia, T.; Kawamura, Y.; Uemura, M.; Extracellular freezing-induced mechanical stress and surface area regulation on the plasma membrane in cold-acclimated plant cells. *Plant Signal. Behav.* **2009,** *4(3),* 231–3.

24. Corpas, F.J; Chaki, M.; Fernandez-Ocana, A.;Valderrama, R.; Palma, J.M.; et al. Metabolism of reactive nitrogen species in pea plants under abiotic stress conditions. *Plant Cell Physiol.* **2008,** *49(11),* 1711–22.

25. Saddhe, A.A.; Malvankar, M.R.; Karle, S.B.; Kumar, K.; Reactive nitrogen species: paradigms of cellular signaling and regulation of salt stress in plants. *Environ. Exp. Bot.* **2019,** *161,* 86–97.

26. Roach, T.; Krieger-Liszkay, A.; Regulation of photosynthetic electron transport and photoinhibition. *Curr. Protein Pept. Sci.* **2014,** *15(4),* 351–62.

27. Nishiyama, Y.; Allakhverdiev, S.I.; Murata, N.; A new paradigm for the action of reactive oxygen species in the photoinhibition of photosystem II. *BBA-Bioenergetics.* **2006,** *1757(7),* 742–9.

28. Kavamura, V.N.; Esposito, E.; Biotechnological strategies applied to the decontamination of soils polluted with heavy metals. *Biotechnol. Adv.* **2010,** *28(1),* 61–9.

29. Panda, S.K.; Baluska, F.; Matsumoto, H.; Aluminium stress signaling in plants. *Plant Signal. Behav.* **2009,** *4(7),* 592–597.

30. Zhang, W.; Long, Y.; Huang, J.; Xia, J.; Molecular Mechanisms for Coping with Al Toxicity in Plants. *Int. J. Mol. Sci.* **2019,** *20(7),* 1551.

31. Cuypers, A.; Plusquin, M.; Remans, T.; Jozefczak, M.; Keunen, E.; et al. Cadmium stress: an oxidative challenge. *Biometals.* **2010,** *23(5),* 927–40.

32. Apel, K.; Hirt, H.; Reactive oxygen species: metabolism, oxidative stress, and signal transduction. *Annu. Rev. Plant Biol.* **2004,** *55,* 373–99.

33. Nabi, R.B.; Tayade, R.; Hussain, A.; Kulkarni, K.P.; Imran, Q.M.; et al. Nitric oxide regulates plant responses to drought, salinity, and heavy metal stress. *Environ. Exp. Bot.* **2019,** *161,* 120–33.

34. Astier, J.; Gross, I.; Durner, J.; Nitric oxide production in plants: an update. *J. Exp. Bot.* **2018,** *69(14),* 3401–11.

35. Turkan, I.; Emerging roles for ROS and RNS–versatile molecules in plants. *J. Exp. Bot.* **2017,** *68(16),* 4413–4416.

36. Sharma, P.; Jha, A.B.; Dubey, R.S.; Pessarakli, M.; Reactive oxygen species, oxidative damage, and antioxidative defense mechanism in plants under stressful conditions. *J. Bot.* **2012,** 2012.

37. Zolla, L.; Rinalducci, S.; Involvement of active oxygen species in degradation of light-harvesting proteins under light stresses. *Biochem.* **2002,** *41(48),* 14391–402.

38. Krieger-Liszkay, A.; Singlet oxygen production in photosynthesis. *J. Exp. Bot.* **2005,** *56(411),* 337–46.

39. Luis, A.; Sandalio, L.M.; Corpas, F.J.; Palma, J.M.; Barroso, J.B.; Reactive oxygen species and reactive nitrogen species in peroxisomes. Production, scavenging, and role in cell signaling. *Plant Physiol.* **2006,** *141(2),* 330–5.

40. Cao, S.S.; Kaufman, R.J.; Endoplasmic reticulum stress and oxidative stress in cell fate decision and human disease. *Antioxid. Redox Sign.* **2014,** *21(3),* 396–413.

41. Del Río, L.A.; ROS and RNS in plant physiology: an overview. *J. Exp. Bot.* **2015,** *66(10),* 2827–37.

42. Kapoor, D.; Singh, S.; Kumar, V.; Romero, R.; Prasad, R.; et al. Antioxidant enzymes regulation in plants in reference to reactive oxygen species (ROS) and reactive nitrogen species (RNS). *Plant Gene.* **2019,** *19,* 100182.

43. Corpas, F.J.; Barroso, J.B.; Palma, J.M.; Rodriguez-Ruiz, M.; Plant peroxisomes: a nitro-oxidative cocktail. *Redox Biol.* **2017,** *11,* 535–42.

44. Frungillo, L.; Skelly, M.J.; Loake, G.J; Spoel, S.H; Salgado, I.; S-nitrosothiols regulate nitric oxide production and storage in plants through the nitrogen assimilation pathway. *Nat. Commun.* **2014,** *5(1),* 1–0.

45. Foresi, N.; Correa-Aragunde, N.; Parisi, G.; Caló, G.; Salerno, G.; et al. Characterization of a nitric oxide synthase from the plant kingdom: NO generation from the green alga Ostreococcus tauri is light irradiance and growth phase dependent. *Plant Cell.* **2010,** *22(11),* 3816–30.

46. Guo, F.Q.; Crawford, N.M.; *Arabidopsis* nitric oxide synthase1 is targeted to mitochondria and protects against oxidative damage and dark-induced senescence. *Plant Cell.* **2005,** *17(12),* 3436–50.

47. Chamizo-Ampudia, A.; Sanz-Luque, E.; Llamas, A.; Galvan, A.; Fernandez, E.; Nitrate reductase regulates plant nitric oxide homeostasis. *Tren. Plant Sci.* **2017,** *22(2),* 163–74.

48. Vandelle, C.E.; Delledonne, M.; Peroxynitrite formation and function in plants. *Plant Sci.* **2011,** *181(5),* 534–9.

49. Planchet, E.; Jagadis Gupta, K.; Sonoda, M.; Kaiser, W.M.; Nitric oxide emission from tobacco leaves and cell suspensions: rate limiting factors and evidence for the involvement of mitochondrial electron transport. *Plant J.* **2005,** *41(5),* 732–43.

50. Gupta, K.J.; Igamberdiev, A.U.; Reactive nitrogen species in mitochondria and their implications in plant energy status and hypoxic stress tolerance. *Front. Plant Sci.* **2016,** *7,* 369.

51. Procházková, D.; Wilhelmová, N.; Pavlík, M.; Reactive nitrogen species and nitric oxide. In *Nitric Oxide Action in Abiotic Stress Responses in Plants,* Springer, Cham, **2015,** p 3–19.

52. Hasanuzzaman, M.; Fotopoulos, V.; Nahar, K.; Fujita, M.; *Reactive Oxygen, Nitrogen and Sulfur Species in Plants: Production, Metabolism, Signaling and Defense Mechanisms,* John Wiley & Sons, **2019.**

53. Fich, E.A.; Segerson, N.A.; Rose, J.K.; The plant polyester cutin: biosynthesis, structure, and biological roles. *Annu. Rev. Plant Biol.* **2016,** *67,* 207–33.

54. He, M.; He, C.Q.; Ding, N.Z.; Abiotic stresses: general defenses of land plants and chances for engineering multistress tolerance. *Front. Plant Sci.* **2018,** *9,* 1771.

55. Wang, W.; Vinocur, B.; Shoseyov, O.; Altman, A.; Role of plant heat-shock proteins and molecular chaperones in the abiotic stress response. *Tren. Plant Sci.* **2004,** *9(5),* 244–52.

56. Shafi, A.; Chauhan, R.; Gill, T.; Swarnkar, M.K.; Sreenivasulu, Y.; et al. Expression of SOD and APX genes positively regulates secondary cell wall biosynthesis and promotes plant growth and yield in *Arabidopsis* under salt stress. *Plant Mol. Biol.* **2015,** *87,* 615–631.

57. Hossain, M.A.; Hoque, M.A.; Burritt, D.J.; Fujita, M.; Proline protects plants against abiotic oxidative stress: biochemical and molecular mechanisms. In *Oxidative Damage to Plants* Academic Press. **2014,** pp. 477–522.

58. Szabados, L.; Savoure, A.; Proline: a multifunctional amino acid. *Trends Plant Sci.* **2010,** *15(2),* 89–97.

59. Signorelli, S.; Coitiño, E.L.; Borsani, O.; Monza, J.; Molecular mechanisms for the reaction between• OH radicals and proline: insights on the role as reactive oxygen species scavenger in plant stress. *J. Phys. Chem. B.* **2014,** *118(1),* 37–47.

60. Soares, C.; Carvalho, M.E.; Azevedo, R.A.; Fidalgo, F.; Plants facing oxidative challenges—A little help from the antioxidant networks. *Environ. Exp. Bot.* **2019,** *161,* 4–25.

61. Gill, S.S.; Tuteja, N.; Reactive oxygen species and antioxidant machinery in abiotic stress tolerance in crop plants. *Plant Physiol. Biochem.* **2010**, *48(12)*, 909–30.

62. Foyer, C.H.; Noctor, G.; Oxidant and antioxidant signalling in plants: a re-evaluation of the concept of oxidative stress in a physiological context. *Plant Cell Environ.* **2005**, *28(8)*, 1056–71.

63. Munné-Bosch, S.; Queval, G.; Foyer, C.H.; The impact of global change factors on redox signaling underpinning stress tolerance. *Plant Physiol.* **2013**, *161(1)*, 5–19.

64. Szymańska, R.; Ślesak, I.; Orzechowska, A.; Kruk, J.; Physiological and biochemical responses to high light and temperature stress in plants. *Environ. Exp. Bot.* **2017**, *139*, 165–77.

65. Halder, T.; Agarwal, T.; Ray, S.; Isolation, cloning, and characterization of a novel Sorghum dehydrin (SbDhn2) protein. *Protoplasma.* **2016**, *253(6)*, 1475–88.

66. Shafi, A.; Gill, T.; Zahoor, I.; Ahuja, P.S.; Sreenivasulu, Y.; Kumar, S.; Singh, A.K.; Ectopic expression of SOD and APX genes in *Arabidopsis* alters metabolic pools and genes related to secondary cell wall cellulose biosynthesis and improve salt tolerance. *Mol. Bio. Rep.* **2019**, *46(2)*, 1985–2002.

67. Weydert, C.J.; Cullen, J.J.; Measurement of superoxide dismutase, catalase and glutathione peroxidase in cultured cells and tissue. *Nat. Protoc.* **2010**, *5(1)*, 51–66.

68. Mhamdi, A.; Queval, G.; Chaouch, S.; Vanderauwera, S.; Van Breusegem, F.; et al. Catalase function in plants: a focus on *Arabidopsis* mutants as stress-mimic models. *J. Exp. Bot.* **2010**, *61(15)*, 4197–220.

69. Bela, K.; Horváth, E.; Gallé, Á.; Szabados, L.; Tari, I.; Csiszár, J.; Plant glutathione peroxidases: emerging role of the antioxidant enzymes in plant development and stress responses. *J. Plant Physiol.* **2015**, *176*, 192–201.

70. Asada, K.; The water-water cycle in chloroplasts: scavenging of active oxygens and dissipation of excess photons. *Annu. Rev. Plant Physiol. Plant Mol. Biol.* **1999**, *50*, 601–639.

71. Foyer, C.H.; Noctor, G.; Redox regulation in photosynthetic organisms: signaling, acclimation, and practical implications. *Antioxid. Redox Sign.* **2009**, *11(4)*, 861–905.

72. Choudhary, A.; Kumar, A.; Kaur, N.; ROS and oxidative burst: Roots in plant development. *Plant Divers.* **2020**, *42(1)*, 33–43.

73. Stern, R.; Kogan, G.; Jedrzejas, M.J.; Šoltés, L.; The many ways to cleave hyaluronan. *Biotechnol. Adv.* **2007**, *25(6)*, 537–57.

74. Oracz, K.; Karpiński, S.; Phytohormones signaling pathways and ROS involvement in seed germination. *Front. Plant Sci.* **2016**, *7*, 864.

75. Diaz-Vivancos, P.; Barba-Espín, G.; Hernández, J.A.; Elucidating hormonal/ROS networks during seed germination: insights and perspectives. *Plant Cell Rep.* **2013**, *32(10)*, 1491–502.

76. Lavenus, J.; Goh, T.; Roberts, I.; Guyomarc'h, S.; Lucas, M.; et al. Lateral root development in *Arabidopsis*: fifty shades of auxin. *Tren. Plant Sci.* **2013**, *18(8)*, 450–8.

77. Potocký, M.; Pejchar, P.; Gutkowska, M.; Jiménez-Quesada, M.J.; Potocká, A.; et al. NADPH oxidase activity in pollen tubes is affected by calcium ions, signaling phospholipids and Rac/Rop GTPases. *J. Plant Physiol.* **2012**, *169(16)*, 1654–63.

78. Mhamdi, A.; Van Breusegem, F.; Reactive oxygen species in plant development. *Development.* **2018**, *145(15)*, dev164376.

79. Huang, H.; Ullah, F.; Zhou, D.X.; Yi, M.; Zhao, Y.; Mechanisms of ROS regulation of plant development and stress responses. *Front. Plant Sci.* **2019**, 10.

80. Schippers, J.H.; Foyer, C.H.; van Dongen, J.T.; Redox regulation in shoot growth, SAM maintenance and flowering. *Curr. Opin. Plant Biol.* **2016**, *29,* 121–8.

81. Lu, D.; Wang, T.; Persson, S.; Mueller-Roeber, B.; Schippers, J.H.; Transcriptional control of ROS homeostasis by KUODA1 regulates cell expansion during leaf development. *Nat. Commun.* **2014**, *5(1),* 1–9.

82. Mangano, S.; Juárez, S.P.; Estevez, J.M.; ROS regulation of polar growth in plant cells. *Plant Physiol.* **2016**, *171(3),* 1593–605.

83. Duan, Q.; Kita, D.; Johnson, E.A.; Aggarwal, M.; Gates, L.; et al. Reactive oxygen species mediate pollen tube rupture to release sperm for fertilization in *Arabidopsis*. *Nat. Commun.* **2014**, *5(1),* 1–0.

84. Petrov, V.; Hille, J.; Mueller-Roeber, B.; Gechev, T.S.; ROS-mediated abiotic stress-induced programmed cell death in plants. *Front. Plant Sci.* **2015**, *6,* 69.

85. Kim, Y.; Mun, B.G.; Khan, A.L.; Waqas, M.; Kim, H.H.; et al. Regulation of reactive oxygen and nitrogen species by salicylic acid in rice plants under salinity stress conditions. *PLos One.* **2018**, *13(3),* e0192650.

86. Anjum, N.A.; Sofo, A.; Scopa, A.; Roychoudhury, A.; Gill, S.S; et al. Lipids and proteins—major targets of oxidative modifications in abiotic stressed plants. *Environ. Sci. Pollut. Res.* **2015**, *22(6),* 4099–121.

87. Shafi, A.; Pal, A.K.; Sharma, V.; Kalia, S.; Kumar, S.; Ahuja, P.S.; Singh, A.K.; Transgenic potato plants overexpressing SOD and APX exhibit enhanced lignification and starch biosynthesis with improved salt stress tolerance. *Plant Mol. Biol. Rep.* **2017**, *35,* 504–518.

88. Banerjee, A.; Roychoudhury, A.; *Reactive Oxygen Species in Plants: Boon or Bane—Revisiting the Role of ROS.* Vijay Pratap Singh, Samiksha Singh, Durgesh Kumar Tripathi, Sheo Mohan Prasad, and Devendra Kumar Chauhan (Editors). John Wiley & Sons. **2018**, 23–50.

89. Singh, V.P.; Singh, S.; Tripathi, D.K.; Prasad, S.M.; Chauhan, D.K.; *Reactive Oxygen Species in Plants: Boon or Bane-reVisiting the Role of ROS.* John Wiley & Sons; **2017**.

90. Garcia-Mata, C.; Gay, R.; Sokolovski, S.; Hills, A.; Lamattina, L.; et al. Nitric oxide regulates K+ and Cl-channels in guard cells through a subset of abscisic acid-evoked signaling pathways. *P.N.A.S.* **2003**, *100(19),* 11116–21.

91. Agurla, S.; Gayatri, G.; Raghavendra, A.S.; Nitric oxide as a secondary messenger during stomatal closure as a part of plant immunity response against pathogens. *Nitric Oxide.* **2014**, *43,* 89–96.

92. Nieves-Cordones, M.; López-Delacalle, M.; Ródenas, R.; Martínez, V.; Rubio, F.; et al. Critical responses to nutrient deprivation: A comprehensive review on the role of ROS and RNS. *Environ. Exp. Bot.* **2019**, *161,* 74–85.

93. Zhang, L.; Zhou, S.; Xuan, Y.; Sun, M.; Zhao, L.; Protective effect of nitric oxide against oxidative damage in *Arabidopsis* leaves under ultraviolet-B irradiation. *J. Plant Biol.* **2009**, *52(2),* 135.

94. Zhao, L.; Zhang, F.; Guo, J.; Yang, Y.; Li, B.; et al. Nitric oxide functions as a signal in salt resistance in the calluses from two ecotypes of reed. *Plant Physiol.* **2004**, *134(2),* 849–57.

95. Zhao, H.; Jin, Q.; Wang, Y.; Chu, L.; Li, X.; et al. Effects of nitric oxide on alleviating cadmium stress in Typha angustifolia. *Plant Growth Regul.* **2016**, *78(2),* 243–51.

96. Ismail, G.S.; Protective role of nitric oxide against arsenic-induced damages in germinating mung bean seeds. *Acta Physiol. Plant.* **2012**, *34(4),* 1303–11.

97. Mostofa, M.G.; Seraj, Z.I.; Fujita, M.; Exogenous sodium nitroprusside and glutathione alleviate copper toxicity by reducing copper uptake and oxidative damage in rice (Oryza sativa L.) seedlings. *Protoplasma.* **2014,** *251(6),* 1373–86.

98. Singh, H.P.; Kaur, S.; Batish, D.R.; Sharma, V.P.; Sharma, N.; et al. Nitric oxide alleviates arsenic toxicity by reducing oxidative damage in the roots of Oryza sativa (rice). *Nitric Oxide.* **2009,** *20(4),* 289–97.

99. Begara-Morales, J.C.; Chaki, M.; Valderrama, R.; Sánchez-Calvo, B.; Mata-Pérez, C.; et al. Nitric oxide buffering and conditional nitric oxide release in stress response. *J. Exp. Bot.* **2018,** *69(14),* 3425–38.

100. Ortega-Galisteo, A.P.; Rodríguez-Serrano, M.; Pazmiño, D.M.; Gupta, D.K.; Sandalio, L.M.; et al. S-Nitrosylated proteins in pea (*Pisum sativum* L.) leaf peroxisomes: changes under abiotic stress. *J. Exp. Bot.* **2012,** *63(5),* 2089–103.

101. Greenacre, S.A.; Ischiropoulos, H.; Tyrosine nitration: localisation, quantification, consequences for protein function and signal transduction. *Free Radic. Res.* **2001,** *34(6),* 541–81.

102. Fancy, N.N.; Bahlmann, A.K.; Loake, G.J.; Nitric oxide function in plant abiotic stress. *Plant Cell Environ.* **2017,** *40(4),* 462–72.

103. Radi, R.; Peroxynitrite, a stealthy biological oxidant. *J. Biol.* **2013,** *288(37),* 26464–72.

104. González-Pérez, S.; Quijano, C.; Romero, N.; Melø, T.B.; Radi, R.; et al. Peroxynitrite inhibits electron transport on the acceptor side of higher plant photosystem II. *Arch. Biochem.* **2008,** *473(1),* 25–33.

105. Pacher, P.; Beckman, J.S.; Liaudet, L.; Nitric oxide and peroxynitrite in health and disease. *Physiol. Rev.* **2007,** *87(1),* 315–424.

106. Sewelam, N.; Kazan, K.; Schenk, P.M.; Global plant stress signaling: reactive oxygen species at the cross-road. *Front. Plant Sci.* **2016,** *7,* 187.

107. Chan, Z.; Yokawa, K.; Kim, W.Y.; Song, C.P.; ROS regulation during plant abiotic stress responses. *Front. Plant Sci.* **2016,** *7,* 1536.

108. Jalmi, S.K.; Sinha, A.K.; ROS mediated MAPK signaling in abiotic and biotic stress-striking similarities and differences. *Front. Plant Sci.* **2015,** *6,* 769.

109. Xing, Y.; Chen, W.H.; Jia, W.; Zhang, J.; Mitogen-activated protein kinase kinase 5 (MKK5)-mediated signalling cascade regulates expression of iron superoxide dismutase gene in *Arabidopsis* under salinity stress. *J. Exp. Bot.* **2015,** *66(19),* 5971–81.

110. Asai, S.; Ohta, K.; Yoshioka, H.; MAPK signaling regulates nitric oxide and NADPH oxidase-dependent oxidative bursts in Nicotiana benthamiana. *Plant Cell.* **2008,** *20(5),* 390–406.

111. Adachi, H.; Nakano, T.; Miyagawa, N.; Ishihama, N.; Yoshioka, M; et al. WRKY transcription factors phosphorylated by MAPK regulate a plant immune NADPH oxidase in Nicotiana benthamiana. *Plant Cell.* **2015,** *27(9),* 2645–63.

112. Suharsono, U.; Fujisawa, Y.; Kawasaki, T.; Iwasaki, Y.; Satoh, H.; et al. The heterotrimeric G protein α subunit acts upstream of the small GTPase Rac in disease resistance of rice. *P.N.A.S.* **2002,** *99(20),* 13307–12.

113. Mittler, R.; Vanderauwera, S.; Suzuki, N.; Miller, G.A.; Tognetti, V.B.; et al. ROS signaling: the new wave?. *Tren. Plant Sci.* **2011,** *16(6),* 300–9.

114. Gilroy, S.; Białasek, M.; Suzuki, N.; Górecka, M.; Devireddy, A.R.; et al. ROS, calcium, and electric signals: key mediators of rapid systemic signaling in plants. *Plant Physiol.* **2016,** *171(3),* 1606–15.

115. Bowler, C.; Fluhr, R.; The role of calcium and activated oxygen as signals for controlling cross-tolerance. *Tren. Plant Sci.* **2000,** *5(6),* 241–6.

116. Jayakannan, M.; Bose, J.; Babourina, O.; Shabala, S.; Massart, A.; et al. The NPR1-dependent salicylic acid signalling pathway is pivotal for enhanced salt and oxidative stress tolerance in *Arabidopsis*. *J. Exp. Bot.* **2015**, *66(7)*, 1865–75.

117. Cheng, F.; Yin, L.L.; Zhou, J.; Xia, X.J; Shi, K.; et al. Interactions between 2-Cys peroxiredoxins and ascorbate in autophagosome formation during the heat stress response in Solanum lycopersicum. *J. Exp. Bot.* **2016**, *67(6)*, 1919–33.

118. Qiao, B.; Zhang, Q.; Liu, D.; Wang, H.; Yin, J.; et al. A calcium-binding protein, rice annexin OsANN1, enhances heat stress tolerance by modulating the production of H_2O_2. *J. Exp. Bot.* **2015**, *66(19)*, 5853–66.

119. You, J.; Chan, Z.; ROS regulation during abiotic stress responses in crop plants. *Front. Plant Sci.* **2015**, *6*, 1092.

120. Wang, P.; Du, Y.; Li, Y.; Ren, D.; Song, C.P.; Hydrogen peroxide–mediated activation of MAP kinase 6 modulates nitric oxide biosynthesis and signal transduction in *Arabidopsis*. *Plant Cell.* **2010**, *22(9)*, 2981–98.

121. Kohli, S.K.; Khanna, K.; Bhardwaj, R.; Abd-Allah, E.F.; Ahmad, P.; et al. Assessment of subcellular ROS and NO metabolism in higher plants: multifunctional signaling molecules. *Antioxidants.* **2019**, *8(12)*, 641.

122. Tanou, G.; Job, C.; Rajjou, L.; Arc, E.; Belghazi, M.; et al. Proteomics reveals the overlapping roles of hydrogen peroxide and nitric oxide in the acclimation of citrus plants to salinity. *Plant J.* **2009**, *60(5)*, 795–804.

123. Yun, B.W.; Feechan, A.; Yin, M.; Saidi, N.B.; Le Bihan, T.; et al. S-nitrosylation of NADPH oxidase regulates cell death in plant immunity. *Nature.* **2011**, *478(7368)*, 264–8.

124. De Abreu Neto, J.B.; Frei, M.; Microarray meta-analysis focused on the response of genes involved in redox homeostasis to diverse abiotic stresses in rice. *Front. Plant Sci.* **2015**, *6*, 1260.

125. Wei, Y.; Shi, H.; Xia, Z.; Tie, W.; Ding, Z.; Yan, Y.; Wang, W.; Hu, W.; Li, K.; Genome-wide identification and expression analysis of the WRKY gene family in cassava. *Front. Plant Sci.* **2016**, *7*, 125.

126. Suzuki, N.; Mittler, R.; Reactive oxygen species and temperature stresses: a delicate balance between signaling and destruction. *Physiol. Plant.* **2006**, *126(1)*, 45–51.

127. Zhang, Q.; Wang, M.; Hu, J.; Wang, W.; Fu, X.; et al. PtrABF of Poncirus trifoliata functions in dehydration tolerance by reducing stomatal density and maintaining reactive oxygen species homeostasis. *J. Exp. Bot.* **2015**, *66(19)*, 5911–27.

128. Fang, Y.; Liao, K.; Du, H.; Xu, Y.; Song, H.; et al. A stress-responsive NAC transcription factor SNAC3 confers heat and drought tolerance through modulation of reactive oxygen species in rice. *J. Exp. Bot.* **2015**, *66(21)*, 6803–17.

CHAPTER 4

Physiological Mechanisms of Plants Involved in Phosphorus Nutrition and Deficiency Management

ZAHOOR AHMAD[1], CELALEDDIN BARUTÇULAR[1],
EJAZ AHMAD WARAICH[2], ADEEL AHMAD[3],
MUHAMMAD ASHAR AYUB[3*], RANA MUHAMMAD SABIR TARIQ[4],
MUHAMMAD AAMIR IQBAL[5] and AYMAN EL SABAGH[6,7]

[1]*Department of Field Crops, Faculty of Agriculture, Cukurova University, Adana, Turkey*

[2]*Department of Agronomy, University of Agriculture, Faisalabad, Pakistan*

[3]*Institute of Soil and Environmental Sciences, University of Agriculture, Faisalabad, Pakistan*

[4]*Department of Agriculture and Agribusiness Management, University of Karachi, Karachi, Pakistan*

[5]*Department of Agronomy, University of Poonch, Rawalakot, Azad Kashmir, Pakistan*

[6]*Department of Field Crops, Faculty of Agriculture, Siirt University, Siirt, Turkey*

[7]*Department of Agronomy, Faculty of Agriculture, Kafrelsheikh University, Kafr El-Sheikh, Egypt*

Corresponding author. E-mail: muhammadasharayub@gmail.com

ABSTRACT

Nutrients play an important role for the better growth and production of the field crops. In fulfilling the requirements of food along with the globe, through an improvement in crop productivity, balanced nutrition plays a vital role. Food crops require phosphorus (P) as a macronutrient for several functions like transfer of energy, cell division, and storage. Phosphorus improves forage, fiber, root growth, and grain yield. It not only strengthens stalk but also improves the early maturity of plants. In resistance against cold injury and root rot disease, P plays its role. Phosphorus is also crucial for cell differentiation and energy transactions. In the plant body, as a part of nucleic acids, phosphor-proteins, and phospholipids, P is a critical constituent of plant cells. The deficiency of P creates a negative impact on the morphology as well as the physiological process of the plants. Many researchers worked on the enhancement of crop productivity and assessed the role of P in plants. This chapter elaborated on three critical aspects of phosphorus; first is the P uses and role in crop plants while the second is the physiological mechanism of P in the plant. The third section discussed the management of P deficiency or toxicity in plants.

4.1 INTRODUCTION

Pakistan has a 22.2 m ha cropped area in total. Food grain crops occupied its significant portion having 54% share, after that sugar cane (20% share) and cotton (20% share); fruit and vegetables have 4% share; oilseed crops 3%; pulses 6%; and other crops have 13% share to total cropped area. In food crops, wheat occupied 36.3% of the total cropped area as a major food crop in Pakistan, followed by cotton with 14% share, paddy having 9.5% share, sugarcane (4.5%), maize (4.5%), and other crops with 20.8% share to the total cropped area [1].

In modern farm technology, fertilizers through improvement in the soil fertility status help to get higher crop production. As high yielding cultivars of cereal crops, with more requirements for nutrients, started in 1966–1967 under the green revolution; thus fertilizer era was started in Pakistan. Fertilizer use was negligible before that period [2]. During 1952–1953, there was a start of commercial fertilizer consumption in Pakistan, and on that time offtake of nitrogen was only one thousand tones. In 1959–60, P fertilizers were introduced to the farmers, and initially their consumption was hundred nutrient tones only. Worldwide, phosphate fertilizers consume about 30 million metric ton P_2O_5 every year and rock phosphate supplied about 99% of this P_2O_5 [3]. In India, imitative fertilizer demand function predicted the long- and short-run elasticity in price [4]. They concluded that the demand for fertilizers in India is inelastic of price in both the long and short run. The total offtake of di-ammonium phosphate (DAP) fertilizer during Kharif 2019 was 974,000 tones in Pakistan. While during Rabi 2019–2020, the offtake of DAP was 1094,000 tones [5].

4.2 ROLE OF P IN THE GROWTH AND DEVELOPMENT OF PLANTS

Phosphorus (P) is of much importance to major fertilizer for all living organisms, but for plants, it is a fundamental element. For metabolism as well as reproductive and vegetative growth of plants, P is a leading and most crucial element for the better production of crops. P concentration affects several plant processes like photosynthesis, enzyme regulation, nucleic acids and membranes synthesis, nitrogen fixation, respiration, and metabolism [6]. Adequate P nutrition accelerates plant developmental processes including root growth, early flowering, and fruiting [7]. P availability determines the transfer and storage of energy in the form of ADP and ATP. Photosynthetic activity of plants depends on P, as it is a structural component of nucleic acids, essential nutrient, part of phospholipids, part of coenzymes, forms nucleotides and proteins [8]. In developing meristematic tissues and cell division, P has a vital role, as it is an integral part of cell nuclei [9]. Many researchers elaborated on a correlation between N fertilization and basal P [10]. In legumes, the requirement of P is substantial for biological N-fixation [11]; it also showed a strong relationship with N [12]. At the crop's early stages, P supply is vital as P supplied at planting maintains vigor and early growth of plants [13].

One of the most critical factors responsible for seed germination and seedling growth is the presence of P content in seed. At the time of germination and early seedling growth, the only P available to plants is seed P contents. This P pool has a pivotal role for faster establishment and better nutrition of young seedling; however, it is of very little importance for mature plants. Phosphorus requirement of plants after seed germination is achievable through roots from growing media [14]. An increased uptake from soil P was witnessed by high P seeds than low P seeds of wheat, as root development in high P seeds was more prominent [15]. After seedling establishment, increase in the biomass of roots is an adoptive measure by plants to explore more P in P-deficit conditions. But if these P deficit conditions prevail for a more extended period then it leads to a decrease in growth rate due to the low concentration of ATP in plant roots [16].

Phosphorus is also much needed for the better reproductive growth (including flower and seed formation) of plants; as P is involved in anthocyanins production in flower stalks, in P-deficit, there was a decrease in its production [17]. Where P is essential for the formation and development of seeds, P quantities were also found higher in seeds as well as in fruit [14]. Most of the P taken up moves to seeds in cereal crops like wheat and rice. Therefore, inadequate P supply reduces the size, number, and viability of seeds. Still, the optimum supply of P by soil enhances seed yield, seed number, seed dry matter, and ultimately harvest index of crop plants. Seeds coated with mono-ammonium phosphate improved the yield and growth parameters of soybean [18].

4.3 CHEMICAL P FERTILIZER IN THE SOIL AND P UPTAKE

Human and agriculture activities are dominant in the modern terrestrial P cycle [19]. Though Pakistani soils are rich in total P contents (0.005%–0.15%) yet lowest in available P (0.01–1 mg L^{-1}) [20]. In Pakistani soils, there is an issue of P availability, as these soils are low in organic matter, alkaline, and calcareous. Extreme temperature and moisture regimes further aggravate the problem of less P availability. All the above factors are creating conducive conditions for fixation and precipitation of P [21]. While in acid soils there is an issue of P fixation due to Al and Fe. Therefore, farmers apply chemical phosphatic fertilizers like nitro phosphate, single super-phosphate, DAP, and triple super-phosphate (TSP) to supply P to fulfill the requirement of the crop [22]. There is an immediate conversion of added phosphatic fertilizers to orthophosphates, $H_2PO_4^-$, and HPO_4^{2-}, as plant roots absorb P in these both forms [23]. When the soil pH is less than 7.0, $H_2PO_4^-$ is the predominant form in the soil. These forms of phosphorus are anions (having negative charge) and therefore are not attracted and retained by the negatively charged soil. So, by all logic, phosphorus anions *should* be mobile in the soil, as is the case with nitrogen anions; but the opposite is true. Added P becomes unavailable to the plants due to some reactions like precipitation as dicalcium phosphate, surface precipitation on solid-phase calcium carbonate, and retention by clays [24, 25] as explained in Figure 4.1.

4.4 PLANT RESPONSE UNDER PHOSPHORUS DEFICIENCY

A most common nutritional factor responsible for limiting agricultural production is P deficiency around the globe [26]. An optimum amount of P is not present in 40% soils in the world field area [7]. After nitrogen, P is the second most lacking nutrient in many agricultural lands worldwide for the better production of crops [27]. Due to P deficiency, certain changes occur in plants such as leaf senescence [28], decreased number of leaves [29], and reduced photosynthetic efficiency [30]. P deficiency symptoms first occur in older leaves, as it is a mobile nutrient in plants, while P deficiency occurs

P remobilized from older leaves to younger ones [31]. There is extensive work for checking the impact of P fertilization on wheat grain yield and pasture by many researchers [32]. There was a 50% decrease in the yield of lowland rice under P deficiency [33]. The growth of cotton is restricted by P deficiency in the early stages of growth [34].

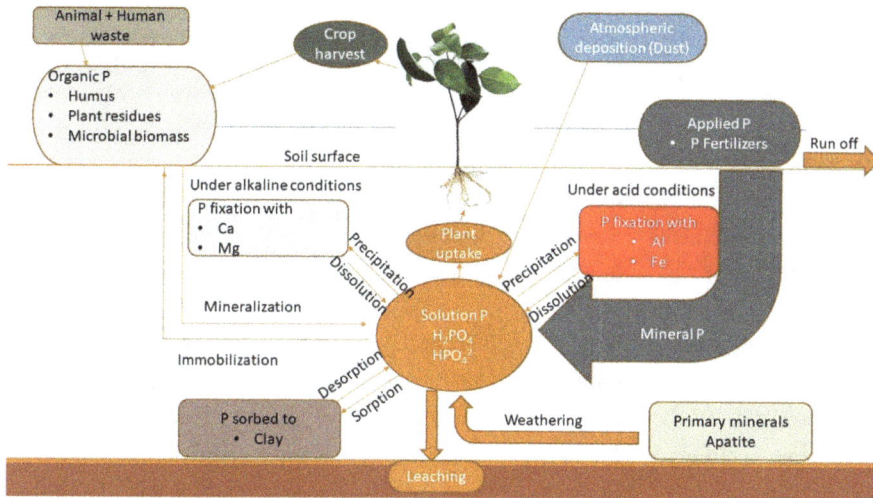

FIGURE 4.1 Phosphorous fate and dynamics in soil.

Under P deficient conditions to maintain growth and metabolic activities, plants have adapted a range of mechanisms responsible for P uptake under such P deficient conditions [35, 36]. Table 4.1 illustrated some of the common responses of plants to P deficiency. These plant adaptations are utilization and acquisition efficiencies. The acquisition efficiency of plants is related to the absorption of the sparingly soluble nutrients. The utilization efficiency of plants is related to the production of more biomass per unit absorption of nutrients [35]. In acquisition efficiency, strategies adapted by plants to enhance P uptake are the release of organic acids from roots, exudation of phosphatases [37–39], enhanced expression of Pi transporters, development of root hair [40], and improved root growth with changed root architecture [38, 41, 42]. In utilization efficiency, strategies adapted by plants to reduce P use with improved growth are altered respiratory pathways, modified carbon metabolism, remobilization of internal P, decreased growth rate, and modified membrane biosynthesis requiring less P [36, 43]. Besides, improved growth per unit of P absorbed [43] is also a type of utilization efficiency.

Due to differences in adapting these mechanisms, crop varieties and plant species differ widely in phosphorus use efficiency [26, 44].

TABLE 4.1   Mechanisms Adopted by Plants to Cope with P Deficiency

Trait	P Sufficiency	P Deficiency	Type of Adaptation	Reference
Release of organic acids by roots	Normal	Increased	Acquisition efficiency	[37–39]
Biosynthesis of membrane	Normal	Modified	Utilization efficiency	[36, 43]
Expression of Pi transporters	Normal	Enhanced	Acquisition efficiency	[40]
Growth rate	Normal	Decreased	Utilization efficiency	[36, 43]
Root hair development	Normal	Accelerated	Acquisition efficiency	[40]
Remobilization of internal P	Normal	Increased	Utilization efficiency	[36, 43]
Root hair architecture	Normal	Changed	Acquisition efficiency	[38, 41, 42]
Carbon metabolism	Normal	Modified	Utilization efficiency	[36, 43]
Root growth	Normal	Improved	Acquisition efficiency	[38, 41, 42]
Respiratory pathway	Normal	Altered	Utilization efficiency	[36, 43]
Exudation of phosphatases	Normal	Increased	Acquisition efficiency	[37–39]
Growth per unit P absorbed	Normal	Improved	Utilization efficiency	[43]

4.5   INTERACTION OF P WITH OTHER NUTRIENTS AND PHOSPHORUS USE EFFICIENCY (PUE)

To maintain the sufficiency level of P in the soil, there is a concept of excessive P fertilization. The concept of sufficiency is only for soil immobile nutrients and P is a soil immobile nutrient. The total nutrient amount in the soil does not determine the response of a plant to immobile nutrients like P, but its concentration in root surface sorption zone does matter. Two factors are responsible for the uptake of P by plant roots; first, the root surface area and second, P availability in the area surrounding the plant roots. The bigger surface area of roots explores greater soil volume and greater P concentrations ensure higher amounts of available P for plant uptake, as soil immobile nutrients are taken up mainly via root interception and diffusion [45]. Anyhow, P fertilizer should be applied in an adequate amount to ensure P availability to plants because in the soil more than 80% of it is lost through P fixation, in acid soils with Al and Fe while in alkaline soils with Ca and Mg [46]. When P fertilizer was knifed to the soil or banded with seed, PUE was about 16% in wheat [47]. In corn when P availability from fertilizer drains out and plant development is in progress, the plant takes up some amount of P from the organic pool in the soil but P availability from organic source is very slow. There was a higher P uptake at the later vegetative stages of maize. P uptake was dropped during pollination and then again increased continuously at the grain filling stage [48]. Many studies indicated that there was an increase in the availability and mobility of P with frequent irrigations and high moisture maintenance [49]. Several researchers demonstrated a movement of P up to 3 to 4 cm from the point where fertilizer was applied [50]. Plants uptake the readily available form of P through diffusion either in the particulate form or dissolved form (soil solution) [51]. P diffusion rate affects both the P sorption capacity of soil and concentration of P in soil solution. There was a decrease in mobility of P, as radii of water-filled pores lowered because of a reduction in soil water contents. Thus, P absorption by plants and P availability to plants is low due to lower soil moisture content [52].

To prevent P losses like fixation, there are different approaches like the addition of organic acids, band placement, acid application, bio-fertilizers, and microbial inoculations of the P [53]. But PUE is not exceeding 25% under all circumstances besides the use of these techniques. However, the use of controlled-release fertilizers or polymer-coated fertilizers could be one of the best strategies to enhance the use efficiency of phosphorus, as these fertilizers release P slowly according to the need of the plant. Hence, the interaction of P with soil will be minimum leaving a tiny P for fixation which in turn improves PUE and reduce the rate of application by controlling the adverse effects linked to over-usage [54, 55].

Phosphorus presence in soil affects the availability of other nutrients also, as it has synergistic and antagonistic interactions with some macro- and micronutrients [14]. Phosphorus and nitrogen (N) have a synergistic interaction. The ammoniacal form of N enhances the availability of P to plants. Phosphorus is one of the nutrients required for N fixation and for ensuring efficient use of N by plants. Where alone N gave 71 bushels per acre in sorghum, the combined application of N and P gave 93 bushels per acre [56]. P and potassium (K) also have synergistic interaction, as their combined application gave 64 bushels per acre in comparison to 38–41 bushel per acre with the alone application of each nutrient in corn [57]. In moong seeds, there is an antagonistic interaction between P and sulfur (S), as 40 ppm S declined 18% P in the vegetative portion and 12% in grains of moong in a greenhouse experiment [58]. Micronutrients interaction of P is widely reported. Synergistic interaction between P and boron (B) was observed in maize [59]. There was a synergistic interaction between B and P in lettuce [60]. Similarly, in lettuce, a synergistic interaction was observed between copper (Cu) and P [61] and molybdenum (Mo) and P in *Brassica napus* [62]. At the same time, an antagonistic interaction was there between iron (Fe) and P [63] and zinc (Zn) and P [64].

4.6   PLANT RESPONSE TO FOLIAR APPLICATION OF P

For about 40 years or even more, foliar fertilization (FF) is being studied and used as an agricultural practice. There was an increase in the yield and quality of several crops with the foliar application of some micronutrients like potassium (K), nitrogen (N), and phosphorus (P) [65]. Earlier work described that almost all plants via foliage obtain various solutes, gases, and water from the environment [66]. To determine the factors affecting foliar uptake of nutrients, a considerable research work was done [67–69]. At lower pH (about 2–3), absorption of foliar-applied P was much faster [66], especially absorption of mono-ammonium phosphates was at a much higher rate [70].

As foliar P application lowers the contact between P fertilizer and soil, it is an excellent substitute to lower the problems related to soil fertilization. Moreover, it also suppresses the drawbacks of P management like low P use efficiency, high costs of fertilization, eutrophication, and low crop productivity and quality. Foliar P fertilization is a recommendation for dry top soils to slightly P deficient soils of the semiarid region and its recommendation is not as a substitute for soil fertilization; indeed, it is a supplemental source. By itself, foliar P fertilization is not enough to meet the nutrient demand of crop; however, it acts as a supplement for adequate soil-applied P fertilizers [71]. When there is P depletion from the root zone, plants are unable to take up more P while it is needed because of the very low diffusion coefficient of P [72]. Thus, in that condition, foliar P application is increasingly important. Foliar applied nutrients enter the plant body either through leaf stomata [73] or through leaf cuticle, which contains hydrophilic pores in it [27].

In an experiment, for comparison of the foliar P fertilization and soil-applied P, the conclusion noted that the metabolism and absorption of P applied as foliar was superior as compared to P applied to the soil. Still, the sole foliar application of P was not enough to meet the P requirements of the plant over the plant cycle [74]. In the case of maize, when soil-applied NPK fertilizers are not enough then their foliar application can supplement the plants but alone foliar application is not an alternate of soil fertilization. Thus, it seems plausible to correct the deficiency of P to plants by foliar application [75]. In a pot experiment, it was noticed that by the foliar application of P as ammonium triple phosphate on young leaves of maize, P absorption was 66% within 10 days after sowing (DAS) and from that absorbed P the translocation was 87% [76]. However, somewhat contradicting results were presented by showing decreased maize grain yield when P supplied as foliar after two weeks of sulking [77]. Though there was no impact on the production of maize dry matter cob index and starch contents were showing intensification by applying foliar Zn and P [78]. With the foliar application of P as KH_2PO_4 at 5–10 kg ha^{-1}, an increase in the grain yield of wheat was noticed [79]. Application of P at critical wheat stages as foliar (0.19%–0.23% P concentration) showed an increase in the grain yield up to 90% [80].

When initial P deficiency symptoms appeared 25 DAS in wheat, higher doses of ammonium phosphate as a foliar spray gave the most significant reduction in P deficiency, and highest yields [81]. There was an increase in P uptake and yield at Freekes 7 in comparison to no foliar applied P. While they applied 1, 2, 4, 8, 12, 16, and 20 kg P ha^{-1} with and without pre-plant rates of P (30 kg P ha^{-1}); however, at Freaks 10.5 highest P use efficiency was noted [47]. A substantial improvement in grain concentration and forage in corn, but a small yield increases with foliar application of 8 kg of P ha^{-1}. An increase of 1 Mg ha^{-1} in wheat grain yield was observed with applying foliar of 1.1 to 2.2 kg P ha^{-1} as KH_2PO_4 after anthesis, in another experiment conducted at Morocco [82]. This increase in yield could be an attribute of the delay in senescence, which is stimulated by fertilization under heat and water stress [79].

At a P concentration of 0.31% and grain yield level of 4.24 to 8.83 Mg ha^{-1}, the total P taken up by corn grains was 13.8 to 26.4 kg P ha^{-1} [50]. While at P concentration 0.30% and yield level 8.10 to 14.47 Mg ha^{-1}, total P has taken up was 21.4 to 47.4 kg P ha^{-1} in corn grain [83]. At late reproductive stages, the spray of liquid N-P-K-S fertilizers enhanced the yield up to 27 to 31%. During pod filling, root activity reduces; thus, nutrient uptake reduces even this uptake is not enough to meet nutrient demand of seed [84]. Thus, during the pod fill stage of soybean, the application of N-P-K-S as foliar spray enhanced the growth and yield of soybean [85]. But, as the duration of photosynthesis does not extend with foliar application of N-P-K-S fertilizers, there was no noteworthy impact on the yield of soybean; anyhow, there is an increase in the concentration of nutrients in soybean. The yield of foliar treated soybean was 3617 kg ha^{-1}, while in control yield was 3825 kg ha^{-1} [86].

4.7 PHYSIOLOGICAL MECHANISM OF PHOSPHORUS IN PLANTS

The existence of P is in different forms like in energy-rich pyrophosphate or through hydroxyl group (phosphate ester) in the carbon cycle or as an inorganic phosphate [87]. P through constituting RNA and DNA molecules participate in the translation and transportation of genetic information; thus, it is a principal constituent of nucleic acids in the plant [88]. Energy-rich phosphates and phosphate esters manufacture ADP and ATP molecules that play a role in the synthesis of starch. Furthermore, to control some enzymatic reactions, a regulatory role is played by inorganic phosphates [87].

The process of photosynthesis is highly dependent on the P availability, as H_2O, CO_2, and Pi are primary substrates in this process which form sugars and ATP by utilizing light energy through chlorophyll. The ATP formed by this way acts as a driving force for regulating several metabolic processes in plants. Sugars formed in photosynthesis participate in the production of storage and structural constituents. During the photosynthesis process, for the fixation of every three molecules of CO_2 in three molecules of O_2, nine Pi is consumed. Through polymerase chain reaction cycle, eight of these nine Pi released into the chloroplast and in the form of triose phosphate remaining one is exported to cytosol. This triose phosphate by releasing and recycling Pi converted to sucrose. The released Pi further moves back to chloroplast to form more triose phosphates. Pi translocators carry the movement of triose phosphate (triose-P), 3-phosphoglyceric acid (3-PGA), and Pi across the membrane [89, 90]. By the alteration in the function of Pi translocators, the external P level regulates photosynthesis. Any alteration in the availability of Pi in cytoplasm changes the enzyme activation (fructose-1,6-bisphosphatase, sedoheptulose-1,7-bisphosphatase, and RuBisCO) and number of intermediates in the Calvin cycle. Anyhow, in cytosol Pi concentration remains stable due to vacuolar P pool [91–93].

For the maintenance of photosynthetic machinery (PSI, PSII cyt-b, cyt-f, LHCP, and antenna mobility) phosphorus is an essential element [94]. A decrease in Pi concentration in the stroma of the membrane induced by Pi deficiency resulted in reduced photophosphorylation; thus, photosynthesis is reduced as a consequence [95]. Due to long-term P deficiency photoinhibition of PSII occurs, resulting in reduced CO_2 fixation [96]. Similarly, due to the high concentration of Pi in the stroma, the breakdown of starch occurs.

From the degradation of starch glucose-1-phosphate forms, that is through oxidative pentose pathway or phosphofructokinase converted to triose-P or PGA, respectively. Various kinetic studies verified these findings [97, 98]. Like photosynthesis, in respiratory processes of plants, P plays a pivotal role. Plant roots tend to respire through the nonphosphorylating pathway under P deficient conditions. Under such a cyanide resistant pathway, a reduced number of ADP and ATP produced due to which energy-dependent processes of plants got affected [99, 100]. During P deficiency, reduction in levels of respiratory intermediates like PGA and hexose phosphates occurs. Concentrations of Pi and adenylate determine the activities of several glycolytic enzymes like 3-PGA kinase, NAD-G3P-DH, PK, and PFK. Under P deprived conditions, activities of nonphosphorylating NAD-G3P-DH and PFP, PEP phosphatase, and PEP carboxylase found to be increased [14, 101–104].

In the human diet, legumes are a good source of protein [105]. These are also important in maintaining soil fertility because these can carry atmospheric nitrogen fixation. In the nodules of leguminous plants, *Rhizobium* bacteria are residing, which are responsible for carrying out the conversion of atmospheric nitrogen (N_2) to ammonia (NH_3) [106]. Energy is needed by these bacteria to perform their function; they obtain energy from a P rich molecule that is ATP. For the reduction of each N_2 molecule, at least 16 ATP molecules are needed to be hydrolyzed. Thus, P deficiency causes both reduction in plant growth and a reduction in root nodules, ultimately affects the rate of nitrogen fixation [107]. P supply increases the biomass of root nodules in pea plants [108].

4.8 MANAGEMENT OF PHOSPHORUS DEFICIENCY OR TOXICITY IN PLANTS

Reduced supply of available P in the soil causes a reduced uptake of P by plants, thus due to P deficiency, overall growth and development of the plants are severely affected [14]. To enhance the acquisition and use of P, two broad strategies adopted by plants are: first, to improve uptake or acquisition by plants and second, to conserve P use which is already available in the soil [109]. Enhanced root hair development and root growth (expanded root surface area) [40, 110], enhanced expression of phosphate transporters [6], mycorrhizal associations [111], and enhanced exudation and organic acid synthesis [112, 113] are primary responses to improve uptake or acquisition. To conserve P, plants have adopted a decreased growth rate, internal remobilization of P [6], a modified carbon metabolism [114], and more growth per unit P taken up [110]. However, to cope with P deficiency in plants and to enhance the adaptability of plants to P deficiency, a collaboration between breeders, physiologists, and geneticists is required. In future, studies should be conducted to enhance the understanding of P uptake, transport, and utilization in the P stress environment [14].

Phosphorus availability may affect the availability of one or more other nutrients, as it has synergistic and antagonistic interactions with other macro- and micronutrients [14]. In subterranean clovers, the P fertilizer application enhanced the symptom-resembling Zn deficiency. They tested three levels of Zn and P fertilizers at siliceous (low capacity to react with P) and ferruginous (high capacity to react with P) sand. First, excessive P application depressed Zn absorption-inducing Zn deficiency symptoms in ferruginous sand but not on siliceous sand. Second, additional P supply where there was P limited growth also depressed concentrations of Zn, thus by promoting the plant growth it induced Zn deficiency by diluting Zn supplies. Third, an additional P supply where P did not limit growth can increase its concentration to a toxic level, and old leaves show necrotic symptoms, ultimately plant growth is compromised [115]. Thus, by managing the supply of other nutrients (which have synergistic and antagonistic interactions with P) in a balanced amount it could manage the deficiency and toxicity of P in plants and also be helpful for improving the growth and yield of crop plants [114, 115].

KEYWORDS

- **phosphorus**
- **phosphor-proteins**
- **grain yield**
- **root growth**

REFERENCES

1. Govt. of Pakistan; Agricultural Statistics of Pakistan, Ministry of Food Agricultural and Livestock, Food and Agriculture, Islamabad, **2014**.
2. Ahmad, N.; Rashid, M.; Fertilizer and their use in Pakistan, Government of Pakistan, Planning and Development Division, NFDC, Islamabad, Pakistan, **2004**.
3. IFA; Summary report—World Agriculture and Fertilizer Demand, Global Fertilizer Supply and Trade, International Fertilizer Industry Association 2005–2006, **2005**. http://www.

fertilizer.org/ifa/publicat/PDF/2005_counsil_sevilla_ifa_summary.pdf www.fertilizer.org/ifa/publicat/PDF/2005_counsil_sevilla_ifa_summary.pdf

4. Dholakia, RH; Majumdar, J.; Estimation of price elasticity of fertiliser demand at macro level in India, *Indian J. Agric. Econ.*, **1995**, 50, 36–46.

5. GOP; Economic Survey of Pakistan 2019–20, Economic Adviser's Wing, Finance Division, Government of Pakistan, Islamabad, **2020**.

6. Raghothama, K.G.; Phosphate acquisition, *Ann. Rev. Plant Physiol. Plant Mol. Biol.*, **1999**, 50, 665–693.

7. Vance, C.P.; Symbiotic nitrogen fixation and phosphorus acquisition, Plant nutrition in a world of declining renewable resources, *Plant Physiol.*, **2001**, 127: 390–397.

8. Taiz, L.; Zeiger, E.; *Plant Physiology*, 4th ed., Sinauer Associates, Inc., Sunderland, Massachusetts, USA, **2006**.

9. Russell, E.W.; *Soil Condition and Plant Growth*, The English Language Book Society and Longman, London, **1973**, pp. 68.

10. Israel, D.W.; Investigation of the role of phosphorus in symbiotic dinitrogen fixation, *Plant Physiol*, **1987**, 84, 835–840.

11. Graham, P.H.; Vance, C.P.; Nitrogen fixation in perspective: An overview of research and extension needs, *Field Crops Res.*, **2000**, 65, 93–106.

12. Sanginga, N.; Lyasse, O.; Singh, B.B.; Phosphorus use efficiency and nitrogen balance of cowpea breeding lines in a low P soil of the derived savanna zone in West Africa, *Plant Soil*, **2000**, 220, 119–128.

13. Grant, C.A.; Flaten, D.N.; Tomasiewicz, D.J.; Sheppard, S.C.; The importance of early season phosphorus nutrition, *Canadian J. Soil Sci.*, **2001**, 81, 211–224.

14. Malhotra, H.; Sharma, S.; Pandey, R.; *Phosphorus Nutrition: Plant Growth in Response To Deficiency and Excess*, Springer Nature, Singapore, **2018**.

15. Zhu, Y.G.; Smith, S.E.; Seed phosphorus (P) content affects growth, and P uptake of wheat plants and their association with arbuscular mycorrhizal (AM) fungi, *Plant Soil*, **2001**, 231, 105–112.

16. Gniazdowska, A.; Mikulska, M.; Rychter, A.M.; Growth, nitrate uptake and respiration rate in bean roots under phosphate deficiency, *Biol. Plant*, **1998**, 41, 217–226.

17. Chen, H.C.H.; Jiang, G.J.J.; Osmotic adjustment and plant adaptation to environmental changes related to drought and salinity, *Environ. Rev.*, **2010**, 18, 309–319.

18. Soares, M.M.; Sediyama, T.; Neves, J.C.L.; dos Santos Junior, H.C.; da Silva, L.J.; Nodulation, growth and soybean yield in response to seed coating and split application of phosphorus, *J. Seed Sci.*, **2016**, 38, 30–40.

19. Oelkers, E.H.; Valsami-Jones, E.; Phosphate mineral reactivity and global sustainability, *Elements*, **2008**, 4, 83–87

20. Zaida, A.; Khan, M.S.; Amil, MD; Interactive effect of rhizotrophic microorganisms on yield and nutrient uptake of chickpea (*Cicer arietinum* L.). *Eur. J. Agron.*, **2003**, 19, 15–21.

21. Sharif, M.; Sabir, M.S.; Rabi, F.; Biological and chemical transformation of phosphorus in some important soil series of NWFP, *Sarhad J. Agric.*, **2000**, 16, 587–592.

22. Rehman, O.; Zaka, M.A.; Rafa, H.U.; Hassan, N.M.; Effect of balanced fertilization on yield and phosphorus uptake in wheat rice rotation, *J. Agric. Res.*, **2006**, 44, 105–113.

23. Ai, P.; Sun, S.; Zhao, J.; Fan, X.; Xin, W.; Guo, Q.; Yu, L.; Shen, Q.; Wu, P.; Miller, A.J.; Two rice phosphate transporters, OsPht1;2 and OsPht1;6, have different functions and kinetic properties in uptake and translocation, *Plant J.*, **2009**, 57, 798–809.

24. Rashid, A.; Memon, K.S.; Bashir, E.L.; Bantel, R.; *Soil Science*, National Book Foundation, Islamabad, **1996**.
25. Shen, J.; Yuan, L.; Zhang, J.; Li, H.; Bai, Z.; Chen, X.; Zhang, W.; Zhang, F.; Phosphorus dynamics: From soil to plant, *Plant Physiol.*, **2011**, 156, 997–1005.
26. Aziz, T.; Sabir, M.; Farooq, M.; Maqsood, M.A.; Ahmad, H.R.; Warraich, E.A.; Phosphorus deficiency in plants: responses, adaptive mechanisms, and signaling, In: Hakeem K., Rehman R., Tahir I. (eds), *Plant Signaling: Understanding the Molecular Crosstalk*, Springer, New Delhi, **2014**.
27. Tyree, M.T.; Scherbatskoy, T.D.; Tabor, C.A.; Leaf cuticles behave as asymmetric membranes: Evidence from the measurement of diffusion potentials, *Plant Physiol.*, **1990**, 92, 103–109.
28. Berchtold, H.; Reshetnikova, L.; Reiser, C.O.; Schirmer, N.K.; Sprinzl, M.; Hilgenfeld, R.; Crystal structure of active elongation factor Tu reveals major domain rearrangements, *Nature*, **1993**, 365, 126–132.
29. Lynch, J.; Läuchli, A.; Epstein, E.; Vegetative growth of the common bean in response to phosphorus nutrition, *Crop Sci.*, **1991**, 31, 380–387.
30. Lauer, M.J.; Blevins, D.G.; Sierzputowska-Gracz, H.; 31P-Nuclear magnetic resonance determination of phosphate compartimentation in leaves of reproductive soybeans (Glycine max Merr.) as affected by phosphate nutrition, *Plant Physiol.*, **1989**, 89, 1331–1336.
31. Bernardo, R.; Prediction of maize single-cross performance using RFLPs and information from related hybrids, *Crop Sci.*, **1994**, 34, 20–25.
32. Smart, CM; Gene expression during leaf senescence, *New Phytol.*, **1994**, 126, 419–448.
33. Saleque, M.A.; Abedin, M.J.; Panaullah, G.M.; Bhuiyan, N.I.; Yield and phosphorus efficiency of some lowland rice varieties at different levels of soil-available phosphorus, *Commun. Soil Sci. Plant Anal.*, **1998**, 29, 2905–2916.
34. Hearn, A.B.; Cotton nutrition, *Field Crop Abstracts*, **1981**, 34, 11–34.
35. Rengel, Z.; Marschner, P.; Nutrient availability and management in the rhizosphere: Exploiting genotypic differences, *New Phytol.*, **2005**, 168, 305–312.
36. Lambers, H.; Michael, W.S.; Michael, D.C.; Stuart, J.P.; Erik, J.V.; Root structure and functioning for efficient acquisition of phosphorus: Matching morphological and physiological traits, *Ann. Bot.*, **2006**, 98, 693–713.
37. Vance, C.P.; Uhde-Stone, C.; Allan, D.L.; Phosphorus acquisition and use: critical adaptations by plants for securing a nonrenewable resource, *New Phytol.*, **2003**, 157, 423–447.
38. Gahoonia, T.S.; Nielsen, N.E.; Barley genotypes with long root hairs sustain high grain yields in low-P field, *Plant Soil*, **2004**, 262, 55–62.
39. Johnson, S.E.; Loeppert, R.H.; Role of organic acids in phosphate mobilization from iron oxide, *Soil Sci. Soc. Am. J.*, **2006,** 70, 222–234.
40. Gilroy, S.; Jones, D.L.; Through form to function: Root hair development and nutrient uptake, *Trends Plant Sci.,* **2000**, 5, 56–60.
41. Raghothama, K.G.; Karthikeyan, AS; Phosphate acquisition, *Plant Soil*, **2005**, 274, 37–49.
42. Bucher, M.; Functional biology of plant phosphate uptake at root and mycorrhiza interfaces, *New Phytol.*, **2006**, 173, 11–26.
43. Uhde-Stone, C.; Zinn, K.E.; Ramirez-Yáñez, M.; Li, A.; Vance, C.P.; Allan, D.L.; Nylon filter arrays reveal differential gene expression in proteoid roots of white lupin in response to phosphorus deficiency, *Plant Physiol.*, **2003**, 131, 1064–1079.

44. Aziz, T.; Rahmatullah, MA; Maqsood, M.; Tahir, I.A.; Cheema, M.A.; Phosphorus utilization by six Brassica cultivars (*Brassica juncea* L.) from tri-calcium phosphate; a relatively insoluble P compound, *Pak. J. Bot.*, **2006**, 38, 1529–1538.

45. Bray, R.H.; A nutrient mobility concept of soil-plant relationships, *Soil Sci.*, **1954**, 78, 9–22.

46. Sanyal, SK; De Datta, S.K.; Chemistry of phosphorus transformations in soil, *Adv. Soil Sci.*, **1991**, 16, 1–120.

47. Mosali, J.; Desta, K.; Teal, R.K.; Freeman, K.W.; Martin, K.L.; Lawles, J.W.; Raun, W.R.; Effect of foliar application of phosphorus on winter wheat grain yield, phosphorus uptake, and use efficiency, *J. Plant Nutri.*, **2006,** 29, 2147–2163.

48. Karlen, D.L.; Flannery, R.L.; Sadler, E.J.; Aerial accumulation and partitioning of nutrients by corn, *Agron. J.*, **1988**, 80, 232–242.

49. Kargbo, D.; Skopp, J.; Knudsen, D.; Control of nutrient mixing and uptake by irrigation frequency and relative humidity, *Agron. J.*, **1991**, 83, 1023–1028.

50. Eghball, B.; Sander, D.H.; Distance and distribution effects of phosphorus fertilizer on corn, *Soil Sci. Soc. Am. J.*, **1989**, 53, 282–287.

51. Haygarth, P.M.; Sharpley, A.N.; Terminology for phosphorus transfer, *J. Environ. Qual.*, **2000**, 29, 10–15.

52. Nye, P.H.; Tinker, P.B.; Solute exchange between solid, liquid and gas phases in the soil, In: Solute movement in the soil–root system, Studies of Ecological Blackwell Sciences, D.J. Anderson, P. Greig Smith, and Frank A. Pitelka (Eds.), Oxford, UK, **1977**, pp. 33–68.

53. Trolove, S.; Hedley, M.J.; Kirk, G.; Bolan, N.S.; Loganathan, P.; Progress in selected areas of rhizosphere research on P acquisition, *Soil Res.*, **2003**, 41, 471–499.

54. Yaseen, M.; Aziz, M.Z.; Manzoor, A.; Naveed, M.; Hamid, Y.; Noor, S.; Khalid, M.A.; Promoting growth, yield and phosphorus use efficiency of crops in maize-wheat cropping system by using polymer coated di-ammonium phosphate, *Commun. Soil Sci. Plant Anal.*, **2017**, 48, 646–655.

55. Noor, S.; Yaseen, M.; Naveed, M.; Ahmad, R.; Use of controlled release phosphatic fertilizer to improve growth, yield and phosphorus use efficiency of wheat crop, *Pak. J. Agric. Sci.*, **2017**, 54, 541–547.

56. Schlegel, A.; Bond, H.D.; Long-term nitrogen and phosphorus fertilization of irrigated grain sorghum, *Kansas Agric. Exp. Station Res. Rep.*, **2017,** 3, 1–8.

57. Usherwood, N.R.; Segars, W.I.; Nitrogen interactions with phosphorus and potassium for optimum crop yield, nitrogen use effectiveness and environmental stewardship, *Sci. World*, **2001**, 1, 57–60.

58. Aulakh, M.S.; Pasricha, NS; Interaction effect of sulphur and phosphorus on growth and nutrient content of moong (Phaseolus aureus L.), *Plant Soil*, **1977**, 47, 341–350.

59. Chatterjee, C.; Sinha, P.; Agarwala, SC; Interactive effect of boron and phosphorus on growth and metabolism of maize grown in refined sand, *Can. J. Plant Sci.*, **1990**, 70, 455–460.

60. Chowdhury, S.Z.; Sobahan, M.A.; Shamim, AHM; Akter, N.; Hossain, M.M.; Interaction effect of phosphorus and boron on yield and quality of lettuce, *Azarian J. Agric.*, **2015,** 2, 147–154.

61. De Iorio, A.F.; Gorgoschide, L.; Rendina, A.; Barros, M.J.; Effect of phosphorus, copper, and zinc addition on the phosphorus/copper and phosphorus/zinc interaction in lettuce, *J. Plant Nutr.*, **1996**, 19, 481–491.

62. Liu, H.; Hu, C.; Hu, X.; Nie, Z.; Sun, X.; Tan, Q.; Hu, H.; Interaction of molybdenum and phosphorus supply on uptake and translocation of phosphorus and molybdenum by Brassica napus, *J. Plant Nutr.*, **2010**, 33, 1751–1760.

63. Zheng, L.; Huang, F.; Narsai, R.; Wu, J.; Giraud, E.; He, F.; Cheng, L.; Wang, F.; Wu, P.; Whelan, J.; Shou, H.; Physiological and transcriptome analysis of iron and phosphorus interaction in rice seedlings, *Plant Physiol.*, **2009**, 151, 262–274.

64. Drissi, S.; Houssa, A.A.; Bamouh, A.; Coquant, J.M.; Benbella, M.; Effect of zinc-phosphorus interaction on corn silage grown on sandy soil, *Agriculture*, **2015**, 5, 1047–1059.

65. Romheld, V.; El-Fouly, M.M.; Foliar nutrient application: Challenges and limits in crop production, In: Suwanarit A., (ed), *Proceedings of 2nd Intl. Workshop on Foliar Fertilization*, Bangkok, Thailand, **1999**, pp. 1–32.

66. Wittwer, S.H.; Teubner, F.G.; Foliar absorption of mineral nutrients, *Ann. Rev. Plant Physiol.*, **1959**, 10, 13–30.

67. Swanson, C.A.; Whitney, J.B.; Studies on the translocation of foliar applied P-32 and other isotopes in bean plants, *Am. J. Botany*, **1953**, 40, 816–823.

68. Fisher, E.G.; Walker, D.R.; The apparent absorption of P and Mg from sprays applied to the lower surface of 'McIntosh' apple leaves, *Proc. Amer. Soc. Hort. Sci.*, **1955**, 65, 1724–1734.

69. Koontz, H.; Biddulph, O.; Factors affecting absorption and translocation of foliar applied phosphorus, *J. Plant Physiol.*, **1957**, 32, 463.

70. Wittwer, S.H.; Teubner, F.G.; McCall, W.W.; Comparative absorption and utilization by beans and tomatoes of phosphorus applied to the soil and foliage, *Proc. Am. Soc. Hort. Sci.*, **1957**, 69, 302–308.

71. Fritz, A.; Foliar fertilization—a technique for improved crop production, *Acta Hort. (ISHS)*, **1978**, 84, 43–56. Available at http://www.actahort.org/books/84/84_5.htm (verified November 2009).

72. Clarkson, D.T.; Nutrient interception and transport by root systems, In: Johnson C. B. (ed), *Physiological Processes Limiting Plant Productivity*, Butterworth's, London, **1981**, pp. 307–330.

73. Eichert, T.J.; Burkhardt, J.; A novel model system for the assessment of foliar fertilizer efficiency, In: Technology and Applications of foliar fertilizers. Proceedings of the Second International Workshop on Foliar Fertilization, Bangkok, The Soil and Fertilizer Society of Thailand, **1999**, pp. 41–54.

74. Boynton, D.; Nutrition by foliar application, *Ann. Rev. Plant Physiol.*, **1954**, 5, 31–54.

75. Ling, F.; Silberbush, M.; Response of maize to foliar vs. soil application of nitrogen–phosphorus–potassium fertilizers, *J. Plant Nutr.*, **2002**, 25, 2333–2342.

76. Barel, D.; Black, C.A.; Foliar application of P. II. Yield response of corn and soybeans sprayed with various condensed phosphates and P–N compounds in greenhouse and field experiments, *Agron. J.*, **1979**, 71, 21–24.

77. Harder, H.J.; Carlson, R.E.; Shaw, R.H.; Corn grain yield and nutrient response to foliar fertilizer applied during grain fill, *Agron. J.*, **1982**, 74, 106–110.

78. Leach, K.A.; Hameleers, A.; The effects of a foliar spray containing phosphorus and zinc on the development, composition and yield of forage maize, *Grass Forage Sci.*, **2001**, 56, 311–315.

79. Benbella, M.; Paulsen, G.M.; Efficacy of treatments for delaying senescence of wheat leaves: II. Senescence and grain yield under field conditions, *Agron. J.*, **1998**, 90, 332–338.

80. Elliott, D.E.; Reuter, D.J.; Reddy, G.D.; Abbott, R.J.; Phosphorus nutrition of spring wheat (*Triticum aestivum* L.) 1. Effects of phosphorus supply on plant symptoms, yield, components of yield, and plant phosphorus uptake, *Aus. J. Agric. Res.*, **1997**, 48, 855–867.

81. Haloi, B.; Effect of foliar application of phosphorus salt on yellowing of wheat seedlings, *J. Res. Assam. Agric. Uni.,* **1980**, 1, 108–109.

82. Girma, K.; Martin, K.L.; Freeman, K.W.; Mosali, J.; Teal, R.K.; Raun, W.R.; Moges, S.M.; Arnall, B.; Determination of the optimum rate and growth stage for foliar applied phosphorus in corn, *Commun. Soil Sci. Plant Anal.*, **2007**, 38, 1137–1154.

83. Raun, W.R.; Sander, D.H.; Olson, R.A.; Phosphorus fertilizer carriers and their placement for minimum till corn under sprinkler irrigation, *Soil Sci. Soc. Am. J.*, **1987**, 51, 1055–1062.

84. Garcia, R.; Hanway, J.J.; Foliar fertilization of soybeans during the seed-filling period, *Agron. J.*, **1976**, 68, 653–657.

85. Poole, W.D.; Randall, G.W.; Ham, G.E.; Foliar fertilization of soybean: I. Effect of fertilizer sources, rates and frequency of application, *Agron. J.*, **1983**, 75, 195–200.

86. Boote, K.J.; Gallaher, R.N.; Robertson, W.K.; Hinson, K.; Hammond, L.C.; Effect of foliar fertilization on photosynthesis, leaf nutrition, and yield of soybeans 1. *Agron. J.*, **1978**, 70, 787–791.

87. Marschner, H.; *Mineral nutrition of higher plants*, Academic Press, London, **1995**, pp. 430–433.

88. Schönknecht, G.; Plant mineral nutrition class, **2009**, [Online]. Available at http://botany courses.okstate.edu/schoenk/bot5423/15_Phosphorus/index.html (verified 15 Jan. 2010).

89. Heber, U.; Heldt, H.W.; The chloroplast envelope: Structure, function, and role in leaf metabolism, *Ann. Rev. Plant Physiol.*, **1981**, 32, 139–168.

90. Flugge, U.I.; Heldt, H.W.; The phosphate-triose phosphate-phosphoglycerate translocator of the chloroplast, *Trends Biochem. Sci.,* **1984**, 9, 530–533.

91. Heldt, H.W.; Chon, C.J.; Lorimer, H.; Phosphate requirement for the light activation of ribulose-1,5-biphosphate carboxylase in intact spinach chloroplasts, *FEBS Lett.*, **1978**, 92, 234–240.

92. Bhagwat, AS; Activation of spinach ribulose 1,5-bisphosphate carboxylase by inorganic phosphate, *Plant Sci. Lett.*, **1981**, 23, 197–206.

93. Plaxton, WC; Podesta, FE; The functional organization and control of plant respiration, *Crit. Rev. Plant Sci.*, **2006**, 25, 159–198.

94. Rychter, A.M.; Rao, R.M.; Role of phosphorus in photosynthetic carbon metabolism, In: Pessarakali M (ed), *Handbook of Photosynthesis*, 2nd edn. CRC Press, Boca Raton, FL, USA, **2005**, pp. 1–27.

95. Robinson, S.P.; Giersch, C.; Inorganic-phosphate concentration in the stroma of isolated-chloroplasts and its influence on photosynthesis, *Aust. J. Plant Physiol.*, **1987**, 14, 451–462.

96. Nielsen, T.H.; Krapp, A.; Roper-Schwarz, U.; Stitt, M.; The sugar-mediated regulation of genes encoding the small subunit of Rubisco and the regulatory subunit of ADP glucose pyrophosphorylase is modified by phosphate and nitrogen, *Plant Cell Environ.*, **1998**, 21, 443–454.

97. Pettersson, G.; Ryde-Pettersson, U.; Metabolites controlling the rate of starch synthesis in chloroplast of C3 plants, *Eur. J. Biochem.*, **1989**, 179, 169–172.

98. Preiss, J.; Regulation of the C3 reductive cycle and carbohydrate synthesis, In: Tolbert NE (ed), *Regulation of atmospheric CO$_2$ and O$_2$ by photosynthetic carbon metabolism,* Oxford University Press, New York, **1994**, pp. 93–102.

99. Rychter, A.M.; Mikulska, M.; The relationship between phosphate status and cyanide-resistant respiration in bean roots, *Physiol. Plant*, **1990**, 79, 663–667.

100. Rychter, A.M.; Chauveau, M.; Bomsel, J.L.; Lance, C.; The effect of phosphate deficiency on mitochondrial activity and adenylate levels in bean roots, *Physiol. Plant*, **1992**, 84, 80–86.

101. Fukuda, T.; Saito, A.; Wasaki, J.; Shinano, T.; Osaki, M.; Metabolic alterations proposed by proteome in rice roots grown under low P and high Al concentration under low pH, *Plant Sci.*, **2007**, 172, 1157–1165.

102. Li, K.; Xu, C.; Li, Z.; Zhang, K.; Yang, A.; Zhang, J.; Comparative proteome analyses of phosphorus responses in maize (*Zea mays* L.) roots of wildtype and a low-P-tolerant mutant reveal root characteristics associated with phosphorus efficiency, *Plant J.*, **2008**, 55, 927–939.

103. Fang, Z.Y.; Shao, C.; Meng, Y.J.; Wu, P.; Chen, M.; Phosphate signaling in *Arabidopsis* and *Oryza sativa*, *Plant Sci.*, **2009**, 176, 170–180.

104. Nilsson, L.; Muller, R.; Nielsen, T.H.; Dissecting the plant transcriptome and the regulatory responses to phosphate deprivation, *Physiol. Plant*, **2010**, 139, 129–143.

105. Qamar, S.; Yady, J.; Manrique; Parekh, H.; Falconer, J.R.; Nuts, cereals, seeds and legumes proteins derived emulsifiers as a source of plant protein beverages: A review. *Critic. Rev. Food Sci. Nutr.*, **2019**, DOI: 10.1080/10408398.2019.1657062

106. Iannetta, P.P.; Young, M.; Bachinger, J.; Bergkvist, G.; Doltra, J.; Lopez-Bellido, R.J.; Walker, R.L.; A comparative nitrogen balance and productivity analysis of legume and non-legume supported cropping systems: the potential role of biological nitrogen fixation. *Front. Plant Sci.*, **2016**, 7, 1700.

107. Bonetti, R.; Montanheiro, M.; Saito, S.; The effects of phosphate and soil moisture on the nodulation and growth of Phaseolus vulgaris, *J. Agric. Sci.*, **1984**, 103, 95–102.

108. Jakobsen, I.; The role of phosphorus in nitrogen fixation by young pea plants (*Pisum sativum*), *Plant Physiol.*, **1985**, 64, 190–196.

109. Lajtha, K.; Harrison, A.F.; Strategies of phosphorus acquisition and conservation by plant species and communities, In: Tiessen H (ed), *Phosphorus in the Global Environment*, John Wiley and Sons, Chichester, UK, **1995**, pp. 140–147.

110. Lynch, J.P.; Brown, K.M.; Regulation of root architecture by phosphorus availability, *Am. Soc. Plant Physiol.*, Rockville, MD, **1998**.

111. Marschner, H.; Dell, B.; Nutrient uptake in mycorrhizal symbiosis, *Plant Soil*, **1994**, 159, 89–102.

112. Marschner, H.V.; Romheld, V.; Horst, W.J.; Martin, P.; Root-induced changes in the rhizoshpere: Importance for the mineral nutrition of plants. *Zeitschrift fur Pflanzenernaehrung und Bodenkunde.*, **1986**, 149, 441–456.

113. Gilbert, G.A.; Vance, C.P.; Allan, D.L.; Regulation of white lupin metabolism by phosphorus availability. In: Lynch JP, Deikman J. (eds), *Phosphorus in Plant Biology: Regulatory Roles in Molecular, Cellular, Organismic and Ecosystem Processes*. American Society of Plant Physiology, Rockville, MD., **1998.**

114. Plaxton, WC; Carswell, M.C.; Metabolic aspects of the phosphate starvation response in plants. In: Lerner HR (ed), *Plant Responses to Environmental Stress: From Phytohormones To Genome Reorganization*, Marcel-Dekker, New York, NY, **1999**, pp. 350–372.

115. Loneragan, J.F.; Grove, T.S.; Robson, A.D.; Snowball, K.; Phosphorus toxicity as a factor in zinc-phosphorus interactions in plants. *Soil Sci. Soc. Am. J.*, **1979**, 43, 966–972.

CHAPTER 5

Responses and Adaptation of Photosynthesis and Respiration under Challenging Environments

KARUNA SHARMA[1], FARAN SALIK[1], and SHABIR A. RATHER[2*]

[1]Department of Botany, University of Delhi, Delhi, India

[2]School of Integrative Plant Sciences, Section of Plant Biology, Cornell University, Ithaca, New York, USA

*Corresponding author. E-mail: rathershabir100@gmail.com

ABSTRACT

Every organism is continuously subjected to challenges of both biotic and abiotic origin and struggles to maintain its survival with minimum damage to its development and physiology. Plants are no exception to this, rather have an additional disadvantage of being unable to locomote and avert the unprecedented abiotic stresses. Therefore, over time, they have evolved inherent adaptive mechanisms by which they cope up with stress in a better way. Plants exhibit very intricate and complex stress responses. Plant's two vital processes of photosynthesis and respiration are deemed as the most crucial physiological functions that directly govern the overall growth of the plant and are also very sensitive to environmental stresses. Abiotic stress including salinity, elevated CO_2, anoxia, drought, high light, heavy metalloids, and extreme temperature impede photosynthesis and respiration by affecting fundamental photosynthetic and respiratory apparatus, structural and regulatory proteins of chloroplast and mitochondria, carrier proteins and complexes of electron transport chain (ETC), photo- and oxidative phosphorylation, and CO_2 fixation. Inhibition of photosynthesis further afflicts respiration, thereby compromising plant growth. However, with varying degree of tolerance, plants manifest diverse adaptive responses to an array of abiotic stresses. Henceforth, it is exigent to delve deeper into the mechanistic effects of abiotic stresses on plants and to unravel the details of molecular reprogramming employed by the plants to mitigate the potential harms that any kind of stress beholds. This chapter entails the impact of various environmental stresses on photosynthesis and respiration and how these processes get regulated under such unfavorable conditions. A comprehensive awareness of plant responses to stress has realistic and practical implications for designing effective stress management techniques to ensure proper crop management and food security.

5.1   INTRODUCTION

Abiotic stresses, including drought, salinity, metalloid toxicity, high carbon dioxide content, extreme temperature, among others, imperil the normal physiology of plants resulting in a great deal of crop damage, globally. These stresses mentioned above, commonly referred to as environmental or abiotic stress, influence the geographic spread of plants, restrict agricultural yield, and endanger food security. Due to sessility, plants are incapable of averting the environmental challenges they are confronted with constantly. So, in order to sustain such a hostile environment, they have to develop adaptations in order to avert or tolerate such stresses and thrive through the challenging environment. These stresses trigger in plants a plethora of molecular, cellular, and physiological responses to help plants endure through unfavorable conditions. Because plants undergo specific alterations in gene expression, protein abundance, metabolic pathways, and physiology in response to stresses, it is explicit that plants can sense, perceive, and transduce the environmental signals [1], generating suitable cellular responses. Environmental cues sensed by the membrane or cytoplasmic sensors are conveyed to the transcriptional machinery via an intricate signal cascade, culminating in differential transcriptional response that renders the plant tolerant to stress. This differential transcriptional response causes altered expression of an array of genes that are directly or indirectly related to defense response. Hence, identification of key responsive genes against environmental stress is necessary to have a thorough acquaintance with the stress responsive and adaptive mechanisms in crop plants. Plants' response and retaliation to stress, however, is an intricate process that varies with the type and combination of stress along with the host species experiencing stress.

All the environmental challenges that plants face in their natural environment hamper the photosynthetic efficiency of plants by paralyzing the structural and functional integrity of photosynthetic apparatus, decreasing stomatal conductance, and resulting in oxidative damage [2–4]. Photosynthesis is the plant's most crucial physiological function that directly influences the growth of the plant and is also extremely sensitive to stress-induced damage [5, 6]. Any degree of inhibition in photosynthesis will be consequential in compromising plants' optimum growth and yield [7]. On the other hand, another very important process, that is, respiration, which reflects the overall plant metabolism [8], has been shown to get affected by several factors. Many environmental stresses, such as drought, heavy metals (HMs), salinity, etc. affect electron partitioning and distribution betwixt the respiratory pathways (cytochrome and alternative oxidase pathway), consequently impeding ATP synthesis [9]. Structural impairment of enzymes such as cytochrome oxidase (marker enzyme of mitochondria) under stress has also been reported [10]. Based on the degree of tolerance exhibited by plants, their responses to abiotic stress differ.

Hence, in light of that, this chapter explores the frontiers of the under-standing of consequences of stress in plant processes, photosynthesis, and respiration, in particular, to gain finer cognizance of the physiology of stress response and tolerance in plants.

5.2 PHOTOSYNTHESIS

Photosynthesis is a complex biochemical mechanism operating in all land plants, higher and lower, that transduce photons of light energy into chemical energy (NADPH and ATP). Complex carbohydrates are synthesized with the evolution of oxygen, all happening in the specialized organelles, called chloroplast that harbors photosynthetic apparatus. Photosynthesis occurs in two phases; the light reaction in thylakoid membrane succeeded by the dark reaction in the stroma. While the light-dependent reaction is a light-driven electron transport chain (ETC), dark or light-independent phase involves carbon fixation, popularly named as Calvin–Benson cycle or C_3 cycle.

In light reaction, light photons are absorbed by an array of pigments harbored in the light-harvesting complex (LHC) and centrally located reaction center of two multisubunit photosystems—PSI (P680) and PSII (P700). Light energy absorbed by LHC pigments (carotene and xanthophylls, majorly) gets transferred to the reaction center (key chlorophyll molecules), whose electrons, on the absorption of light, get promoted to the higher, unstable state of energy. This inaugurates the transfer of electrons through various electron carriers (pheophytin, plastoquinone, cytochrome complex, plastocyanin) from PSII to PSI. Toward the end, the shuttling electron is passed on to its final acceptor $NADP^+$ to form NADPH [11–13]. This transport of electrons is very sensitive to toxic metalloids like Na, Cr, Al, Cd among others [14–16].

The energy products of light-driven electron-transport chain, that is, ATP and NADPH, are employed in the plants to drive CO_2 fixation in dark reaction, resulting in the formation of complex carbohydrates. C_3 cycle of photosynthesis commences in three stages:

1. *Carboxylation*: It begins with carboxylation of ribulose 1,5-bisphosphate (RuBP), the primary CO_2 acceptor in a dark reaction resulting in the synthesis of 3-phosphoglycerate (3-PGA).
2. *Reduction of 3-PGA*: A two-step process of reduction of 3-PGA into glyceraldehyde 3-phosphate, a triose-phosphate.
3. *Regeneration of RuBP*: The 3-PGA generated in the second phase is either commuted to the cytosol for sucrose biosynthesis or invested in regeneration [17–19].

The structural and functional integrity of enzymes involved in all the three phases of dark reaction is influenced by different stresses, especially osmotic stress, water stress, and elevated CO_2 concentrations [20, 21].

5.3   ENVIRONMENTAL STRESSES: A PERIL ON PHOTOSYNTHESIS

Stress, either of biological or physiological origin, is the response of the organism subjected to a stressor-like environmental conditions [22]. It is an opposing force that cripples the normal functioning and wellness of an organism such as plants [23]. Abiotic stresses of different origins account to a substantial crop loss every year by majorly inhibiting the crucial physiological functions such as photosynthesis [24]. PSII is the most vulnerable component of the photosynthetic apparatus that faces the onus of environmental stress. Surfeit amount of reactive oxygen species (ROS) are generated in response to stress meddles with the functioning of PSII causing photodamage to the machinery [25]. Abiotic stresses majorly influence the PSII functioning by not directly resulting in photoinhibition, rather promoting inhibition of PSII damage repair process [25]. Several studies have suggested that ROS furnished as a result of extreme temperature, high light, high salt, and limited CO_2 fixation [26, 27, 28] is potent enough to impede the translation process of *PsbA* mRNA, hence hampering the repair mechanism of PSII [28]. An outline of a variety of abiotic stresses, their impacts on photosynthesis, and the photosynthetic adaptation to the stress is discussed in the following section.

5.3.1 SALT STRESS AND ITS IMPACT ON PHOTOSYNTHESIS

Salinization, resulting from the accumulation of salts like sulphates and chlorides of sodium, calcium, magnesium, etc., refers to the gradual accrual of salt level in the soil that is toxic for the plants beyond a certain threshold. Salinization increases the total dissolved solids and decreases the soil fertility. Salinity, which is emerging as one of the most excruciating abiotic stresses of the time, can be classified as (1) primary salinity that results from gradual accrual of soluble salts through weathering of rocks that contain salts, that gets subsequently carried away by wind and rain flow [115]; (2) secondary salinity is a consequence of anthropogenic activities that result in mobilization of soluble salts, accumulated from primary salinization, on the surface of soil brought up by rising groundwater level due to excessive exploitation of land [29].

Salt ions present in the soil severely impact plant by interfering with water absorption, disturbing ion equilibrium and resulting in toxicity. Salinity may have a short-term or long-term outcome on photosynthesis [30]. Since it interferes with water absorption by plant roots, it results in the conditions of physiological drought in plants, which in severe cases might result in wilting. In order to maintain hydro homeostasis, plants execute abscisic acid (ABA)-mediated stomatal closure to shun loss of water through transpiration [31]. However, salinity triggered the closure of stomata compromises CO_2 uptake by plants, which directly affects the photosynthetic processes. These outcomes have been validated through several studies where photosynthetic rate and stomatal conductance were decreased in salt-affected plants, as observed in *Kandelia candel* at 430 mM salt concentration [32]. Similarly, salinity-mediated decline in the stomatal conductance of *Gossypium hirsutum* (more pronounced at 250 mM NaCl) and *Phaseolus vulgaris* (more pronounced at 150 mM NaCl) was observed, with the effect being more dramatic in the latter [33]. A substantial decline in the stomatal conductance and 30% reduction in intracellular CO_2 concentration [Ci] were observed in salt-stressed *P. vulgaris* grown over a range of NaCl concentration, that is, 0–150 mM [34]. Similarly, reduction in intracellular CO_2, photosynthetic rate (fv/fm), water use efficiency (WUE), stomatal conductance was reported in *Spinacia oleracea* stressed with 200 mM NaCl concentration [35].

Salt also affects the structure and abundance of photosynthetic pigments and proteins. The decrease in total content of chlorophyll and carotene in tomato plants challenged by salinity was observed [36]. Similar reduction in chlorophyll content has been observed in several plants namely, tomato, potato, mung bean and pea, okra, wheat, castor bean among others [37–42].

A decrease in carotenoid content was also reported in *Nigella sativa*, sugarcane, and pistachio under salt stress [43–45]. However, several studies report upregulation of carotenoid synthesis under stress owing to the photoprotective and signaling roles of carotenoid under stress [46, 47]. The salt-induced reduction in chlorophyll, attributed to impaired biosynthesis or rapid degradation, has been found accompanied by the ultra-structural aberrations in the chloroplast and impaired function of photosynthetic machinery [48]. Accumulation of toxic ions in the chloroplast's damages thylakoid membrane. Various electron microscopic studies have displayed the inhibitory effect of salt on thylakoid structure, resulting in thylakoid disorganization and swelling, unstacking of grana, and variation in starch content in cells [49]. Swollen thylakoids, reduced grana stacking, and accumulation of starch were obtained in potato subjected to long-term NaCl treatments (100 and 200 mM) [50] and *Thellungiella salsuginea* [51] under 400 mM NaCl dose. Aberration of thylakoid and chloroplast membrane and accumulation of plastoglobule were some of the salt-induced adverse effects of salt on the architecture of photosynthetic apparatus of *Cucumis* seedlings [52]. Starch accumulation in response to toxic salt ions is attributed to the impairment of cytosolic sucrose phosphate synthase enzyme, directing triose phosphate pathway toward starch biosynthesis or obliteration of starch-degrading enzymes [53]. Membrane damage is a typical consequence of oxidative stress, downstream stress resulting from salinity, explaining the structural alterations of the chloroplast envelope and thylakoids observed in these studies.

Inhibition of plant growth in response to salinity is often ascribed to compromised photosynthetic performance [47]. Since PSII is the fulcrum of photosynthetic performance, the salt-induced outcome on PSII has been well studied, and reports suggest that salinity results in compromised quantum yield of electron transport at PSII, the quantity of light reaching the reaction center as well as interferes with the activity of oxygen-evolving complex (OEC) associated with PSII [54, 55]. Salt-triggered reduction in fv/fm ratio, an indicator of quantum efficiency of PSII, has been reported in *Vicia faba* [56], wheat [57], maize [58] among other significant crops. Impairment of OEC capacity in response to salt has been accredited to changes in the protein profile of thylakoid membrane that peter the energy conduction from LHC to PSII. It has been proposed that salt stress represses electron transfer rates at PSII donor site [57]. Radioactive labeling and molecular studies in *Synechocystis* have also revealed salt-mediated inhibition of translation and transcription of D1-encoding *psbA* gene. D1 protein is the reaction center protein that represents the key subunit of PSII and its phosphorylation confers immunity from oxidative damage [59]. Salinity stress can also wreck activities of crucial stromal enzymes that participate in CO_2 reduction. *In vitro* RuBisCO activity was found inhibited by high levels of salts in barley [60], *Vigna unguiculate* [61], soybean [62], *Kalidium foliatum* [63]. Other than RuBisCO, enzymes participating in the regeneration of RuBP are also critical targets of salinity stress [47]. A substantial reduction in RuBP repertoire was attributed to salt influence on RuBP regeneration in *P. vulgaris* [64], fructose-1,6-bisphosphatase enzyme (FBPase), committed to RuBP regeneration, was reported to be negatively influenced by salinity in *Helianthus annuus* [65], *Oryza sativa* [66], and so forth. Reduction intracellular pool of another enzyme crucial for photosynthesis, phosphoenol pyruvate carboxylase (PEPC) was reported under salt stress [67] in wheat and maize.

5.3.2 *DROUGHT AND ITS IMPACT ON PHOTOSYNTHESIS*

Reports on climate change anticipate a continuous increase in atmospheric temperature resulting in frequent drought conditions worldwide. Water stress commences in plants when there is a lag between the amount of water absorbed and the amount transpired, when the latter is higher than the former, plants experience physiological drought. Apart from disturbing plant water status, water deficit conditions also exert damaging influence on important physiological processes, like stomatal movements, gaseous exchange, photosynthesis, and respiration [68]. The initial reduction in photosynthesis under drought conditions is due to closure of stomata, restricting CO_2 influx, and as drought becomes sustained it severely impairs the metabolism accounting for severe downregulation of photosynthesis effectuating in decreased yield and biomass production [20]. Early and the foremost response of plants under drought conditions is restraining water loss by stomatal regulation. This restricts water loss but also reduces CO_2 absorption and circulation of nonstructural carbon [69]. Drought-induced reduction in stomatal conductance (g_s) and intracellular CO_2 (C_i) has been suggested by several studies, to state a few, drought decreased g_s and photosynthetic efficiency in *Chrysanthemum* [70], soybean [71], sunflower [72], and kidney bean [73]. All these studies pointed at the close correlation between water availability, stomatal conductance, leaf CO_2 concentration, and photosynthetic efficiency. Stomatal close down is not the exclusive factor that accounts for photosynthetic lapse under drought. Additionally, several other nonstomatal mechanisms influence the photosynthetic processes under "low-water" environmental conditions [74].

In the quest of establishment of drought-responsive processes, local pool of carbohydrates and energy [ATP/NADPH] are used up leaving a dearth of carbohydrates and energy for other physiological processes of significance to plants, therefore, resulting in carbon starvation [4]. Reduced synthesis of ATP, along with altered conformation of CO_2-fixing enzymes of dark reaction, potentially inhibits RuBisCo activity by impeding RuBP regeneration. For example, in soybean, wheat, sunflower, and *Trifolium subterraneum* drought resulted in reduced RuBisCO activity [75, 76, 77]. Similar implications were put forth by Marques and Arrabica [78] in *Setaria sphacelota* wherein the rate of RuBisCO catalyzed reaction declined with severe drought stress owing to (1) changes in CO_2 availability in the chloroplast, (2) alterations in RUPB (substrate) availability or, (3) RuBisCO deactivation [78]. Along similar lines, rapid abatement in the transcript abundance of the smaller subunit of RuBisCO (rbcS) under drought stress was observed in tomato, *Arabidopsis*, and rice [79, 80, 81]. The negative influence of drought stress on various other enzymes of C4 cycle of photosynthesis, namely PEPcase, FBPase, NADP-malic enzyme, phosphopyruvate dikinase (PPDK), have been obtained in sugarcane [82]. Significant reduction in chlorophyll pigments in drought-stressed plants has been reported in wheat [83], *Vigna radiata* [84], and many other species. The reduction in chlorophyll pigment with fluctuations in Chl a/b ratio under drought may, chiefly, result from chloroplast damage triggered by ROS generated in response to stress [85].

Large body of evidence imputes the progressive suppression of photosynthesis under drought stress to perturbation of processes that involves the participation of PSII reaction center and accompanying uncoupling of noncyclic light reaction that disturbs the balance between the generation-and-exploitation dynamics of electrons, thereby, accounting for reduced electron transport rate [84]. Drought mainly affects the functionality of PSII; the transcript levels of *psbA, psbD*, and *cab* genes and abundance of D1, D2, and LHCII got markedly reduced due to drought triggered accelerated degradation of proteins and mRNA transcripts [86]. The process of phosphorylation of proteins associated with PSII has been widely acknowledged as a post-translational modification that regulates the stability and turnover of the proteins of the reaction center. Drought stress stimulates dephosphorylation of PSII proteins rapidly coupled with phosphorylation of barley LHCIIb4 and CP29 proteins [87]. While CP29 phosphorylation may result in uncoupling of LHCII from the P680 complex and cleavage of LHCII trimer, the dephosphorylated state of PSII may be significant for PSII repair process and signal transmission during stress [87]. Numerous ad rem in vivo studies have demonstrated the effect of drought on OEC associated with PSII, reduced quantum efficiency of PSII photochemistry and degradation of D1 protein bringing about inactivation of the P680 reaction center ([88] in pea; [89] in wheat; [90] in *Arabidopsis*; [91] in *Brassica*).

5.3.3 HEAVY METALS STRESS AND ITS IMPACT ON PHOTOSYNTHESIS

The vulnerability of plants to HMs present in the soil or irrigation water has become a climacteric environmental stress due to both natural and anthropogenic origins [4]. Anthropogenic activities, such like abuse of inorganic agrochemicals, atmospheric deposition of automobile exhausts and fossil fuel combustion, the release of domestic effluents, waste discharge from industries like refining, mining, dyeing, pyrometallurgy, etc. contribute to the prevalence of HMs in the soil, water bodies, and groundwater [92]. Crops growing in such soil substratum, or irrigated with water contaminated with HM, acquire these metalloids that impose excruciating repercussions on their growth and metabolism. The crucial process of photosynthesis is very sensitive to toxicity by HMs and sustained exposure to HMs can affect photosynthetic machinery at many levels of organization and framework that may invoke consequences like reduced content of photosynthetic pigments, damaged the structural integrity of chloroplast, and attenuated activity of enzymes involved in CO_2 fixation [92, 93, 94]. HMs have been reported to impede chlorophyll biosynthesis by inhibiting the activity of protochlorophyllide reductase and delta-aminolevulinic acid dehydrogenase by interacting with their sulphydryl groups [95]. Cadmium (Cd) that has been reported to inhibit the HMs can alter cellular and molecular processes by blocking the active sites of the enzymes, rendering them inactive, substituting essential cofactors from important biomolecules, occluding the functional groups of enzymes, cause membrane damage and alter the native transcription patterns in plants [4]. The resulting ROS generated from HM stress causes lipid peroxidation of the membrane and oxidative damage of many several physiologically important biomolecules as well as meddles with the electron transport system (ETS) in the chloroplast and mitochondrial membrane [96]. Chloroplast also undergoes some ultrastructural alterations under HM stress. Cd is widely known as one of the major inhibitors of photosynthesis and has been reported to alter the chloroplast architecture and results in thylakoid inflation. Cd inhibits the complex formation between protochlorophyllide reductase and its substrates. The number, size, and architecture of chloro-plasts, accumulation of starch and plastoglobuli, stacking and orientation of grana and lamellar arrangement in thylakoids—all these traits are negatively affected by treatment of plants with HMs like chromium (Cr), Cd, lead (Pb), aluminum (Al), nickel (Ni), etc. HMs also increase the activity of lipoxy-genase enzyme that functions in oxidation of polyunsaturated fatty acids forming free radicals.

Furthermore, HMs also influence the biosynthesis and accumulation of chlorophyll [97] and encourage the degradation of the former by upregulation of chlorophyllase activity [98]. Many HMs such as mercury (Hg), Cd, Ni, zinc (Zn), Pd, etc. oust magnesium (Mg) ions from the chlorophyll head making it inactive to perform photosynthetic processes [99]. Besides, HMs also impede ATPase activity and damage membrane integrity. HMs also hinder the structural components of ETS, especially PSII, the principal component which absorbs light, OEC, and electron carrier intermediates. PSII is juxtaposed to OEC that harbors three extrinsic polypeptides—PsbO, PsbP, and PsbQ along with Mn_4O_3Ca cluster. HMs (Cu, Ni, Pb, Al, Hg) have been reported to replace the ions (Mn, Ca, Cl) from the inorganic cluster and also dissociate the extrinsic polypeptide, thereby sabotaging O_2-evolution at OEC, a process responsible for the generation of electrons that are circulated through ETS [4, 100]. HMs have also been shown to disrupt the flow of electrons from Q_A (primary acceptor) to Q_B (secondary acceptor) [15]. It has been demonstrated through fluorescence techniques in tobacco that Al impairs the reduction of Q_B acceptor by interacting with nonheme iron localized between Q_A and Q_B acceptor molecules [101]. Aluminum also perturbs the electron flow between PSII and redox-active TyrZ, which under normal physiological conditions, receives electrons from OEC of PSII and transfers it to Q_B via Q_A [16]. HMs effectively occlude electron flow at PSII, both the donor and acceptor site and have also been reported to deplete the number of reaction centers associated with PSII [102]. Conformational changes of the trimeric LHC in the photosynthetic machinery have been reported under Cd and Pb toxicity. Plants experiencing Cd stress have yielded disordered arrangement of LHC monomers—lhcb1 and lhcb2—resulting in a decreased number of actively assembled LHC complexes [97]. Through FTIR spectroscopic study, it has been observed that under Pb stress, the assembly of LHCII components is ravaged, wreaking damage to the photosynthetic machinery [103]. Moreover, the enzymology of plants is also influenced due to HM toxicity. Many enzymes involved in CO_2 reduction are irreversibly damaged. Detrimental effects of Pb, Cu, Zn, Cd, etc. on the activity of enzymes RuBisCO and phosphoribulokinase have been reported by many authors (Table 5.1).

TABLE 5.1 Inhibitory Effect of HMs on the Structural and Regulatory Components of Photosynthesis

Heavy Metals	Structural/Regulatory Component Affected	References
Cd, Pb, Cu	Photosystem II	[98, 100, 104–106]
Cu, Ni, Al, Hg, Zn, Cr	Oxygen-evolving complex	[106–109]
Cu, Zn, Pb	Photosystem I	[110, 111]
Cd, Hg, Pb	Light-harvesting complex	[97, 103, 112]
Zn, Cd, Cu, Pb, Mn, Ni	RuBisCo	[113–117]
Cd, Zn, Ni, Zn, Pb, Cu	PEP carboxylase and 3-phosphoglyceric acid kinase	[116, 118, 119]
Cd, Ni	Fructose-1,6-bisphosphate	[115, 119]
Cd	δ-Aminolevulinic acid dehydratase Porphobilinogen deaminase Protochlorophyllide	[120–122]
	Chloroplast architecture	
Cd	Decreased number and size of chloroplasts, higher number of plastoglobuli, lower content of starch in the stroma	[123]
	Disordered structure of chloroplasts	[124]
	Reduced grana stacking, swollen grana	[125]
Cr	Poorly arranged stroma lamella, fewer grana, widely placed thylakoids	[126]
	Loss of starch grain, disruption of chloroplast, and thylakoid membrane	[127]
Pb	Altered composition of lipids constituting the thylakoid membrane	[118]
Hg, Ni, Zn, Cd, Pb, Cu	Chlorophyll molecule inactivation	[99, 128]

5.3.4   *HEAT STRESS AND IMPACT ON PHOTOSYNTHESIS*

Temperature is a prominent environmental variable that governs the distribution of plants. In most plants, fluctuations in the photosynthetic rate are reversible over a specific range of temperature (10–35 °C). However, exposure of plants to temperature conditions below and above this range irreversibly damages the photosynthetic apparatus. Extreme temperatures can perturb photosynthesis by fracturing the integrity of the photosynthetic machinery [129]. Innumerable reports claim the reduction in chlorophyll biosynthesis in plants experiencing high temperature or heat stress. This lapse in chlorophyll content of the plants under regimes of high temperature can be attributed to either impaired activity of chlorophyll synthesizing enzymes or accelerated disintegration of the chlorophyll molecule, or even a combination of both the possibilities [47]. The synthesis of enzymes 5-aminolevulinate dehydratase, an enzyme committed to the first step of pyrrole biosynthesis pathway, and protochlorophyllide was found compromised in temperature stressed cucumber and wheat seedlings, respectively, contributing to reduced chlorophyll accumulation [130]. Metabolic reprogramming of plants under temperature stress also includes upregulated activity of chlorophyll degrading enzymes such as chlorophyllase or peroxidases [131]. This might be a strategy to promote photosynthetic efficiency under stress by maintaining homeostasis between chlorophyll synthesis and degradation. Deduced from the ultrastructural studies of heat-stressed thylakoids, temperature beyond the normal range inflicts dissociation of Chla/b proteins of LHCII that results in ensuing destacking of appressed part of grana [132, 133].

Since photosynthetic components present in the chloroplast are liable to thermal damages, these organelles render a crucial role in the emanation of heat-triggered responses. The direct impact of heat stress on photosystem components (PSI and PSII), RuBisCO enzyme, Cytb6f complex, etc. has been reported [134]. The principal target site of heat stress in photosynthetic machinery is PSII which is most thermolabile of all the membrane components. During heat stress, thylakoid leakiness is accelerated and due to the enhanced fluidity, PSII falls off the membrane, thereby affecting the transport of electrons [135, 136]. The OEC component associated with PSII is also damaged due to the heat stress resulting in the disrupted transfer of electrons from OEC to the PSII complex [137, 138]. Heat-mediated inactivation of OEC results from either disintegration of the extrinsic polypeptides that are found in the complex [139] or release of manganese (Mn) ions from the inorganic cluster [140]. Dephosphorylation of PSII core proteins, catalyzed by membrane protein phosphates, was observed in heat-stressed spinach thylakoids. Temperature-mediated activation of the phosphatases facilitates accelerated repair of photodamaged PSII which behaves as a cue for other heat responses in the chloroplasts [141].

High temperature also influences the solubility of CO_2 within the leaf tissue; it fosters the solubility of O_2 over CO_2, reducing the accessibility of CO_2 as a substrate for RuBisCO. Temperature beyond optimum thermal results in dissociation of RuBisCO activase enzyme in both C_3 and C_4 plants, thereby lowering the photosynthetic rates [142]. A marked reduction in the activities of RuBisCO and PEP carboxylase were observed in wheat subjected to 12 day-heat stress regime [143] and RuBisCO in tobacco and cotton [144]. Moreover, as temperature escalates, the propensity of RuBisCO diverts from CO_2 toward O_2, this then catalyzes oxygenation reaction. This causes an increase in photorespiration reaction and reduced efficiency of photosynthesis, accounting to compromised crop yield.

Regulation of root hydraulic conductivity is a principle control mechanism of stomatal closure in dry soil conditions. Heat stress noticeably influences the WUE, leaf stomatal, and mesophyll conductance and intracellular concentration of CO_2. Closure of stomatal pores adds another mechanism contributing to impaired photosynthesis as it limits the CO_2 abundance in the tissues [135]. Influence of high temperature on stomatal conductance has been studied in several crops of great economic importance, like *Zea mays* [145] and many others.

5.3.5 IMPACT OF HIGH CO₂ LEVELS ON PHOTOSYNTHESIS

It is widely established that anthropogenic activities have substantially contributed to an accelerated increase in greenhouse gases concentrations in the atmosphere, mainly CO_2 due to rapidly increasing rates of urbanization and industrialization. The elevated levels of atmospheric CO_2 may have drastic ramifications on the structural and functional architecture of natural ecosystems [146]. Responses of plants to overwhelming levels of CO_2 in the environment are essentially mediated by photosynthetic processes which are coupled with alterations in the structural and metabolic framework of leaf tissues [147]. Moreover, the influence of CO_2 on structural and biochemical attributes of the leaf may affect photosynthetic rates. It has been documented that plants grown under elevated CO_2 possess thick leaves and greater mesophyll size and cross-sectional area owing to the increased availability of carbohydrate substrate under high rate of internal CO_2 assimilation [148, 149, 150]. CO_2 serves as the substrate for dark photosynthetic reaction, and how escalating atmospheric concentration of CO_2 is influencing crops and the ecosystem is the area of major interest and significance. Many past studies have shown the positive correlation between elevated CO_2

levels and photosynthetic efficiency, especially in C3 plants, famed as "CO_2-fertilization effect" owing to the magnified carboxylation efficiency of RuBisCo, reduced photorespiration, and inhibition of dark respiration [151, 152]; however, in contrariety, some plants exhibit a decline in photosynthetic efficiency with increasing CO_2 concentration due to observed decline in stomatal conductance [146, 153, 154]. Zheng et al. [146] found a negative quadratic relationship betwixt photosynthetic rate and CO_2 fixation in soybean with the least rate of photosynthesis occurring at 1200 ppm of CO_2. A similar trend was observed with g_s and WUE under elevated CO_2 conditions. Also, a decline in the stomatal index, both on adaxial and abaxial sides of leaves, was observed. The authors accredited the downregulation of photosynthesis beyond certain thresholds of CO_2 to reduced stomatal density and conductance, excessive accumulation of carbohydrates in leaves that negatively feedback the photosynthetic CO_2 fixation, and reduced concentration and activity of RuBisCO. The maximum rate of carboxylation of RuBisCO (abbreviated as Vcmax) and a maximum RUBP regeneration capacity (abbreviated as Jmax), were also remarkably reduced under elevated CO_2, implicating the influence of the latter under both light-dependent and independent (dark) reactions of photosynthesis [146]. A marked reduction in the abundance and activity of enzyme RuBisCO, accompanied by a drop in chlorophyll levels and thylakoid membrane proteins (D1, D2, and Cytf), was observed in *Prunus avium* [155]. Downregulation of RuBisCO and thereby photosynthesis under high CO_2 may be attributed to the carbohydrate feedback inhibition hypothesis: an inordinate amount of photosynthate synthesized in chloroplasts under high CO_2 may trigger downregulation of sugar content through the sugar-signal network [156].

However, plant behavior toward CO_2 enrichment is specific to species and also dependent on environmental factors, especially nutrient conditions, water, and temperature status. Net photosynthetic rate of *Caragana microphylla* was found upregulated in the presence of N augmentation under elevated CO_2 conditions with beneficial effects on shoot biomass [139]. Thus, photosynthetic efficiency in plants is boosted under enriched CO_2 conditions only when plants are not limited by nitrogen. In complete harmony with this was the study carried out by Ainsworth et al. [151] which demonstrated that photosynthetic capacity of soybean plants decreased when they were N-limited and not when they had optimum N availability and sink strength [151]. Under heat stress, elevated CO_2 improved RuBisCo carboxylation and RUBP regeneration as well as photosynthetic rates in tomato, while had no significant effect under normal conditions. CO_2 enrichment also resulted in reduced accumulation of H_2O_2

under heat stress, suggesting its protective role in preventing ROS-induced damage to PSII [157]. Rising temperature and CO_2 levels thus differentially regulate photosynthetic processes in plants.

5.4 PHOTOSYNTHETIC ADAPTIVE RESPONSES OF PLANTS TO ENVIRONMENTAL CHALLENGES

Depending on the magnitude and combination of stress regimes, plants exhibit reprogramming of physiology and metabolism and regulation of gene expression to retaliate stress. Almost all the abiotic stresses damage the architectural and functional integrity of PSII and compromise its photochemical efficiency. Compromised quantum efficiency of PSII can result in absorption of light beyond the competence of photosynthetic machinery. This excess light that is absorbed by stressed PSII can be dispelled by nonphotochemical quenching whereby xanthophyll cycle furnishes an essential photoprotective role. As a defensive response to stress, xanthophyll cycle gets activated wherein two carotenoids—violaxanthin and zeaxanthin—interconvert via antheraxanthin [47]. When stressed photosystems absorb excess light, xanthophyll cycle is activated wherein violaxanthin de-epoxidized to zeaxanthin via antheraxanthin, dissipating excess light as heat [158].

Many abiotic stress-inducible genes are activated under stressful conditions that can be bracketed into majorly three categories:

1. Genes participating in direct protection of plants under stress such as chaperons (Heat shock proteins [HSPs]), LEA proteins, ROS-scavengers, and osmoprotectants [131].
2. Genes participating in signal transduction and transcriptional regulation, such as transcription factors (TFs) (DREB and HSF), SOS kinases, MAPK, and CDPK [159–161].
3. Genes encoding functional proteins that are involved in uptake and conduction of water and ions, that is, ion transporters (NHX1, HKT) and aquaporins (PIPs, TIPs, NIPs, etc.) [162].

HSPs are molecular chaperons that are invested in the protection of proteins by preventing their disintegration and aggregation [163]. A chaperone, APG6, is involved in information of the thylakoid membrane and provides thermotolerance to chloroplasts under high temperature [164]. Small HSPs, like HSP21 associated with chloroplast, get activated under various abiotic stresses, especially heat stress to stabilize PSII by protecting OEC [165, 166]. Another holdase chaperone, Or (orange) protein, is responsible for carotenoids biosynthesis and gauging environmental stresses in plants, for instance, IbOr of sweet potato stabilizes PSII under heat stress [167]. A group of proteins, called late embryogenesis abundance (LEA), accumulates in plants under desiccation and protects the structure of proteins, vesicles, mitochondria, and chloroplast membrane and mediates functioning of chaperons [168]. In *Arabidopsis*, LEA-1 family proteins have been found to enhance plants' endurance to salinity and drought stress. COR15A is a chloroplast LEA protein, encoded by nuclear genes under cold stress, which provides cold acclimation to plants [169].

Genes contributing to the accumulation of osmoprotectants, such as proline encoding P5CS gene, are also upregulated under osmotic stress. Another compatible osmolyte glycine betaine is accumulated under stress. It confers cellular protection under hyperionic, drought, and heat stress by stabilizing macromolecules and preventing damage to PSII by stabilizing the association of PSII extrinsic proteins with thylakoid membrane [170]. Other osmolytes such as trehalose, polyhydric alcohols, polyamines, tertiary sulphonium compounds, etc., enhance stress tolerance in plants by protecting the architecture and functionality of enzymes involved in CO_2 fixation as well as scavenging ROS that potentially damage thylakoid and chloroplast membrane. ROS like O_2^{-}, H_2O_2, $OH^{.}$, $HO_2^{.}$, etc. are often accrued in plants when they experience abiotic stress. They cause lipid peroxidation of membranes resulting in disintegration of membrane protein complexes involved in ETS. Plants have evolved antioxidant defense machinery with enzymatic (GSR, APX, CAT, DHAR, MDHAR, SOD) and nonenzymatic (GSH, ascorbate, carotenoids, etc.) drivers to shield the plant against oxidative injury (Figure 5.1).

Closure of stomata is a common response observed in plants experiencing drought, salinity, and heat stress. More than just a response, it is an adaptation in plants to maintain water homeostasis in disturbed environments. Being the gateways that control CO_2 uptake and rates of transpiration, they are chief determinants of photosynthetic process. Among many signals that govern stomatal movements, ABA is the most widely known along with other secondary messengers, like H_2O_2, NO, and Ca^{2+} [171] (Figure 5.1).

It has been established that stressed plants trigger sulfur assimilatory pathway to ensure an optimum supply of glutathione for phytochelatin biosynthesis. Phytochelatins, HM ATPases, and metallothionein genes furnish crucial roles in acquisition, accumulation, detoxification, and signaling of metalloids. Their combined role prevents hyperaccumulation and toxic effects of HMs in plants [173].

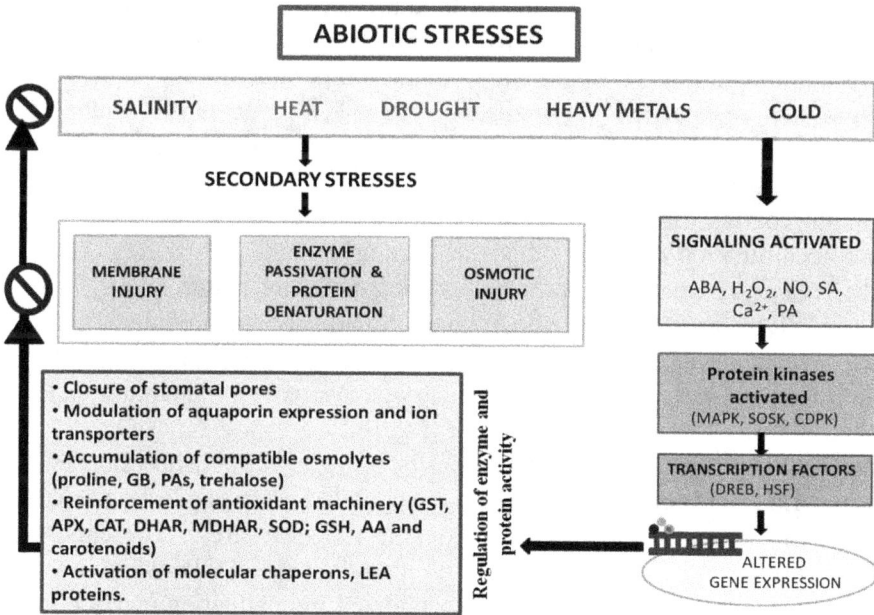

FIGURE 5.1 Once triggered, different signaling molecules (Ca^{2+}, H_2O_2, NO, ABA, etc.) undergo extensive cross-talk at different levels, cooperatively to orchestrate signaling by downstream effectors (TFs, protein kinases) that interact with the defense-responsive genes to launch a fortified defense response.

Of many signaling cascades activated in plants to reinforce defense response against stress, mitogen-activated protein kinases (MAPK) cascades have extraordinary significance. They are conserved signaling cascades in all eukaryotes which transponder environmental cues into intracellular responses for optimum cellular adjustments. MAPK cascade, constituted by three connected components—MAPKKK (MAPK kinases kinases), MAPKK (MAPK kinases), and MAPK—are involved in activation of defense response by regulation of TFs of defense-responsive genes (LEA, DREB, etc.) and antioxidant defense machinery, thus mediating tolerance in plants against abiotic stress [174] (Figure 5.1). Calcium-dependent protein kinases (CDPK), another principal intracellular protein kinases, transduce calcium signals into events of phosphorylation. Several calcium sensors, such as calmodulin-like proteins, regulate the activity and abundance of CDPK in facilitating plant stress tolerance [175]. Ectopic overexpression of CDPK1 gene mediated salt and drought tolerance in tobacco, with an increment in photosynthetic efficiency, chlorophyll content, and water content [176].

Dehydration responsive element binding (DREB) factors partake in the ABA-independent pathway of stress tolerance and activate stress-responsive genes under salinity; drought and extreme temperature stress [177]. Over-expression of AtDREB1.A in *Arabidopsis* [178], rice [179], and potato [180] resulted in enhanced tolerance and better endurance of transformed plants to various abiotic stresses. The transgenic plants revealed higher chlo-rophyll content, better herbage and yield, a higher rate of photosynthesis, and accumulation of proline a sugar. Similarly, many TFs such as Bzip62 of *Arabidopsis* regulate the SOS signaling pathway under salinity stress. SOS1, SOS2, and SOS3 constitute the SOS pathway that modulates cellular responses and facilitates ion homeostasis under salinity. They extrude toxic Na^+ ions from the cell [181] which otherwise decrease the quantum yield of electron transport and activity of PSII [54, 182].

Through a stream of evidence, it has been widely established that several aquaporin (AQP)-encoding genes are upregulated when plants are subjected to abiotic stress, mainly salinity, drought, and heat stress. The enhanced expression of AQP genes results in an abundance of AQPs which are the conserved transmembrane proteins specialized in conduction of H_2O, CO_2, and other neutral solutes [183]. Increasing the bioavailability of CO_2 to photosynthetic machinery correlates with the enhanced photosynthetic rate and yield of the plant. AQPs influence the photosynthetic rate also because of their marked effect on mesophyll conductance [184]. Tobacco plants having overexpressed AQP gene (McMIPB) registered a higher rate of photosyn-thesis, yield, growth rate, and biomass than their wild counterparts under conditions of water deficit [185].

It is widely known that acquisition, translocation, extrusion, and compart-mentation of toxic metal ions (Na^+, Cl^-, etc.) provide fundamental grounds for salinity tolerance in plants. Plants adapt to these stresses by augmenting the expression levels of ion transporters, Na^+/H^+ antiporter NHX1, SOS1, and K^+/Na^+ symporter HKT1 that mediate vacuolar sequestration and long-distance translocation of Na^+, respectively [186, 187]. The Na^+/H^+ antiporter, NhaD, localized in the chloroplast membrane mediates efflux of Na^+ ions from the stroma [188]. These transporters are found differentially upregulated under salt stress to mediate stress tolerance. Inorganic phosphate transporters such as PHT4;1, PHT4;4, etc., localized in the thylakoid membrane, are also involved in modulating chloroplastic Na^+ accumulation [189].

Apart from responses and adaptation of plants to the aforementioned stresses, mediation of photosynthetic responses by elevated CO_2, alone or in concert with other abiotic stresses, is gaining importance. Many elevated CO_2-triggered responses have been studies in C_3 and C_4 plants, such as suppression of photorespiration manifested as lower Gly:Ser ratio; higher carboxylation of RuBisCO accounting to higher photosynthetic efficiency, higher reinforcement of antioxidant defense machinery under other abiotic stresses, and higher accumulation of secondary metabolites, among others. CO_2 enrichment doesn't influence photosynthesis in C_4 plants, much, under unstressed conditions, but promotes it under stressful conditions. This offers a potential benefit for C_4 species under ensuing climate change scenarios [156].

Together with photosynthesis, respiration is also (un)favorably impacted by environmental conditions. Respiration involved in the biological oxidation of food prepared by the process of photosynthesis explicitly indicates that both processes are connected at the various substrate and metabolic levels. Fixation of CO_2 (photosynthesis) on one the side and release of the same (respiration) on the other side are interconnected at the level of substrate. Factors like light, oxygen, temperature, CO_2, and water have potential influence on these mechanisms; thereby, their imbalance turns out to be a major cause of stress for both photosynthesis as well as respiration. The following subset of the chapter highlights the impact of various abiotic stresses on the process of respiration and what plants resort to in order to avert damage to the integrity of the process, and hence ultimate yield of the plant.

5.5 RESPIRATION

Respiration in plants is a complex process of chemical events that involves bioconversion of stored food, chiefly starch and sucrose, into the chemical energy via glycolysis and mitochondrial aerobic respiration, facilitating growth and development of plants [190]. In mitochondrial aerobic respiration, ATP molecules are synthesized in the terminal step via oxidative phosphorylation, which is carried out when organic acids are oxidized resulting in the release of CO_2 and reduction of O_2 to water. The intermediates of TCA cycle also serve as precursors for the formation of various nitrogenous compounds involved in defense mechanisms. Hence, mitochondria participate in numerous other cellular mechanisms and form an important link between N- and C-metabolism in plant cells [191].

5.6 ENVIRONMENTAL FACTORS AFFECTING RESPIRATION IN PLANTS

Plants growing in extreme conditions experience various types of environmental stresses. For example, alpine plants face icy and dry winds, on the other hand, farm crop may encounter fluctuating temperatures along with drought conditions. These extremities in environmental factors create a burden over plants' life cycle, having a notable influence on their cellular metabolism and thereby, affecting their normal physiology, growth and development, and survival [190]. The process of respiration is influenced by the prevailing environmental factors with significant effects on metabolism and respiration rates [192]. Hereby, we will analyze the roles of oxygen, carbon dioxide, light, temperature, and water in respiration and how the latter gets affected by fluctuations thereof.

5.6.1 OXYGEN AND ITS IMPACT

The availability of oxygen is a significant factor responsible for controlling the respiration rate because oxygen serves as the final acceptor of electrons in the mitochondrial ETC. Under normal conditions, in air saturated (21%) aqueous solution, at 25 °C, the equilibrium constant of oxygen is 250 μM. Direct impact on the respiration rate due to external oxygen concentration is rare because the Km value for O_2 is below 1 μM in the reaction catalyzed by the cytochrome *c* oxidase [192]. However, there can be some scenarios where the concentration of oxygen has direct influence on respiration rate [190]. For instance, the respiration rate is reduced when the concentration of oxygen is below 5% for the whole organ and 2–3% for tissue of a plant [192].

Oxygen availability to the plant can be affected because of its diffusion rate in the tissue. A high diffusion rate is required for oxygen to go through the tissues and meet the rate of oxygen utilization. Therefore, the concentration of oxygen may fall due to a decrease in the diffusion gradient. The slow diffusion rate is observed in the heavy storage tissues like tubers of potato, pea and the bean seeds. The lowering of internal O_2 level ultimately results in the reduction of ATP/ADP ratio along with adenylate energy charge within these tissues, which reflects the lowering of respiration rate [193]. Second, the plant may experience the fluctuated oxygen concentration in waterlogged conditions as well (flooding). Though plants usually experience deficit of O_2 at the time of flooding, waterlogging, or due to actions of microbes in the soil [193], however, some studies suggest that flooding does not necessarily result in lower O_2 level, particularly at the time of active photosynthesis [194]. These conditions result in lower O_2 supply to the roots due to blockage of air spaces in the soil with water molecules. For example, in hydroponics, where plants are grown in a soilless culture, regular aeration must be maintained to ensure the continuous supply of O_2 to the roots [190]. Plants growing in the soilless culture depend on the nutrient media solution for meeting their needs of micro- and macronutrients. At times when condition becomes anoxic, the plants are unable to take nutrients from the solution as proper oxygen supply in the nutrient solution is imperative for the uptake and transport of nutrients [195]. In conditions where oxygen level is significantly lower, so much so that plants completely lose the access to it, the activity of cytochrome oxidase gets severely compromised (Km $[O_2]$ = 14 mM, which is equivalent to 0.013% oxygen) and result in the inhibition of ATP production via oxidative phosphorylation. This indicates that paucity of ATP molecules in the cell would sabotage the crucial cellular biochemical reactions. This also signals that ATP formation must be done via fermentation to meet the need of energy in a plant cell [193].

5.6.2 CARBON DIOXIDE AND ITS IMPACT

Carbon dioxide has been increasing globally since the onset of industrial revolution. It can probably reach up toa range of 550–1000 ppm by the end of 21st century, resulting in the rise of 1–3.7 °C in the global temperature [196]. According to one view, increase in the CO_2 concentration does not significantly affect the rate of respiration. For instance, a simulated concentration of 3%–5%, CO_2 has a limited effect on the respiration rate [192]. However, continuous application of CO_2 to soybean plants resulted in lowering of respiration rate, reflecting the continuous and direct effects of CO_2 levels on the process of respiration [197]. A twofold increase in the CO_2 concentration has shown to restrict the O_2 uptake along with a decrease in the cytochrome c oxidase and succinate dehydrogenase enzymes activity in an isolated mitochondrion [198]. The studies suggest that the processes like translocation, nitrate reduction, and others may also get affected at high CO_2 concentration whose rate depends on respiration [199].

Other prevailing view suggests the positive correlation between elevated CO_2 concentration and respiration rate can be attributed to the contribution of CO_2 in photosynthesis. Higher CO_2 levels imply higher CO_2 fixation by photosynthetic enzymes, thus higher production of metabolites and other substrates which can be used up in the process of respiration. Increase in the number of mitochondria has been observed in leaves grown under elevated

CO_2 level. The rise in respiratory mechanism in vegetation is also observed which can be a result of increase in leaf mass per unit area. This increase in respiratory mechanism is linked to the upregulation of respiratory genes under higher levels of carbon dioxide [196]. However, the exact effects of CO_2 concentration on the respiration mechanism are yet to be revealed entirely. Therefore, in current scenario it is difficult to predict the significance of plants as a storehouse of anthropogenic CO_2 and its potential impacts on respiration and yield [192, 196, 198].

5.6.3   *LIGHT AND ITS IMPACT*

Effects of light on respiration are mainly evident during the night hours. Heliophytes generally have higher dark respiratory rates than sciophytes. Similarly, the rate in fully developed leaves that show intolerance to shade is higher than the shade-tolerant leaves emphasizing that shade-loving plants have a lower respiration rate in response to low light. This effect relates to the use of substrate in the process of dark respiration in case of plants that are shade intolerant [190]. The idea explains the dependence of respiration rate on sugar concentration level in the leaves and roots [200]. Similarly, the rate of respiration after few hours of photosynthesis is higher than that after sustained period of darkness [190].

Light has direct effects on respiration rate in dark. The conversion of the nonphosphorylated (activate) form of pyruvate dehydrogenase (PDH) complex to phosphorylated (inactive) form, in presence of light, is stimulated by ammonium ions which are produced from glycine during photorespiration in mitochondria (Figure 5.2) [190]. Photophosphorylation-induced inactivation of PDH reduces the conversion of pyruvate into acetyl-CoA along with production of CO_2 [201]. The inactivation of PDH complex in turn affects the turnover of tricarboxylic acid cycle (TCA) enzymes and production of metabolic intermediates [202]. Light is a vital factor that regulates the expression of genes encoding essential enzymes that are involved in the respiratory pathway, like cytochrome *c* oxidase. However, the effect of light on respiration remains an elusive subject and demands more attention of the researchers [200].

5.6.4   *TEMPERATURE AND ITS IMPACT*

Temperature and respiration rate share a direct relationship till the range of temperature lies between 0 °C and 30 °C. However, respiration rate becomes constant beyond the temperature range of 40–50 °C [192]. Beyond the temperature range of 50–60 °C, respiratory rate shows a drastic lapse due to denaturation of enzymes involved in the respiratory mechanism. Temperature coefficient or Q_{10} is a general applied measure which is used to describe the temperature effect on respiration [190]. The respiration rate increases with an increase in every 10 °C [192]. This can be given by the equation [190]:

$$Q_{10} = \frac{\text{rate at}(t+10)^\circ \text{C}}{\text{rate at } t \,^\circ \text{C}}$$

where Q_{10} values are used to study temperature response of plant respiration rate for short term measurement [203]. At higher temperatures (<30 °C), the Q_{10} value drops as the availability of substrate becomes restrained due to denaturation. Notably, solubility of O_2 also decreases with rising temperature and to balance the lower accessibility of O_2 to the plant tissues, diffusion rates of gases are modulated resulting in a plunge in respiratory rates [190].

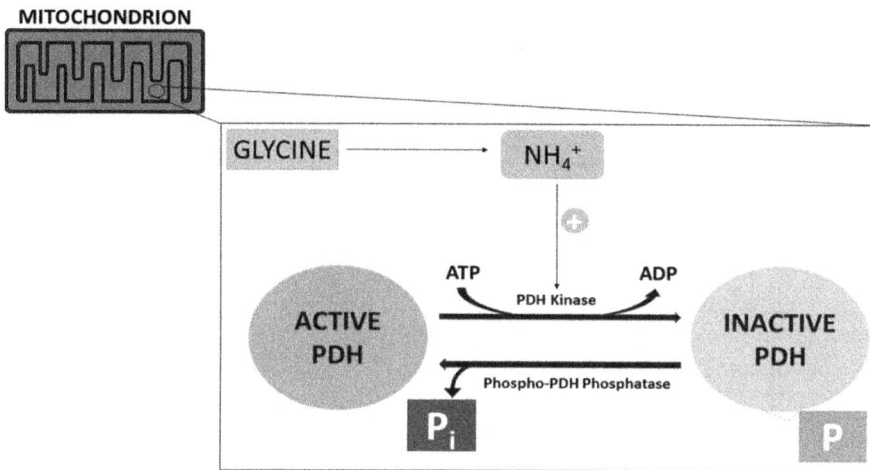

FIGURE 5.2 Mitochondrial PDH regulation. PDH kinase has a role in the interconversion of PDH enzyme in active and inactive forms by utilizing ATP to phosphorylate PDH and phospho-PDH phosphatase has a role in dephosphorylation of PDH. Inactivation of PDH complex is done by ammonium ion which is by triggering PDH kinase.

The rise in night temperature is comparatively quicker than that of day temperature in tropical regions [204]. The respiration rate in the dark fluctuates with different growth temperatures and is found to be lower in plants grown at a higher temperature over those grown at lower temperature [202]. Plants acclimatize to lower temperature for more extended periods by enhancing the respiration activity so as to ensure continuous generation of ATP that is used up to mediate other defensive mechanisms, such as overproduction of osmolytes, reinforcement of ROS scavenging machinery, and development of physical barriers (waxes, cuticles, etc.), among others [192]. It has been reported in several studies that plants growing under higher temperatures experience reduced respiratory rate when compared to the control plants growing at normal temperature range [196]. The production of metabolites in TCA cycle, like amino acid and organic acids, is also affected by fluctuations in temperature [202]. Chilling stress causes potential damage to the electron movement via the cytochrome pathway and results in reduction of subunit II of cytochrome c oxidase along with enhancement in H_2O_2 generation [205].

5.6.5   *WATER AND ITS IMPACT*

One of the most important and relevant factors that restricts the plant yield is water stress. Water stress results in alteration of plant behavior which involves different signaling communications, genetic pathways, growth metabolism, and photosynthesis [9]. It is considered that there is significant reduction in the respiration rate under water-deficit conditions, like that of drought. Under conditions of water deficit, the available free form of cellular water is reduced drastically, leaving behind only the bound form of cellular water, thus causing effective reduction in the respiration rate. It has been documented that extreme conditions of water deficit alter the membrane functions of mitochondria as they are the regions where respiratory ETS operates. Hence, stress-triggered generation of ROS and RNS would result in lipid peroxidation of membrane causing disintegration and release of membrane proteins involved in ETS; this apparently reflects the interdependency of membrane structure of mitochondria and respiration function. The lowered water potential under drought was also reported to impede the oxidation rate of enzymes—succinate and malate-pyruvate and potential of NADH oxidation in the isolated mitochondria [206]. Water stress also impairs ATP production in the mitochondria, as reported in case of soybean [9].

5.7   RESPIRATORY ADAPTIVE RESPONSES AGAINST ENVIRONMENTAL CHALLENGES

If the mechanism of respiration is affected, it would in turn alter the cellular redox status and result in the generation of ROS [207]. One of the major organelles involved in ROS generation is mitochondria which is enrolled in the metabolism where plant cells generate reactive oxidants. Superoxide formation takes place at the sites of ROS production in the complexes I and III of ETC while the process of respiration is taking place in mitochondria resulting in subsequent dismutation of superoxide to H_2O_2 [208]. In natural conditions, the ROS and RNS are maintained at a balanced amount. On the onset of stress, there is a drastic upsurge in their level, serving as a marker of stressed plants, resulting in oxidative imbalance as well as nitrosative stress [209].

Based on their availability, the oxygen levels can be normoxic (normal or standard O_2 level), hypoxic (lesser O_2 level), or anoxic (O_2 absence). Under conditions of oxygen deficiency, the respiration shifts to anaerobic conditions and ATP formation takes place through the process of alcoholic fermentation. This is the first metabolic response to a low level of oxygen, followed by responses at the genetic level [210]. Due to the lack of "final electron acceptor", that is, O_2 in the ETS, the Krebs cycle operates partially and in both directions. The cytoplasm acidifies and pyruvate is converted into ethanol and lactate which signifies major fermentation reactions in the plant cells [211]. The enzymes of the glycolytic and alcoholic (ethanolic) fermentation pathway, like hexokinase, fructokinase, pyruvate kinase and others, show specific activity in hypoxia. This incident results in the activation of glycolytic flux, ATP generation along with lactate production which acidifies the cytoplasm of the cell. This has been evidenced that different metabolisms and signaling pathways are related to the ROS production; however, the mechanism is still not understood. This transfer of aerobic to anaerobic mechanism quickly activates various genetic level of expressions that are controlled at both transcription and post-transcription level. Decreased O_2 level interferes with the activity of several genes. It has been reported that the genes that encode enzymes participating in the breakdown of disaccharide sucrose and starch, glycolytic pathway, and ethanolic fermentation are upregulated. In aerobic and anaerobic cells, the alcohol dehydrogenase isoforms are expressed differentially due to different promoter sequences. Anaerobic response elements are localized in the promoters of many such hypoxia-stimulated genes which relate with hypoxia-responsive transcriptional regulators of ERF family [210]. The conditional responses of lower O_2 level have been summarized in Table 5.2.

The diurnal temperature fluctuations to varying degrees have a corresponding effect on the biosynthesis of Krebs cycle intermediates. It has been demonstrated in many plants that fumarate and malate preferentially accumulate during the day and citrate, aconitate, and succinate get selectively accumulated during the night. Additionally, the Krebs cycle can be enhanced via the activity of light induced PEPC that generates malate and oxaloacetate in the Hatch–Slack pathway. In this mechanism, PEPC usually replenishes the Krebs cycle intermediates (malate or oxaloacetate) which are formed in scanty amount when α-ketoglutarate is utilized for the synthesis of amino acids [202]. Furthermore, alternative oxidases (AOX) activity increases under various stress conditions. Principally, water stress, together with light, increases the threat of oxidative stress by enhancing the ROS production in the cell. In this situation, the alternate respiratory pathway, AOX, becomes operational and restricts the adverse effects of ROS entities [9].

TABLE 5.2 Effects of Oxygen Conditions on Plant Metabolism

Status of Oxygen	Condition	Effect on Metabolism	ATP Status
Normoxia	Normal O_2 level or aerobic condition	The regular respiratory mechanism takes place where all ATP is generated through oxidative phosphorylation.	Normal energy level
Hypoxia	Low level of O_2	Low level of ATP generation by oxidative phosphorylation. ATP production via the glycolytic pathway is more in comparison to normal conditions. Variation in metabolism and development are triggered once the plant gets acclimatization to a reduced level of oxygen.	Low energy level
Anoxia	Absence of O_2 or anaerobic condition	Glycolytic pathways remain as the only source of energy where cells face the reduced level of ATP, decreased protein manufacturing along with damaged division and elongation. Numerous cells can die if this situation continues.	Least energy

The mechanism of plant respiration is flexible and can detour to alternate pathway, such as AOX pathway, under conditions of stress that impair normal cytochrome-mediated respiratory pathways. [212]. The distribution of AOX has been reported in organisms like many algae, fungi, some protozoa, and land plants. The AOX is firmly bonded to the inner mitochondrial membrane and diverts the normal movement of electrons from the typical cytochrome-mediated ETS pathway (Figure 5.3) and results in ATP production, lesser than the cytochrome pathway, along with additional free energy that is dissipated as heat [210]. AOX mechanism is an efficient alternative for plants under various unfavorable physiological conditions as it smoothens the metabolism and reduces the amount of ROS produced in mitochondria. AOX pathways are activated under conditions of drought, high temperature, elevated CO_2, high light, and salt stress [212]. AOX pathway offers a mechanism by which over reduction of mitochondrial ETS is prevented which results in low ATP production and lowers the oxidative stress which engenders from extreme temperature or others [190]. AOX displays buffering effect when there is fluctuation in the abundance of respiratory substrates and ATP production via mitochondrial ETS under stressful situations. AOX pathway helps plant manage the anoxic or hypoxic conditions without disturbing the turnover of oxidative phosphorylation, since the affinity of AOX for oxygen is quite below than that of COX (cytochrome c oxidase) [194]. Krebs cycle intermediates are found to have principal role in regulation of AOX activity, the fact that can be exploited to design novel AOXs with desirable regulatory characteristics [212].

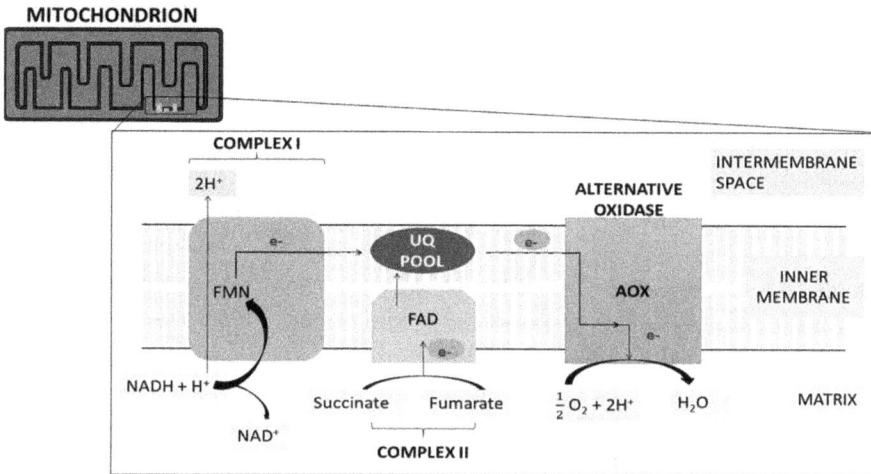

FIGURE 5.3 The alternative respiratory pathway. Electrons which get captured to AOX pass through one or no phosphorylating sites.

The functionality of alternative pathways has also been observed to increase when CO_2 concentrations get doubled, which results in direct inhibiting the cytochrome oxidase [198]. Plants experiencing sustained high CO_2 concentrations exhibit partitioning of electrons between COX and AOX pathways, closely associated with alteration in energy requirement in the tissue. Additionally, electron partitioning in mitochondria to AOX linked to the enhancement of leaf carbon balance and respiration efficiency under fluctuating levels of carbon dioxide. Indeed, overexpression of AOX has been revealed in the leaves of tobacco subjected to elevated CO_2 level that helped in maintaining carbohydrate and energy homeostasis [212]. In *Arabidopsis*, five families of AOX genes have been reported-*AOX1a*, *AOX1b*, *AOX1c*, *AOX1d*, and *AOX2*, all encoding mitochondria-localized proteins Amongst hundreds of genes that code for different proteins in mitochondria, the most expressive genes for stress response were found to be *AOX1a* and *AOX1d* in *Arabidopsis* [213]. Insertional inactivation of *AOX1a* resulted in accumulation of ROS and production of different anthocyanin phenotype under conditions of combined drought and light stresses [191]. It is, thus, quite evident that AOX is an extremely important adaptive respiratory response of plants toward an array of abiotic stresses that challenge respiratory mechanism at various levels. From the studies revealing the extensive participation of AOX enzyme in plant defense mechanism against various stresses, it can be claimed that AOX delivers a beneficial in plants that are challenged by a combination of stresses induced by the alteration in the climate [212]. Thus, overexpression of AOX gene is a potential strategy that can go a long way in imparting blanket tolerance to plants against broad spectrum of abiotic stresses.

5.8 CONCLUSION

Increasing temperature and carbon dioxide (due to global warming), salinization, drought, heavy metal(loids), and limited oxygen are some of the very significant environmental challenges that plants confront in their entire life cycle. These stresses of variable nature, extent, and tenure incite stress-responses in plants, many of which trigger defense pathways to help mitigate the adverse effects of these stresses on the two most sensitive mechanisms in plants, that is, photosynthesis and respiration. All these stresses impose structural and functional deformities, some overlapping while some discrete, in the photosynthetic and respiratory machinery which has detrimental impacts on the overall growth and yield of crop plants that imperils the global food security. To recapitulate, this chapter provides an elaborated overview of the impacts of different abiotic stresses on the structural and functional integrity of photosynthesis apparatus, for instance, disruption of photosystems and OEC, blockage of ETC, the disintegration of chlorophyll and carotenoids, inactivation of the pathway and biosynthetic enzymes (those participating in Calvin cycle), the disintegration of chloroplast ultrastructure, and so on, among others. The chapter also reiterates how all these environmental abnormalities (oxygen, carbon dioxide, temperature, light, and water) afflict respiratory machinery of the plants. These environmental stresses result in fluctuation of respiratory rate and accordingly affect the ATP production also. Alternation in the Krebs cycle intermediates, decrease in enzyme activities, and hindrance in normal electron flow via the cytochrome pathway are some of the repercussions of these stresses on respiration. However, respiration is not a discrete process, it is rather connected to other mechanisms like photosynthesis, photorespiration, and lipid metabolism at various levels of substrate and metabolites. It has been reported that entirely operative mitochondria are required for optimum photosynthesis. The environmental stresses trigger a responsive behavior and defense mechanism in plants. The alternative pathway through AOX has been considered mostly as an adaptation of the plant toward stress, aimed at completing the mitochondrial ETC which gets paralyzed under stress. Therefore, advanced genetic analysis on AOX can prove to be a supremely practical approach to know its regulatory and behavioral contribution and crosstalk to other pathways. Also, studying plant responses under combined abiotic stresses is a more practical approach to unravel and understand the universal standard and exclusive defense mechanisms, as multiple stresses challenge plants in the environment at a time.

Furthermore, how plants acclimate to challenges of such nature by incorporating functional and structural reprogramming is a matter of utmost significance as it deepens the understanding of "plant-abiotic stress" arm-race and offers new insights on how lessons from nature can be incorporated into the in vitro studies to head toward a future of best quality of transgenic plants with blanket resistance toward a broader range of stresses in the environment.

KEYWORDS

- **photosynthesis**
- **respiration**
- **salt stress**
- **drought stress**
- **heavy metal stress**
- **heat stress**
- **elevated CO_2**
- **light stress**
- **anoxia**

REFERENCES

1. Zhu, J.K. Abiotic stress signaling and responses in plants. *Cell*, **2016**, *167*(2), 313–324.
2. Allen, J.F.; Forsberg, J. Molecular recognition in thylakoid structure and function. *Trends in Plant Science*, **2001**, *6*(7), 317–326.
3. Efeoglu, B.; Terzioglu, S. Photosynthetic responses of two wheat varieties to high temperature. *Eurasian Journal of Biosciences*, **2009**, *3(1)*, 97–106.
4. Sharma, A.; Kumar, V.; Shahzad, B.; Ramakrishnan, M.; Sidhu, G.P.S.; Bali, A.S.; Handa, N.; Kapoor, D.; Yadav, P.; Khanna, K.; Bakshi, P. Photosynthetic response of plants under different abiotic stresses: a review. *Journal of Plant Growth Regulation*, **2019**, *39*, 1–23.
5. Chaves, M.M.; Maroco, J.P.; Pereira, J.S. Understanding plant responses to drought—from genes to the whole plant. *Functional Plant Biology*, **2003**, *30*, 239–264.
6. Lawlor, D.W.; Tezara, W. Causes of decreased photosynthetic rate and metabolic capacity in water-deficient leaf cells: a critical evaluation of mechanisms and integration of processes. *Annals of Botany*, **2009**, *103*, 561–579.
7. Sudhir, P.; Murthy, S.D.S. Effects of salt stress on basic processes of photosynthesis. *Photosynthetica*, **2004**, *42*(2), 481–486.
8. Flexas, J.; Galmes, J.; Ribas-Carbo, M.; Medrano, H. The effects of water stress on plant respiration, In *Plant respiration*, Springer, Dordrecht, **2005**, pp. 85–94.
9. Ribas-Carbo, M.; Taylor, N.L.; Giles, L.; Busquets, S.; Finnegan, P.M.; Day, D.A.; Lambers, H.; Medrano, H.; Berry, J.A.; Flexas, J. Effects of water stress on respiration in soybean leaves. *Plant Physiology*, **2005**, *139*(1), 466–473.
10. Peñuelas, J.; Ribas-Carbo, M.; Giles, L. Effects of allelochemicals on plant respiration and oxygen isotope fractionation by the alternative oxidase. *Journal of Chemical Ecology*, **1996**, *22*(4), 801–805.
11. Fajer, J.; Davis, M.S.; Forman, A.; Klimov, V.V.; Dolan, E.; Ke, B. Primary electron acceptors in plant photosynthesis. *Journal of the American Chemical Society*, **1980**, *102* (23), 7143–7145.
12. Hasni, I.; Yaakoubi, H.; Hamdani, S.; Tajmir-Riahi, H.A.; Carpentier, R. Mechanism of interaction of Al^{3+} with the proteins composition of photosystem II. *PLoS One*, **2015**, *10*(3), e0120876.
13. Klimov, V.V.; Klevanik, A.V.; Shuvalov, V.A.; Krasnovsky, A.A. Reduction of pheophytin in the primary light reaction of photosystem II. *FEBS Letters*, **1977**, *82*, 183–186.
14. Dixit, V.; Pandey, V.; Shyam, R. Chromium ions inactivate electron transport and enhance superoxide generation in vivo in pea (*Pisum sativum* L. cv. Azad) root mitochondria. *Plant, Cell & Environment*, **2002**, *25*(5), 687–693.
15. Geiken, B.; Masojidek, J.; Rizzuto, M.; Pompili, M.L.; Giardi, M.T. Incorporation of (35S) methionine in higher plants reveals that stimulation of the D1 reaction centre II protein turnover accompanies tolerance to heavy metal stress. *Plant, Cell & Environment*, **1998**, *21*(12), 1265–1273.
16. Hasni, I.; Hamdani, S.; Carpentier, R. Destabilization of the oxygen evolving complex of photosystem II by Al^{3+}. *Photochemistry and Photobiology*, **2013**, *89*(5), 1135–1142.
17. Buchanan, B.B. Role of light in the regulation of chloroplast enzymes. *Annual Review of Plant Physiology*, **1980**, *31*(1), 341–374.
18. Hoober, J. K. The process of photosynthesis—The light reactions; In *Chloroplasts*, Springer, Boston, MA, **1984**, pp. 79–109.

19. Heineke, D. Photosynthesis: dark reactions, **2001**, *e LS.* https://doi.org/10.1038/npg. els.0001291
20. Sharkey, T. D. Water-stress effects on photosynthesis, *Photosynthetica*, 24, **1990**, 651–651.
21. Li, X.; Zhang, G.; Sun, B.; Zhang, S.; Zhang, Y.; Liao, Y.; Zhou, Y.; Xia, X.; Shi, K.; Yu, J. Stimulated leaf dark respiration in tomato in an elevated carbon dioxide atmosphere. *Scientific Reports*, **2013**, *3*, 3433.
22. Muthukumar, K.; Nachiappan, V. Cadmium-induced oxidative stress in *Saccharomyces cerevisiae. Indian Journal of Biochemistry and Biophysics*, **2010**, *47*(6), 383–387.
23. Jones, H.G.; Jones, M.B. Introduction: some terminology and common mechanisms, In: *Plants under Stress*, Jones, H.., Flowers, T.J., Jones, M.B. (Eds.), Society for Experimental Biology Seminar Series. Cambridge University Press, Cambridge, **1989**, pp. 1–10.
24. Mahajan, S.; Tuteja, N. Cold, salinity and drought stresses: an overview. *Archives of Biochemistry and Biophysics*, **2005**, *444*(2), 139–158.
25. Murata, N.; Takahashi, S.; Nishiyama, Y.; Allakhverdiev, S. I. Photoinhibition of photosystem II under environmental stress. *Biochimica et Biophysica Acta (BBA)-Bioenergetics*, **2007**, *1767*(6), 414–421.
26. Ohnishi, N.; Murata, N. Glycinebetaine counteracts the inhibitory effects of salt stress on the degradation and synthesis of D1 protein during photoinhibition in *Synechococcus* sp. PCC 7942. *Plant Physiology*, **2006**, *141*(2), 758–765.
27. Yang, Y.; Xu, S.; An, L.; Chen, N. NADPH oxidase-dependent hydrogen peroxide production, induced by salinity stress, may be involved in the regulation of total calcium in roots of wheat. *Journal of Plant Physiology*, **2007**, *164*(11), 1429–1435.
28. Takahashi, S.; Murata, N. Glycerate-3-phosphate, produced by CO_2 fixation in the Calvin cycle, is critical for the synthesis of the D1 protein of photosystem II. *Biochimica et Biophysica Acta (BBA)-Bioenergetics*, **2006**, *1757*(3), 198–205.
29. Manchanda, G.; Garg, N. Salinity and its effects on the functional biology of legumes. *Acta Physiologiae Plantarum*, **2008**, *30*(5), 595–618.
30. Parida, A.K.; Das, A.B. Salt tolerance and salinity effects on plants: a review. *Ecotoxicology and Environmental Safety*, **2005**, *60*(3), 324–349.
31. Munns, R.; Tester, M. Mechanisms of salinity tolerance. *Annual Review of Plant Biology*, **2008**, *59*, 651–681.
32. Kao, W.Y.; Tsai, H.C.; Tsai, T.T. Effect of NaCl and nitrogen availability on growth and photosynthesis of seedlings of a mangrove species, *Kandelia candel*(L.) Druce. *Journal of Plant Physiology*, **2001**, *158*(7), 841–846.
33. Brugnoli, E.; Lauteri, M. Effects of salinity on stomatal conductance, photosynthetic capacity, and carbon isotope discrimination of salt-tolerant (*Gossypium hirsutum* L.) and salt-sensitive (*Phaseolus vulgaris* L.) C3 non-halophytes. *Plant Physiology*, **1991**, *95*(2), 628–635.
34. Seemann, J. R.; Critchley, C. Effects of salt stress on the growth, ion content, stomatal behaviour and photosynthetic capacity of a salt-sensitive species, *Phaseolus vulgaris* L. *Planta*, **1985**, *164*(2), 151–162.
35. Downton, W. J. S.; Grant, W. J. R.; Robinson, S. P. Photosynthetic and stomatal responses of spinach leaves to salt stress. *Plant Physiology*, **1985**, *78*(1), 85–88.
36. Khavari-Nejad, R.A.; Mostofi, Y. Effects of NaCl on photosynthetic pigments, saccharides, and chloroplast ultrastructure in leaves of tomato cultivars. *Photosynthetica*, **1998**, *35*(1), 151–154.

37. Lapina, L. P.; Popov, B. A; The effect of sodium chloride on the photosynthetic apparatus of tomatoes. *FiziologiyaRastenii*, **1970**, *17*, 580–584.

38. Abdullah, Z.; Ahmad, R. Effect of pre-and post-kinetin treatments on salt tolerance of different potato cultivars growing on saline soils. *Journal of Agronomy and Crop Science*, **1990**, *165*(2-3), 94–102.

39. Hamada, A. M.; El-Enany, A. E. Effect of NaCl salinity on growth, pigment and mineral element contents, and gas exchange of broad bean and pea plants. *Biologia Plantarum*, **1994**, *36*(1), 75–81.

40. Shahbaz, M.; Ashraf, M. Does exogenous application of 24-epibrassinolide ameliorate salt induced growth inhibition in wheat (*Triticum aestivum* L.). *Plant Growth Regulation*, **2008**, *55*(1), 51–64.

41. Pinheiro, H.A.; Silva, J.V.; Endres, L.; Ferreira, V.M.; de Albuquerque Câmara, C.; Cabral, F.F.; Oliveira, J.F.; de Carvalho, L.W.T.; dos Santos, J.M.; dos Santos Filho, B.G. Leaf gas exchange, chloroplastic pigments and dry matter accumulation in castor bean (*Ricinus communis* L) seedlings subjected to salt stress conditions. *Industrial Crops and Products*, **2008**, *27*(3), 385–392.

42. Saleem, A.; Ashraf, M.;Akram, N.A.Salt(NaCl)-induced modulation in some key physio-biochemical attributes in okra (*Abelmoschus esculentus*L.). *Journal of Agronomy and Crop Science*, **2011**, *197*(3), 202–213.

43. Hajar, A. S.; Zidan, M. A.; AlZahrani, H. S. Effect of salinity stress on the germination, growth and some physiological activities of black cumin (*Nigella sativa* L). *Arab Gulf Journal of Scientific Research*, **1996**, *14*(2), 445–454.

44. Gomathi, R.; Rakkiyapan, P. Comparative lipid peroxidation, leaf membrane thermostability, and antioxidant system in four sugarcane genotypes differing in salt tolerance. *International Journal of Plant Physiology and Biochemistry*, **2011**, *3*(4), 67–74.

45. Shamshiri, M.H.; Fattahi, M. Effects of arbuscular mycorrhizal fungi on photosystem II activity of three pistachio rootstocks under salt stress as probed by the OJIP-test. *Russian Journal of Plant Physiology*, **2016**, *63*(1), 101–110.

46. Ziaf, K.; Amjad, M.; Pervez, M.A.; Iqbal, Q.;Rajwana, I.A.; Ayyub, MU. Evaluation of different growth and physiological traits as indices of salt tolerance in hot pepper (*Capsicum annuum* L.). *Pakistan Journal of Botany*, **2009**, *41*(4), 1797–1809.

47. Ashraf, M..; Harris, P.J. Photosynthesis under stressful environments: an overview. *Photosynthetica*, **2013**, *51*(2), 163–190.

48. Shu, S.; Guo, S.R.; Sun, J.; Yuan, L.Y.Effects of salt stress on the structure and function of the photosynthetic apparatus in *Cucumis sativus* and its protection by exogenous putrescine. *Physiologia Plantarum*, **2012**, *146*(3), 285–296.

49. Bruns, S.; Hecht-Buchholz, C. Light and electron microscope studies on the leaves of several potato cultivars after application of salt at various development stages. *Potato Research*, **1990**, *33*(1), 33–41.

50. Fidalgo, F.; Santos, A.; Santos, I.; Salema, R. Effects of long-term salt stress on antioxidant defense systems, leaf water relations and chloroplast ultrastructure of potato plants. *Annals of Applied Biology*, **2004**, *145*(2), 185–192.

51. Goussi, R.; Manaa, A.; Derbali, W.; Cantamessa, S.; Abdelly, C; Barbato, R. Comparative analysis of salt stress, duration and intensity, on the chloroplast ultrastructure and photosynthetic apparatus in *Thellungiella salsuginea*. *Journal of Photochemistry and Photobiology B: Biology*, **2018**, *183*, 275–287.

52. Shu, S.; Yuan, L.Y.; Guo, S.R.; Sun, J.; Yuan, Y.H. Effects of exogenous spermine on chlorophyll fluorescence, antioxidant system and ultrastructure of chloroplasts in *Cucumis sativus* L. under salt stress. *Plant Physiology and Biochemistry*, **2013**, *63*, 209–216.

53. Rahman, M.S.; Miyake, H.; Takeoka, Y. Effects of exogenous glycinebetaine on growth and ultrastructure of salt-stressed rice seedlings (*Oryza sativa* L.). *Plant Production Science*, **2002**, *5*(1), 33–44.

54. Xia, J.; Li, Y.; Zou, D. Effects of salinity stress on PSII in *Ulva lactuca* as probed by chlorophyll fluorescence measurements. *Aquatic Botany*, **2004**, *80*(2), 129–137.

55. Sudhir, P.; Pogoryelov, D.; Kovács, L.; Garab, G.; Murthy, S.D. The effects of salt stress on photosynthetic electron transport and thylakoid membrane proteins in the cyanobacterium *Spirulina platensis*. *Journal of Biochemistry and Molecular Biology*, **2005**, *38*(4), 481.

56. Tavakkoli, E.; Rengasamy, P.; McDonald, G.K. High concentrations of Na^+ and Cl^- ions in soil solution have simultaneous detrimental effects on growth of faba bean under salinity stress. *Journal of Experimental Botany*, **2010**, *61*(15), 4449–4459.

57. Mehta, P.; Jajoo, A.; Mathur, S.; Bharti, S. Chlorophyll a fluorescence study revealing effects of high salt stress on Photosystem II in wheat leaves. *Plant Physiology and Biochemistry*, **2010**, *48*(1), 16–20.

58. Khan, W.U.D.; Aziz, T.; Hussain, I.; Ramzani, P.M.A.; Reichenauer, T.G. Silicon: A beneficial nutrient for maize crop to enhance photochemical efficiency of photosystem II under salt stress. *Archives of Agronomy and Soil Science*, **2017**, *63*(5), 599–611.

59. Chen, L.; Jia, H.; Tian, Q.; Du, L.; Gao, Y.; Miao, X.; Liu, Y. Protecting effect of phosphorylation on oxidative damage of D1 protein by down-regulating the production of superoxide anion in photosystem II membranes under high light. *Photosynthesis Research*, **2012**, *112*(2), 141–148.

60. Miteva, T.S.; Zhelev, N.Z.; Popova, L.P. Effect of salinity on the synthesis of ribulose-1, 5-bisphosphate carboxylase/oxygenase in barley leaves. *Journal of Plant Physiology*, **1992**, *140*(1), 46–51.

61. Aragão, M.E.F.D.; Guedes, M.M.; Otoch, M.D.L.O.; Guedes, M.I.F.; Melo, D.F.D.; Lima, M.D.G.S. Differential responses of ribulose-1, 5-bisphosphate carboxylase/ oxygenase activities of two *Vigna unguiculata* cultivars to salt stress. *Brazilian Journal of Plant Physiology*, **2005**, *17*(2), 207–212.

62. He, Y.; Yu, C.; Zhou, L.; Chen, Y.; Liu, A.; Jin, J.; Hong, J.; Qi, Y.; Jiang, D. Rubisco decrease is involved in chloroplast protrusion and Rubisco-containing body formation in soybean (*Glycine max.*) under salt stress. *Plant Physiology and Biochemistry*, **2014**, *74*, 118–124.

63. Gong, D.H.; Wang, G.Z.; Si, W.T.; Zhou, Y.; Liu, Z.; Jia, J. Effects of salt stress on photosynthetic pigments and activity of ribulose-1,5-bisphosphate carboxylase/oxygenase in *Kalidiumfoliatum*. *Russian Journal of Plant Physiology*, **2018**, *65*(1), 98–103.

64. Seemann, J.R.; Sharkey, T.D. The effect of abscisic acid and other inhibitors on photosynthetic capacity and the biochemistry of CO_2 assimilation. *Plant Physiology*, **1987**, *84*(3), 696–700.

65. Gimenez; Carmen; Valerie J. Mitchell; David W. Lawlor. Regulation of photosynthetic rate of two sunflower hybrids under water stress. *Plant Physiology*, **1992**, *98*(2), 516–524.

66. Ghosh, S.; Bagchi, S.; Majumder, A.L. Chloroplast fructose-1, 6-bisphosphatase from *Oryza* differs in salt tolerance property from the *Porteresia* enzyme and is protected by osmolytes. *Plant Science*, **2001**, *160*(6), 1171–1181.

67. Abdel-Latif, A. Phosphoenolpyruvate carboxylase activity of wheat and maize seedlings subjected to salt stress. *Australian Journal of Basic Applies Sciences*, **2008**, *2*, 37–41.

68. Cornic, G. Drought stress and high light effects on leaf photosynthesis. In: Baker, N.B., Bowyer, J.R. (eds.) *Photoinhibition of Photosynthesis: From Molecular Mechanisms to the Field*, Bios Scientific Publishers, Oxford 1994, pp. 297–313.

69. McDowell, N.G.; Sevanto, S. The mechanisms of carbon starvation: how, when, or does it even occur at all? *The New Phytologist*, **2010**, *186*(2), 264–266.

70. Sun, J.; Gu, J.; Zeng, J.; Han, S.; Song, A.; Chen, F.; Fang, W.; Jiang, J.; Chen, S. Changes in leaf morphology, antioxidant activity and photosynthesis capacity in two different drought-tolerant cultivars of chrysanthemum during and after water stress. *Scientia Horticulturae*, **2013**, *161*, 249–258.

71. Anjum, S.A.; Xie, X.Y.; Wang, L.C.; Saleem, M.F.; Man, C.; Lei, W. Morphological, physiological and biochemical responses of plants to drought stress. *African Journal of Agricultural Research*, **2011**, *6*(9), 2026–2032.

72. Ashraf, M.; O'Leary, J.W. Effect of drought stress on growth, water relations, and gas exchange of two lines of sunflower differing in degree of salt tolerance. *International Journal of Plant Sciences*, **1996**, *157*(6), 729–732.

73. Miyashita, K.;Tanakamaru, S.; Maitani, T.; Kimura, K. Recovery responses of photosynthesis, transpiration, and stomatal conductance in kidney bean following drought stress. *Environmental and Experimental Botany*, **2005**, *53*(2), 205–214.

74. Hajiboland, R.; Cheraghvareh, L.; Poschenrieder, C. Improvement of drought tolerance in tobacco (*Nicotiana rustica* L.) plants by silicon. *Journal of Plant Nutrition*, **2017**, *40* (12), 1661–1676.

75. Majumdar, S.; Ghosh, S.; Glick, B.R.; Dumbroff, E.B. Activities of chlorophyllase, phosphoenolpyruvate carboxylase and ribulose-1,5-bisphosphate carboxylase in the primary leaves of soybean during senescence and drought. *Physiologia Plantarum*, **1991**, *81*(4), 473–480.

76. Tezara W; Lawlor DW. Effects of water stress on the biochemistry and physiology of photosynthesis in sunflower, In: Mathis P (ed). *Photosynthesis: From light to biosphere IV*. Kluwer Academic Publishers, Dordrecht, **1995**, 625–628.

77. Medrano, H.; Parry, M.A.J.; Socias, X.D.W.L.; Lawlor, D.W. Long term water stress inactivates Rubisco in subterranean clover. *Annals of Applied Biology*, **1997**, *131*(3), 491–501.

78. Marques da Silva, J.; Arrabica, M.C. Effect of water stress on Rubisco activity of *Setariasphacelota*. *Photosynthesis: From light to Biosphere*, **1995**, 545–548.

79. Bartholomew, D.M.; Bartley, G.E.; Scolnik, P.A. Abscisic acid control of rbcS and cab transcription in tomato leaves. *Plant Physiology*, **1991**, *96*(1), 291–296.

80. Williams, J.; Bulman, M.P.; Neill S.J. Wilt-induced ABA biosynthesis, gene-expression and down-regulation of *rbc*S messenger-RNA levels in *Arabidopsis thaliana*. *Physiologia Plantarum*, **1994**, *91*, 177–182.

81. Vu J.C.V., Gesch RW, Allen LH, Boote KJ, Bowes G. CO_2 enrichment delays a rapid, drought-induced decrease in Rubisco small subunit transcript abundance. *Journal of Plant Physiology*, **1999**, *155*, 139–142.

82. Du, Y.C.; Nose, A.; Wasano, K.; Uchida, Y. Responses to water stress of enzyme activities and metabolite levels in relation to sucrose and starch synthesis, the Calvin cycle and the C4 pathway in sugarcane (*Saccharum* sp.) leaves. *Functional Plant Biology*, **1998**, *25*(2), 253–260.

83. Nikolaeva, M.K.; Maevskaya, S.N.; Shugaev, A.G.; Bukhov, N.G. Effect of drought on chlorophyll content and antioxidant enzyme activities in leaves of three wheat cultivars varying in productivity. *Russian Journal of Plant Physiology*, **2010**, *57*(1), 87–95.

84. Batra, N.G.; Sharma, V.; Kumari, N. Drought-induced changes in chlorophyll fluorescence, photosynthetic pigments, and thylakoid membrane proteins of *Vigna radiata*. *Journal of Plant Interactions*, **2014**, *9*(1), 712–721.

85. Smirnoff, H. Antioxidant systems and plant response to the environment. In: Smirnoff, V. (Ed), Environment *and Plant Metabolism: Flexibility and Acclimation*, BIOS Scientific Publishers, Oxford, 217–243, **1995**.

86. Duan, H.G.; Yuan, S.; Liu, W.J.; Xi, D.H.; Qing, D.H.; Liang, H.G.; Lin, H.H. Effects of exogenous spermidine on photosystem II of wheat seedlings under water stress. *Journal of Integrative Plant Biology*, **2006**, *48*(8), 920–927.

87. Liu, X.; Wang, Z.; Wang, L.; Wu, R.; Phillips, J.; Deng, X. LEA 4 group genes from the resurrection plant *Boeahygrometrica* confer dehydration tolerance in transgenic tobacco. *Plant Science*, **2009**, *176*(1), 90–98.

88. Giardi, M.T.; Cona, A.; Geiken, B.; Kučera, T.; Masojidek, J.; Mattoo, A.K. Long-term drought stress induces structural and functional reorganization of photosystem II. *Planta*, **1996**, *199*(1), 118–125.

89. Zlatev, Z. Drought-induced changes in chlorophyll fluorescence of young wheat plants. *Biotechnology & Biotechnological Equipment*, **2009**, *23*(sup1), 438–441.

90. Moustakas, M.; Sperdouli, I.; Kouna, T.; Antonopoulou, C.I.; Therios, I. Exogenous proline induces soluble sugar accumulation and alleviates drought stress effects on photosystem II functioning of *Arabidopsis thaliana* leaves. *Plant Growth Regulation*, **2011**, *65*(2), 315.

91. Mittal, S.; Kumari, N.; Sharma, V. Differential response of salt stress on *Brassica juncea*: photosynthetic performance, pigment, proline, D1 and antioxidant enzymes. *Plant Physiology and Biochemistry*, **2012**, *54*, 17–26.

92. Prasad, M.N.V.; de Oliveira Freitas, H.M. Feasible biotechnological and bioremediation strategies for serpentine soils and mine spoils. *Electronic Journal of Biotechnology*, **1999**, *2*, 7–8.

93. Bishnoi, N.R.; Sheoran, I.S.; SINGH, R. Influence of cadmium and nickel on photosynthesis and water relations in wheat leaves of different insertion level. *Photosynthetica (Praha)*, **1993**, *28*(3), 473–479.

94. Sheoran, I.S.; Singh, R., Effect of heavy metals on photosynthesis in higher plants, In: *Photosynthesis: Photoreactions to plant productivity*, Springer, Dordrecht, **1993**, 451–468.

95. Ouzounidou, G.; Čiamporová, M.; Moustakas, M.;Karataglis, S. Responses of maize (*Zea mays* L.) plants to copper stress—I. Growth, mineral content and ultrastructure of roots. *Environmental and Experimental Botany*, **1995**, *35*(2), 167–176.

96. Guo, H.; Hong, C.; Chen, X.; Xu, Y.; Liu, Y.; Jiang, D.; Zheng, B. Different growth and physiological responses to cadmium of the three *Miscanthus* species. *PLoS One*, **2016**, *11*(4), e0153475.

97. Parmar, P.; Kumari, N.; Sharma, V. Structural and functional alterations in photosynthetic apparatus of plants under cadmium stress. *Botanical Studies*, **2013**, *54*(1), 45.

98. Drazkiewicz, M. Chlorophyllase: occurrence, functions, mechanism of action, effects of external and internal factors. *Photosynthetica(Czech Republic)*, **1994**, *30*(3), 321–331.

99. Shahzad, B.; Tanveer, M.; Rehman, A.; Cheema, S.A.; Fahad, S.; Rehman, S.; Sharma, A.Nickel; whether toxic or essential for plants and environment-A review. *Plant Physiology and Biochemistry*, **2018**, *132*, 641–651.

100. Rashid, A.; Bernier, M.; Pazdernick, L.; Carpentier, R. Interaction of Zn^{2+} with the donor side of photosystem II. *Photosynthesis Research*, **1991**, *30*(2–3), 123–130.

101. Li, Z.; Xing, F.; Xing, D. Characterization of target site of aluminum phytotoxicity in photosynthetic electron transport by fluorescence techniques in tobacco leaves. *Plant and Cell Physiology*, **2012**, *53*(7), 1295–1309.

102. Appenroth, K.J.; Stöckel, J.; Srivastava, A.; Strasser, R.J. Multiple effects of chromate on the photosynthetic apparatus of Spirodelapolyrhiza as probed by OJIP chlorophyll a fluorescence measurements. *Environmental Pollution*, **2001**, *115*(1), 49–64.

103. Ahmed, A.; Tajmir-Riahi, H.A. Interaction of toxic metal ions Cd2+, Hg2+, and Pb2+ with light-harvesting proteins of chloroplast thylakoid membranes. An FTIR spectroscopic study. *Journal of Inorganic Biochemistry*, **1993**, *50*(4), 235–243.

104. Faller, P.; Kienzler, K; Krieger-Liszkay, A. Mechanism of Cd^{2+} toxicity: Cd^{2+} inhibits photoactivation of Photosystem II by competitive binding to the essential Ca^{2+} site. *Biochimica et Biophysica Acta (BBA)-Bioenergetics*, **2005**, *1706*(1–2), 158–164.

105. Pagliano, C.; Raviolo, M.; Dalla Vecchia, F.; Gabbrielli, R.; Gonnelli, C.; Rascio, N.; Barbato, R.; La Rocca, N. Evidence for PSII donor-side damage and photoinhibition induced by cadmium treatment on rice (*Oryza sativa* L.). *Journal of Photochemistry and Photobiology B: Biology*, **2006**, *84*(1), 70–78.

106. Yruela, I.; Alfonso, M.; Barón, M.; Picorel, R. Copper effect on the protein composition of photosystem II. *Physiologia Plantarum*, **2000**, *110*(4), 551–557.

107. Bernier, M.; Carpentier, R. The action of mercury on the binding of the extrinsic polypeptides associated with the water oxidizing complex of photosystem II. *FEBS Letters*, **1995**, *360*(3), 251–254.

108. Boisvert, S; Joly, D.; Leclerc, S.; Govindachary, S.; Harnois, J.; Carpentier, R. Inhibition of the oxygen-evolving complex of photosystem II and depletion of extrinsic polypeptides by nickel. *Biometals*, **2007**, *20*(6), 879–889.

109. Boucher, N.; Carpentier, R. Hg^{2+}, Cu^{2+}, and Pb^{2+}-induced changes in photosystem II photochemical yield and energy storage in isolated thylakoid membranes: A study using simultaneous fluorescence and photoacoustic measurements. *Photosynthesis Research*, **1999**, *59*(2–3), 167–174.

110. Miles, C.D.; Brandle, J.R.; Daniel, D.J.; Chu-Der, O.; Schnare, P.D.; Uhlik, D.J. Inhibition of photosystem II in isolated chloroplasts by lead. *Plant physiology*, **1972**, *49*(5), 820–825.

111. Radmer, R.; Kok, B. Kinetic observation of the System II electron acceptor pool isolated by mercuric ion. *Biochimica et Biophysica Acta (BBA)-Bioenergetics*, **1974**, *357*(2), 177–180.

112. Krupa, Z.; Siedlecka, A.; Mathis, P. Cd/Fe interaction and its effects on photosynthetic capacity of primary bean leaves. In: *Proceed. X Intern. Photosynthesis Congress, Montpellier, France*, **1995**, 20–25.

113. Angelov, M.; Tsonev, T.; Uzunova, A.; GAIDARJIEVA, K. Cu^{2+} effect upon photosynthesis, chloroplast structure, RNA and protein synthesis of pea plants. *Photosynthetica(Praha)*, **1993**, *28*(3), 341–350.

114. Houtz, R.L.; Nable, R.O.; Cheniae, G.M. Evidence for effects on the in vivo activity of ribulose-bisphosphate carboxylase/oxygenase during development of Mn toxicity in tobacco. *Plant Physiology*, **1988**, *86*(4), 1143–1149.

115. Malik, D., Sheoran, I.S.; Singh, R. Carbon metabolism in leaves of cadmium treated wheat seedlings. *Plant Physiology and Biochemistry (Paris)*. **1992**, *30*(2), 223–229.

116. Stiborová, M., Doubravová, M.; Leblová, S.A comparative study of the effect of heavy metal ions on ribulose-1, 5-bisphosphate carboxylase and phosphoenolpyruvate carboxylase. *Biochemie und Physiologie der Pflanzen*, **1986**, *181*(6), 373–379.

117. Van Assche, F.; Clijsters, H. Inhibition of photosynthesis in Phaseolus vulgaris by treatment with toxic concentration of zinc: effect on ribulose-1,5-bisphosphate carboxylase/oxygenase. *Journal of Plant Physiology*, **1986**, *125*(3–4), 355–360.

118. Sharma, P.; Dubey, R.S. Lead toxicity in plants. *Brazilian Journal of Plant Physiology*, **2005**, *17*(1), 35–52.

119. Sheoran, I.S.; Singal, H.R.; Singh, R. Effect of cadmium and nickel on photosynthesis and the enzymes of the photosynthetic carbon reduction cycle in *pigeonpea* (*Cajanuscajan* L.). *Photosynthesis Research*, **1990**, *23*(3), 345–351.

120. Myśliwa-Kurdziel, B.; Strzałka, K., Influence of metals on biosynthesis of photosynthetic pigments. In: *Physiology and biochemistry of metal toxicity and tolerance in plants*, Springer, Dordrecht, **2002**, 201–227.

121. Skrebsky, E.C.; Tabaldi, L.A.; Pereira, L.B.; Rauber, R.; Maldaner, J.; Cargnelutti, D.; Gonçalves, J.F.; Castro, G.Y.; Shetinger, M.R.; Nicoloso, F.T. Effect of cadmium on growth, micronutrient concentration, and δ-aminolevulinic acid dehydratase and acid phosphatase activities in plants of *Pfaffia glomerata*. *Brazilian Journal of Plant Physiology*, **2008**, *20*(4), 285–294.

122. Stobart, A.K.; Griffiths, W.T.; Ameen-Bukhari, I.; Sherwood, R.P. The effect of Cd^{2+} on the biosynthesis of chlorophyll in leaves of barley. *Physiologia Plantarum*, **1985**, *63*(3), 293–298.

123. Ying, R.R.;Qiu, R.L.; Tang, Y.T.; Hu, P.J.; Qiu, H.; Chen, H.R.; Shi, T.H.; Morel, J.L. Cadmium tolerance of carbon assimilation enzymes and chloroplast in Zn/Cd hyperaccumulator *Picris divaricata*. *Journal of Plant Physiology*, **2010**, *167*(2), 81–87.

124. Remans, T.; Opdenakker, K.; Smeets, K.; Mathijsen, D.; Vangronsveld, J.; Cuypers, A. Metal-specific and NADPH oxidase dependent changes in lipoxygenase and NADPH oxidase gene expression in *Arabidopsis thaliana* exposed to cadmium or excess copper. *Functional Plant Biology*, **2010**, *37*(6), 532–544.

125. Wang, F.; Chen, F.; Cai, Y.; Zhang, G.; Wu, F. Modulation of exogenous glutathione in ultrastructure and photosynthetic performance against Cd stress in the two barley genotypes differing in Cd tolerance. *Biological Trace Element Research*, **2011**, *144*(1–3), 1275–1288.

126. Garg, P.; Chandra, P.; Devi, S. Chromium (VI) induced morphological changes in *Limnanthemum cristatum* Griseb.: A possible bioindicator. *Phytomorphology*, **1994**, *44*(3–4), 201–206.

127. Appenroth, K.J.; Keresztes, A.; Sárvári, É.; Jaglarz, A.; Fischer, W. Multiple effects of chromate on *Spirodelapolyrhiza*: Electron microscopy and biochemical investigations. *Plant Biology*, **2003**, *5*(3), 315–323.

128. Bertrand, M.; Poirier, I. Photosynthetic organisms and excess of metals. *Photosynthetica*, **2005**, *43*(3), 345–353.

129. Berry, J.; Bjorkman, O. Photosynthetic response and adaptation to temperature in higher plants. *Annual Review of Plant Physiology*, **1980**, *31*(1), 491–543.

130. Tewari, A.K.; Tripathy, B.C. Temperature-stress-induced impairment of chlorophyll biosynthetic reactions in cucumber and wheat. *Plant Physiology*, **1998**, *117*(3), 851–858.

131. Wang, Q.L.; Chen, J.H.; He, N.Y.; Guo, F.Q. Metabolic reprogramming in chloroplasts under heat stress in plants. *International Journal of Molecular Sciences*, **2018**, *19*(3), 849.

132. Fristedt, R.; Willig, A.; Granath, P.; Crèvecoeur, M.; Rochaix, J.D.; Vener, A.V. Phosphorylation of photosystem II controls functional macroscopic folding of photosynthetic membranes in *Arabidopsis*. *The Plant Cell*, **2009**, *21*(12), 3950–3964.

133. Tang, Y.; Wen, X.; Lu, Q.; Yang, Z.; Cheng, Z.; Lu, C. Heat stress induces an aggregation of the light-harvesting complex of photosystem II in spinach plants. *Plant Physiology*, **2007**, *143*(2), 629–638.

134. Hu, S.; Ding, Y.; Zhu, C. Sensitivity and Responses of Chloroplasts to Heat Stress in Plants. *Frontiers in Plant Science*, **2020**, *11*, 375.

135. Mathur, S.; Agrawal, D.; Jajoo, A. Photosynthesis: response to high temperature stress. *Journal of Photochemistry and Photobiology B: Biology*, **2014**, *137*, 116–126.

136. Sharkey, T.D. Effects of moderate heat stress on photosynthesis: importance of thylakoid reactions, rubisco deactivation, reactive oxygen species, and thermotolerance provided by isoprene. *Plant, Cell & Environment*, **2005**, *28*(3), 269–277.

137. Allakhverdiev, S.I.; Kreslavski, V.D.; Klimov, V.V.; Los, D.A.; Carpentier, R.; Mohanty, P. Heat stress: An overview of molecular responses in photosynthesis. *Photosynthesis Research*, **2008**, *98*(1–3), 541.

138. Wang, L.J.; Fan, L.; Loescher, W.; Duan, W.; Liu, G.J.; Cheng, J.S.; Luo, H.B.; Li, S.H. Salicylic acid alleviates decreases in photosynthesis under heat stress and accelerates recovery in grapevine leaves. *BMC Plant Biology*, **2010**, *10*(1), 34.

139. Zhang, L.T.; Zhang, Z.S.; Gao, H.Y.; Xue, Z.C.; Yang, C.; Meng, X.L.; Meng, Q.W. Mitochondrial alternative oxidase pathway protects plants against photoinhibition by alleviating inhibition of the repair of photodamaged PSII through preventing formation of reactive oxygen species in Rumex K-1 leaves. *Physiologia Plantarum*, **2011**, *143*(4), 396–407.

140. Nash, D.; Miyao, M.; Murata, N. Heat inactivation of oxygen evolution in photosystem II particles and its acceleration by chloride depletion and exogenous manganese. *Biochimica et Biophysica Acta (BBA)-Bioenergetics*, **1985**, *807*(2), 127–133.

141. Rokka, A.; Aro, E.M.; Herrmann, R.G.; Andersson, B.; Vener, A.V. Dephosphorylation of photosystem II reaction center proteins in plant photosynthetic membranes as an immediate response to abrupt elevation of temperature. *Plant Physiology*, **2000**, *123*(4), 1525–1536.

142. Sage, R.F.; Zhu, X.G. Exploiting the engine of C4 photosynthesis. *Journal of Experimental Botany*, **2011**, *62*(9), 2989–3000.

143. Xu, W.; Zhou, Y.; Chollet, R. Identification and expression of a soybean nodule-enhanced PEP-carboxylase kinase gene (NE-PpcK) that shows striking up-/down-regulation in vivo. *The Plant Journal*, **2003**, *34*(4), 441–452.

144. Crafts-Brandner, S.J.; Salvucci, M.E. Rubisco activase constrains the photosynthetic potential of leaves at high temperature and CO_2. *Proceedings of the National Academy of Sciences*, **2000**, *97*(24), 13430–13435.

145. Crafts-Brandner, S.J; Salvucci, M.E. Sensitivity of photosynthesis in a C4 plant, maize, to heat stress. *Plant Physiology*, **2002**, *129*(4), 1773–1780.

146. Zheng, Y.; Li, F.; Hao, L.; Yu, J.; Guo, L.; Zhou, H.; Ma, C.; Zhang, X.; Xu, M. Elevated CO_2 concentration induces photosynthetic down-regulation with changes in leaf structure, non-structural carbohydrates and nitrogen content of soybean. *BMC Plant Biology*, **2019**, *19*(1), 255.

147. Drake, B.G.; Gonzàlez-Meler, M.A.; Long, S.P. More efficient plants: a consequence of rising atmospheric CO2? *Annual Review of Plant Biology*, **1997**, *48*(1), 609–639.

148. Pritchard, S.G.; Peterson, C.M.; Prior, S.A.; Rogers, H.H. Elevated atmospheric CO2 differentially affects needle chloroplast ultrastructure and phloem anatomy in Pinus palustris: interactions with soil resource availability. *Plant, Cell & Environment*, **1997**, *20*(4), 461–471.

149. Ranasinghe, S.; Taylor, G. Mechanism for increased leaf growth in elevated CO2. *Journal of Experimental Botany*, **1996**, *47*(3), 349–358.

150. Taylor, G.; Ranasinghe, S.; Bosac, C.; Gardner, S.D.L.; Ferris, R. Elevated CO_2 and plant growth: cellular mechanisms and responses of whole plants. *Journal of Experimental Botany*, **1994**, *45*, 1761–1774.

151. Ainsworth, E.A.; Davey, P.A.; Bernacchi, C.J.; Dermody, O.C.; Heaton, E.A.; Moore, D.J.; Morgan, P.B.; Naidu, S.L.; Yoo Ra, H.S.; Zhu, X.G.; Curtis, P.S. A meta-analysis of elevated (CO2) effects on soybean (*Glycine max*) physiology, growth and yield. *Global Change Biology*, **2002**, *8*(8), 695–709.

152. Lee, T.D.; Tjoelker, M.G.; Ellsworth, D.S.; Reich, P.B. Leaf gas exchange responses of 13 prairie grassland species to elevated CO_2 and increased nitrogen supply. *New Phytologist*, **2001**, *150*(2), 405–418.

153. Erice, G.; Irigoyen, J.J., Pérez, P.; Martínez-Carrasco, R.; Sánchez-Díaz, M. Effect of elevated CO_2, temperature and drought on photosynthesis of nodulated alfalfa during a cutting regrowth cycle. *Physiologia Plantarum*, **2006**, *126*(3), 458–468.

154. Wheeler, R.M.;Mackowiak, C.L.; Siegriest, L.M.; Sager, J.C. Supraoptimal carbon dioxide effects on growth of soybean (*Glycine max* (L.)Merr.). *Journal of Plant Physiology*, **1993**, *142*(2), 173–178.

155. Wilkins, D.; Van Oosten, J.J.; Besford, R.T. Effects of elevated CO_2 on growth and chloroplast proteins in *Prunus avium*. *Tree Physiology*, **1994**, *14*(7–8–9), 769–779.

156. Xu, Z.; Jiang, Y.; Zhou, G. Response and adaptation of photosynthesis, respiration, and antioxidant systems to elevated CO_2 with environmental stress in plants. *Frontiers in Plant Science*, **2015**, *6*, 701.

157. Pan, C.; Ahammed, G.J.; Li, X.; Shi, K. Elevated CO_2 improves photosynthesis under high temperature by attenuating the functional limitations to energy fluxes, electron transport and redox homeostasis in tomato leaves. *Frontiers in Plant Science*, **2018**, *9*, 1739.

158. Misra, A.N.; Latowski, D.; Strzalka, K. The xanthophyll cycle activity in kidney bean and cabbage leaves under salinity stress. *Russian Journal of Plant Physiology*, **2006**, *53*(1), 102–109.

159. Choi, H.I.; Hong, J.H.; Ha, J.O.; Kang, J.Y.; Kim, S.Y. ABFs, a family of ABA-responsive element binding factors. *Journal of Biological Chemistry*, **2000**, *275*(3), 1723–1730.

160. Ludwig, A.A.; Romeis, T.; Jones, J.D. CDPK-mediated signaling pathways: specificity and cross-talk. *Journal of Experimental Botany*, **2004**, *55*(395), 181–188.

161. Zhu, J.K. Cell signaling under salt, water and cold stresses. *Current Opinion in Plant Biology*, **2001**, *4*(5), 401–406.

162. Blumwald, E., Sodium transport and salt tolerance in plants. *Current Opinion in Cell Biology*, **2000**, *12*(4), 431–434.

163. Barua, D.; Downs, C.A.; Heckathorn, S.A. Variation in chloroplast small heat-shock protein function is a major determinant of variation in thermotolerance of photosynthetic electron transport among ecotypes of *Chenopodium album*. *Functional Plant Biology*, **2003**, *30*(10), 1071–1079.

164. Myouga, F.; Motohashi, R.; Kuromori, T.; Nagata, N.; Shinozaki, K. An *Arabidopsis* chloroplast-targeted Hsp101 homologue, APG6, has an essential role in chloroplast development as well as heat-stress response. *The Plant Journal*, **2006**, *48*(2), 249–260.

165. Heckathorn, S.A.; Downs, C.A.; Sharkey, T.D.; Coleman, J.S. The small, methionine-rich chloroplast heat-shock protein protects photosystem II electron transport during heat stress. *Plant Physiology*, **1998**, *116*(1), 439–444.

166. Heckathorn, S.A.; Ryan, S.L.; Baylis, J.A.; Wang, D.; Hamilton III, E.W.; Cundiff, L.; Luthe, D.S. In vivo evidence from an *Agrostis stolonifera* selection genotype that chloroplast small heat-shock proteins can protect photosystem II during heat stress. *Functional Plant Biology*, **2002**, *29*(8), 935–946.

167. Kang, L.; Kim, H.S.; Kwon, Y.S.; Ke, Q.; Ji, C.Y.; Park, S.C.; Lee, H.S.; Deng, X; Kwak, S.S. IbOr regulates photosynthesis under heat stress by stabilizing *IbPsbP* in sweet potato. *Frontiers in Plant Science*, **2017**, *8*, 989.

168. Close, T.J. Dehydrins: Emergence of a biochemical role of a family of plant dehydration proteins. *Physiologia Plantarum*, **1996**, *97*(4), 795–803.

169. Thalhammer, A; Hincha, D.K., The function and evolution of closely related COR/LEA (cold-regulated/late embryogenesis abundant) proteins in *Arabidopsis thaliana*. In: *Plant and microbe adaptations to cold in a changing world*, Springer, New York, NY, **2013**, 89–105.

170. Slama, I.; Abdelly, C.; Bouchereau, A.; Flowers, T.; Savoure, A. Diversity, distribution and roles of osmoprotective compounds accumulated in halophytes under abiotic stress. *Annals of Botany*, **2015**, *115*(3), 433–447.

171. Zhang, X.; Zhang, L.; Dong, F.; Gao, J.; Galbraith, D.W.; Song, C.P. Hydrogen peroxide is involved in abscisic acid-induced stomatal closure in *Vicia faba*. *Plant Physiology*, **2001**, *126*(4), 1438–1448.

172. He, M.; He, C.Q.; Ding, N.Z. Abiotic stresses: General defenses of land plants and chances for engineering multistress tolerance. *Frontiers in Plant Science*, **2018**, *9*, 1771.

173. Chaudhary, K.; Agarwal, S.; Khan, S. Role of phytochelatins (PCs), metallothioneins (MTs), and heavy metal ATPase (HMA) genes in heavy metal tolerance. In: *Mycoremediation and environmental sustainability*, Springer, Cham, **2018**, 39–60.

174. Sinha, A.K.; Jaggi, M.; Raghuram, B.; Tuteja, N. Mitogen-activated protein kinase signaling in plants under abiotic stress. *Plant Signaling & Behavior*, **2011**, *6*(2), 196–203.

175. Mittal, S.; Mallikarjuna, M.G.; Rao, A.R.; Jain, P.A.; Dash, P.K.; Thirunavukkarasu, N. Comparative analysis of cdpk family in maize, arabidopsis, rice, and sorghum revealed potential targets for drought tolerance improvement. *Frontiers in Chemistry*, **2017**, *5*, 115.

176. Vivek, P.J.; Tuteja, N.; Soniya, E.V. CDPK1 from ginger promotes salinity and drought stress tolerance without yield penalty by improving growth and photosynthesis in Nicotiana tabacum. *PLoS One*, **2013**, *8*(10), e76392.

177. Lata, C.; Prasad, M. Role of DREBs in regulation of abiotic stress responses in plants. *Journal of Experimental Botany*, **2011**, *62*(14), 4731–4748.

178. Liu, Q.; Kasuga, M.; Sakuma, Y.; Abe, H.; Miura, S.; Yamaguchi-Shinozaki, K.; Shinozaki, K. Two transcription factors, DREB1 and DREB2, with an EREBP/AP2 DNA binding domain separate two cellular signal transduction pathways in drought-and low-temperature-responsive gene expression, respectively, in *Arabidopsis*. *The Plant Cell*, **1998**, *10*(8), 1391–1406.

179. Oh, S.J., Song, S.I., Kim, Y.S., Jang, H.J., Kim, S.Y., Kim, M., Kim, Y.K., Nahm, B.H. and Kim, J.K., 2005. *Arabidopsis* CBF3/DREB1A and ABF3 in transgenic rice increased tolerance to abiotic stress without stunting growth. *Plant Physiology*, *138*(1), 341–351.

180. Behnam, B.; Kikuchi, A.; Celebi-Toprak, F.; Yamanaka, S.; Kasuga, M.; Yamaguchi-Shinozaki, K.; Watanabe, K.N. The *Arabidopsis* DREB1A gene driven by the stress-inducible

rd29A promoter increases salt-stress tolerance in proportion to its copy number in tetrasomic tetraploid potato (*Solanum tuberosum*). *Plant Biotechnology*, **2006**, *23*(2), 169–177.

181. Zhu, J.K.; Liu, J.; Xiong, L.; Genetic analysis of salt tolerance in *Arabidopsis*: evidence for a critical role of potassium nutrition. *The Plant Cell*, **1998**, *10*(7), 1181–1191.

182. Lu, C.; Vonshak, A. Effects of salinity stress on photosystem II function in cyanobacterial Spirulina platensis cells. *Physiologia Plantarum*, **2002**, *114*(3), 405–413.

183. Bramley, H.; Turner, D.W.; Tyerman, S.D.; Turner, N.C. Water flow in the roots of crop species: the influence of root structure, aquaporin activity, and waterlogging. *Advances in Agronomy*, **2007**, *96*, 133–196.

184. Sade, N.; Gallé, A.; Flexas, J.; Lerner, S.; Peleg, G.; Yaaran, A.; Moshelion, M. Differential tissue-specific expression of NtAQP1 in *Arabidopsis thaliana* reveals a role for this protein in stomatal and mesophyll conductance of CO_2 under standard and salt-stress conditions. *Planta*, **2014**, *239*(2), 357–366.

185. Kawase, M.; Hanba, Y.T.; Katsuhara, M. The photosynthetic response of tobacco plants overexpressing ice plant aquaporin McMIPB to a soil water deficit and high vapor pressure deficit. *Journal of Plant Research*, **2013**, *126*(4), 517–527.

186. Brini, F.; Masmoudi, K. Ion transporters and abiotic stress tolerance in plants. *ISRN Molecular Biology*, *2012*, **2012**. https://doi.org/10.5402/2012/927436

187. Rus, A.; Lee, B.H.; Muñoz-Mayor, A.; Sharkhuu, A.; Miura, K.; Zhu, J.K.; Bressan, R.A.; Hasegawa, P.M. AtHKT1 facilitates Na^+ homeostasis and K^+ nutrition in planta. *Plant Physiology*, **2004**, *136*(1), 2500–2511.

188. Huber, S.C.; Maury, W. Effects of magnesium on intact chloroplasts: I. Evidence for activation of (sodium) potassium/proton exchange across the chloroplast envelope. *Plant Physiology*, **1980**, *65*(2), 350–354.

189. Bose, J.; Munns, R.; Shabala, S.; Gilliham, M.; Pogson, B.; Tyerman, S.D., Chloroplast function and ion regulation in plants growing on saline soils: lessons from halophytes. *Journal of Experimental Botany*, **2017**, *68*(12), 3129–3143.

190. Hopkins, W. G.; Huner, N. P. A. (Eds.) *Introduction to Plant Physiology*. 4th Edition; John Wiley & Sons, Inc., USA, **2008**, 187, 192–193, 223.

191. Millar, A.H.; Whelan, J.; Soole, K.L.; Day, D.A. Organization and regulation of mitochondrial respiration in plants. *Annual Review of Plant Biology*, **2011**, *62*, 79–104.

192. Taiz, L.; Zeiger, E.; Moller, I.M.; Murphy, A.(Eds.). *Plant physiology and development*. 6th Edition; Sinauer Associates, Massachusetts. USA, **2015**; 342–343.

193. Geigenberger, P. Response of plant metabolism to too little oxygen. *Current Opinion in Plant Biology*, **2003**, *6*(3), 247–256.

194. Gupta, K.J.; Zabalza, A.; Van Dongen, J.T. Regulation of respiration when the oxygen availability changes. *Physiologia Plantarum*, **2009**, *137*(4), 383–391.

195. Nguyen, N.T.; McInturf, S.A.; Mendoza-Cózatl, D.G. Hydroponics: A versatile system to study nutrient allocation and plant responses to nutrient availability and exposure to toxic elements. *JoVE(Journal of Visualized Experiments)*, **2016**, (113), e54317.

196. Dusenge, M.E.; Duarte, A.G.; Way, D.A. Plant carbon metabolism and climate change: elevated CO_2 and temperature impacts on photosynthesis, photorespiration and respiration. *New Phytologist*, **2019**, *221*(1), 32–49.

197. Bunce, J.A.; Ziska, L.H. Responses of respiration to increases in carbon dioxide concentration and temperature in three soybean cultivars. *Annals of Botany*, **1996**, *77*(5), 507–514.

198. Gonzalez-Meler, M.A.; Taneva, L.I.N.A.; Trueman, R.J. Plant respiration and elevated atmospheric CO2 concentration: Cellular responses and global significance. *Annals of Botany*, **2004**, *94*(5), 647–656.

199. Bunce, J.A. A comparison of the effects of carbon dioxide concentration and temperature on respiration, translocation and nitrate reduction in darkened soybean leaves. *Annals of Botany*, **2004**, *93*(6), 665–669.

200. Ribas-Carbo, M.; Robinson, S.A.; Gonzalez-Meler, M.A.; Lennon, A.M.; Giles, L.; Siedow, J.N.; Berry, J.A. Effects of light on respiration and oxygen isotope fractionation in soybean cotyledons. *Plant, Cell & Environment*, **2000**, *23*(9), 983–989.

201. Shapiro, J.B.; Griffin, K.L.; Lewis, J.D.; Tissue, D.T. Response of Xanthium strumarium leaf respiration in the light to elevated CO_2 concentration, nitrogen availability and temperature. *New Phytologist*, **2004**, *162*(2), 377–386.

202. Ahmad Rashid, F.A.; Scafaro, A.P.; Asao, S.; Fenske, R.; Dewar, R.C.; Masle, J.; Taylor, N.L.; Atkin, O.K. Diel and temperature temperature-driven variation of leaf dark respiration rates and metabolite levels in rice. *New Phytologist*, **2020**. https://doi.org/10.1111/nph.16661

203. Frantz, J.M.; Cometti, N.N.; Bugbee, B. Night temperature has a minimal effect on respiration and growth in rapidly growing plants. *Annals of Botany*, **2004**, *94*(1), 155–166.

204. Peraudeau, S.; Lafarge, T.; Roques, S.; Quiñones, C.O.; Clement-Vidal, A.; Ouwerkerk, P.B.; Van Rie, J.; Fabre, D.; Jagadish, K.S.; Dingkuhn, M. Effect of carbohydrates and night temperature on night respiration in rice. *Journal of Experimental Botany*, **2015**, *66*(13), 3931–3944.

205. Fiorani, F.; Umbach, A.L.; Siedow, J.N. The alternative oxidase of plant mitochondria is involved in the acclimation of shoot growth at low temperature. A study of *Arabidopsis AOX1a* transgenic plants. *Plant Physiology*, **2005**, *139*(4), 1795–1805.

206. Bell, D.T.; Koeppe, D.E.; Miller, R.J. The effects of drought stress on respiration of isolated corn mitochondria. *Plant Physiology*, **1971**, *48*(4), 413–415.

207. Van Dongen, J.T.; Gupta, K.J.; Ramírez-Aguilar, S.J.; Araújo, W.L.; Nunes-Nesi, A.; Fernie, A.R. Regulation of respiration in plants: A role for alternative metabolic pathways. *Journal of Plant Physiology*, **2011**, *168*(12), 1434–1443.

208. Tiwari, B.S.; Belenghi, B.; Levine, A. Oxidative stress increased respiration and generation of reactive oxygen species, resulting in ATP depletion, opening of mitochondrial permeability transition, and programmed cell death. *Plant Physiology*, **2002**, *128*(4), 1271–1281.

209. Dumont, S.; Rivoal, J. Consequences of oxidative stress on plant glycolytic and respiratory metabolism. *Frontiers in Plant Science*, **2019**, *10*, 166.

210. Jones, R.; Ougham, H.; Thomas, H.; Waaland, S. (Eds.). *The Molecular Life of Plants.* 1st Edition; John Wiley & Sons, Ltd., UK, **2013**; 236, 564–565.

211. Sousa, C.A.F.D.; Sodek, L. The metabolic response of plants to oxygen deficiency. *Brazilian Journal of Plant Physiology*, **2002**, *14*(2), 83–94.

212. Florez-Sarasa, I.; Fernie, A.R.; Gupta, K.J. Does the alternative respiratory pathway offer protection against the adverse effects resulting from climate change? *Journal of Experimental Botany*, **2020**, *71*(2), 465–469.

213. Zhang, D.W.; Xu, F.E.I.; Zhang, Z.W.; Chen, Y.E.; Du, J.B.; Jia, S.D.; Yuan, S.; Lin, H.H. Effects of light on cyanide-resistant respiration and alternative oxidase function in *Arabidopsis* seedlings. *Plant, Cell & Environment*, **2010**, *33*(12), 2121–2131.

CHAPTER 6

Forward and Reverse Genetic Approaches for Improving Abiotic Stress Tolerance in Crop Plants

SHAKRA JAMIL[1†], RAHIL SHAHZAD[1†], SHAKEEL AHMAD[2†], AMINA NISAR[3], MUQADAS ALEEM[3,4], JAVARIA TABASSUM[2], MUHAMMAD MUNIR IQBAL[5], AAMAR SHEHZAD[6], ABDELHALIM I. GHAZY[7], AQIB ZEB[2], SHAMSA KANWAL[1], and MUHAMMAD AFZAL[7]

[1]Agricultural Biotechnology Research Institute, Ayub Agricultural Research Institute, Faisalabad, 38000, Punjab, Pakistan

[2]State Key Laboratory of Rice Biology, China National Rice Research Institute, Hangzhou, 310006, China

[3]Department of Plant Breeding and Genetics, University of Agriculture, Faisalabad, 38040, Punjab, Pakistan

[4]National Centre for Soybean Improvement, Key Laboratory of Biology and Genetics and Breeding for Soybean Ministry of Agriculture, State Key Laboratory of Crop Genetics and Germplasm Enhancement, Nanjing Agricultural University, Nanjing 210095, China.

[5]Centre for Plant Genetics and Breeding, The University of Western Australia, Perth, 6009, WA, Australia

[6]Maize Research Station, Ayub Agricultural Research Institute, Faisalabad, 38000, Punjab, Pakistan

[7]Plant Production Department, Food Science and Agricultural College, King Saud University, Riyadh, Saudi Arabia

*Corresponding author. E-mail: shakeelpbg@gmail.com

†These authors have contributed equally.

ABSTRACT

The climate change is adversely affecting the crop production and increasing the intensity of biotic and abiotic stresses. The abiotic stresses such as drought, heat, salinity, waterlogging, flooding, heavy metals, and lodging are hampering the crop productivity of all major crops of economic importance around the globe. The plants are sessile, hence cannot escape the stress by moving away from the stressed environment. Therefore, plants have to adjust their metabolic activities to respond to stress and survive under unfavorable climatic conditions. Wild plants have evolved different stress responsive mechanisms by switching on the stress responsive genes through signal transduction pathways. However, the cultivated plants have lost the stress tolerance mechanism during focused selection for improved yield. Now scientists are using different forward and reverse genetic approaches to equip the cultivated plant species with stress tolerance genes to enable them better survive in the changing climate. The current book chapter summarizes the existing knowledge about different forward and reverse genetic approaches and their use in plant breeding for development of abiotic stress tolerant crop varieties. Further, it highlights the role of different online crop improvement tools which are helping plant scientists in developing climate resilient crops through different crop improvement techniques.

6.1 INTRODUCTION

Climate change is putting pressure on agriculture through altered rain patterns, geographic shift in the global temperatures which have exposed the crops to face drought, heat, waterlogging, lodging, heavy metals stresses, and new insects and pests. Recent climate change trends and future predictions showed that the global mean temperature is gradually increasing and predicted to increase by 1.5–5 °C by 2100. Moreover, the extreme weather conditions have prevailed more frequently in last three decades. These abiotic stresses are hampering the crop productivity and sometime leads to total crop failure [1]. As crops are sessile hence, they cannot move away from the stressed area and have to face the stress by adapting to the surrounding environment and by changing their metabolic activities. To face these unforeseen circumstances, crops need to be tolerant to abiotic stresses to avoid yield losses.

Different forward and reverse genetic approaches possess the potential to equip crops with abiotic stress tolerance genes which will help crops in combating abiotic stress tolerance. Quantitative trait locus (QTL) mapping is helping the plant scientists in exploring the genetic loci which govern the quantitative traits, that is, yield and stress tolerance. The markers are designed from the flanking regions of these QTLs for their introgression into stress susceptible varieties from stress tolerant varieties [2]. Similarly, phenomics approaches are helping the breeders in precise phenotyping of the crops to identify the different plant traits/attributes which are affected by abiotic stresses [3]. Marker-assisted breeding (MAB) of crops is helping in easy stacking of multiple stress responsive genes and their confirmation at early seedling stage in genotypes to develop multiple stress tolerant crop varieties [4]. Similarly, the reverse genetic approaches, that is, mutagenesis, RNA silencing, genetically modified organisms, and genome editing are helping to develop stress tolerant varieties through targeted mutagenesis of the stress responsive genes, by changing them from susceptible (S genes) to tolerant (T genes) [5–7] as shown in Figure 6.1.

This book chapter summarizes the understanding about the role of different forward and reverse genetics approaches in the development of abiotic stress tolerant crop varieties. Further, we provide a future outlook about the use of different forward and reverse genetic approaches in crop improvement.

6.2 FORWARD GENETIC APPROACHES TO COMBAT ABIOTIC STRESSES

Plants face various abiotic stresses such as salinity, heat, cold, heavy metals, and waterlogging during their growth and development cycle. All these stresses result in the activation or repression of different genetic pathways. The forward genetic approaches, that is, microarrays, phenomics, QTL mapping, MAB, and genome sequencing have played a significant role in developing abiotic stress tolerant crops as described below.

6.2.1 ROLE OF MICROARRAYS APPROACH FOR ABIOTIC STRESS TOLERANCE

There is a dire need to explore the functions of stress responsive genes in order to enhance the stress tolerance in crops. The microarray-based expression profiling in different crops has led the foundation for characterization of the plant stress responsive genes. Microarrays are effective for

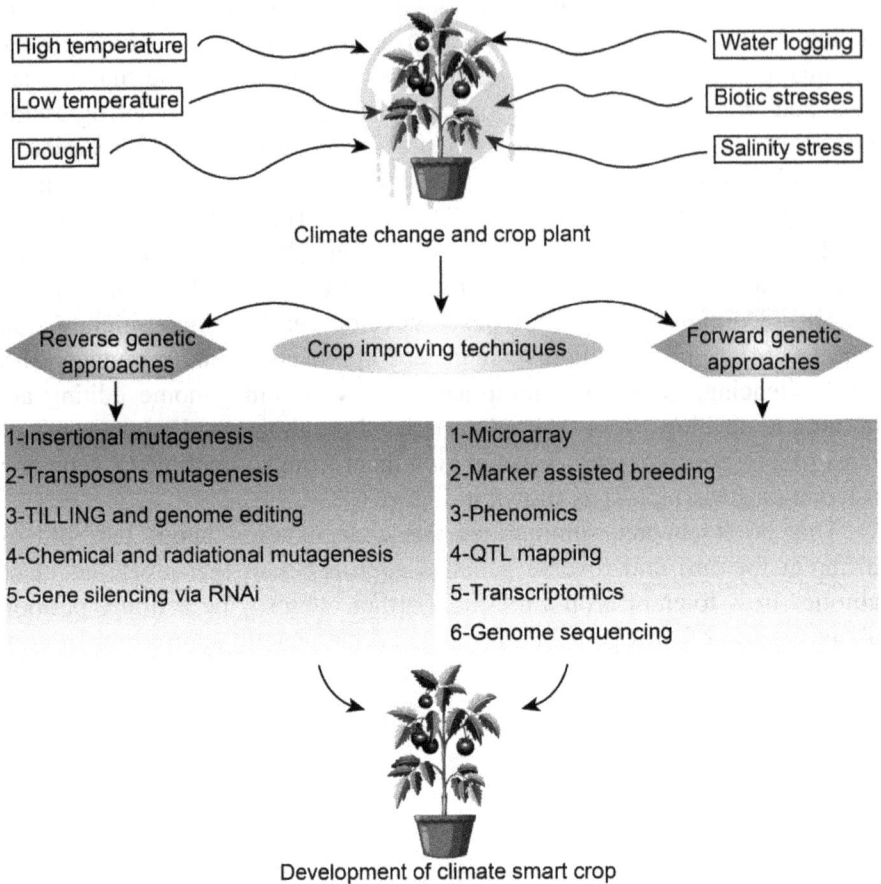

FIGURE 6.1 The flowchart for the development of climate resilient crops by the use of forward and reverse genetic approaches.

identification of stress responsive genes at the transcriptional level [8, 9]. Microarray technology not only identifies the stress responsive genes but is also effective for recognition of cis-acting elements and understanding of the molecular pathways during stress responses. Specific DNA microarrays store the genomic sequence data along with expression data for identification of downstream genes of stress-associated transcriptional factors (TFs) and cis-acting elements [10, 11]. This forward genetic approach is extensively used to identify the several knock-out mutants having stress sensitive and tolerant phenotypes. Microarrays are helpful for gene identification and understanding of their responsive mechanisms during abiotic stresses. These are worthwhile for the recognition of multiple stress stimulated genes and stress responsive TFs and transcriptome analysis during abiotic stress.

Several hundred genes which respond to different abiotic stresses, that is, drought, heat, cold, heavy metals stresses, salinity, and waterlogging have been identified through microarray analysis in different crops [12]. Similarly, microarray analysis was also helpful in the identification of different TFs, cis-regulatory elements in the promotor region of different genes, coregulatory genes, downstream, or regulatory genes [13]. The microarray approach was used for the identification of abiotic stress responsive genes in many crops, that is, *Arabidopsis thaliana* [14], rice [15], sorghum [16] cotton, potato [17], and maize [18].

6.2.2 ROLE OF PHENOMICS APPROACH FOR ABIOTIC STRESS TOLERANCE IN CROPS

The plants are grown in field or controlled conditions and observations are recorded on the phenotypic changes undergone by crop plant in order to assess the impact of different abiotic stresses. Even in the post genomic era, phenotypic observations are very necessary because of their crucial role in genome wide association mapping studies, QTL mapping, fine mapping of the genes, and genomic selection [19]. Phenomics approaches encompass infrared thermography (IRT), fluorescence imaging, spectroscopic techniques, and integrated imaging techniques. IRT is widely adaptive technology to measure the differences of temperature with respect to emission of infrared waves. It distinguishes the canopy, leaf, and air temperature under abiotic stresses in different plant species. Moreover, this technology was applied to estimate the osmotic level during salinity stress. Similarly, it also detects the transpiration rate and leaf temperature during drought stress [20, 21].

Under fluorescence imaging methods, the most precise, affordable, and widely used technique is chlorophyll fluorescence. In *Arabidopsis thaliana*, to observe different chemical reactions during cold and drought stress, researchers have linked the digital imaging (2D) of plant development with the chlorophyll fluorescence. Chlorophyll fluorescence has been used to observe the photosynthetic rate or processing in *Arabidopsis thaliana* during abiotic stresses [3, 22]. Spectroscopic technique is a powerful tool to investigate the rate of photosynthesis at canopy and leaves level. Photosynthesis-related parameters have been observed in *Beta vulgaris* and *Pinus* by employing this technology [23]. Similarly, integrated imaging technique include X-ray computed tomography and magnetic resonance imaging applications to analyze the anatomical organizations and morphological features under multiple abiotic and also biotic stresses [24].

The brief understanding of the interrelationship of different plant processes under stress condition demands the precise phenotyping of crops subjected to different stresses under control and stress conditions. Precise phenotyping of crops under stress conditions will help to elucidate the different stress responsive mechanisms of crops. Advanced phenotyping techniques make use of visible to near-infrared spectrum light sources to visualize the images for the development of plant phenotyping database in nondestructive manners [25]. The different imaging platforms, that is, visible light imaging, infrared imaging, fluorescence imaging, X-ray computed tomography, and hyperspectral imaging make use of robust software systems to generate unique multilevel phenotyping data. The precise phenotyping and imaging techniques which are based on measurement of interaction between light and plant surfaces, such as photons (reflected absorbed or transmitted) are playing key role in reaching the desired efficacy for accurate measurement of quantitative phenotypic traits. For in-depth discussion about the plant phenotyping and its significance readers may refer Ref. [19].

6.2.3 *ROLE OF QUANTITATIVE TRAIT LOCI (QTL) MAPPING FOR ABIOTIC STRESS TOLERANCE IN CROPS*

A QTL is a genetic locus which correlates with phenotypic variation observed for a particular quantitative trait in a population. QTL mapping is statistical procedure which makes use of both the genotypic and phenotypic data to assign a genetic locus to an observed phenotypic trait. QTL mapping plays a key role to enhance the understanding of genetic makeup of quantitative traits by using diverse mapping populations such as near isogenic lines, recombinant inbred lines, F2 generation, double haploids, and backcross plants [2].

Two prerequisites for QTL mapping are (1) two or more genotypes of genetically different organism and (2) molecular markers (restriction fragment length polymorphism, simple sequence repeat markers, and single nucleotide polymorphism [SNPs]) to distinguish between genotypes. The genotypes are crossed to develop heterozygous (F1) individuals and then these heterozygous individuals are crossed with each other following different cross-combinations to harness maximum genetic diversity [26]. Genetic loci or molecular markers which are linked to trait of interest will segregate more frequently with trait value than the unlinked markers. The major use of QTL analysis is to figure out whether the phenotypic differences are controlled by few loci having large effects or many loci having smaller effects. Generally phenotypic variations are governed by few loci with great effects [27]. Sample size also have a significant impact on the downstream results as low samples size fails to detect the QTLs effect [28].

This technique has recognized numerous abiotic stresses tolerant QTLs such as flooding, heat, low and high temperature, salt, and drought. In *Oryza sative*, QTL mapping has identified QTL 12.1 at 12th chromosome, which is effective for yield of rice grains during drought stress. Moreover, this locus also enhanced the plant height, yield, and biomass in stressed environment [29]. Similarly, in pearl millet, the linkage group 2 (LG2) have terminal drought stress tolerant QTL that has been recognized through QTL mapping. QTL mapping has been effective for enhancing the yield of pearl millet via marker-assisted selection (MAS) of traits of economic importance during the terminal drought stress [30].

Highly effective QTLs qSTIY5.1/qSSIY5.2 and qSTIPSS9.1 under heat stress in *Oryza sativa* were detected and used in MAS programs for crop advancement. In *Lycopersicum esculentum*, four cold stress responsive QTLs were identified. Those QTLs involved qCTS-12 on chromosome 12, Qcts-1 and qSPA-1 on chromosome 1, and qCTS-2 chromosome 2 [29, 31]. Submergence tolerant QTL, Sub1, was identified in indicia rice (FR13A cultivar) with the help of QTL mapping. This QTL was introduced into extensively grown Indian "sawaran" variety of rice, where it shown the effective stress tolerance [32].

6.2.4   ROLE OF MAB IN ABIOTIC STRESS TOLERANCE IN CROPS

Marker-assisted breeding is used to select plants with desirable attributes for farmers and consumers. It uses DNA markers for selection of desirable traits/plant for inclusion in a plant breeding program at the early stages of its development. MAB is the most precise, efficient, and widely used technique for effective incorporation of QTLs and genes to other genotypes. The different steps included in MAB are (1) selection of plants with desirable phenotypic traits, (2) breeding population development, (3) MAS of desirable plants, (4) marker validation in improved plants, (5) field trials of improved plants expressing desired traits, and (6) enhanced crop growth, yield, and stress tolerance [4] as demonstrated in Figure 6.2.

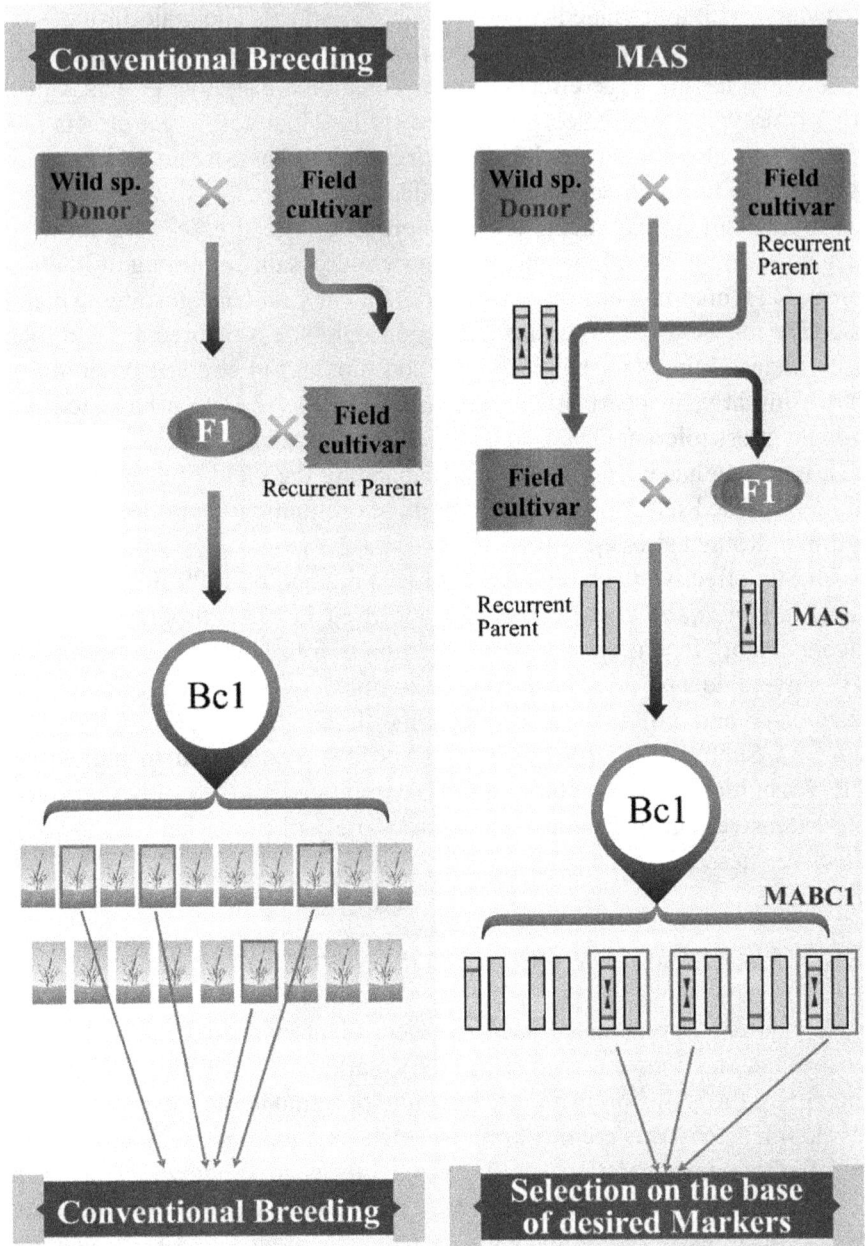

FIGURE 6.2   Strategy followed by conventional breeding and marker assisted breeding for abiotic stress tolerant breeding programs. The flowchart indicates that in F1, MAS favors the selection of desirable plants using gene linked DNA markers whereas in conventional breeding selection is accomplished phenotypically which is time consuming and unreliable.

First, this technique was applied in rice variety (MAS 946–1), where it proved fruitful for incorporation of drought tolerance. Similarly, this was effectively applied in rice to develop the varieties for submergence tolerance using *Sub1* gene, bacterial blast resistance using *OsXa5*, *OsXa13*, and *OsXa21* genes [33], rice quality improvement, and abiotic stress tolerance [34]. Moreover, Marker Assisted Back Crossing scheme has been adopted for transferring the salinity tolerant QTL, Saltol, of FL478 rice variety into extensively grown, well adapted and high yielding ASS996 rice variety [35]. Moreover, MAS was also applied in *Cicer arietinum* at International Crop Research Institute for Semi-Arid Tropics (ICRISAT), in order to introduce the QTLs for increased root size. A deeper root system has better efficacy for up taking the nutrients as well as soil's moisture during terminal drought stress situations [36].

Similarly, MAB was used for introgression of drought tolerance in wheat using 13 drought responsive QTLs derived from the cross between drought sensitive variety (Yecora Rojo) and drought tolerant variety (Pavon 76) [37]. MAB was used for heat stress tolerance [38], and salinity tolerance [39] in wheat. Efficacious implementations of this technique lead toward the pyramiding of both biotic and abiotic stress tolerant QTLs into other species for improving the crops' architecture. To combat the prevailing situation of changing climate, MAS along with "Omics" techniques will be a powerful tool [40].

6.2.5 ROLE OF GENOME SEQUENCING IN ABIOTIC STRESS TOLERANCE IN CROPS

During last decade, genome sequencing was revolutionized by the latest discoveries [41]. The journey of genome sequencing was started with the traditional sequencing techniques back in 1970 with Sanger based technique while later on it was advanced having the capability to sequence the whole plants' genome. Currently, modern genome sequencing techniques such as Next Generation Sequencing (NGS) are widely used. The NGS setup was revealed via 454 techniques, which were the most precise and fast genome sequencing techniques having the efficacy to sequence more than 20 million bases pairs within 4 hours. With the passage of time, the efficacy of sequencing technology was enhanced with sequencing ability for sequencing complete organism genome within hours [42].

Genomics studies provide highly reliable and the most appropriate evidence of crop's genetic architecture. The Expressed Sequence Tags (ESTs) have been utilized at large scale for gene detection. At present, National Centre for Biotechnological Information (NCBI) carries more than 1 million ESTs in its EST-related databases. Most of these sequenced tags are of major crops like rice, maize, wheat, and soybean and a number of ESTs of other crops are also present (http://www.ncbi.nlm.nih.gov/dbEST/). The sequencing of crop genomes has provided us an opportunity to compare the ESTs of stress tolerant and resistant genotypes to identify stress responsive locus/genes [43]. cDNA libraries from various tissues, developmental stages, and various treatment serve as the source for EST analysis to identify differentially expressed genes [44].

An alternate sequencing approach to EST sequencing is Serial Analysis of Gene expression (SAGE), which was developed to quantitate the abundance of thousands of transcripts. In SAGE approach, short sequence tags from transcripts are sequenced, giving an absolute measure of gene expression. The ability of SAGE to identify genes is dependent on the availability of the EST databases for the respective species. The use of SAGE remained limited in plant sciences however SuperSAGE and DeepSAGE have been explored heavily in plant species to identify and tag stress responsive genes and loci. The SAGE was used for the first time in rice plants for the identification of function of novel genes of rice seedlings genes. Likewise, another tag-linked approach is Massively Parallel Signature Sequencing (MPSS), which has the ability to analyze more than million transcripts instantaneously. However, MPS sequencing records of soybean, maize and rice are available at MPSS database (http://mpss.udel.edu/). Further improvements in genome sequencing will be significantly helpful in development of multiple abiotic stress responsive varieties [29].

6.3 REVERSE GENETIC APPROACHES TO COMBAT ABIOTIC STRESSES

Reverse genetics manipulates the genetic regions within the gene to understand the function and role of a gene. The reverse genetics techniques include mutagenesis (physical, chemical, and Targeting Induced Local Lesions in Genomes [TILLING]), genetically modified organisms, gene silencing, and genome editing [45]. The following section discusses the role of these reverse genetics approaches for development of abiotic stress tolerant crops.

6.3.1 ROLE OF MUTAGENESIS IN COMBATING ABIOTIC STRESSES

A mutagen is defined as a physical or chemical agent which changes the genetic architecture of a plant by targeting the hereditary material, usually DNA, of an organism and increase the mutation frequency of a population above the natural background mutation rate. Mutagenesis bears huge potential to help plant breeders to develop climate resilient crops under changing climate. Mutagenesis creates novel genetic variations which do not exist in the genetic makeup of a species and produce novel plants which are then screened against particular stress to identify potent stress tolerant plants. There are different types of mutagenesis, that is, insertional, chemical, physical radiations and transposons. Mutagenesis along with whole genome sequencing provide platform for numerous reverse and forward genetic techniques [46]. Following section discusses in brief about the role of different types of mutagenesis utilized in crop improvement for abiotic stress tolerance.

6.3.1.1 INSERTIONAL MUTAGENESIS (TRANSPOSONS MUTAGENESIS)

Trans-positioning or transposons mutagenesis is a biological process which provides platform for genes to be transferred to a host genome, modifying or interrupting the function of distant host genes causing mutation or disruption of function. Insertional mutagenesis with the help of transposons, retro-transposons, T-DNA, and activators/dissociation (Ac/Ds) plays the main role in gene identification. Insertional mutagenesis has an advantage over physical and chemical mutagenesis in recognizing the functions of specific genes [47]. Advancement of transformation methods in rice and existence of wide variety of transformation vectors are the main drivers for insertional mutagenesis as well as to study the rice's genetic architecture. In *Oryza sativa*, this approach was extensively used due to presence of efficient *Agrobacterium Tumefaciens*-Mediated Transformation system [48, 49]. The study of transposable elements in *Arabidopsis thaliana* has indicated their role in abiotic stress tolerance [50]. The transposable elements contribute in the activation of various stress responsive genes and play their role in development of stress tolerant crop varieties [51]. However, low-mutation rate and inefficiency to develop large mutant populations are some drawbacks of this type of mutagenesis [46].

6.3.1.2   RADIATION AND CHEMICAL MUTAGENESIS

Chemical and radiation mutagenesis are helpful tools for the development of climate smart varieties of different crops. The physical mutagens used for mutation breeding are X-rays, Gamma rays, alpha particles, ultraviolet radiations, and radioactive decays. Similarly, chemical mutagens used for mutation breeding are alkylating agent, deamination agents, polycyclic aromatic hydrocarbons, aromatic amines, alkaloid, bromine, sodium azide, and benzene [52].

The mutation breeding was used to develop abiotic stress-tolerant crop varieties in various crops. The procedure for development of abiotic stress tolerant varieties through mutation breeding includes (1) development of mutant population, (2) screening of mutant population against abiotic stresses, (3) identification of stress tolerant mutants, and (4) use of tolerant genotypes as parents in the breeding programs for the development of stress tolerant varieties. A number of tolerant varieties were developed against various abiotic stresses, that is, salt, drought, alkalinity, low and high temperature, and lodging resistance using mutation breeding (Table 6.1). More than 50% of these abiotic stress tolerant mutants are developed through mutations induced by gamma rays [52, 53].

6.3.1.3   TILLING (TARGETING INDUCED LOCAL LESIONS IN GENOMES)

TILLING is a reverse genetics approach of molecular biology which allows direct identification of mutant genes. The term was first coined and used in 2000, in model plant *Arabidopsis thaliana* [54]. TILLING has emerged as a nontransgenic reverse genetics approach which is applicable to all plants and animals which can be mutagenized, regardless of their genome size, ploidy level, and mating/pollinating system. The prerequisite for TILLING is prior DNA sequencing information which is used to detect the mismatch endonuclease to locate and detect induced mutations. As a result, an allelic series with altered DNA sequences is observed which contains missense, nonsense, splice, and silent mutations for examination of the effect of mutation in gene. TILLING is proving as a practical, efficient, and an effective tool for functional genomics studies in plant species. Previous studies have indicated the use of mutation breeding in rice [55], soybean [7], sorghum [56], barley [57], and maize [58] to develop abiotic stress resistance varieties. TILLING was used in legumes to identify the genetic regions which are responsible for salt stress response [59].

TABLE 6.1 List of Mutant Varieties Developed for Different Abiotic Stress Tolerance across the Globe

Stress	Crop	Variety/Country
Cold	Barely	**Austria;** Robin **Bulgaria;** Jubiley (Yubilei 100), Diana, IZ Bori **Greece;** Grammos **Russain Federation;** Novator, Accord, Radical, Shyrokolystnii, Dobrynia-3 **Turkey;** Akdeniz M-Q-54, **United States;** Pennrad
	Bermuda Grass	**United States;** Tifway II, Tift 94
	Broom Grass	**Russian Federation;** Fakel 89
	Crape Myrtle	**United States;** Centennial Spirit
	Crested Wheatgrass	**United States;** CD-II
	Eggplant	**Japan;** Daijiro
	Groundnut	**China;** Changhua 4,
	Iris	**Russian Federation;** Belyi Karlik, Chistoe Pole, Marina Raskova, Marshal Pokryshkin, Podmoskownaya Osen
	Lentil	**India;** Rajendra Masoor 1
	Maize	**China;** Longfuyu 1 (hybrid)
	Mandarin	**China;** Hongju 420
	Oat	**United States;** Bob, Ozark
	Rapeseed	**China;** Huyou 4, Xinyou 1, Xiuyou 1, Ganyou 5
	Raspberry	**Russian Federation;** Colocolchik
	Red Clover	**Belgium;** Rotra, R.v.P.
	Rice	**China;** Fu 709, Fuzhu, Hongnan, M 114, Wanfu 33, Xiangfudao, Fuxiang 1, M 112, Nongshi 4, Wanhongfu, Zhongmounuodao, Qiufu 1, Xindao 1, Qikesui, Xiushui 48, Ailiutiaohong, Zhefu 7, Changwanxian, Guiwanfu, Zhong 156, Fuxuan 8,
		Chuba; IACuba 22 India; Pusa-NR-162 **Japan;** Hanahikari, Houhai, Ibukiwase, Tsugaruotome, Hyogo-Kitanishiki, Hanabusa, Yume-akari, Fusa-no-mai, Oborozuki, Dewanosato **United States;** S-301, M-204 **Viet Nam;** DB-2

TABLE 6.1 *(Continued)*

Stress	Crop	Variety/Country
	Rye	**Finland;** Hankkija's Jussi (Hja 6900)
	Soybean	**Viet Nam;** S-31
	Wheat	**China;** Yuanfeng 1, Yuanfeng 2, Yuanfeng 3, 1161, Changwei 51503, Emai 9, Fuer, Jiaxuan 1, Zhonga 1, 503, Jimai 28, YF188, **Russian Federation;** Albidum 12, Omskaya ozimaya, Meshenskaya, Inma, Kazanskaya 84, Sibirskaya niva **United States;** Stadler
Drought	Amarant	**Russian Federation;** Sterke
	Barely	**Czech Republic;** Safir **Finland;** Balder J. **India;** RD-137 **Jordan;** Madaba-1 **Russian Federation;** Kharkovskii 84 **Ukarine;** Phenix **United Kingdom;** Goldmarker, Goldspear
	Chickpea	**Pakistan;** Thal-2006,
	Chinese Garlic	**China;** Ningsuan 1
	Cotton	**India;** MCU 7, M.A.9, MCU 10,
	Eggplant	**India;** PKM 1
	Grass Plavine	**Russian Federation;** Poltavskaya 2
	Grass Pea	**Moldova, Republic;** Bogdan
	Groundnut	**China;** Huayu 32
	Lentil	**Bulgaria;** Mutant 17 MM **Moldova, Republic;** Verzuie, Aurie
	Maize	**Bulgaria;** Kneja 682, Kneja 509, Kneja 570 **Hungary;** De 2205 SC
	Millet	**China;** Nunxuan 14, Fugu 6, Yugu 6, Nunxuan 12, Chigu 4, Longgu 28 **Russian Federation;** Cheget
	Moth Bean	**India;** RMO 40
	Mulberry	**China;** Ji 7681, Fuzaofeng
	Mung Bean	**India;** Co 4
	Pigeon Pea	**India;** Co 5

TABLE 6.1 *(Continued)*

Stress	Crop	Variety/Country
	Rice	**China;** 202 **Cuba;** IACuba 23, INCA LP-7, INCA LP-10 **India;** CNM 6 (Lakshmi), **Indonesia;** Danau atas **Philippines;** Rc346 (Sahod Ulan 11), Rc272 (Sahod Ulan 2) **Thailand;** RD-15 **United States;** Calmochi 201, Calmochi 202
	Sorghum	**Russian Federation;** Donetskaya 5
	Soybean	**Algeria;** Cerag No.1 **China;** Heinong 16, Heinong 6, Heinong 31, Heinong 32, Jiyuan 1, Wendou 79012, Liaodou 7, **Moldova, Republic;** Albisoara, Amelina, Clavera **Viet Nam;** DT96, DT2010, DT2012
	Sudan Grass	**Russian Federation;** Mironovskaya 8
	Sunflower	**India;** TAS-82
	Tomato	**Cuba;** MAGINE, Maybel, Domi
	Wheat	**Bulgaria;** Zlatostrui **China;** Jienmai 2, Taifu 1, Taifu 23, Zhengliufu, Chuanfu 1, Heichun 2, Qichun 1, Taifu 10, Taifu 15, Taifu 22, Wuchun 3, Yunfu 2, Longfumai 3, Xinchun 2, Jinmai 22, Xifu 4, H6765, Longfumai 17, Shaannong 138, Longfumai 19 **Kenya;** Njoro-BW1 **Pakistan;** Tatara **Ukraine;** Giant, Leana, Deada
Heat	Cotton	**Pakistan;** NIAB Karishma, NIAB 999, NIAB 111, NIAB-846, NIAB 777, NIAB-Kiran
	Creeping Bent Grass	**Japan;** Springs
	Lettuce	**Japan;** Evergreen, Giantgreen
	Rice	**China;** Zaoyeqing, Guifu 3, Shuangke 1 **Japan;** Norin PL 12, Nijihikari
	Tomato	**Mauritius;** Summer Star, Summer King
	Wheat	**China;** Emai 6
Lodging	Barely	**Austria;** Berta, Vienna, Carmen, Jutta, Amalia, **Bulgaria;** Krasi 2 **Canada;** Atlanta **Czech Republic;** Ametyst, Atlas, Diamant, Favorit, Hana, Rapid, Koral, Mars, Opal, Perun, Zenit **Estonia;** Liisa, Anni, Leelo **France;** Betina **Germany;** Comtesse, Acclaim, Nadja, Trumpf, Consista, Defra, Delita, Dera, Derkado, Dorina, Femina, Gerlinde, Grit, Ilka, Lada, Lenka, Maresi, Nebi, Salome, Spirit, Tamina, Defia, Arena, Beate, Cheri, Dorett, Korinna, Larissa, Jutta **India;** RDB-1, Karan-3, Karan-4, Karan-265 **Iraq;** Tuwaitha, Shua **Japan;** Fuji Nijo II **Korea Republic;** Radiation, **Russian Federation;** Secret, Debut, Fakel, Araraty 7,

TABLE 6.1 (Continued)

Stress	Crop	Variety/Country
		Kaskad, Minsk, Taeler, Maksim, Vavilon, Scorohod, VITIM, Bastion, **Slovakia;** Bonus, Fatran, Horal, Orbit, Novum, Profit, **Sweden;** Hellas, Kristina, Lina, Troja **Syrian Arab Republic;** Furat 3 **Ukraine;** Exotic, Badjory **United Kingdom;** Minak, Camargue, Corniche, Ayr, Beauly, Cromarty, Donan, Esk, Nairn, Tyne **United States;** Boyer, Luther, Advance, Hesk, Mal
	Buckwheat	**Russian Federation;** Chernoplodnaya,
	Chamomile	**Russian Federation;** Podmoskovnaya
	Chinese Matgrass	**Japan;** Toyomidori
	Cowpea	**India;** V240,
	Durum	**Austria;** Attila, Grandur, Probstdorfer Miradur, Unidur, **Bulgaria;** Lozen 76, Gergana, Impuls, Yavor, Vuzhod, Beloslava, Progress **Greece;** G-0367 **Italy;** Castel del Monte, Castelfusano, Castelnuovo, Castelporziano, Creso, Mida, Tito, Icaro, Febo, Giano, Peleo, Ulisse
	Faba Bean	**Germany;** Ti-Nova
	Fibre Flex	**Russian Federation;** M-5, Baltyuchai
	Flax/Linseed	**China;** Heiya 4, Ningya 10, Heiya 6, Luhua 7
	Job's Tears	**Japan;** Hato-yutaka
	Lentil	**Pakistan;** NIAB MASOOR-2006
	Lod	**Hungary;** Mutashali **Poland;** Agra
	Groundnut	**China;** Luhua 7
	Maize	**Russian Federation;** Krasnodarskii 303 VK, Collectivnii 225 MV (H) **China;** Jidan 101, Luyuan SC 4 **Viet Nam;** DT-8
	Marrow Pea	**Russian Federation;** Talovets 60
	Oat	**United States;** Belle
	Opium Poppy	**India;** BC-28/9/4 (Vivek)

TABLE 6.1 *(Continued)*

Stress	Crop	Variety/Country
	Pea	**Poland;** Hamil, Sum, Wasata, Milewska, Mihan, Ramir, Heiga, Jaran, Bosman, Legenda **Russian Federation;** Nemchynovskii 85, Streletchkii 11
	Rice	**Bangladesh;** Iratom 24, Binadhan 5, Binadhan 4, Binadhan-13, **Brazil;** SCS121 CL **Burkina Faso;** IRAT 144 **China;** Dongting 3, Fushe 31, Shuangchengnuo, Fubao 201, Guangfen 1, Suiwan 3, 7404, Hangfeng, Hu 2205, Yangfuxian 9850, Yangfujing4901, Zhe 101 **Cote D'Ivoire;** IRAT 13 **Guyana;** IRAT 194 (IREM 194), IRAT 256 (IREM 46–2), IRAT 257 (IREM 4113), IRAT 258 (IREM 4114) **Hungary;** Oryzella **India;** Jagannath (BSS-873), Prabhavati **Iran Islamic Republic;** Pooya, Tabesh **Iraq;** Amber-Baghdad, Amber-Manathera **Japan;** Hayahikari, Mineasahi, Miyanishiki, Musashikogane, Mutsukaori, Mutsukomachi, Niigatawase, Reimei, Iwate 21, Heiseimochi, Chuukan-bohon Nou-14, Kinuhikari, Koihime, Suzutakara, Yume-minori, Doman-naka, Hareyaka, Sakata-mezuru, Minami-yutaka, Yukimi-mochi, Hae-nuki, Hashiri-aji, Yume-tsukushi, Fukuhibiki, Tsukushi-wase, Tsugaru-roman, Sakitamahime, Yume-izumi, Aki-geshiki, Gin-ginga, Yume-no-kaori, Itadaki, Mine-hibiki, Milky Princess, Shun-you, Silky Pearl, Kusa-yutaka, Churahikari, Sai-no-kirabiyaka, Fuku-izumi, Akineiro, Yumeaoba, Bekoaoba, Kusayutaka, Hoshiaoba, Moretsu, Nishiaoba **Korea Republic;** Milyang 10, Wonpyongbyeo, Wonkwangbyeo, Wonmibyeo, Woncheongbyeo **Romania;** Oltenita **Russian Federation;** Madjan, Nucus 2, Mutant 428 **United States;** Calpearl, Calrose 76, M-101, M-301, M-302, M-401, M7, S 201, M-202, Calmochi-101, M-102, Mercury, M-203, S2-Calpearl, Valencia 87, S-102, A-201 **Viet Nam;** DB 250, DT-10, MT-4, MT-6, THDB, DT36, DT38
	Rye	**Finland;** HJA 6902 **Germany; Donar,** Pollux
	Sesame	**China;** Zhongzhi11 **Korea Republic;** Suwonkkae **Pakistan;** NIAB-Pearl
	Sorghum	**Mali;** Fambe, Sofin, Gnome
	Soybean	**Bulgaria;** Bisser, **China;** Tiefeng 18, Fengshou 11, Liaodou 3, Ludou 9, Liaodou 10, Beinong 103 **Indonesia;** Muri, **Japan;** Raiden, Raikou, Kosuzu, Tamaurara, Ryuhou, Suzukaori, Suzu-honoka **Mexico;** Hector, Esperanza **Russian Federation;** Universal I **Slovakia;** Aida **Viet Nam;** M-103
	Spring Barely	**Russian Federation;** BIOS-1, Moskovskii 2, Veras, Tuteishy, Perelom
	Sugarcane	**India;** Co 85035

TABLE 6.1 *(Continued)*

Stress	Crop	Variety/Country
	Tossa Jute	**Bangladesh;** Atompat-38
	Wheat	**China;** Jingfen 1, Luten 1, Nanjing 3, Yuanfeng 4, Yuyuan 1, 092, Henong 1, Wanyuan 75–6, 77 L15, Longfumai 2, Lumai 4, Lumai 5, Lumai 8, Qicheng 115, Weimai 6, Xinchun No.30 **Finland;** Hankkija's Taava **Germany;** Sirius, Els **Hungary;** Mv 8 **Italy;** Spinnaker **Japan;** Zenkoji Komugi (Zenkouzi-Komugi) **Mexico;** Centauro, Bajio Plus **Russian Federation;** Kiyanka, Novosibirskaya 67, Odesskaja Polukarlykovaja, Odesskaja 75, Polukarlykovaja-49, Deda, Kormovaya 30, Schedraja Polesja, Erytrospermum 103, Pitikul, Progress, Polukarlik 3, Yunnat odesskii, Birlik, Dnestryanka, Kharkovskaya 90, Nemchinovskaya 86, Mriya Khersona, Nemchinovskaya 52, Spartanka, Moskovskaya 70, Khersonskaya 86, Belchanka, Skifyanka, Moskovskaya nizkostebelnaya, Lutestsens 7 **United States;** Lewis, Payne
	White Jute	**India;** Hyb 'C' (Padma)
	White Lupin	**Russian Federation;** Gorizont, Sinii parus
Salinity	Chickpea	**India;** Kiran
	Common Bean	**Tunisia;** CIAT 899
	Cucumber	**Russian Federation;** Altaj
	Groundnut	**India;** TKG-19A, TG-22
	Mung Bean	**Indonesia;** Camar
	Rapeseed	**Bangladesh;** Binasarisha-6, Binasarisha-5 **China;** Xiangyou 11
	Rice	**China;** Fuxuan 1, Liaoyan 2, Yuanjing 11 **India;** Rasmi, Mohan (=CSR4), Lunisree **Indonesia;** Atomita 2, **Korea Republic;** Wonhaebyeo **Pakistan;** Shadab, Shua 92, NIAB-IRRI-9 **Viet Nam;** 6 B, A-20, DT17, VND 95–20, VND99–3
	Wheat	**China;** Changwei 19, Weifu 6757
Water-logging	Sesame	**Bangladesh;** Binatil-2, Binatil-4
	Groundnut	**China;** Yueyou 551–116

TABLE 6.1 *(Continued)*

Stress	Crop	Variety/Country
	Maize	**China**; Yuan 79–418 (inbred)
	Millet	**China**; Zhangnong 11
	Soybean	**China**; Heinong 26
Heavy Metal	Crown Vetch	**China**; Xifuxiaoguanhua
Multiple Stresses	Alfalfa	**China**; Xinmu 1
	Durum	**Bulgaria**; Sredetz
	Maize	**Bulgaria**; KNEJA-HP-556 (hybrid)
	Millet	**China**; Nunxuan 11
	Rice	**Cuba**; José LP-20
	Soybean	**Viet Nam**; DT99
	Wheat	**Bulgaria**; Fermer, Guinness/1322 China; Yuandong 3, Neimai 5

EcoTILLING, a variant of the TILLING, examines the natural genetic variation of a population for its utilization in crop improvement and discovery of SNPs [54, 60, 61]. EcoTILLING, like TILLING, can be implemented on polyploids species and it distinguishes the alleles from paralogous and homologous genes. Targeting of specific drought stress responsive transcription factors (TFs) in rice varieties has been achieved through EcoTILLING approach. Detection of stress responsive genes through the abovementioned techniques is dependent on available EST databases of different crops. However, functional markers are also helpful and efficient in the discovery of naturally existing variations in comparison to random DNA markers [62, 63].

6.3.2 ROLE OF GENETICALLY MODIFIED CROPS FOR ABIOTIC STRESS TOLERANCE IN CROPS

The adverse climatic conditions prevailing due to climate change demand the development of climate resilient crop varieties. One possible approach for the development of stress tolerant crop varieties is the development of genetically modified crop plants. The development of GMOs offer the best alternative approach where different stress responsive genes and TFs are reported in various crop wild relatives and nonplant species and they cannot be transferred to the crops through conventional breeding approaches and wild hybridization. Hence, nonconventional ways are used for their transfer to the cultivates species. Genetically modified organisms have shown better adaptability in the face of environmental stresses in comparison to ordinary plants [64, 65]. Wheat (*Triticum aestivum*) was improved for cold stress through transfer of *TaPIE1*. Moreover, the introduction of *AtMYB44*, *AtMYB60*, and *ATMYB61* in *Arabidopsis thaliana* was effective for drought tolerance [6]. *GmWRKY54*, *GmWRKY21*, and *GmWRKY76* of soybean (*Glycine max)* were introduced into *Arabidopsis thaliana* for resistance development in response to drought, low temperature, and salt stress, respectively [66].

ZmMYB30 enhanced salt stress tolerance in *A. thaliana* [67], *MdMYB1* enhanced drought, salinity, and chilling tolerance in apple and tobacco [68]. Correspondingly, *TaWRKY33* and *TaWRKY1* genes of wheat (*Triticum aestivum*) were highly effective in heat and drought stresses of genetically modified *A. thaliana* [69]. Likewise, over expression of specific maize *ZmWRKY33* gene is responsible for salt stress tolerance in *A. thaliana* [70]. *TaNHX2* showed enhanced slat tolerance in tobbaco plant by showing high antiporter activity in 200 mM NaCL stress [71]. *AlNHX1* showed enhanced salt tolerance in by accumulation of increased Na^+ deposition in roots and increase Na$^+$/K$^+$ ration in the shoots. All these examples highlights that huge potential exists in utilization of the genetically modified organisms for development of climate resilient and stress tolerant crops.

6.3.3 ROLE OF GENOME SILENCING THROUGH RNAI IN COMBATING ABIOTIC STRESSES

RNA interference is a biological process which makes use of the small RNA molecules to inhibit gene translation and expression through targeting of mRNA molecules. RNAi was known historically by other names, that is, posttranscriptional gene silencing, quelling, and co-suppression. RNA interference provides platform for the development of climate resilient crops. This approach has the ability to recognize and functionally characterize several stress responsive genes from diverse genomes [72].

MicroRNA (miRNA) plays significant role in enhancing drought tolerance of *Oryza sativa, Populus trichorpa*, and *A. thaliana*. The experimentation has proved that several microRNAs, that is, *miR158, miR169, miR156, miR396, miR394, miR171, miR168, miR165*, and *miR167* are drought responsive [73]. Similarly, rice drought stress response was induced by *miR393* and *miR169* genes [74]. In *Medicago truncatula*, downregulation of *miR169* was observed in roots under drought stress. The microRNA, that is, *miR398a, miR398b*, and *miR408* were upregulated in response to drought stress both in shoots and roots in various crops [75].

Similarly, the microRNAs, that is, *miR394, miR398, miR156, miR167, miR171*, and *miR393* are upregulated in response to salt stress in *Arabidopsis thaliana* [76]. Moreover, the role of microRNA was observed under chilling stress in various species, that is, *Populus, Brachypodium*, and *Arabidopsis thaliana* [76–78]. In the abovementioned species, miR172, miR397, and miR169 were upregulated in response to chilling [76]. These examples highlighted the role of microRNAs for stress tolerance in crops as was illustrated by Ding et al. [79].

6.3.4 ROLE OF GENOME EDITING IN COMBATING ABIOTIC STRESSES

The genome editing tools make use of guided nucleases to cut the DNA at a specific site followed by cellular DNA repair system [80]. Basically, a synthetic nuclease is designed for a specific sequence to recognize the target site and to create a double-strand DNA break (DSB). The DSB is repaired by nonhomologous end joining (NHEJ), causing mutation and knocking out the gene of interest. Presence of donor DNA fragment having homology with target site initiate homology-directed repair (HDR) leading to the integration of donor DNA template at the DSB site. NHEJ and HDR are integral component of all living cells; therefore, selection of nuclease having reprogrammable recognition site is a critical step in all genome editing experiments [81].

There are four types of nucleases, that is, meganucleases, transcription activator-like effector nucleases (TALENs), zinc finger nucleases (ZFNs), and RNA-guided nucleases (RGNs) from CRISPR/Cas system [82]. The simplest and most popular among all these nucleases is Cas9 nuclease, which uses Watson–Crick base pairing system for recognition of target DNA by single stranded guide RNA (sgRNA). The advanced versions of CRISPR such as Cpf1, Cas13, Base Editors, and Prime Editors are helping to address the biological questions in depth, along with the development of practical and innovative applications of biology [83, 84].

Base editing (BE) is a modified version of CRISPR/Cas9 tools for precise genome editing. The BE provides an additional benefit of irreversible base conversion at the target site. The BE apparatus comprises of a complex of nucleobase deaminase domain (having the ability of base conversion) and Cas protein, sgRNA. BE is a precise and simple tool for base conversion without the involvement of DSBs at target site [85]. Anzalone et al. [86] devised a genome editing method known as prime editing (PE). PE provides platform for base to base transitions, transversions, and introduction of indels without causing DSBs. For PE, Cas9 endonucleases are guided by Prime editing guide RNA (pegRNA). The pegRNA contains spacer (which is complementary to one DNA strand) and primer binding site that is specific for target gene. The PBS region creates a reverse transcriptase that is linked to the Cas9 (H840A) nickase [86].

Abiotic stress tolerance can be incorporated in crops by modification of structural genes, regulatory genes, and cis-regulatory elements. Structural genes have important regulatory role to play in the development of any crop plant and also serve as targets for stress tolerance. Several genes encoding antioxidant enzymes, that is, superoxide dismutase (SOD), glutathione-*S*-transferases (GST), catalases (CAT), peroxidases (POD), and many glutathione reductases (GR) play crucial role of scavenging of reactive oxygen species (ROS) molecules produced during abiotic stresses. The genes which work to scavenge the ROS are termed as Tolerance genes (T genes). There is another class of genes termed as "Sensitivity genes" (S genes) which favors the enhanced programmed cell death, reduced antioxidant activity, and excessive ROS production upon exposure to different abiotic stresses [87].

Different regulatory genes, that is, phosphatases, TFs, and kinases are another class of genes which can be modulated to regulate the expression of downstream-located genes to initiate stress response. For example, *ANAC069* (a NAC TFs) regulates the expression of several S genes during the onset of abiotic stresses by binding to the core motif sequence of C[A/G]CG[T/G] in the promotor region of various S genes and render plants susceptible to various abiotic stresses. On contrary, overexpression of *AtMYB44* (T gene) confers salt and drought tolerance via ABA-induced dependent pathway through closure of stomata [6]. Similarly, overexpression of *ZmWRKY106* confers heat and drought tolerance by reducing the ROS production, increased activities of antioxidant enzymes, and regulation of the stress-related genes [88].

The role of Cis-regulatory elements in abiotic stress tolerance is inevitable as these have key role in regulating the expression of genes by facilitating the attachment of stress responsive TFs [89]. Several cis regulatory elements, that is, GCC box (AGCCGCC) and W-box (TTGACC) counter abiotic stresses by favoring the attachment of particular TFs. Potential exists for targeting of the Cis-regulatory elements to modify them using the advance genome editing tools for BE and providing the altered attachment site for TFs to regulate the abiotic stresses. This has been accomplished in model organism *Arabidopsis thaliana* by modulating the attachment site for *ANAC069* which inhibits expression of many T genes, for example, Pyrroline-5-carboxylate synthase, P5CS GST, POD, and SOD. Remarkably, CRISPR-Cas9 drove cis-categorizations mutagenesis in the promoters' region of many genes which has generated a variety of phenotypic and genotypic mutations that are responsible for the creation of new QTLs and boosted the tomato (*Lycopersicum esculentum*) yield [90].

6.4 ONLINE RESOURCES FOR CROP IMPROVEMENT

The technological advancements have brought us to the era of information and technology. The role of information technology and online tools in crop improvement have become inevitable. Many online tools and databases have been built through the use of information technology which stores large biological data sets which are efficiently being utilized in different crop improvement programs. The most highlighted examples of online database which is extensively used in genetic studies are National Centre for Biotechnology Information (NCBI) database. The NCBI stores the genomics, transcriptomics, and proteomics database of different crops whose genomes are sequenced. This database is used for comparative genetic studies, gene ontology studies, and functional genomics studies. Similarly, Phytozome is another highlighted example of plant database. It is different from NCBI in a sense that it stores the genomics, proteomics, transcriptomics, and metabolomics information of plants only while NCBI contains the information for both plants and animals [91].

There are certain other online portfolios which are used in crop improvement, that is, Plant Transcriptional Factors Data Base (Plant TFDB) which stores the information about different groups of TFs gene families reported in plants and provides a platform for genome wide expression studies of different TFs gene families in various stresses and different plant processes [92]. There other widely used online genomics tools which help in crop improvement include MEME suit, EMBL-EBI, ISAAA, E-CRISP, E-RNAi, and many others as described in detail in Table 6.2.

6.5 CONCLUSION AND FUTURE PROSPECTS

The abiotic stresses, that is, drought, heat, waterlogging, chilling, flooding, lodging, and heavy metals are hampering the crop productivity worldwide. The changing climate is further amplifying these abiotic stresses with elevated temperatures, unpredicted rainfalls, glaciers melting at rapid pace, and unforeseen climatic threats like emerging of new crop pests and diseases [70]. In this book chapter, we have summarized how different forward and reverse genetics approaches are helping the plant scientists/breeders in developing the abiotic stress tolerant crop varieties to combat the climate change.

Thanks to CRISPR technology which has helped in the modification of plant genome in a very precise way as compared to random mutagenesis using physical and chemical mutagenesis. It has helped in gene annotations and transcriptome profiling of the different genes. The multiplex genome editing has equipped the 21st century scientists with a tool to target and mutate multiple genes governing a metabolic pathway in a single attempt to explore the role of each gene in that pathway [93]. The latest developments in genome editing, that is, BE [94] and PE [95] have opened the new horizons for plant breeders and geneticists and provided them freedom to manipulate single base pair efficiently without disturbing the rest of the genome. This allows repairing the genes with single base insertion/deletion/and substitution to repair the faulty S genes and converting them to T genes [96].

The future of crop improvement for abiotic stress tolerance lies in the advancement of genome sequencing techniques (providing platform for

TABLE 6.2 The List of Online Crop Improvement Tools which Helps in Genomics, Proteomics, Transcriptomics, Metabolomics, and Phenomics Studies

Sr. No.	Name of Tool	Weblink	Description	PD/AD	Keywords
1.	Illumina	https://www.illumina.com	Sequencing Platform	Both	Genome, Metagenome sequencing, associated metadata
2.	Solid	http://www.appliedbiosystems.com/technologies	Next Generation DNA sequencing Platform	Both	Genome, Metagenome sequencing, associated metadata
3.	PacBio RS	http://www.pacificbiosciences.com	Sequencing of Genome, Transcriptome and epigenome	Both	Genome, Metagenome sequencing, associated metadata
4.	Helicos	http://www.helicosbio.com	Sequencing Platform	Both	Genome, Metagenome sequencing, associated metadata
5.	EMBL-EBI	http://www.ebi.ac.uk/embl	Genomic Database	Both	Genome, Metagenome sequencing, associated metadata
6.	Genbank	http://www.ncbi.nlm.nih.gov/genbank/	General public sequence repository	Both	Genome, Metagenome sequencing, associated metadata
7.	EMBL	http://www.ebi.ac.uk/embl/	General public sequence repository	Both	Genome, Metagenome sequencing, associated metadata
8.	DDBJ	http://www.ddbj.nig.ac.jp	General public sequence repository	Both	Genome, Metagenome sequencing, associated metadata
9.	UniProt	http://www.uniprot.org/	Protein sequences and functional information	Both	Genome, Metagenome sequencing, associated metadata
10.	NCBI	http://www.ncbi.nlm.nih.gov/	Biomedical and genomical information	Both	Genome, Metagenome sequencing, associated metadata
11.	Gene Index Project	http://compbio.dfci.harvard.edu/tgi/	Transcriptome repository	Both	Genome, Metagenome sequencing, associated metadata

TABLE 6.2 *(Continued)*

Sr. No.	Name of Tool	Weblink	Description	PD/AD	Keywords
12.	GOLD	http://genomesonline.org/cgi-bin/GOLD/bin/gold.cgi	Repository of genomes databases	Both	Genome, Metagenome sequencing, associated metadata
13.	Phytozome	http://www.phytozome.net/	Genomic plant database	PD	Gene sequence, gene structure, gene family
14.	MaizeGDB	http://www.maizegdb.org/	Maize information resource	PD	Maize, Genetics, Genomics
15.	Tair	http://www.arabidopsis.org/	*Arabidopsis* information resource	PD	Genetic, molecular biology, *Arabidopsis* thaliana
16.	CPGR	http://cpgr.plantbiology.msu.edu/	Phytopathogen genomic resource	PD	Suit of Genomics, Proteomics platforms
17.	MASWheat	https://maswheat.ucdavis.edu	Wheat Genomic information for MAS	PD	Marker Assisted Selection, Wheat, Markers Data
18.	PlantTFDB	http://planttfdb.gao-lab.org	Transcriptional Factors Database of Plants	PD	Portal, functional, evolutionary study of TFs
19.	ISAAA	http://www.isaaa.org/default.asp	GM Approval Database	PD	GM Crops, Events, Traits
20.	CottonGen	https://www.cottongen.org	Cotton Database Resources	PD	Genomics, Genetics, Breeding, Cotton
21.	GrainGenes	https://wheat.pw.usda.gov/GG3	A Database for Triticeae and Avena	PD	Central data repository, Wheat, Barely, Oat
22.	Gramene	https://www.gramene.org	Integrated data resource for comparative functional genomics in crops and model plant species	PD	Browse genomes, annotations, variation, comparative tools
23.	Sol Genomics Network	https://solgenomics.net	Genomics Database of Solanaceae family	PD	Web Portal, Genomics Data, Phenotypic Data, Analysis tools

TABLE 6.2 *(Continued)*

Sr. No.	Name of Tool	Weblink	Description	PD/AD	Keywords
24.	Spud DB	http://solanaceae.plantbiology.msu.edu/	Genomics Database for Potato Crop	PD	Annotation, Genome, Transcriptome
25.	Integrated Breeding Platform	https://www.integratedbreeding.net	Helps manage your breeding data across all phases of the crop improvement cycle, from program planning to decision making	PD	Genotyping and Phenotyping Data Management, Plant Breeding,
26.	AraPheno	https://arapheno.1001genomes.org	Phenomics Database	PD	Collect, Store, Analyze data from images of plants
27.	Plant Genomics & Phenomics Research Data Repository	https://edal-pgp.ipk-gatersleben.de	Phenomics Database	PD	Collect, Store, Analyze data from images of plants
28.	Phenome Networks	https://phenome-networks.com	Phenomics Database	PD	Collect, Store, Analyze data from images of plants
29.	Phenopsis DB	http://bioweb.supagro.inra.fr/phenopsis	*Arabidopsis* Thaliana Phenotyping Database	PD	Collect, Store, Analyze data from images of plants
30.	PHIDIAS	https://cordis.europa.eu/article/id/150886-global-phenotype-database-for-plant-scientists	Global phenotype database for plant scientists	PD	Collect, Store, Analyze data from images of plants
31.	Plant phenotyping Datasets	https://www.plant-phenotyping.org/datasets-home	Global phenotype database for plant scientists	PD	Scan plants, Analyze Data, Predicts yield
32.	International Plant Phenotyping Network	https://www.plant-phenotyping.org	IPPN aims to provide all relevant information about plant phenotyping	PD	Scan plants, Analyze Data, Predicts yield

TABLE 6.2 (*Continued*)

Sr. No.	Name of Tool	Weblink	Description	PD/AD	Keywords
33.	TERRA Phenotyping Reference Platform	https://terraref.org	Global phenotype database for plant scientists	PD	Remote sensing, Quantifying of traits, Prediction of yield
34.	DRYAD	https://datadryad.org/stash	Contains phenotype database for plant scientists	PD	Plant genotyping, Plant phenotyping, Research and Education Data
35.	The Triticeae Tool Box	https://triticeaetoolbox.org	Contains phenotype database for plant scientists	PD	Wheat, Oat, Barely
36.	Swiss-Prot	http://www.expasy.ch	Protein Data Base	Both	Bioinformatics, Databases, Repositories
37.	TrEMBL	http://www.expasy.ch	Protein Data Base	Both	Bioinformatics, Databases, Repositories
38.	PIR	http://pir.georgetown.edu/pirwww/pirhome.html	Protein Data Base	Both	Bioinformatics, Databases, Repositories
39.	OWL	http://www.leeds.ac.uk/bmb/owl/owl.html	Protein Data Base	Both	Bioinformatics, Databases, Repositories
40.	Washington University	http://genome.wustl.edu/est/esthmpg.html	EST Database	Both	Bioinformatics, Databases, Repositories
41.	PDB	http://www.rcsb.org/pdb	Protein Data Base	Both	Bioinformatics, Databases, Repositories
42.	ProteomicsDB	https://www.proteomicsdb.org	Protein Data Base	Both	Bioinformatics, Databases, Repositories
43.	AgBase	https://www.hsls.pitt.edu/obrc/index.php?page=URL1174595451	Search and analyze functional genomics datasets in agricultural species.	Both	Plants, Animals, Microbes

TABLE 6.2 *(Continued)*

Sr. No.	Name of Tool	Weblink	Description	PD/AD	Keywords
44.	General plant databases	https://www.hsls.pitt.edu/obrc/index.php?page=general	A hub of different Plant Databases	Both	AgBase, AutoSNPdb, BarleyBase
45.	MEME Suite	http://meme-suite.org	Motif Based Sequence Analysis	Both	Motif analysis, nucleotide, protein
46.	BLAST	https://blast.ncbi.nlm.nih.gov/Blast.cgi	To search homologous sequences	Both	Identity, Homology, DNA, RNA, Protein
47.	Mutant variety DB	https://www.iaea.org/resources/databases/mutant-varieties-database	Holds information on induced mutations suitable for breeding programme and genetic analysis	PD	Mutagens used, Characters Improved, varieties developed
48.	Transcriptome	https://molbiol-tools.ca/Transcriptome.htm	Contains tools for normalization, preprocessing, viewing, clustering, differential expression, supervised classification, and data mining	Both	GEPAS, Cyber-T, GEMS
49.	GMOMETHODS	https://gmo-crl.jrc.ec.europa.eu/gmomethods/entry?db=gmometh&id=QT-EVE-GM-012	GMO Detection Methods Database	Both	CRMs, Nucleic acid and protein based testing
50.	GMDD	http://gmdd.sjtu.edu.cn	GMO Detection Methods Database	Both	CRMs, Nucleic acid and protein based testing
51.	CRISPR-Genome Editing	http://skl.scau.edu.cn/	A toolkit for CRISPR-based genome editing	Both	Primer design, Target design, Off target finder, Genome editing

TABLE 6.2 *(Continued)*

Sr. No.	Name of Tool	Weblink	Description	PD/AD	Keywords
52.	CRISPR-P	http://crispr.hzau.edu.cn/CRISPR2/	A web tool for synthetic single-guide RNA design of CRISPR-system in plants	PD	sgRNA design, CRISPR-system, Genome editing
53.	E-CRISP	http://www.e-crisp.org/E-CRISP/	A tool for designing CRISPR constructs	Both	Gene editing, homology peak detection, sgRNA design
54.	E-TALEN	www.e-talen.org/E-TALEN	To design TALENs for introducing knock-out mutations, for endogenous tagging and targeted excision repair.	Both	TALENs, Gene editing, De-novo
55.	E-RNAi	www.dkfz.de/signaling/e-rnai3	E-RNAi is a tool for the design and evaluation of RNAi reagents for a variety of species. It can be used to design and evaluate long dsRNAs (including esiRNAs) as well as siRNAs.	AD	RNAi construct, dsRNAs, esiRNAs, siRNAs.
56.	GenomeRNAi	www.genomernai.org/GenomeRNAi	GenomeRNAi is a database containing phenotypes from RNA interference (RNAi) screens in Drosophila and Homo sapiens. In addition, the database provides an updated resource of RNAi reagents and their predicted quality.	AD	RNAi reagents, RNAi, Gene silencing
57.	CRISPR Library Designer	https://github.com/boutroslab/cld	a software for the multispecies design of sgRNA libraries	-	sgRNA, CRISPR, Genome editing

TABLE 6.2 *(Continued)*

Sr. No.	Name of Tool	Weblink	Description	PD/AD	Keywords
58.	CHOPCHOP	https://chopchop.cbu.uib.no/	A web tool for selecting target sites for CRISPR/Cas9, CRISPR/Cpf1, CRISPR/Cas13 or NICKASE/TALEN-directed mutagenesis.	Both	CRISPR system, Genome editing, Nikase, Mutagenesis
59.	CRISPy-web	https://crispy.secondarymetabolites.org	Designing of gRNA for CRISPR	Both	sgRNAs, CRISPR, Genome Editing
60.	IGTRCN	https://igtrcn.org/knowledgebase/crisprcas-web-resources	Web Resources for design and planning of CRISPR/Cas9 studies	Both	sgRNAs, CRISPR, Genome Editing
61.	CRISPR-PLANT	https://www.genome.arizona.edu/crispr	A portal for CRISPR/Cas based genome editing	PD	sgRNAs, CRISPR, Genome Editing
62.	Benchling	https://www.benchling.com	Web Resources for design and planning of CRISPR/Cas9 studies	Both	sgRNAs, CRISPR, Genome Editing

cost-effective genome sequencing), Phenomics (for the identification of plant traits which are most affected by any stress) and genome editing with CRISPR/Cas system. Although genetically modified organisms, RNAi silencing and TILLING have the potential for crop improvement but there are certain limitations associated with each technique. On the other hand, modern gene editing techniques and their products face certain regulatory road locks which hamper their utilization in crop improvement especially in the European Union (EU). If the products of genome editing obtain the regulatory status as non-GMOs from EU, it would be a great development and would lead to the cultivation of genome editing crops at a vast area as compared to their present cultivation which is confined to USA, China, Canada, and some countries of the EU [97].

Recently, the European Sustainable Agriculture through Genome Editing (EU-SAGE) network and its members from 132 European research institutes and associations urged the European Council, European Parliament, and the European Commission to reconsider their stance on genome editing, which is one of the tools needed to achieve the Sustainable Development Goals. In an open statement, the EU-SAGE network said that developing new crop varieties needs tools that are safe, easy, and fast, and the latest addition to these tools is precision breeding or genome editing. If EU reconsider its definition of genome edited crops, it would be a great development and will increase the impact of this technology in crop improvement [98–103].

KEYWORDS

- **crop improvement**
- **forward genetics**
- **reverse genetics**
- **stress resistant**
- **climate change**

REFERENCES

1. Lane, A.; Jarvis, A., Changes in climate will modify the geography of crop suitability: agricultural biodiversity can help with adaptation. *SAT eJournal* **2007**, 4, (1), 1–12.

2. Ahmad, H.; Azeem, F.; Tahir, N.; Iqbal, M., QTL mapping for crop improvement against abiotic stresses in cereals. *JAPS: Journal of Animal & Plant Sciences* **2018**, 28, (6), 1558–1573.

3. Rungrat, T.; Awlia, M.; Brown, T.; Cheng, R.; Sirault, X.; Fajkus, J.; Trtilek, M.; Furbank, B.; Badger, M.; Tester, M., Using phenomic analysis of photosynthetic function for abiotic stress response gene discovery. *The Arabidopsis Book/American Society of Plant Biologists* **2016**, 14.

4. Wani, S. H.; Choudhary, M.; Kumar, P.; Akram, N. A.; Surekha, C.; Ahmad, P.; Gosal, S. S., Marker-assisted breeding for abiotic stress tolerance in crop plants. In *Biotechnologies of Crop Improvement*, Volume 3, Springer: 2018; pp 1–23.

5. Sikora, P.; Chawade, A.; Larsson, M.; Olsson, J.; Olsson, O., Mutagenesis as a tool in plant genetics, functional genomics, and breeding. *International Journal of Plant Genomics* **2011**, 2011, Article ID 314829, 1–13.

6. Jung, C.; Seo, J. S.; Han, S. W.; Koo, Y. J.; Kim, C. H.; Song, S. I.; Nahm, B. H.; Do Choi, Y.; Cheong, J.-J., Overexpression of AtMYB44 enhances stomatal closure to confer abiotic stress tolerance in transgenic *Arabidopsis*. *Plant Physiology* **2008**, 146, (2), 623–635.

7. Cooper, J. L.; Till, B. J.; Laport, R. G.; Darlow, M. C.; Kleffner, J. M.; Jamai, A.; El-Mellouki, T.; Liu, S.; Ritchie, R.; Nielsen, N., TILLING to detect induced mutations in soybean. *BMC Plant Biology* **2008**, 8, (1), 9.

8. Xiong, L.; Schumaker, K. S.; Zhu, J.-K., Cell signaling during cold, drought, and salt stress. *The Plant Cell* **2002**, 14, (suppl 1), S165–S183.

9. Lee, B.-H.; Henderson, D. A.; Zhu, J.-K., The *Arabidopsis* cold-responsive transcriptome and its regulation by ICE1. *The Plant Cell* **2005**, 17, (11), 3155–3175.

10. Seki, M.; Narusaka, M.; Abe, H.; Kasuga, M.; Yamaguchi-Shinozaki, K.; Carninci, P.; Hayashizaki, Y.; Shinozaki, K., Monitoring the expression pattern of 1300 *Arabidopsis* genes under drought and cold stresses by using a full-length cDNA microarray. *The Plant Cell* **2001**, 13, (1), 61–72.

11. Maruyama, K.; Sakuma, Y.; Kasuga, M.; Ito, Y.; Seki, M.; Goda, H.; Shimada, Y.; Yoshida, S.; Shinozaki, K.; Yamaguchi-Shinozaki, K., Identification of cold-inducible downstream genes of the *Arabidopsis* DREB1A/CBF3 transcriptional factor using two microarray systems. *The Plant Journal* **2004**, 38, (6), 982–993.

12. Seki, M.; Okamoto, M.; Matsui, A.; Kim, J.-M.; Kurihara, Y.; Ishida, J.; Morosawa, T.; Kawashima, M.; To, T. K.; Shinozaki, K., Microarray analysis for studying the abiotic stress responses in plants. In *Molecular Techniques in Crop Improvement*, Springer: 2010; pp 333–355.

13. Kimotho, R. N.; Baillo, E. H.; Zhang, Z., Transcription factors involved in abiotic stress responses in Maize (*Zea mays* L.) and their roles in enhanced productivity in the post genomics era. *PeerJ* **2019**, 7, e7211.

14. Ghorbani, R.; Alemzadeh, A.; Razi, H., Microarray analysis of transcriptional responses to salt and drought stress in *Arabidopsis* thaliana. *Heliyon* **2019**, 5, (11), e02614.

15. Sirohi, P.; Yadav, B. S.; Afzal, S.; Mani, A.; Singh, N. K., Identification of drought stress-responsive genes in rice (*Oryza sativa*) by meta-analysis of microarray data. *Journal of Genetics* **2020**, 99, 1–10.

16. Devnarain, N.; Crampton, B. G.; Olivier, N.; Van der Westhuyzen, C.; Becker, J. V.; O'Kennedy, M. M., Transcriptomic analysis of a Sorghum bicolor landrace identifies a role for beta-alanine betaine biosynthesis in drought tolerance. *South African Journal of Botany* **2019**, 127, 244–255.

17. Kim, C.-K.; Oh, J.-H.; Na, J.-K.; Cho, C.; Kim, K.-H.; Yu, G. E.; Kim, D.-Y., The genes associated with drought tolerance by multi-layer approach in potato. *Plant Breeding and Biotechnology* **2019**, 7, (4), 405–414.

18. Liu, X.; Zhang, X.; Sun, B.; Hao, L.; Liu, C.; Zhang, D.; Tang, H.; Li, C.; Li, Y.; Shi, Y., Genome-wide identification and comparative analysis of drought-related microRNAs in two maize inbred lines with contrasting drought tolerance by deep sequencing. *PLoS One* **2019**, 14, (7), e0219176.

19. Singh, B.; Mishra, S.; Bohra, A.; Joshi, R.; Siddique, K. H., Crop phenomics for abiotic stress tolerance in crop plants. In *Biochemical, Physiological and Molecular Avenues for Combating Abiotic Stress Tolerance in Plants*, Elsevier: 2018; pp 277–296.

20. Sirault, X. R.; James, R. A.; Furbank, R. T., A new screening method for osmotic component of salinity tolerance in cereals using infrared thermography. *Functional Plant Biology* **2009**, 36, (11), 970–977.

21. Wedeking, R.; Mahlein, A.-K.; Steiner, U.; Oerke, E.-C.; Goldbach, H. E.; Wimmer, M. A., Osmotic adjustment of young sugar beets (Beta vulgaris) under progressive drought stress and subsequent rewatering assessed by metabolite analysis and infrared thermography. *Functional Plant Biology* **2017**, 44, (1), 119–133.

22. Jansen, M.; Gilmer, F.; Biskup, B.; Nagel, K. A.; Rascher, U.; Fischbach, A.; Briem, S.; Dreissen, G.; Tittmann, S.; Braun, S., Simultaneous phenotyping of leaf growth and chlorophyll fluorescence via GROWSCREEN FLUORO allows detection of stress tolerance in *Arabidopsis* thaliana and other rosette plants. *Functional Plant Biology* **2009**, 36, (11), 902–914.

23. Siebke, K.; Ball, M. C., Non-destructive measurement of chlorophyll b: a ratios and identification of photosynthetic pathways in grasses by reflectance spectroscopy. *Functional Plant Biology* **2009**, 36, (11), 857–866.

24. Fatangare, A.; Gebhardt, P.; Saluz, H.; Svatoš, A., Comparing 2-[18F] fluoro-2-deoxy-D-glucose and [68Ga] gallium-citrate translocation in *Arabidopsis* thaliana. *Nuclear Medicine and Biology* **2014**, 41, (9), 737–743.

25. Rahaman, M.; Chen, D.; Gillani, Z.; Klukas, C.; Chen, M., Advanced phenotyping and phenotype data analysis for the study of plant growth and development. *Frontiers in Plant Science* **2015**, 6, 619.

26. Svischeva, G. R., Quantitative trait locus analysis of hybrid pedigrees: variance-components model, inbreeding parameter, and power. *BMC Genetics* **2007**, 8, (1), 50.

27. Mäki-Tanila, A.; Hill, W. G., Influence of gene interaction on complex trait variation with multilocus models. *Genetics* **2014**, 198, (1), 355–367.

28. Belonogova, N. M.; Svishcheva, G. R.; van Duijn, C. M.; Aulchenko, Y. S.; Axenovich, T. I., Region-based association analysis of human quantitative traits in related individuals. *PLoS One* **2013**, 8, (6), e65395.

29. Taunk, J.; Rani, A.; Singh, R.; Yadav, N. R.; Yadav, R. C., Genomic strategies for improving abiotic stress tolerance in crop plants. In *Genetic Enhancement of Crops for Tolerance to Abiotic Stress: Mechanisms and Approaches, Volume I*, Springer: 2019; pp 205–230.

30. Yadav, R. S.; Sehgal, D.; Vadez, V., Using genetic mapping and genomics approaches in understanding and improving drought tolerance in pearl millet. *Journal of Experimental Botany* **2011**, 62, (2), 397–408.

31. Zhang, S.; Zheng, J.; Liu, B.; Peng, S.; Leung, H.; Zhao, J.; Wang, X.; Yang, T.; Huang, Z., Identification of QTLs for cold tolerance at seedling stage in rice (*Oryza sativa* L.) using two distinct methods of cold treatment. *Euphytica* **2014**, 195, (1), 95–104.

32. Gonzaga, Z. J. C.; Carandang, J.; Sanchez, D. L.; Mackill, D. J.; Septiningsih, E. M., Mapping additional QTLs from FR13A to increase submergence tolerance in rice beyond SUB1. *Euphytica* **2016,** 209, (3), 627–636.

33. Chukwu, S. C.; Rafii, M. Y.; Ramlee, S. I.; Ismail, S. I.; Oladosu, Y.; Okporie, E.; Onyishi, G.; Utobo, E.; Ekwu, L.; Swaray, S., Marker-assisted selection and gene pyramiding for resistance to bacterial leaf blight disease of rice (*Oryza sativa* L.). *Biotechnology & Biotechnological Equipment* **2019,** 33, (1), 440–455.

34. Jena, K.; Mackill, D., Molecular markers and their use in marker-assisted selection in rice. *Crop Science* **2008,** 48, (4), 1266–1276.

35. Luu TN, H.; Luu M, C.; Abdelbagi M, I.; Le H, H., Introgression the salinity tolerance QTLs Saltol into AS996, the elite rice variety of Vietnam. *American Journal of Plant Sciences* **2012,** 3, (7), 981–987.

36. Crouch, J. H.; Serraj, R. In *5.1 DNA Marker Technology as a Tool for Genetic Enhancement of Drought Tolerance at ICRISAT*, Field Screening for Drought Tolerance in Crop Plants with Emphasis on Rice: Proceedings of an International Workshop on Field Screening for Drought Tolerance in Rice, 11–14 Dec 2000, ICRISAT, Patancheru, India. Patancheru 502 324, Andhra Pradesh, India, and the Rockefeller Foundation, New York, NY 10018-2702, USA. 208 pp. Order code CPE 139, 2002; Abstract: 2002; p 155.

37. Barakat, M.; Saleh, M.; Al-Doss, A.; Moustafa, K.; Elshafei, A.; Zakri, A.; Al-Qurainy, F., Mapping of QTLs associated with abscisic acid and water stress in wheat. *Biologia Plantarum* **2015,** 59, (2), 291–297.

38. i Azam, F.; Chang, X.; Jing, R., Mapping QTL for chlorophyll fluorescence kinetics parameters at seedling stage as indicators of heat tolerance in wheat. *Euphytica* **2015,** 202, (2), 245–258.

39. Rana, M. M.; Takamatsu, T.; Baslam, M.; Kaneko, K.; Itoh, K.; Harada, N.; Sugiyama, T.; Ohnishi, T.; Kinoshita, T.; Takagi, H., Salt tolerance improvement in rice through efficient SNP marker-assisted selection coupled with speed-breeding. *Internationa Journal of Molecular Sciences* **2019,** 20, (10), 2585.

40. Gantait, S.; Sarkar, S.; Verma, S. K., Marker-assisted Selection for Abiotic Stress Tolerance in Crop Plants. *Molecular Plant Abiotic Stress: Biology and Biotechnology*, Wiley Online Library. **2019,** 335–368.

41. Shendure, J.; Aiden, E. L., The expanding scope of DNA sequencing. *Nature Biotechnology* **2012,** 30, (11), 1084.

42. Edwards, D., The impact of genomics technology on adapting plants to climate change. In *Plant Genomics and Climate Change*, Springer: 2016; pp 173–178.

43. Akpınar, B. A.; Lucas, S. J.; Budak, H., Genomics approaches for crop improvement against abiotic stress. *The Scientific World Journal* **2013,** 2013.

44. Peace, C. P.; Bianco, L.; Troggio, M.; Van de Weg, E.; Howard, N. P.; Cornille, A.; Durel, C.-E.; Myles, S.; Migicovsky, Z.; Schaffer, R. J., Apple whole genome sequences: recent advances and new prospects. *Horticulture Research* **2019,** 6, (1), 1–24.

45. Bahuguna, R. N.; Gupta, P.; Bagri, J.; Singh, D.; Dewi, A. K.; Tao, L.; Islam, M.; Sarsu, F.; Singla-Pareek, S. L.; Pareek, A., Forward and reverse genetics approaches for combined stress tolerance in rice. *Indian Journal of Plant Physiology* **2018,** 23, (4), 630–646.

46. Li, C.; Zhang, R.; Meng, X.; Chen, S.; Zong, Y.; Lu, C.; Qiu, J.-L.; Chen, Y.-H.; Li, J.; Gao, C., Targeted, random mutagenesis of plant genes with dual cytosine and adenine base editors. *Nature Biotechnology* **2020,** 38, (7), 875–882.

47. Springer, P. S., Gene traps: tools for plant development and genomics. *The Plant Cell* **2000,** 12, (7), 1007–1020.

48. Wu, J.-L.; Wu, C.; Lei, C.; Baraoidan, M.; Bordeos, A.; Madamba, M. R. S.; Ramos-Pamplona, M.; Mauleon, R.; Portugal, A.; Ulat, V. J., Chemical-and irradiation-induced mutants of indica rice IR64 for forward and reverse genetics. *Plant Molecular Biology* **2005**, 59, (1), 85–97.

49. Ram, H.; Soni, P.; Salvi, P.; Gandass, N.; Sharma, A.; Kaur, A.; Sharma, T. R., Insertional mutagenesis approaches and their use in rice for functional genomics. *Plants* **2019**, 8, (9), 310.

50. Joly-Lopez, Z.; Forczek, E.; Vello, E.; Hoen, D. R.; Tomita, A.; Bureau, T. E., Abiotic stress phenotypes are associated with conserved genes derived from transposable elements. *Frontiers in Plant Science* **2017**, 8, 2027.

51. Makarevitch, I.; Waters, A. J.; West, P. T.; Stitzer, M.; Hirsch, C. N.; Ross-Ibarra, J.; Springer, N. M., Transposable elements contribute to activation of maize genes in response to abiotic stress. *PLoS Genetics* **2015**, 11, (1), e1004915.

52. Suprasanna, P.; Mirajkar, S.; Patade, V.; Jain, S. M., Induced mutagenesis for improving plant abiotic stress tolerance. *Mutagenesis: Exploring Genetic Diversity of Crops*. Wageningen Academic Publishers, Wageningen **2014**, 345–376.

53. Human, S.; Indriatama, W. M. In *Sorghum improvement program by using mutation breeding in Indonesia*, IOP Conference Series: Earth and Environmental Science, 2020; IOP Publishing: 2020; p 012003.

54. McCallum, C. M.; Comai, L.; Greene, E. A.; Henikoff, S., Targeting induced locallesions in genomes (TILLING) for plant functional genomics. *Plant Physiology* **2000**, 123, (2), 439–442.

55. Cooper, J. L.; Henikoff, S.; Comai, L.; Till, B. J., TILLING and ecotilling for rice. In *Rice Protocols*, Springer: 2013; pp 39–56.

56. Xin, Z.; Wang, M. L.; Barkley, N. A.; Burow, G.; Franks, C.; Pederson, G.; Burke, J., Applying genotyping (TILLING) and phenotyping analyses to elucidate gene function in a chemically induced sorghum mutant population. *BMC Plant Biology* **2008**, 8, (1), 103.

57. Caldwell, D. G.; McCallum, N.; Shaw, P.; Muehlbauer, G. J.; Marshall, D. F.; Waugh, R., A structured mutant population for forward and reverse genetics in Barley (Hordeum vulgare L.). *The Plant Journal* **2004**, 40, (1), 143–150.

58. Till, B. J.; Reynolds, S. H.; Weil, C.; Springer, N.; Burtner, C.; Young, K.; Bowers, E.; Codomo, C. A.; Enns, L. C.; Odden, A. R., Discovery of induced point mutations in maize genes by TILLING. *BMC Plant Biology* **2004**, 4, (1), 12.

59. De Lorenzo, L.; Merchan, F.; Laporte, P.; Thompson, R.; Clarke, J.; Sousa, C.; Crespi, M., A novel plant leucine-rich repeat receptor kinase regulates the response of Medicago truncatula roots to salt stress. *The Plant Cell* **2009**, 21, (2), 668–680.

60. Chen, L.; Huang, L.; Min, D.; Phillips, A.; Wang, S.; Madgwick, P. J.; Parry, M. A.; Hu, Y.-G., Development and characterization of a new TILLING population of common bread wheat (Triticum aestivum L.). *PLoS One* **2012**, 7, (7), e41570.

61. Barkley, N.; Wang, M., Application of TILLING and EcoTILLING as reverse genetic approaches to elucidate the function of genes in plants and animals. *Current Genomics* **2008**, 9, (4), 212–226.

62. Bagge, M.; Xia, X.; Lübberstedt, T., Functional markers in wheat. *Current Opinion in Plant Biology* **2007**, 10, (2), 211–216.

63. Garg, B.; Lata, C.; Prasad, M., A study of the role of gene TaMYB2 and an associated SNP in dehydration tolerance in common wheat. *Molecular Biology Reports* **2012**, 39, (12), 10865–10871.

64. Nejat, N.; Mantri, N., Plant immune system: crosstalk between responses to biotic and abiotic stresses the missing link in understanding plant defense. *Signal* **2017**, 2, O2.

65. Jamil, S.; Shahzad, R.; Rahman, S. U.; Iqbal, M. Z.; Yaseen, M.; Ahmad, S.; Fatima, R., The level of Cry1Ac endotoxin and its efficacy against *H. armigera* in Bt cotton at large scale in Pakistan. *GM Crops & Food* **2021**, 12, (1), 1–17.

66. Yang, Y.; Zhou, Y.; Chi, Y.; Fan, B.; Chen, Z., Characterization of soybean WRKY gene family and identification of soybean WRKY genes that promote resistance to soybean cyst nematode. *Scientific Reports* **2017**, 7, (1), 1–13.

67. Chen, Y.; Cao, Y.; Wang, L.; Li, L.; Yang, J.; Zou, M., Identification of MYB transcription factor genes and their expression during abiotic stresses in maize. *Biologia Plantarum* **2018**, 62, (2), 222–230.

68. Li, Y.-Y.; Mao, K.; Zhao, C.; Zhao, X.-Y.; Zhang, H.-L.; Shu, H.-R.; Hao, Y.-J., MdCOP1 ubiquitin E3 ligases interact with MdMYB1 to regulate light-induced anthocyanin biosynthesis and red fruit coloration in apple. *Plant Physiology* **2012**, 160, (2), 1011–1022.

69. He, G.-H.; Xu, J.-Y.; Wang, Y.-X.; Liu, J.-M.; Li, P.-S.; Chen, M.; Ma, Y.-Z.; Xu, Z.-S., Drought-responsive WRKY transcription factor genes TaWRKY1 and TaWRKY33 from wheat confer drought and/or heat resistance in *Arabidopsis*. *BMC Plant Biology* **2016**, 16, (1), 116.

70. Raza, A.; Razzaq, A.; Mehmood, S. S.; Zou, X.; Zhang, X.; Lv, Y.; Xu, J., Impact of climate change on crops adaptation and strategies to tackle its outcome: A review. *Plants* **2019**, 8, (2), 34.

71. Cao, D.; Hou, W.; Liu, W.; Yao, W.; Wu, C.; Liu, X.; Han, T., Overexpression of TaNHX2 enhances salt tolerance of 'composite'and whole transgenic soybean plants. *Plant Cell, Tissue and Organ Culture (PCTOC)* **2011**, 107, (3), 541–552.

72. Younis, A.; Siddique, M. I.; Kim, C.-K.; Lim, K.-B., RNA interference (RNAi) induced gene silencing: a promising approach of hi-tech plant breeding. *International Journal of Biological Sciences* **2014**, 10, (10), 1150.

73. Liu, C.; Yang, Z.-J.; Li, G.-R.; Zeng, Z.-X.; Zhang, Y.; Zhou, J.-P.; Liu, Z.-H.; Ren, Z.-L., Isolation of a new repetitive DNA sequence from Secale africanum enables targeting of Secale chromatin in wheat background. *Euphytica* **2008**, 159, (1–2), 249–258.

74. Zhao, B.; Liang, R.; Ge, L.; Li, W.; Xiao, H.; Lin, H.; Ruan, K.; Jin, Y., Identification of drought-induced microRNAs in rice. *Biochemical and Biophysical Research Communications* **2007**, 354, (2), 585–590.

75. Trindade, I.; Capitão, C.; Dalmay, T.; Fevereiro, M. P.; Dos Santos, D. M., miR398 and miR408 are up-regulated in response to water deficit in *Medicago truncatula*. *Planta* **2010**, 231, (3), 705–716.

76. Liu, H.-H.; Tian, X.; Li, Y.-J.; Wu, C.-A.; Zheng, C.-C., Microarray-based analysis of stress-regulated microRNAs in *Arabidopsis thaliana*. *Rna* **2008**, 14, (5), 836–843.

77. Arenas-Huertero, C.; Pérez, B.; Rabanal, F.; Blanco-Melo, D.; De la Rosa, C.; Estrada-Navarrete, G.; Sanchez, F.; Covarrubias, A. A.; Reyes, J. L., Conserved and novel miRNAs in the legume *Phaseolus vulgaris* in response to stress. *Plant Molecular Biology* **2009**, 70, (4), 385–401.

78. Zhang, J.; Xu, Y.; Huan, Q.; Chong, K., Deep sequencing of Brachypodium small RNAs at the global genome level identifies microRNAs involved in cold stress response. *BMC Genomics* **2009**, 10, (1), 449.

79. Ding, Y.; Ding, L.; Xia, Y.; Wang, F.; Zhu, C., Emerging roles of microRNAs in plant heavy metal tolerance and homeostasis. *Journal of Agricultural and Food Chemistry* **2020**, 68, (7), 1958–1965.

80. Voytas, D. F., Plant genome engineering with sequence-specific nucleases. *Annual Review of Plant Biology* **2013**, 64, 327–350.

81. Guha, T. K.; Wai, A.; Hausner, G., Programmable genome editing tools and their regulation for efficient genome engineering. *Computational and Structural Biotechnology Journal* **2017**, 15, 146–160.

82. Ahmad, S.; Wei, X.; Sheng, Z.; Hu, P.; Tang, S., CRISPR/Cas9 for development of disease resistance in plants: recent progress, limitations and future prospects. *Briefings in Functional Genomics* **2020**, 19, (1), 26–39.

83. Doudna, J. A.; Charpentier, E., The new frontier of genome engineering with CRISPR-Cas9. *Science* **2014**, 346, (6213).

84. Hsu, P. D.; Lander, E. S.; Zhang, F., Development and applications of CRISPR-Cas9 for genome engineering. *Cell* **2014**, 157, (6), 1262–1278.

85. Abudayyeh, O. O.; Gootenberg, J. S.; Konermann, S.; Joung, J.; Slaymaker, I. M.; Cox, D. B.; Shmakov, S.; Makarova, K. S.; Semenova, E.; Minakhin, L., C2c2 is a single-component programmable RNA-guided RNA-targeting CRISPR effector. *Science* **2016**, 353, (6299).

86. Anzalone, A. V.; Randolph, P. B.; Davis, J. R.; Sousa, A. A.; Koblan, L. W.; Levy, J. M.; Chen, P. J.; Wilson, C.; Newby, G. A.; Raguram, A., Search-and-replace genome editing without double-strand breaks or donor DNA. *Nature* **2019**, 576, (7785), 149–157.

87. Liu, J.; Shen, J.; Xu, Y.; Li, X.; Xiao, J.; Xiong, L., Ghd2, a CONSTANS-like gene, confers drought sensitivity through regulation of senescence in rice. *Journal of Experimental Botany* **2016**, 67, (19), 5785–5798.

88. Wang, C.-T.; Ru, J.-N.; Liu, Y.-W.; Li, M.; Zhao, D.; Yang, J.-F.; Fu, J.-D.; Xu, Z.-S., Maize WRKY transcription factor ZmWRKY106 confers drought and heat tolerance in transgenic plants. *International Journal of Molecular Sciences* **2018**, 19, (10), 3046.

89. Liu, J.-H.; Peng, T.; Dai, W., Critical cis-acting elements and interacting transcription factors: key players associated with abiotic stress responses in plants. *Plant Molecular Biology Reporter* **2014**, 32, (2), 303–317.

90. Rodríguez-Leal, D.; Lemmon, Z. H.; Man, J.; Bartlett, M. E.; Lippman, Z. B., Engineering quantitative trait variation for crop improvement by genome editing. *Cell* **2017**, 171, (2), 470–480.e8.

91. Goodstein, D. M.; Shu, S.; Howson, R.; Neupane, R.; Hayes, R. D.; Fazo, J.; Mitros, T.; Dirks, W.; Hellsten, U.; Putnam, N., Phytozome: a comparative platform for green plant genomics. *Nucleic Acids Research* **2012**, 40, (D1), D1178-D1186.

92. Pérez-Rodríguez, P.; Riano-Pachon, D. M.; Corrêa, L. G. G.; Rensing, S. A.; Kersten, B.; Mueller-Roeber, B., PlnTFDB: updated content and new features of the plant transcription factor database. *Nucleic Acids Research* **2010**, 38, (suppl_1), D822-D827.

93. Ma, X.; Zhang, Q.; Zhu, Q.; Liu, W.; Chen, Y.; Qiu, R.; Wang, B.; Yang, Z.; Li, H.; Lin, Y., A robust CRISPR/Cas9 system for convenient, high-efficiency multiplex genome editing in monocot and dicot plants. *Molecular Plant* **2015**, 8, (8), 1274–1284.

94. Monsur, M. B.; Shao, G.; Lv, Y.; Ahmad, S.; Wei, X.; Hu, P.; Tang, S., Base editing: The Ever Expanding Clustered Regularly Interspaced Short Palindromic Repeats (CRISPR) Tool Kit for Precise Genome Editing in Plants. *Genes* **2020**, 11, (4), 466.

95. Yan, J.; Cirincione, A.; Adamson, B., Prime editing: precision genome editing by reverse transcription. *Mol. Cell* **2020**, 77, (2), 210–212.

96. Zafar, S. A.; Zaidi, S. S.-e.-A.; Gaba, Y.; Singla-Pareek, S. L.; Dhankher, O. P.; Li, X.; Mansoor, S.; Pareek, A., Engineering abiotic stress tolerance via CRISPR/Cas-mediated genome editing. *Journal of Experimental Botany* **2020**, 71, (2), 470–479.

97. Custers, R., The regulatory status of gene-edited agricultural products in the EU and beyond. *Emerging Topics in Life Sciences* **2017**, 1, (2), 221–229.
98. Eckerstorfer, M. F.; Engelhard, M.; Heissenberger, A.; Simon, S.; Teichmann, H., Plants developed by new genetic modification techniques—comparison of existing regulatory frameworks in the EU and non-EU countries. *Frontiers in Bioengineering and Biotechnology* **2019**, 7, 26.
99. Ahmad S.; Sheng Z.; Jalal R. S.; Tabassum J.; Ahmed F. K.; Hu S.; Shao G.; Wei X.; Abd-Elsalam K. A.; Hu P.; Tang S. CRISPR—Cas technology towards improvement of abiotic stress tolerance in plants. In *CRISPR and RNAi Systems*, Elsevier, **2021a**, pp. 755–772.
100. Ahmad S.; Shahzad R.; Jamil S.; Tabassum J.; Chaudhary M. A.; Atif R. M.; Iqbal M. M.; Monsur M. B.; Lv Y.; Sheng Z.; Ju L. Regulatory aspects, risk assessment, and toxicity associated with RNAi and CRISPR methods. In *CRISPR and RNAi Systems*, Elsevier, **2021b**, pp. 687–721.
101. Monsur M. B.; Shao G.; Lv Y.; Ahmad S.; Wei X.; Hu P.; Tang S. Base editing: the ever expanding clustered regularly interspaced short palindromic repeats (CRISPR) Tool Kit for Precise Genome Editing in Plants. *Genes.* **2020**, 11, (4):466. https://doi.org/10.3390/genes11040466.
102. Jamil S.; Shahzad R.; Ahmad S.; Fatima R.; Zahid R.; Anwar M.; Iqbal M. Z.; Wang X. Role of genetics, genomics and breeding approaches to combat stripe rust of wheat. *Frontiers in Nutrition.* **2020**, 7, 173. https://doi.org/10.3389/fnut.2020.580715.
103. Shahzad R.; Shakra Jamil S. A.; Nisar A.; Amina Z.; Saleem S.; Iqbal M. Z.; Atif R. M.; Wang X. Harnessing the potential of plant transcription factors in developing climate resilient crops to improve global food security: Current and future perspectives. *Saudi Journal of Biological Sciences* **2021**, 28, (4), 2323–2341.

CHAPTER 7

Salinity-Induced Changes on Different Physiological and Biochemical Features of Plants

AADIL RASOOL[1], WASIFA HAFIZ SHAH[1], NAVEED UL MUSHTAQ[1], SEERAT SALEEM[1], KHALID REHMAN HAKEEM[2], and REIAZ UL REHMAN[1*]

[1]*Department of Bioresources, School of Biological Sciences, University of Kashmir, Srinagar 190006, India*

[2]*Department of Biological Sciences, King Abdulaziz University, Jeddah, Saudi Arabia*

Corresponding author. E-mail: rreaizbiores@gmail.com.

ABSTRACT

Salt stress is a severe ecological limitation that hinders the normal functioning of plants. An extensive range of alterations in growth and development are provoked by salt stress. Salt stress is one of the most fierce ecological components constraining the profitability of plants on the grounds that a large portion of the plants are sensitive to excess salt. This circumstance has been additionally exacerbated by anthropogenic implementations. Accordingly, there is a much logical burden on analysts to upgrade crop efficiency under salinity stress so as to adapt to the expanding food security risks. Salinity triggers osmotic, ionic, oxidative stress, and a surge in the production of reactive oxygen species. It hampers plant biomass production, photosynthesis, and antioxidative defense system. The plant capacity to endure salt stress is dictated by different biochemical and physiological systems, ensuring normal functioning of the cell, mainly by managing appropriate water relations and keeping up ion homeostasis. A thorough comprehension on how plants react to salty conditions at various levels and an incorporated methodology of joining molecular biology approaches with physiological and biochemical strategies are basic for the advancement of knowledge toward this hazard. This chapter adds more to the understanding of different biochemical and physiological changes that occur to plants when exposed to salt stress.

7.1   INTRODUCTION

The green revolution in the mid and late 20th century ensured the worldwide supply of food was enough to stay up to date with population increase; it ensured safe, reliable, affordable, and sustainable improvements to food for all segments of society. However, in the 21st century, as a consequence of global climate change and global warming, the trend took a U-turn. Accomplishing global security of food even though the increase in demand due to population explosion, diminishing resources, and the incompetence of the environment to cushion expanding anthropogenic activities is currently observed as the principal challenge within recent history [1, 2]. Deterioration of agricultural land and affording food for the exponentially mounting population is one among a lot of interconnected threats confronted today by humanity [3]. It has been observed that changing climatic conditions severely influence natural systems, human wellbeing, and agricultural production [4]. An increase in population, the world over corresponds to the rise in food demand, which endangers the stability of the global agricultural system. Agricultural efficacy depends upon water accessibility, air contamination, and soil fertility [5]. So, the punitive impacts on agricultural productivity are advancing in immense intensity, which directly results in abiotic stresses. Also, continued deforestation and extreme use of petroleum products have led to an increase in CO_2 concentration from 280 to 400 μmol^{-1} in the air. Moreover, this concentration is set to rise toward the end of this century if it remains unchecked. This CO_2 emission is a leading cause of global warming [6]. Also, industrialization is one of the primary causes of CO_2 emission and temperature rise, which negatively impacts the global environment.

Because of severe climate events, the recurrence of global warming is rising, which will disturb the delicate ecological balance [7]. The two main issues of the 21st century are climate change and food insecurity. It is estimated that ~815 million individuals are affected by malnutrition, hampering to accomplish the objective of eliminating hunger by 2030 [8]. The agricultural production has been diminished around the globe due to the increase in temperature [9]. In the 21st century, the Earth's average temperature is set to rise from 2 °C to 4.5 °C. Likewise, the time between the 19th and the 21st hundreds of years is viewed as the period, which encountered the most warming [10]. These increased temperatures by giving way to various stresses such as drought, high temperature, and salinity affect agricultural production [11]. The impacts of these environmental variations are evaluated by their effect on human health and agriculture, and the yield of crops is mainly endured because of unfavorable ecological conditions, high temperature, and excess of CO_2 [12]. This unevenness can lead to heavy or low precipitation, thus cause floods or drought, respectively [13]. Every living being, for example, plants, animals, fishes, and humans, have been influenced by the changes in the environmental conditions world over. The threat to the world's atmosphere conditions has paved the way for researchers to find new ideas as global food security is at risk [14, 15]. The impacts of environmental change have been exceptionally critical, particularly in cultivated agrarian lands (Figure 7.1).

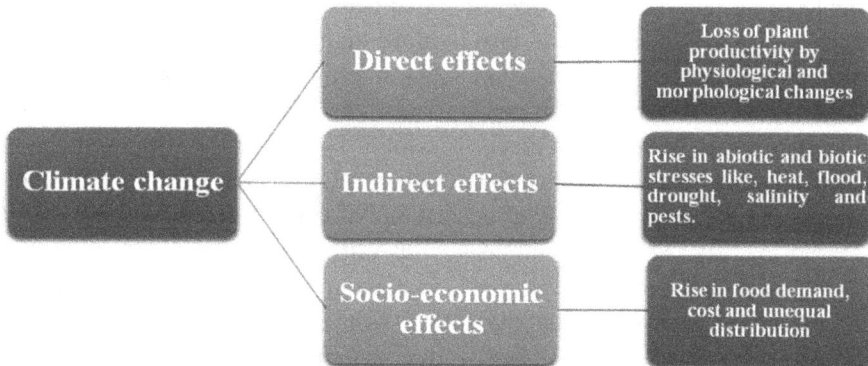

FIGURE 7.1 Overall impact of climate change.

In perspective on the above-discussed facts, the need is to understand how these abiotic stresses affect plant growth and development. An assortment of environmental constraints such as biotic and abiotic stresses hamper the yield efficiency [16, 17]. Abiotic stress has bought the staggering results to our wellbeing and to the agricultural productions, and it is right now being acknowledged as the growing danger to the human presence [18, 19]. Diminished harvest and incompetence of soil, collectively with the aggregation of the harmful components in the food chains as a result of abiotic stress, are a threat to human survival [20]. This chapter provides insights into the adverse effects of salt on overall growth and development.

7.2   SALT STRESS

Anything abiotic or biotic that confines the plant photosynthesis and declines its capacity to grow and develop is characterized as stress to the plant. This may be inappropriate quantities of salt, water, or heat or maybe an attack by disease. Excess salt in the soil is one of the significant stress today's agriculture is facing. Generally, a soil with electrical conductivity (EC) of the saturation extract (EC_e) in the root zone exceeds 4 dS m^{-1} (~40 mM sodium chloride (NaCl)) at 25 °C and has exchangeable sodium of 15% is saline. Most plants show decreased productivity or fail to germinate at this EC_e. Even so, it was observed that several plants show a reduction in yield at lower EC_es. Salinity stress is the gravest constraining element for crop development and food security. About 20% of the entire cultivated and 33% of watered agrarian lands are salt-affected. For many reasons, including both anthropogenic and natural, the salinity stress in agricultural soils is increasing by 10% annually. World over about 6% (800 million ha) of the total land area is salt-affected. Globally, ~230 million ha is irrigated land, of which 45 million ha have been at present influenced by salt stress. Many activities such as low rainfall, high evaporation, inferior social practices, wearing out of rocks, and salt-contaminated irrigation water add to salinity stress. There are predictions that over 50% of the agronomic land would be contaminated with some degree of salinity by 2050 [21]. Adding more to this worsening situation is that the population will increase from 7.7 billion currently to 9.7 billion in 2050, according to a new United Nations report released in 2019. Soils contaminated with salt are comprehended to restrict plant growth and development. Soil salinization is a menace to agriculture globally. Crops grown on salt-contaminated soils experience high osmotic pressure and diminished harvest. The soil degradation by abiotic stresses, mainly soil salinization, is currently the worst hindrances to agricultural efficacy [22]. In many regions around the world, rainfall is inadequate to drain dissolvable salts from the soil, which results in elevated salt levels in soils and hence pose major ecological problems for agricultural production. This problem is set to get worse due to the adverse impacts of climate change [23]. Salts may typically emerge in the subsoil, which is called primary salinization or may be added to the soil by soil amendments, inorganic composites, and in particular water system with brackish water, called secondary salinization [24]. Plants can be grouped generally into two types based on adaptive evolution: the halophytes and the glycophytes. The halophytes can endure salt and thus can be found in saline soils and glycophytes, which cannot survive excess salt, and the majority of the crop plants belong to glycophytes. In this way, salt stress is severe abiotic stress that impedes plant growth and development [25].

Salt stress reduces agricultural yield and negatively affects soil properties (water holding capacity, infiltration ability, bulk density) and ecological symmetry of the area. Salinity stress negatively influences germination, vegetative, and reproductive growth. It causes osmotic stress, ion toxicity, insufficiency of nutrients, and oxidative damage, and along these lines limits water take-up from the soil. Excessive amassing of sodium in the cell can quickly prompt osmotic pressure and cell death [26]. Salinity has a profound effect on plant physiology, as it reduces plant pigment content (Chlorophyll-a, b, carotenoid, and anthocyanin), cell division, and nitrogen absorption. It thereby diminishes the yield and profitability of harvests. Salt stress initially activates osmotic stress in plants. It lessens water take-up by the roots prompting a decrease in relative water content and diminished water potential, which restrains development, stomatal closure, photosynthesis, and transpiration [25]. The second casualty of the salt stress is the ideal K^+/Na^+ proportion in a cell. The build-up of these ions causes an imbalance in Na^+ and Cl^- ideal concentrations [27]. One more antagonistic effect of salt stress is the production of reactive oxygen species (ROS). The ROS is produced by disrupting the essential electron transport chain in mitochondria and chloroplasts, which prompts the generation of harmful singlet oxygen ($^1O_2$), superoxide ($O2^{\cdot-}$), hydroxyl radical ($^{\cdot}OH$), and hydrogen peroxide (H_2O_2), which cause lipid peroxidation, degradation of proteins, lipids, and DNA, thereby hindering basic cell functioning. ROS, if not appropriately quenched, proves detrimental to plants. So the presence of excessive salt in the soil causes various changes such as damage to membranes, nutrient imbalance, improper quenching of ROS, alterations in antioxidant enzymes, and diminished photosynthetic efficacy [28]. The overall detrimental effects of salinity stress are depicted below in Figure 7.2.

7.3 IMPACT OF SALT STRESS ON PLANTS

Salinity disturbs various physiological and biochemical parameters of plants. Some of the features that are hampered by salinity include; however, are not limited to ion balance and compartmentalization, biosynthesis, and activation of antioxidant enzymes and osmoprotectants.

7.3.1 SALINITY STRESS AND IONIC BALANCE

Upholding ion homeostasis in normal conditions and under salinity stress is vital for healthy plant growth and development [29]. The chief variety of

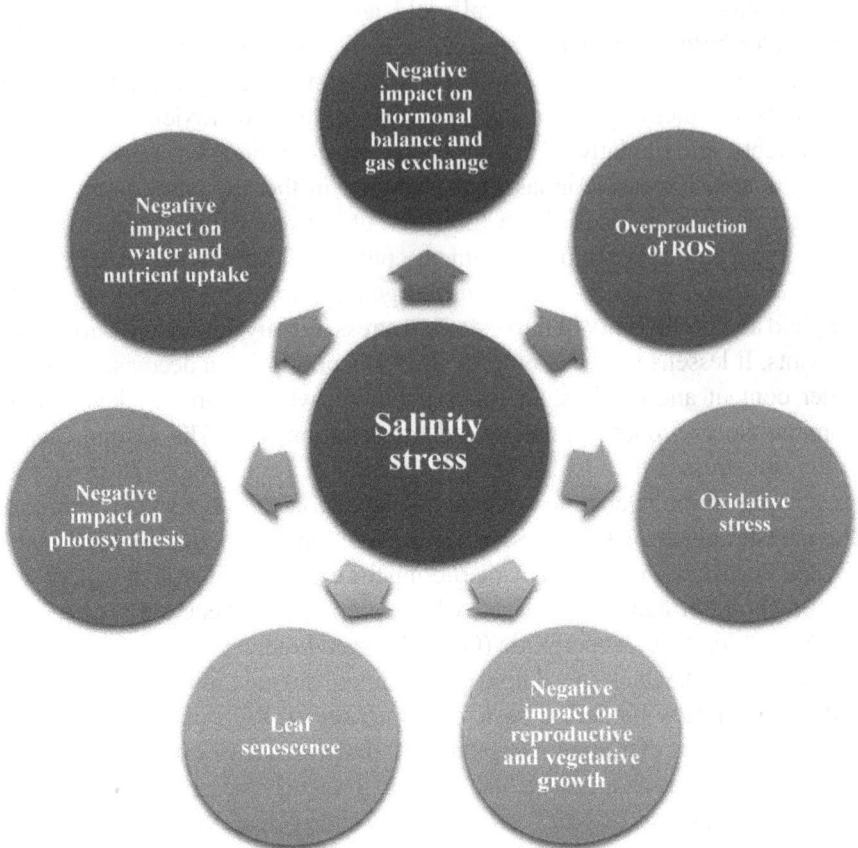

FIGURE 7.2   The effect of salt stress on plants.

salt reported in the soil is NaCl. The net uptake of Na^+ is influenced by its influx, loading into the xylem, exclusion, sequestration, and its recirculation through the phloem [30]. Nonselective cation channels (NSCCs) and high-affinity K^+ transporter (HKT) are the main ways by which Na^+ enters into cells [31]. At the same time, Na^+/H^+ antiporter salt overly sensitive 1 (SOS1) transports excessive Na^+ into the apoplast through the plasma membrane, SOS1 also aids in xylem loading of Na^+ ion [32]. An ideal K^+/Na^+ ratio in the cytoplasm is mandatory in order to survive in high saline conditions. High Na^+ concentration is harmful to plants and endangers the survivability of plants. Plants endure this ionic imbalance by transporting additional Na^+ either to the vacuole or sequestered to older tissues, in this manner, securing the plant from ionic imbalance [33]. After entering the cytoplasm,

Na^+ ion is moved to the vacuole with the help of Na^+/H^+ antiporter. There are two types of transporters in the vacuolar membrane responsible for transport into the vacuole namely, H^+-ATPase (V-ATPase) and the vacuolar pyrophosphatase (V-PPase) [34]. The principal H^+ pump present inside the plant cell is V-ATPase. V-ATPase plays an essential role in both stressed and nonstressed conditions. The survivability of the plants is dedicated to the activity of V-ATPase under stressed conditions [34]. Also, in the nonstress condition, it helps in aiding vesicle fusion, keeping solute homeostasis, and stimulating secondary transport. Salt tolerance of plants directly depends on the net Na^+ efflux capability. Extreme uptake of Na^+ brings the loss of potassium ions (K^+) by depolarizing the cellular membranes. Besides, it augments their abundance in plant cells. Moreover, an increased concentration of Na^+ causes unevenness of the K^+/Na^+ ratio, and these Na^+ ions compete with K^+ for the binding sites in the membrane, lowering the K^+ concentration [30]. Ion homeostasis and transport under salt stress in plants are maintained by the salt overly sensitive (SOS) pathway [35] (Figure 7.3).

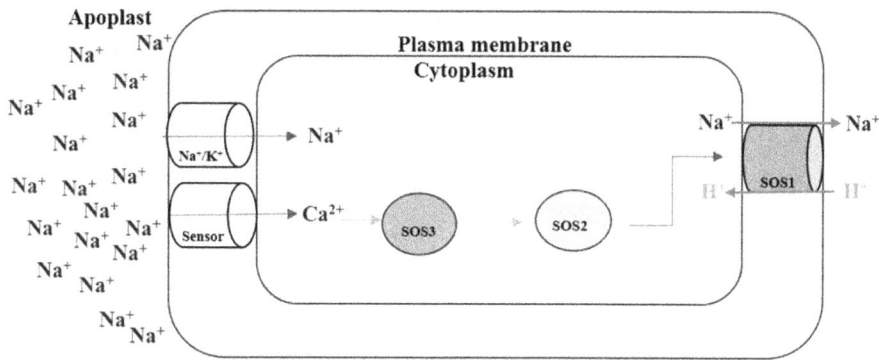

FIGURE 7.3   An overview of the SOS pathway.

After sensing high Na^+, the plant activates Ca^{2+} signals and salt-induced membrane depolarization takes place. However, membrane depolarization is physiologically irrelevant as it do not confer any specificity. However, increments in extracellular NaCl cause quick Ca^{2+} enhancement in the cytosol [36]; the Ca^{2+} signals can be salt specific [37]. The SOS pathway proteins perceive calcium signals. The SOS pathway involves three proteins: SOS1, SOS2, and SOS3. This pathway is essential in managing Na^+ efflux at the cell level. It likewise supports long-distance transport of Na^+ from root to shoot. Overexpression of SOS pathway genes gives salt versatility in plants [32].

The detoxifying process begins with SOS3. SOS3 protein is a myristoylated Ca^+ binding protein and contains a myristoylation site at its N-end. This site has the principal work in salt versatility [27]. When initiated, it collaborates with the SOS2. SOS2 contains a FISL motif, which is around 21 amino acids long arrangements in the C-terminal regulatory domain, which fills in as a site of contact for SOS3. The calcium ties to SOS3 (CBL4), causing dimerization of this protein and expanding the movement of SOS2 (CIPK24) serine/threonine-protein kinase. The subsequent CBL4/CIPK24 (SOS3/SOS2) complex stimulates the Na^+/H^+ SOS1, through phosphorylation [38], which directs Na^+ efflux from the cell. Other than giving salt flexibility, it additionally oversees pH homeostasis and membrane vesicle trafficking [39]. Plants have built up a lucrative system to forestall the efflux of extra ions such as Na^+. Membrane transporters assume a first job in keeping up ion concentration inside the cytoplasm during stress conditions [40]. Keeping up cell Na^+/K^+ homeostasis is pivotal for plant perseverance in saline conditions. Different transporter proteins, channel proteins, antiporters, and symporters carry out the fundamental inward and outward movement of particles.

For normal functioning, plants keep up a raised degree of K^+ (~100 mM) and a low degree of Na^+ (~1 mM or less) inside the cytosol. However, inside the vacuole, K^+ concentration is higher somewhere in the range of 10 and 200 mM. The vacuole fills in as the most significant pool of K^+ inside the plant cell. K^+ plays a vital job in keeping up the turgor inside the cell. It is moved into the plant cell against the concentration gradient utilizing K^+ transporters and membrane channels. When the extracellular K^+ concentration is low, high-affinity K^+ take-up is intervened by K^+ transporters to get K^+ required for normal functioning. At the point when the K^+ level is perfect, take-up is passed on by K^+ channels. So, the take-up of K^+ is chiefly constrained by the concentration of K^+ available in the outside environment.

On the other hand, a shallow convergence of Na^+ ions is kept up in the cytosol. During salt stress, the degree of Na^+ in the soil is increased. As both of the ions, Na^+ and K^+, have a similar vehicle component. Na^+ particle contends with K^+ for the transport, hence declining the K^+ take-up. Numerous transporter encoding genes have been noticed such as HKT and NHX (sodium hydrogen exchanger). Carriers situated on the plasma membrane, like that of the HKT family, play a significant role in salt resistance by dealing with the transportation of Na^+ and K^+ [41]. Class 1 HKT transporters in Arabidopsis shield the plant from the unsafe effects of salt stress by turning away Na^+ particles in leaves. Likewise, in rice class1, the HKT transporter ousts extra Na^+ from xylem and thus protecting the photosynthetic leaf tissues from the destructive effect of Na^+ [42]. Intracellular NHX proteins are Na^+ and antiporters (K^+/H^+), which are related to K^+ homeostasis and endosomal pH regulation. Tonoplast NHX proteins (NHX1 and NHX2) are crucial for dynamic K^+ take-up at the tonoplast. They are, in this way, keeping up the turgor and smooth working of stomata [43] (Figure 7.4).

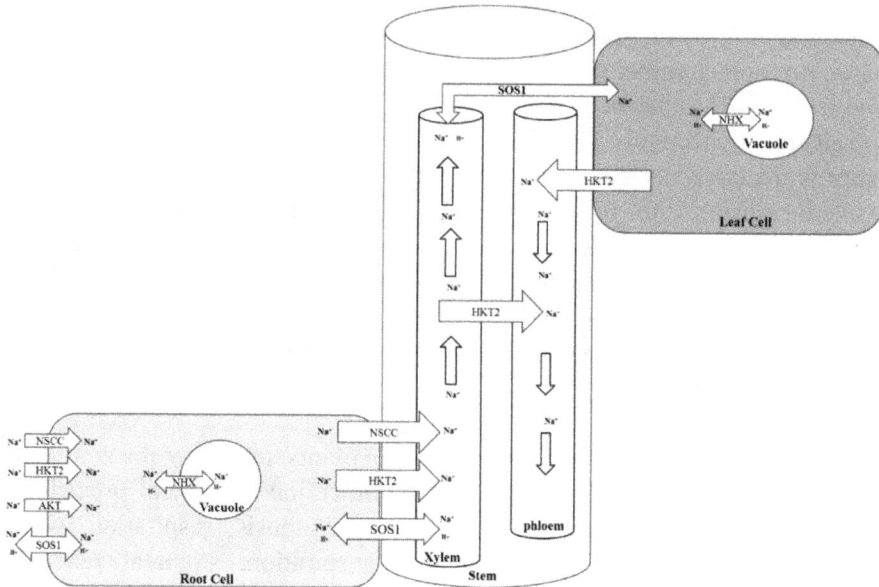

FIGURE 7.4 Overview of Na^+ transport in cells. NSCC: nonselective cation channels; AKT: arabidopsis K^+ transporter; HKT: high-affinity potassium transporters; SOS1: salt overly sensitive1.

7.3.2 *SALINITY STRESS AND PLANT GROWTH*

As already discussed, salt stress harms every aspect of the growth and development of the plant. The degree of restraint brought about by salt stress relies upon different factors such as species, plant's developmental stage, and type of ions. The salt stress causes detrimental changes in the plant from germination until development [26]. The plant cell in the hypertonic environment undergo shrinkage and dehydrate by the outward movement water. If the salt stress persists, it has severe consequences on cell extension and cell division, which in turn decreases root length, shoot length, and leaf development. Prolonged exposure to salt stress alters almost every aspect of the growth and development of plants. This is because of changes in cell–water relations and osmotic changes in the root. The osmotic impact prompts a decrease in the ability to assimilate water, which in turn results in low yields [44]. Excess salt diminishes the plant's capacity to take up water; this limits growth and development by reducing leaf development and root development [45].

Also, when salt enters the plant, it seriously injures leaves causing further reduction in growth [46]. Salt stress considerably diminishes the mass of aerial parts, the number of leaves, the height of plants, and the plant root area and length [47]. It has been observed that different parts of the plant respond to salt stress differently [48]. Under salinity stress, the growth of the upper parts is declined because the plant decreases carbon allocation to shoots, while increasing it for roots [49]. Salt stress results in a decrease in vegetative development such as shoot length, root length, leaf size, and biomass accumulation [50]. The plant spends its photosynthetic assimilates to keep up high water status and for root generation to avoid water loss under saline conditions [51]. In these conditions, it appears that the leaf development is ceased by activation of hormones [25], and the photosynthetic assimilates at that point are diverted to root growth.

Salinity stress induces various changes in plants, including the activation of specific biochemical pathways and modification of critical procedures of energy metabolism. For example, photosynthesis, respiration, water retention, uptake, membrane stability, transpiration, stomatal function, cytoskeleton disruption, etc. [52]. The salt that moves in the plant mainly concentrates on old leaves. However, if the salt continues to transport in the transpiring leaves, the leaves may undergo senescence due to the imbalance of the Na^+ and Cl^- ions. The further inward movement of salt in the cell will be surpassing the capacity of cells to compartmentalize salts in the vacuole. This causes the accumulation of salts in the cytoplasm, restraining the movement of essential compounds, and eventually amass in the cell membrane, which fastens the dehydration and shrinkage of the cell [53]. This ion build-up in the cell leads to the development of osmotic potential, which hinders the capability of the cell to take up water. Both Na^+ and Cl^- ions interfere with the transport of other essential ions such as K^+, which cause severe biochemical and physiological glitches to plants. Na^+ ion interferes with the transport of K^+ ion, which plays a vital role in the regulation of stomata and thus transpiration. While the Cl^-, which is more lethal than Na^+, if present in excess disrupts pH, enzyme activity in the cytoplasm, inhibits the chlorophyll production, and deregulation of turgor [53].

7.3.3 SALINITY STRESS AND PHOTOSYNTHETIC PIGMENTS

Photosynthesis is one of the most significant biochemical pathways by which plants convert solar energy into chemical energy; it is entirely associated with plant productivity and nutrient flow in plants. Salt stress influences the physiological activity of the leaf, especially photosynthesis, which is the fundamental driver of decreased plant efficiency [54]. The decrease in photosynthetic rates and effectiveness in plants under salt stress is primarily because of the reduction in water potential. Besides, photosynthesis is repressed when a high large amount of Na^+ and Cl^- generated from high salt in the soil, are amassed in the chloroplasts. This amassing of these ions causes injury to chlorophyll, which is the leading site of light reaction and hence legitimately downregulates the fitness of the plant [55]. The decline in chlorophyll content is a standard indicator of stress, and chlorophyll content has been utilized as a marker to know the metabolic cell state [56]

Under moderate salt stress, the photosynthetic proficiency can arrive at the qualities comparative to the control; however, under high saltiness, photosynthesis is fundamentally repressed. At high salt concentrations, the photosynthetic quenching and electron transport rate are diminished. The malondialdehyde content increases at high salt concentrations, reporting inordinate amassing of ROS, which is harmful to both the photosystems. Extreme salt concentrations are detrimental to PSII photochemistry and downregulate chloroplast proteins associated with biochemical impediments [57]. Salt stress incites a decrease in biomass because of a reduction in photosynthesis [58]. The reaction of plants to salt stress is intensely reliant on genotype. Salt stress has both short- and long-term impacts on the pattern of photosynthesis. The first consequences can be watched following a couple of hours, and there can be a full discontinuance of carbon take-up for a couple of hours.

The decrease in photosynthesis by salt stress can also be attributed to the decline in leaf water potential, which is responsible for stomata closure [59]. Increased concentrations of salt decrease chlorophyll content [60]. The decrease in photosynthesis is caused additionally by a reduction in stomatal conductance, which confines the entrance of CO_2 for the Calvin–Benson cycle [61]. The diffusion of CO_2 inside the stomata gets restricted, and it is binding at the chloroplast level that diminishes the recovery of RuBP (ribulose bisphosphate), which ultimately disrupts normal metabolic state of the plant [25]. Irregularities in stomatal functioning limit transpiration and influence in energy take-up by chloroplasts and hence hampering the function of these organelles [62]. In short, the reduction in photosynthesis is downregulated by ion build-up, diminishing leaf water potential, and stomatal irregularities under salt stress.

7.3.4   SALINITY STRESS AND ANTIOXIDANTS

Salt stress is accompanied by abundance in ROS [63]. $^1O_2$, H_2O_2, $O2^{\cdot-}$, and $\cdot OH$ are different types of ROS [64]. ROS is formed by the reduction of oxygen (O_2). The first step during oxygen reduction is the formation of $O2^{\cdot-}$ that is quickly changed over to H_2O_2 by the action of superoxide dismutase (SODs). This $O2^{\cdot-}$ is formed at the accessory pigments and by the Mehler reaction at PSI [65]. H_2O_2 thus formed is additionally reduced by the activity of ascorbate peroxidases (APXs) [65]. Transition metals can transform this specie into significantly destructive OH [66]. As cells do not have the means to detoxify OH enzymatically, plants can confine the OH^- provoked damage to the cell membrane by hindering its arrangement, or by detoxifying it by nonenzymatic antioxidants. Another nonradical ROS that can form in plant cells is $^1O_2$. $^1O_2$ can be shaped in PSII by the electron transfer of the energized triplet state of chlorophyll to $^3O_2$ [67], which are as unsafe as $\cdot OH$. $^1O_2$ has a short half-life as it is extinguished by water [65]. H_2O_2 is the steadiest of the ROS. The half-life of $^1O_2$, $O_2^{\cdot-}$, and OH are <1 µs. On the other hand, the half-life of H_2O_2 is in minutes [68]. So, H_2O_2 can diffuse to close by subcellular compartments and cross neighboring cells, which can be encouraged by the unique aquaporin's called peroxiporins [69].

Since aerobic metabolic processes, for example, respiration and photosynthesis, take place in mitochondria, chloroplast, and peroxisome. Therefore, these three organelles are viewed as the primary ROS producers during abiotic stresses [65]. While various other subcellular compartments may add to ROS generation, moreover, ROS can be delivered at the apoplastic space by plant nicotinamide adenine dinucleotide phosphate (NADPH) oxidases, peroxidases (POXs), and oxalate oxidases [70]. The photosynthetic photosystem I (PSI) and PSII reaction centers are the significant destinations of chloroplast ROS generation during light reactions of photosynthesis [65].

Since oxygen is one of the crucial substances in plants, it is involved in essential processes such as metabolism, mitochondrial respiration, and energy-producing processes like oxidative phosphorylation. When it gets converted into ROS during salt stress, it causes various detrimental effects in plant cells and might cause cell death by its oxidative capacity. ROS can damage amino acid residues in proteins such as Met, Phe, Tyr, Trp, and Cys.

Under ideal conditions, intracellular ROS are mainly produced at a low level in organelles. However, ROS are drastically accumulated during stress conditions. Oxidative stress is a critical impact of ROS on plant structures. Impeding effects of ROS are an aftereffect of their capacity to cause lipid peroxidation in cell membranes, DNA damage, protein denaturation, carbohydrate oxidation, pigment breakdown, and inactivation of enzymes (Figure 7.5). ROS generation is known to increase drastically under stress conditions. Oxidative burst is related to plant reactions to pathogens; however, increased ROS generation has been reported in light of an assortment of abiotic stresses including salinity, flooding, heat, cold, UV radiation, drought, and many others [71]. Other than the toxic quality of ROS, ROS additionally acts as signaling molecules that control plant functioning and reaction to biotic and abiotic stress [72]. Numerous studies have concentrated on ROS metabolism [73], ROS sensory and signaling systems [74], and the cross-talk with other signaling molecules and their function in developmental and stress response processes [73, 74]. However, stress like salt stress frequently enhances the production of ROS. The adverse outcomes of ROS build-up such as lipid peroxidation, oxidation of proteins, nucleic acids degradation, enzyme inhibition, and initiation of programmed cell death are likewise connected to signaling. Since these ROS ions acting as signaling molecules transmit information to downstream events, which need to be activated for stress response [75]. The consistent state of ROS levels relies upon the production and decomposition. Once the plant senses surge in ROS, a comprehensive defense system gets activated like system antioxidants, antioxidant enzymes, redox input elements, redox transmitters, redox target proteins, and redox sensors formulate the modification of redox homeostasis and redox-dependent response [76].

Salt stress harms stomatal conductance and brings down transpiration. Stomatal development is connected to ion redistribution, alkalization, and accumulation of abscisic acid [77]. Irregularities in stomatal opening confine gas exchange and thus limits CO_2 take-up, and ultimately brings down intercellular CO_2 fixation, and CO_2 accessibility for the Calvin cycle. As a result, the pool of oxidized nicotinamide adenine dinucleotide phosphate is exhausted, and electron share moved to O_2 to generate $O_2^{\cdot-}$ [78]. The reduction in the CO_2/O_2 proportion in the mesophyll upgrades photorespiration in C3-plants and invigorates H_2O_2 generation in the peroxisome [79]. Photorespiration represents over 70% of the H_2O_2 production under osmotic stress [80]. Similarly, $O_2^{\cdot-}$ is produced when respiratory electron transport is over-reduced. This $O_2^{\cdot-}$ is converted into H_2O_2 which gets reduced to water by catalases (CATs), class I peroxidases (APXs), class III POXs, and thiol POXs. Salt stress activates the cell membrane-bound RBOH [81] and the apoplastic diamine oxidase [82]. The two systems add to the ROS production in the apoplastic space. This leads to the production of $^\cdot$OH, which causes membrane disruption [83]. ROS-producing enzymes comprise oxalate oxidase and amine oxidase. Under salt stress, each sort of organelle utilizes different ways of ROS generation. The signaling activity of ROS depends upon its production, be it cellular or subcellular. Accordingly, ROS is produced individually in chloroplasts, the respiratory electron transport in mitochondria, different oxidases in the peroxisome, and the NADPH oxidase (RBOH) in the plasma membrane. Moreover, the reactions in the endomembrane and the cell wall also contribute toward ROS generation.

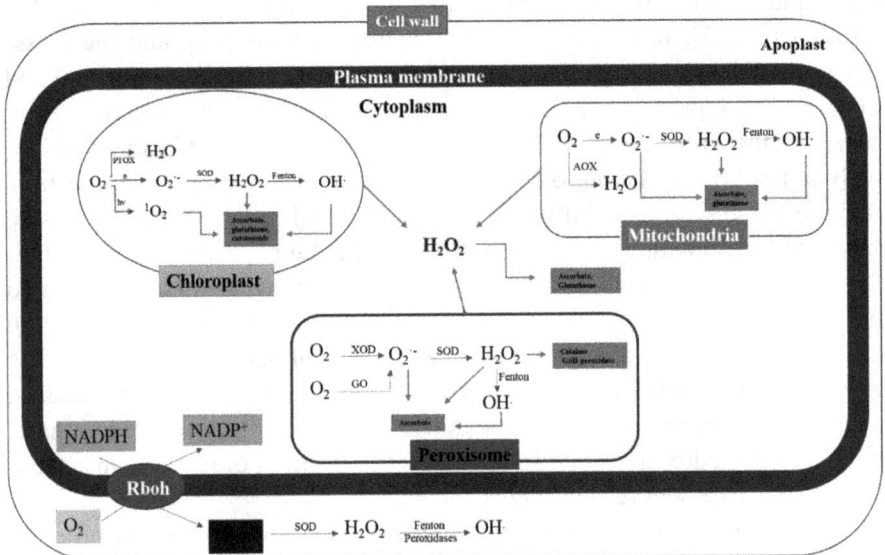

FIGURE 7.5   Different ROS generation sites in plant cell and there scavenging by different antioxidants. H_2O_2: hydrogen peroxide; AOX: alternativeoxidase; GO: glycolate oxidase; PET: photosyntheticelectrontransport; PTOX: plastid terminal oxidase; RBOH: respiratoryburstoxidasehomolog; XOD: xanthineoxidase; Fenton: decomposition of hydrogen peroxide to highly reactive $^\cdot$OH in presence of iron.

Plants have fostered dynamic ROS scavenging systems, including enzymatic or nonenzymatic pathways, to counter the harmful impacts of ROS generation (Figure 7.5). These include both enzymatic antioxidants such as CAT, glutathione reductase (GR), SOD, APX, glutathione peroxidase, guaiacol peroxidase, and glutathione S-transferase (GST), and nonenzymatic antioxidants such as glutathione (GSH), ascorbic acid, and carotenoids, which play an indispensable role in scavenging free radicals [76]. SOD is considered as the first line of antioxidant defense. Being a metalloenzyme, SOD can be either Manganese-SOD (Mn-SOD), Copper/Zinc SOD, and Iron SOD. SOD dismutase $O_2^{\cdot-}$ into H_2O_2 and O_2 [84]. Increased SOD activity frequently seems to upgrade plant resistance to oxidative stress. Among the antioxidant enzymes, CAT was found first, and it dismutase's two atoms of H_2O_2 into water and oxygen. CAT is crucial for ROS detoxification with the potential to change over H_2O_2 into H_2O and O_2 legitimately. CAT has the most elevated turnover rates as one particle of CAT can turnover ~6 million atoms of H_2O_2 into H_2O and O_2 per minute [76]. An upsurge in CAT movement is one of the leading defense strategies in halophytes [85]. Similarly, POXs are grouped as heme or thiol (or selenol) POXs and reduce H_2O_2 at the expense of an alternative electron donor [86]. APX catalyzes the essential first step in the water–water cycle. APX is similar to CAT. However, they catalyze the expulsion of H_2O_2 by utilizing ascorbate as a reductant. It is crucial in the regulation of the intracellular degree of H_2O_2 in higher plants. APX is found in various cell compartments, for example, stroma and thylakoid, microbody including glyoxysomes and peroxisomes, cytosol, and mitochondria. In the same manner, GST and PRX reduce H_2O_2 and organic hydroperoxides through ascorbate-independent thiol-mediated pathways using GSH, thioredoxin, or glutaredoxins [73]. PsTRXo1 and PRXIIF are vital to the cell to protect the cell from oxidative damage along with Mn-SOD and AOX. The connection between salt resistance and enhanced activities of antioxidant enzymes is often recognized [87]. It is well recognized that upon salt stress, plants defend themselves by antioxidative enzymes against various stress, depending on the plant, and it has been observed that there is a difference in the antioxidant enzymes in C3 and C4 plants.

7.3.5   SALINITY STRESS AND OSMOLYTES

Osmolytes are a group of organic compounds that are uncharged, polar, and soluble. Under salinity stress, plants first experience water deficiencies caused by osmotic stress; this nonavailability of water is referred to as physiological dryness because the water is present there; however, plants cannot use it because of the dissolved salts [88]. These osmolytes are hydrophilic and replace water at the surface of proteins, protein complexes, or membranes, and they do not affect the standard biochemical mechanism.

They mitigate the inhibitory effects of increased ion concentrations in the cell. Osmoprotectants accumulate in response to osmotic stress, while some solutes may be already present in the cell, such as trehalose [89]. Plant cells accumulate soluble osmolytes to regulate their osmotic potentials, such as sugars, proline, amino acids, acyclic and cyclic polyols, fructans, ectoine, and sulfonium compounds glycine betaine (GB), glutathione, and proteins, which enable the plants to improve the osmotic stress and keep cell turgor, water uptake, and metabolic activity [25].

Proline is vital for the osmotic modification since it has a low molecular weight, it is extremely soluble in water, it is comparatively nontoxic, and has no net charge in the physiological pH [90]. Proline accumulation is a typical adaptive response to many abiotic stresses [91]. Proline metabolism is vital in stress responses and tolerance to several adverse environmental conditions [92]. Proline is mainly present in the cytosol, chloroplast, cytoplasm, and vacuoles. Its accumulation helps to maintain the osmotic balance under salinity stress, and proline content is used as a physiological indicator of plant resistance to various stresses, including salt stress [93]. Proline can be produced from glutamate or ornithine. Glutamate functions as the primary precursor for proline synthesis in the osmotically stressed cell. The pyrroline carboxylic acid synthetase and pyrroline carboxylic acid reductase comprise two crucial enzymes in the biosynthetic pathway, which help in the overproduction of proline in stress conditions [94]. It quenches ROS, thereby improving the antioxidant capability of plants [95].

GB is another osmolyte found in microorganisms, higher plants, and animals. It is an amphoteric quaternary ammonium compound, nontoxic, easily soluble in water, and is electrically neutral over a wide range of pH. It interacts with both hydrophobic and hydrophilic domains of enzymes and protein complexes as it has a unique structure. GB accumulates in stress conditions and has a vital function in stress tolerance; it advances the osmolarity of the cell during the stress period. GB stabilizes proteins [96] protects the photosynthetic apparatus from injuries [97], helps in scavenging of ROS [98], and safeguards the thylakoid membrane and chloroplast. Accumulation of GB under salt stress is a widespread phenomenon in the plant kingdom. GB is produced within the cell from either choline or glycine. Choline is first oxidized by the enzyme choline monooxygenase to betaine aldehyde and then again oxidized by betaine aldehyde dehydrogenase to form GB [99]. However, some halophytic plants synthesize GB from glycine with the help of enzymes, S-adenosyl methionine dependent methyltransferases, glycine sarcosine N-methyl transferase, and sarcosine dimethylglycine N-methyl transferase [99].

Salinity stress also triggers the accumulation of sugar [100]. Sugars such as glucose, sucrose, and trehalose function as osmoprotectants. They protect the cell membrane and protoplast [101]. Also, they shield enzymes from getting denatured by the high ion concentration. Plants improve their osmotic potential by accumulating sugars to oppose salt stress [100]. Sugars mitigate salt stress by acting as an osmoprotectant, helps in carbon storage, and protection against ROS. In this manner, these sugar-derived osmolytes are a crucial physiological marker of salt stress. The concentration of trehalose, sucrose, and various other sugars is increased under salt stress. They play an essential role in osmoprotection and carbon storage role in physiological responses [99]. During stress, condition plants employ various sucrose metabolizing enzymes. Starch is converted into dextrin's by α-amylases and to maltose by β-amylases [102]. Sucrose is synthesized by Sucrose phosphate synthase and sucrose synthase; however, acid invertase helps in the breakdown of sucrose into glucose [103]. Trehalose (α-D-glucopyranosyl-1,1-α-D-glucopyranoside) is a nonreducing disaccharide that controls carbohydrate metabolism [104]. It protects the plant under stress conditions by keeping up K^+/Na^+ proportion, scavenging ROS, and increasing the concentration of sugars in plants [105].

So salinity stress disturbs the plant's overall growth and development, ion balance, K^+/Na^+ ratio, degrades pigment concentration, leads to the production of ROS, and accumulation of various osmolytes. Thereby hindering the overall growth and development of plants.

7.4   CONCLUSION

This chapter gives incredible information to create robust practices to improve crop resistance and yield in saline conditions. Plants utilize necessary adaptive reactions to antagonize salt stress. To sum up, salt stress in plants is a growing concern for the agricultural industry and overshadows the food security. It triggers a water shortage and ionic toxicity in plants, causing trivial in major plant processes such as photosynthesis and ion balance. This information might be useful to comprehend the physiological, metabolic, and other responses of plants to salt stress, bringing about the plummeting of biomass creation and yield. There is a need for multipronged arrangements for the development of crops and strategies, which will expand resilience to salt stress. Incredible advancement has been made in the most recent decades; however, yet a large number of the fundamental procedures that add to resistance are just incompletely comprehended. Further investigations are direly expected to disentangle the sensitivities of Na^+ and Cl^- ions. There take-up procedures and their connections with one another and with different particles and more prominent comprehension of the role of different minerals, for example, K^+ should empower us to relieve the salt stress by controlling take-up and dissemination of these supplements.

KEYWORDS

- **stress**
- **salinity**
- **reactive oxygen species (ROS)**
- **NaCl**
- **soil salinization**

REFERENCES

1. Beddington JR, Asaduzzaman M, Fernandez A, Clark ME, Guillou M, Jahn MM, Erda L, Mamo T, Bo N and Nobre CA (2012) Achieving food security in the face of climate change: Final Report from the Commission on Sustainable Agriculture and Climate Change.
2. Foley JA, Ramankutty N, Brauman KA, Cassidy ES, Gerber JS, Johnston M, Mueller ND, O'Connell C, Ray DK and West PCJN (2011) Solutions for a cultivated planet. Nature 478:337–342.
3. Pielke Sr RA, Adegoke JO, Chase TN, Marshall CH, Matsui T and Niyogi DJ (2007) A new paradigm for assessing the role of agriculture in the climate system and in climate change. Agricultural Forest Meteorology 142:234–254.
4. Arunanondchai P, Fei C, Fisher A, McCarl BA, Wang W and Yang Y (2018) How does climate change affect agriculture? In: The Routledge Handbook of Agricultural Economics, Routledge, pp. 191–210.
5. Noya I, González-García S, Bacenetti J, Fiala M and Moreira MTJ (2018) Environmental impacts of the cultivation-phase associated with agricultural crops for feed production. Journal of Cleaner Production 172:3721–3733.
6. Vaughan MM, Block A, Christensen SA, Allen LH and Schmelz EAJ (2018) The effects of climate change associated abiotic stresses on maize phytochemical defenses. Phytochemistry Reviews 17:37–49.
7. Kanojia A and Dijkwel PPJ (2018) Abiotic stress responses are governed by reactive oxygen species and age. Annual Plant Reviews 1:295–326.
8. Richardson KJ, Lewis KH, Krishnamurthy PK, Kent C, Wiltshire AJ and Hanlon HMJ (2018) Food security outcomes under a changing climate: Impacts of mitigation and adaptation on vulnerability to food insecurity. Climatic Change 147:327–341.

9. Tito R, Vasconcelos HL and Feeley KJJ (2018) Global climate change increases risk of crop yield losses and food insecurity in the tropical Andes. Global Change Biology 24:e592–e602.

10. Pachauri RK, Allen MR, Barros VR, Broome J, Cramer W, Christ R, Church JA, Clarke L, Dahe Q and Dasgupta P (2014) Climate change 2014: synthesis report. In: Contribution of Working Groups I, II and III to the Fifth Assessment Report of the Intergovernmental Panel on Climate Change. Ipcc.

11. Andy PJ (2016) Abiotic stress tolerance in plants. Plant Science 7:1–9.

12. Rosenzweig C, Elliott J, Deryng D, Ruane AC, Müller C, Arneth A, Boote KJ, Folberth C, Glotter M and Khabarov NJ (2014) Assessing agricultural risks of climate change in the 21st century in a global gridded crop model intercomparison. Proceedings of the National Academy of Sciences 111:3268–3273.

13. Khan A, Ijaz M, Muhammad J, Goheer A, Akbar G and Adnan MJ (2016) Climate change implications for wheat crop in dera ismail khan district of khyber pakhtunkhwa. Pakistan Journal of Meteorology 13, 17–27.

14. Altieri MA and Nicholls CIJ (2017) The adaptation and mitigation potential of traditional agriculture in a changing climate. Climatic Change 140:33–45.

15. Lesk C, Rowhani P and Ramankutty NJ (2016) Influence of extreme weather disasters on global crop production. Nature 529:84–87.

16. Rejeb IB, Pastor V and Mauch-Mani BJ (2014) Plant responses to simultaneous biotic and abiotic stress: molecular mechanisms. Plants 3:458–475.

17. Wani SH and Sah SJJ (2014) Biotechnology and abiotic stress tolerance in rice. Rice Research 2:e105.

18. Pereira AJ (2016) Plant abiotic stress challenges from the changing environment. Frontiers in Plant Science 7:1123.

19. Roberts DP and Mattoo AKJ (2018) Sustainable agriculture—enhancing environmental benefits, food nutritional quality and building crop resilience to abiotic and biotic stresses. Agriculture 8:8.

20. Verstraeten SV, Aimo L and Oteiza PIJ (2008) Aluminium and lead: molecular mechanisms of brain toxicity. Archives of Toxicology 82:789–802.

21. Jamil A, Riaz S, Ashraf M and Foolad MR (2011) Gene expression profiling of plants under salt stress. Critical Reviews in Plant Sciences 30:435–458.

22. Munns R and Gilliham MJ (2015) Salinity tolerance of crops–what is the cost? New Phytologist 208:668–673.

23. Lachhab I, Louahlia S, Laamarti M and Hammani KJ (2013) Effet d'un stress salin sur la germination et l'activité enzymatique chez deux génotypes de Medicago sativa. International Journal of Innovation Applied Studies 2:511–516.

24. Carillo P, Annunziata MG, Pontecorvo G, Fuggi A and Woodrow PJ (2011) Salinity stress and salt tolerance. Abiotic stress in plants–mechanisms adaptations, IntechOpen Limited, London, pp. 1:21–38.

25. Munns R and Tester M (2008) Mechanisms of salinity tolerance. Annual Review of Plant Biology 59:651–681.

26. Munns RJ (2002) Comparative physiology of salt and water stress. Plant, Cell Environmental 25:239–250.

27. Ishitani M, Liu J, Halfter U, Kim C-S, Shi W and Zhu J-KJ (2000) SOS3 function in plant salt tolerance requires N-myristoylation and calcium binding. The Plant Cell 12:1667–1677.

28. Rahnama A, James RA, Poustini K and Munns RJ (2010) Stomatal conductance as a screen for osmotic stress tolerance in durum wheat growing in saline soil. Functional Plant Biology 37:255–263.

29. Hasegawa PM (2013) Sodium (Na+) homeostasis and salt tolerance of plants. Environmental Experimental Botany 92:19–31.

30. Wu HJ (2018) Plant salt tolerance and Na+ sensing and transport. The Crop Journal 6: 215–225.

31. Ward JM, Hirschi KD and Sze HJ (2003) Plants pass the salt. Trends in Plant Science 8:200–201.

32. Shi H, Quintero FJ, Pardo JM and Zhu J-KJ (2002) The putative plasma membrane Na+/H+ antiporter SOS1 controls long-distance Na$^+$ transport in plants. The Plant Cell 14:465–477.

33. Zhu J-K (2003) Regulation of ion homeostasis under salt stress. Current Opinion in Plant Biology 6:441–445.

34. Dietz K-J, Tavakoli N, Kluge C, Mimura T, Sharma S, Harris G, Chardonnens A and Golldack D (2001) Significance of the V-type ATPase for the adaptation to stressful growth conditions and its regulation on the molecular and biochemical level. Journal of Experimental Botany 52:1969–1980.

35. Hasegawa PM, Bressan RA, Zhu J-K and Bohnert HJJ (2000) Plant cellular and molecular responses to high salinity. Annual Review of Plant Biology 51:463–499.

36. Knight H, Trewavas AJ and Knight MRJ (1997) Calcium signalling in Arabidopsis thaliana responding to drought and salinity. The Plant Journal 12:1067–1078.

37. Choi W-G, Toyota M, Kim S-H, Hilleary R and Gilroy SJPotNAoS (2014) Salt stress-induced Ca^{2+} waves are associated with rapid, long-distance root-to-shoot signaling in plants. Proceedings of the National Academy of Sciences of the United States of America 111:6497–6502.

38. Martínez-Atienza J, Jiang X, Garciadeblas B, Mendoza I, Zhu J-K, Pardo JM and Quintero FJJ (2007) Conservation of the salt overly sensitive pathway in rice. Plant Physiology 143:1001–1012.

39. Quintero FJ, Martinez-Atienza J, Villalta I, Jiang X, Kim W-Y, Ali Z, Fujii H, Mendoza I, Yun D-J and Zhu J-KJ (2011) Activation of the plasma membrane Na/H antiporter Salt-Overly-Sensitive 1 (SOS1) by phosphorylation of an auto-inhibitory C-terminal domain. Proceedings of the National Academy of Sciences 108:2611–2616.

40. Sairam R and Tyagi AJ (2004) Physiology and molecular biology of salinity stress tolerance in plants. Current Science, 86:407–421.

41. Emilie Yen H, Wu SM, Hung YH and Yen SKJ (2000) Isolation of 3 salt-induced low-abundance cDNAs from light-grown callus of *Mesembryanthemum crystallinum* by suppression subtractive hybridization. Physiologia Plantarum 110:402–409.

42. Schroeder JI, Delhaize E, Frommer WB, Guerinot ML, Harrison MJ, Herrera-Estrella L, Horie T, Kochian LV, Munns R and Nishizawa NKJ (2013) Using membrane transporters to improve crops for sustainable food production. Nature 497:60–66.

43. Barragán V, Leidi EO, Andrés Z, Rubio L, De Luca A, Fernández JA, Cubero B and Pardo JMJ (2012) Ion exchangers NHX1 and NHX2 mediate active potassium uptake into vacuoles to regulate cell turgor and stomatal function in *Arabidopsis*. The Plant Cell 24:1127–1142.

44. Munns RJ (2005) Genes and salt tolerance: bringing them together. New Phytologist 167:645–663.

45. Munns R (1993) Physiological processes limiting plant growth in saline soils: some dogmas and hypotheses. Plant, Cell Environmental 16:15–24.
46. Flowers TJ and Colmer TDJ (2015) Plant salt tolerance: adaptations in halophytes. Annals of Botany 115:327–331.
47. Mohammad M, Shibli R, Ajlouni M and Nimri LJ (1998) Tomato root and shoot responses to salt stress under different levels of phosphorus nutrition. Journal of Plant Nutrition 21:1667–1680.
48. Negrão S, Schmöckel S and Tester MJ (2017) Evaluating physiological responses of plants to salinity stress. Annals of Botany 119:1–11.
49. López MAH, Ulery AL, Samani Z, Picchioni G and Flynn RJ (2011) Response of chile pepper (Capsicum annuum L.) to salt stress and organic and inorganic nitrogen sources: I. growth and yield. Tropical Subtropical Agroecosystems 14:137–147.
50. Kamiab F, Talaie A, Javanshah A, Khezri M and Khalighi AJ (2012) Effect of long-term salinity on growth, chemical composition and mineral elements of pistachio (Pistacia vera cv. Badami-Zarand) rootstock seedlings. Annals of Biological Research 3:5545–5551.
51. Kafkafi UJ (1991) Root growth under stress: Salinity. Plant Roots.:375–391.
52. Botella MA, Quesada MA, Kononowicz AK, Bressan RA, Pliego F, Hasegawa PM and Valpuesta VJ (1994) Characterization and in situ localization of a salt-induced tomato peroxidase mRNA. Plant Molecular Biology 25:105–114.
53. Parihar P, Singh S, Singh R, Singh VP and Prasad SMJ (2015) Effect of salinity stress on plants and its tolerance strategies: a review. Environmental Science Pollution Research 22:4056–4075.
54. Alem C, Labhilili M, Brahmi K, Jlibene M, Nasrallah N and Filali-Maltouf A (2002) Adaptations hydrique et photosynthétique du blé dur et du blé tendre au stress salin. Comptes Rendus Biologies 325:1097–1109.
55. Zhang M, Qin Z and Liu X (2005) Remote sensed spectral imagery to detect late blight in field tomatoes. Precision Agriculture 6:489–508.
56. Chutipaijit S, Cha-um S and Sompornpailin K (2011) High contents of proline and anthocyanin increase protective response to salinity in'Oryza sativa'L. spp.'indica'. Australian Journal of Crop Science 5:1191.
57. Asrar H, Hussain T, Hadi SMS, Gul B, Nielsen BL and Khan MA (2017) Salinity induced changes in light harvesting and carbon assimilating complexes of *Desmostachya bipinnata* (L.) Staph. Environmental Experimental Botany 135:86–95.
58. Kalaji H, Rastogi A, Živčák M, Brestic M, Daszkowska-Golec A, Sitko K, Alsharafa K, Lotfi R, Stypiński P and Samborska I (2018) Prompt chlorophyll fluorescence as a tool for crop phenotyping: an example of barley landraces exposed to various abiotic stress factors. Photosynthetica 56:953–961.
59. Munns R and Tester MJ (2008) Mechanisms of salinity tolerance. Annual Review of Plant Biology 59:651–681.
60. Rasool A, Shah WH, Tahir I, Alharby HF, Hakeem KR and Rehman R (2020) Exogenous application of selenium (Se) mitigates NaCl stress in proso and foxtail millets by improving their growth, physiology and biochemical parameters. Acta Physiologiae Plantarum 42:1–13.
61. Brugnoli E and Björkman O (1992) Growth of cotton under continuous salinity stress: influence on allocation pattern, stomatal and non-stomatal components of photosynthesis and dissipation of excess light energy. Planta 187:335–347.

62. Iyengar E and Reddy MJHopMD, Baten Rose, USA (1996) Photosynthesis in highly salt tolerant plants. Handbook of photosynthesis. Marshal Dekar, Baten Rose, USA, p. 909.
63. Shah WH, Rasool A, Tahir I and Rehman RU (2020) Exogenously applied selenium (Se) mitigates the impact of salt stress in *Setaria italica* L. and *Panicum miliaceum* L. The Nucleus 63:1–13.
64. Azevedo Neto AD, Gomes Filho E, Prisco JT (2008) Salinity and oxidative Stress. In: Khan NA, Singh S (eds.) Abiotic Stress and Plant Responses, IK International, New Delhi, pp. 58–82.
65. Asada KJ (2006) Production and scavenging of reactive oxygen species in chloroplasts and their functions. Plant Physiology 141:391–396.
66. Rodrigo-Moreno A, Poschenrieder C and Shabala SJ (2013) Transition metals: a double edge sward in ROS generation and signaling. Plant Signaling Behavior 8:e23425.
67. Laloi C, Apel K and Danon AJ (2004) Reactive oxygen signalling: the latest news. Current Opinion in Plant Biology 7:323–328.
68. Pitzschke A, Forzani C and Hirt HJ (2006) Reactive oxygen species signaling in plants. Antioxidants Redox Signaling 8:1757–1764.
69. Bienert GP, Schjoerring JK and Jahn TPJ (2006) Membrane transport of hydrogen peroxide. Biochimica et Biophysica Acta -Biomembranes 1758:994–1003.
70. Kawano TJ (2003) Roles of the reactive oxygen species-generating peroxidase reactions in plant defense and growth induction. Plant Cell Reports 21:829–837.
71. Miller G, Shulaev V and Mittler R (2008) Reactive oxygen signaling and abiotic stress. Physiologia Plantarum 133:481–489.
72. Apel K and Hirt H (2004) Reactive oxygen species: metabolism, oxidative stress, and signal transduction. Annual Review of Plant Biology 55:373–399.
73. Noctor G, Mhamdi A and Foyer CH (2014) The roles of reactive oxygen metabolism in drought: not so cut and dried. Plant Physiology 164:1636–1648.
74. Suzuki N, Koussevitzky S, Mittler R and Miller G (2012) ROS and redox signalling in the response of plants to abiotic stress. Plant, Cell Environmental 35:259–270.
75. Srivastava S and Dubey R (2011) Manganese-excess induces oxidative stress, lowers the pool of antioxidants and elevates activities of key antioxidative enzymes in rice seedlings. Plant Growth Regulation 64:1–16.
76. Gill SS and Tuteja N (2010) Reactive oxygen species and antioxidant machinery in abiotic stress tolerance in crop plants. Plant Physiology Biochemistry 48:909–930.
77. Geilfus CM, Mithöfer A, Ludwig-Müller J, Zörb C and Muehling KH (2015) Chloride-inducible transient apoplastic alkalinizations induce stomata closure by controlling abscisic acid distribution between leaf apoplast and guard cells in salt-stressed Vicia faba. New Phytologist 208:803–816.
78. Mehler AH (1951) Studies on reactions of illuminated chloroplasts: I. Mechanism of the reduction of oxygen and other hill reagents. Archives of Biochemistry Biophysics 33:65–77.
79. Ghannoum O (2009) C4 photosynthesis and water stress. Annals of Botany 103:635–644.
80. Noctor G, Veljovic-Jovanovic S, Driscoll S, Novitskaya L and FOYER CH (2002) Drought and oxidative load in the leaves of C3 plants: a predominant role for photorespiration? Annals of Botany 89:841–850.
81. Rejeb KB, Benzarti M, Debez A, Bailly C, Savouré A and Abdelly C (2015) NADPH oxidase-dependent H2O2 production is required for salt-induced antioxidant defense in Arabidopsis thaliana. Journal of Plant Physiology 174:5–15.

82. Waie B and Rajam MV (2003) Effect of increased polyamine biosynthesis on stress responses in transgenic tobacco by introduction of human S-adenosylmethionine gene. Plant Science 164:727–734.

83. Rodríguez AA, Maiale SJ, Menéndez AB and Ruiz OA (2009) Polyamine oxidase activity contributes to sustain maize leaf elongation under saline stress. Journal of Experimental Botany 60:4249–4262.

84. Tuna AL, Kaya C, Dikilitas M and Higgs D (2008) The combined effects of gibberellic acid and salinity on some antioxidant enzyme activities, plant growth parameters and nutritional status in maize plants. Environmental Experimental Botany 62:1–9.

85. Lokhande VH, Mulye K, Patkar R, Nikam TD and Suprasanna P (2013) Biochemical and physiological adaptations of the halophyte *Sesuvium portulacastrum* (L.) L.,(Aizoaceae) to salinity. Archives of Agronomy Soil Science 59:1373–1391.

86. Dietz K-J (2016) Thiol-based peroxidases and ascorbate peroxidases: why plants rely on multiple peroxidase systems in the photosynthesizing chloroplast? Molecules Cells 39:20.

87. LIU H-x, XIN Z-y and ZHANG Z-y (2011) Changes in activities of antioxidant-related enzymes in leaves of resistant and susceptible wheat inoculated with Rhizoctonia cerealis. Agricultural Sciences in China 10:526–533.

88. Brinker M, Brosché M, Vinocur B, Abo-Ogiala A, Fayyaz P, Janz D, Ottow EA, Cullmann AD, Saborowski J and Kangasjärvi JJ (2010) Linking the salt transcriptome with physiological responses of a salt-resistant Populus species as a strategy to identify genes important for stress acclimation. Plant Physiology 154:1697–1709.

89. Slama I, Abdelly C, Bouchereau A, Flowers T and Savouré AJ (2015) Diversity, distribution and roles of osmoprotective compounds accumulated in halophytes under abiotic stress. Annals of Botany 115:433–447.

90. Liang W, Ma X, Wan P and Liu LJ (2018) Plant salt-tolerance mechanism: a review. Biochemical Biophysical Research Communications 495:286–291.

91. Akula R, and Ravishankar GA (2011) Influence of abiotic stress signals on secondary metabolites in plants. Plant Signaling & Behavior 6:1720–1731.

92. Suprasanna P, Rai AN, HimaKumari P, Kumar SA and Kavi Kishor PJ (2014) Modulation of proline: implications in plant stress tolerance and development. Plant Adaptation to Environmental Change. CABI Publishers, UK, pp. 68–93.

93. Hasanuzzaman M, Alam M, Rahman A, Hasanuzzaman M, Nahar K and Fujita MJ (2014) Exogenous proline and glycine betaine mediated upregulation of antioxidant defense and glyoxalase systems provides better protection against salt-induced oxidative stress in two rice (*Oryza sativa* L.) varieties. BioMed Research International 2014.

94. Ma L, Zhang H, Sun L, Jiao Y, Zhang G, Miao C and Hao FJ (2012) NADPH oxidase AtrbohD and AtrbohF function in ROS-dependent regulation of Na+/K+ homeostasis in Arabidopsis under salt stress. Journal of Experimental Botany 63:305–317.

95. Matysik J, Alia, Bhalu B and Mohanty PJ (2002) Molecular mechanisms of quenching of reactive oxygen species by proline under stress in plants. Current Science:525–532.

96. Mäkelä P, Kärkkäinen J and Somersalo SJBP (2000) Effect of glycinebetaine on chloroplast ultrastructure, chlorophyll and protein content, and RuBPCO activities in tomato grown under drought or salinity. Biologia Plantarum 43:471–475.

97. CHAUM S and Kirdmanee CJ (2010) Effect of glycinebetaine on proline, water use, and photosynthetic efficiencies, and growth of rice seedlings under salt stress. Turkish Journal of Agriculture Forestry 34:517–527.

98. Saxena SC, Kaur H, Verma P, Petla BP, Andugula VR and Majee M (2013) Osmoprotectants: potential for crop improvement under adverse conditions. In: Plant Acclimation to Environmental Stress, Springer, Berlin, Germany, pp. 197–232.

99. Ahmad R, Lim CJ and Kwon S-YJPBR (2013) Glycine betaine: a versatile compound with great potential for gene pyramiding to improve crop plant performance against environmental stresses. Plant Biotechnology Reports 7:49–57.

100. Kerepesi I and Galiba GJ (2000) Osmotic and salt stress-induced alteration in soluble carbohydrate content in wheat seedlings. Crop Science 40:482–487.

101. Guo R, Yang Z, Li F, Yan C, Zhong X, Liu Q, Xia X, Li H and Zhao LJ (2015) Comparative metabolic responses and adaptive strategies of wheat (TriticBMC plant biologyum aestivum) to salt and alkali stress. BMC Plant Biology 15:170.

102. Ballicora MA, Iglesias AA and Preiss JJ (2004) ADP-glucose pyrophosphorylase: a regulatory enzyme for plant starch synthesis. Photosynthesis Research 79:1–24.

103. Peng J, Liu J, Zhang L, Luo J, Dong H, Ma Y, Zhao X, Chen B, Sui N and Zhou ZJ (2016) Effects of soil salinity on sucrose metabolism in cotton leaves. PLoS One 11.

104. Lunn JE, Delorge I, Figueroa CM, Van Dijck P and Stitt MJ (2014) Trehalose metabolism in plants. The Plant Journal 79:544–567.

105. Chang B, Yang L, Cong W, Zu Y, Tang ZJ and Biochemistry (2014) The improved resistance to high salinity induced by trehalose is associated with ionic regulation and osmotic adjustment in *Catharanthus roseus*. Plant Physiology 77:140–148.

CHAPTER 8

Next-Generation Climate-Resilient Agricultural Technology in Traditional Farming for Food and Nutritional Safety in the Modern Era of Climate Change

AKBAR HOSSAIN[1*], SAGAR MAITRA[2], SOURAV GARAI[3], MOUSUMI MONDAL[4], ASGAR AHMED[5], MST. TANJINA ISLAM[6], and JAGAMOHAN NAYAK[7]

[1]*Bangladesh Wheat and Maize Research Institute, Dinajpur-5200, Bangladesh*

[2]*Department of Agronomy, Centurion University of Technology and Management, Paralakhemundi 761211, India*

[3]*Department of Agronomy, Bidhan Chandra KrishiViswavidyalaya, Nadia, West Bengal, India*

[4]*Research Scholar, Department of Agronomy, Bidhan Chandra Krishi Viswavidyalaya, Nadia, West Bengal, India*

[5]*Maize Breeding Division, Bangladesh Wheat and Maize Research Institute, Dinajpur-5200, Bangladesh*

[6]*Department of Agronomy, Hajee Mohammad Danesh Science and Technology University, Dinajpur 5200, Bangladesh*

[7]*Department of Agronomy, Bidhan Chandra KrishiViswavidyalaya, Nadia, West Bengal, India*

Corresponding author. E-mail: akbarhossainwrc@gamil.com.

ABSTRACT

It is predicted that the population across the globe is approaching to 9.5 billion by the year 2050. Therefore, sustainable food and nutritional safety are the foremost encounters of the 21st century. Human activities are largely responsible for rising the ecological complications such as increasing climate change consequences decline the food and nutritional security and consequently mounting the food values. In the current era, global warming as a consequence of the changing climate has turn out to be the extreme fears of agricultural systems, which leads to global food and nutritional security. Since agriculture is amongst the greatly thoughtful systems prejudiced due to the change in ecological condition. These consequences are going to be more serious in countries with agro-based economic and traditional agricultural practices. Therefore to reach the target aims for the eradication of starvation and poverty by the year 2050, and also reducing the GHGs emission, climate-smart agricultural technologies are essential in traditional farmings. This chapter advocates next-generation climate-smart agricultural technologies in traditional farmings for the sustainability of crop production and also deliberates the correlation between climate change and agriculture.

8.1 INTRODUCTION

Climate change is the alteration over normal climatic conditions that are occurred by natural and artificial causes, where natural causes are considered as the revolution of the earth, volcanic eruptions, earthquake, and movement of the earth's crust, but artificial factors including anthropogenic activities are measured as the enhancement of greenhouse gases (GHGs) and aerosol. However, anthropogenic changes occur within a short period of few decades causing harm to the environment and living beings. One of the adverse effects of climate change is global warming, which means the rise of atmospheric temperature that has a huge impact on the earth's ecosystem, particularly on the agroecosystem. The intergovernmental panel on climate change (IPCC) narrated the negative impacts of the changing climate recently due to ever-increasing emissions of GHGs by anthropogenic activities, which are accountable for global warming and interestingly, from the middle of the 20th century it has been caused. In the agriculture production and food value chains, enough of GHGs are emitted and an estimate figures that globally 2.5 million small farmers are responsible for 10%–29% of GHGs caused by agriculture due to their faulty agricultural practices [1, 2]. Further, IPCC [3] estimated that if the same man-made activities go on there will be a chance in an increase of the earth's temperature by 6.4 °C and due to the melting of glaciers, sea level will rise up to 59 cm in the 21st century. In the first half of the 20th century, burning of fossil fuel was comparatively less than the next half, but in the previous century (1906–2005) the earth witnessed a significant increase in temperature by 0.74 °C [4]. The IPCC [3] further projected that there will be an increase in average temperature globally from 1.1 °C to 5.4 °C by 2100, based on anthropogenic influences to the emission of GHGs during the period. Undoubtedly, the enhancement of atmospheric temperature will further create associated effects for rising the sea level water, rainfall patterns and timing, and melting of snow cover [5]. Besides, global warming enhances the possibility of different natural calamities such as flood, drought, and storms as a resultant of the rise of atmospheric temperature and change in the pattern of rainfall.

The cumulative effect of all abnormalities climatic parameters ultimately causes hindrances to normal farming activities. Agriculture is a man-made initiative, which is highly climate dependent and sensitive to agroclimatic situations such as temperature, humidity, and rainfall [6–9]. Climate change directly influences crop productivity in terms of qualitative and quantitative variation. Modification in agronomic management like water and nutrient management and plant protection is also caused due to climate change. Further, alteration in agroecosystem may occur as drought or flood and reduction of cropping intensity, water stagnation, and drainage leading to loss of soil, nutrients, and diversity (Figure 8.1).

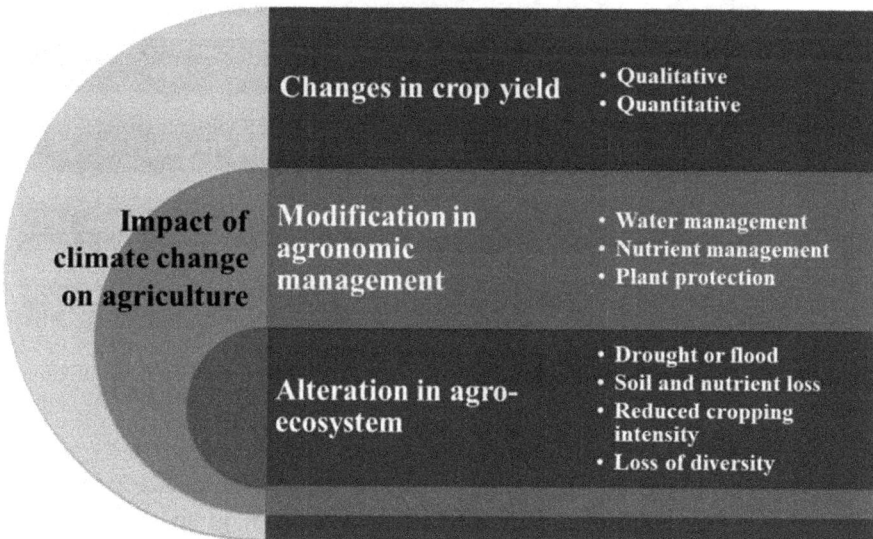

FIGURE 8.1 Adverse effects of global warming on agriculture as a consequence of global changing climate.

The consequence of global changing climate implies some alterations in the natural system and both the human being and natural system face the challenges; thus after assessing the threat for both, adaptation options to be considered [10]. The UNEP [11] well-defined the adaptation options as "The process of alteration of human systems to the real or predictable environment and its effects to reasonable impairment or feat helpful prospects, where the human involvement may simplify for the amendment to projected climate." To combat the hostile concerns of climate change, agronomic manipulations are in practices that in other words are known as adaptation options. The adaptation options include agronomic manipulation with existing crops and cultivars, choice of more abiotic stress cultivars of the existing crop, switching to ecologically sound and stress-tolerant crops, shifting the crop area from vulnerable to the risk-free area, following weather forecast, warning and decision support system, and crop insurance (Figure 8.2).

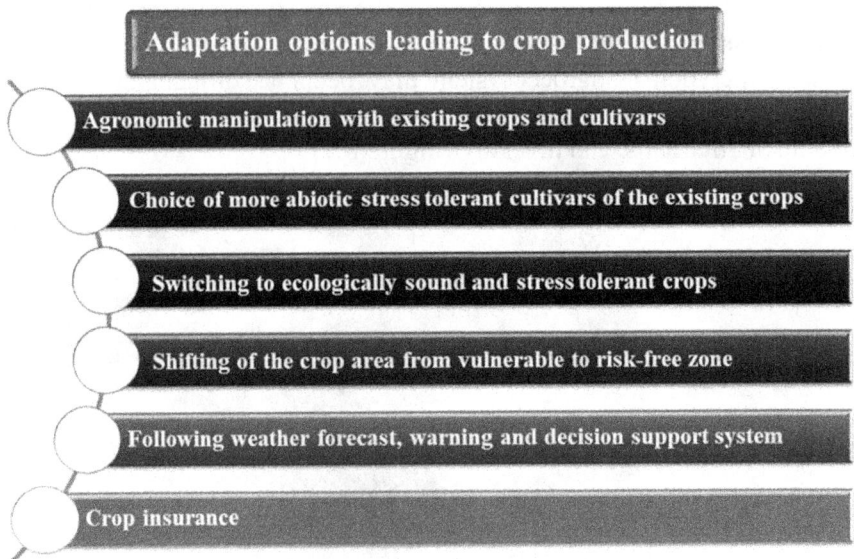

Adaptation options leading to crop production

Agronomic manipulation with existing crops and cultivars

Choice of more abiotic stress tolerant cultivars of the existing crops

Switching to ecologically sound and stress tolerant crops

Shifting of the crop area from vulnerable to risk-free zone

Following weather forecast, warning and decision support system

Crop insurance

FIGURE 8.2 Adaptation options leading to crop production in the modern era of climate change.

Reduction of the adverse impacts of climate change as well as global warming on agriculture and quality production is a great task [12–14]. Adaptation measures have already started in different countries to cope up with the situation. Heat tolerant cultivars have been developed and already have been under cultivation [15, 16]. Moreover, manipulation of sowing date, irrigation, drainage and water management, and soil management including tillage, mulching, and cover cropping is common agronomic management practices generally adopted against climate change [17–19]. The Food and Agricultural Organization (FAO) of United Nations indicated that long-term environmental inconsistency is a vital motorist of global food insecurity [20]. During present times, prediction of climate variability has become easier and considering the climate extremes suitable agronomic measures are adopted [21–23]. In this regard, different agrometeorological tools are useful in weather-related decision support system as farmers can adopt suitable measures [24, 25]. Further, there is a need for more precise information [26] with a proper communication network to the farmers, particularly, smallholders in their local vernacular. Crop insurance is another adaptive measure to safeguard the smallholders from crop failure due to climatic issues. During the present time, elimination of hunger and food security can be achieved by combined application of biotechnology and information technology with eco-friendly adoption of agronomic management [27]. The current chapter focuses on the antagonistic consequences of global warming and their mitigation strategies by the introduction of next-generation climate-resilient agricultural technologies in traditional farming for increasing food and nutritional safety under changing climate.

8.2 ADVERSE EFFECTS AND ADAPTATION OPTIONS FOR THE PRODUCTION OF MAJOR CROPS UNDER CHANGING CLIMATE

Agriculture is the mainstay of several developing countries, whereas it hardly matters in the economy of the developed countries. Agriculture contributes <2% of GDP in developed countries, but it is acting a momentous role in the GDP of low-income countries [28]. It is projected that the world cereal equivalent food demand will be around 14,886 million tons in 2050 [29] to feed 9.7 billion by 2050. Therefore, to meet the demand of the increasing population, food production must be increased in near future. As per the present concept, food security is synonymous to food and nutritional security. To meet the target, the latest and proven technologies are adopted considering the cropping as well as farming systems of various agroecological regions. The climate change effects on agriculture are mainly linked with two key tools. The first approach considers the influence of climate change by applying processed-based crop simulation models [30, 31]. But the second approach deals in econometric models estimated using observational data [32]. Different countries conducted studies based on crop modelings to predict the consequence of changing climate on major crops. The major consequences of changing climate and adaptations practiced in the important crop production are highlighted in the following sections.

8.2.1   RICE

Rice is the furthermost vital cereals of the world and cultivated in diverse areas between the latitudes 53°N and 40°S, that is, the valley of the Amur River (bordering Russia and China) and central Argentina, respectively [33]. The crop is grown in various climatic conditions as it is observed to grow at and altitudes of 2600 MSL in the Himalayan country Nepal and the hot deserts country of Egypt. Rice assures food security of more than half of the world population. An increase in atmospheric temperature and variability in rain are detrimental as they reduce the yield of rice [34, 35]. In rice cultivation, drought has been found more fatal than the flood, because rice area in the world was around 167 million ha in 2018 [36] and rainfed rice contributes 25% of global rice output [37]. In rainfed areas, rice is more subjected to water shortage due to uncertainty and uneven distribution pattern of rain. The congenial climatic condition for rice is the warm temperature (20–30°C) with a rainfall of 1500 mm and above [38]. But when temperature increases (reaches above 35°C), alteration in the physiological and metabolic process may occur. The high temperature causes spikelet sterility, chaffy grains, and low yield [39]. The crop performs well in a congenial temperature condition (Table 8.1). The common symptoms of heat stress of rice during the vegetative stage are chlorotic bands, white bands, specks and white tip in leaves, reduced plant height, and fewer tillering. But heat stress in reproductive stage causes sterility as well as the reduced number of spikelets. However, at the ripening stage of rice abnormal temperature results in poor grain filling [38].

The scanty rain cuts the rate of transpiration resulting in leaf rolling, drying, and reduction of leaf area and ultimately less assimilation of photosynthate [37, 40]. During present times, the unpredicted monsoon in India is creating uncertainty in the main rainy season crop, rice [41], and both the excess and scanty rain adversely affected rice productivity. However, drought has been caused by more damage events [42]. The emission of GHGs may reduce irrigated rice yield by ~10% in 2080, but there will be a reduction of productivity of rainfed rice by ~6% during the period [43]. In Andhra Pradesh, an Indian state, there will be a reduction of 8%–9% yield of rice in 2050 due to the rise of atmospheric temperature [44]. In Uttar Pradesh, India there was a rise of minimum temperature during the monsoon season by 0.06–0.44°C over a decade that adversely affected rice yield, particularly when the temperature prevailed more than 35°C during the reproductive stage of the crop [45], which negatively influenced pollen shading and grain filling [46]. Another study projected a decrease in rice production at *Konkan* region of Maharashtra by 15%–25% in 2040s and 10%–40% by 2080s [47].

TABLE 8.1 Critical Temperature Congenial for Rice Production

Crop Phenology	Life-Threatening Temperature (°C)			References
	High	Low	Optimum	
Gemination/Emergence	45	16–19	18–40	Adopted from sources [33, 39]
Growth of seedling	35	12	25–30	
Formation of root	35	16	25–28	
Elongation of leaf	45	7–12	31	
Tiller formation	33	9–16	25–31	
Initiation panicle primordia	–	15	–	
Panicle differentiation	30	15–20	–	
Anthesis	35–36	22	30–33	
Ripening	>30	12–18	20–29	

In a study, Murdiyarso [48] at IRRI showed that there will be a reduction of rice productivity by 10% for each degree rise in atmospheric temperature. Further, Matsuda [49] expected a reduction in rice yields of 3.8% by the 21st century, as a result of faulty fertilization, water scarcity, and the extreme event of several stresses due to the global warming. In Myanmar, the majority of agricultural land is situated in the central dry zone, where rice productivity is hampered due to rainfall variability [8, 50]. In Japan, already 6.6% area of paddy has been brought under cultivation of heat-tolerant rice varieties [26] along with some other agronomic management like manipulation of transplanting date (for upland rice), water, and nutrient management [16, 19]. There is an indication of the reduction of rice yield due to climate change in Portugal, a leading producer and consumer among EU countries [51].

Bangladesh is highly vulnerable to climate change [52, 53], and rice is the main crop of the country. During the last four decades, Bangladesh witnessed an increase in temperature by 0.103°C per decade [54, 55] and variability in rain [56]. The temperature of the country may increase by 1.4°C and 2.4 °C by middle and end of the century [52], and predicted decrease of rice production by 2050 is 8%–17% [3, 57]. Rice accounts for about 80% of the cropped area with 90% of grain production in Bangladesh [58, 59]. In Bangladesh, rice is cultivated during three seasons, namely, premonsoon (*aus*, March–July), monsoon (*aman*, June–November), and winter (*boro*, November–May) [60]. Majority of monsoon and premonsoon crop is rainfed, and erratic monsoon and other natural calamities cause a yield reduction as well as damage to rice. Rainfed rice is negatively affected by the erratic rainfall and increase of temperature during pre-monsoon and monsoon season [61]. Excess rain, water stagnation, and flood also damage rice. But *boro* rice is irrigated mostly by groundwater and temperature rise above 35°C during the reproductive stage hampers grain filling and productivity. There is an urgent need for the development of climate-resilient varieties of rice for different agroclimatic conditions and cropping systems for sustaining rice yield in Bangladesh [59, 62]. Soil salinity is another issue of rice ecosystem of Bangladesh, which may be further intensified under the climate change scenario and *aus* rice will be affected more [63]. The country has a 700 km long coastline in which 0.83 million ha are subjected to be adversely exaggerated by sea-level rise and rise of sea level by 0.3 m may cause as fatal as a net reduction of 0.5 million tons of farm production [64, 65].

Similar to South Asian countries, rice is also the main cereal of China where above 65% of the population consumes rice [66] and thus playing a pivotal role in food security [67]. China accounts for 27.8% of world rice production from 18.6% of the area [36]. Research evidence indicated that the effect of the rise of temperature due to global warming on rice was ambiguous [67, 68] as Liu et al. [69] mentioned a reduced rice yield in China, but others revealed that there was yield increase in Northeast and South China [70] and high-latitude countries [67]. The Pearl River delta of China is basically a rice-producing region [71]. Guo et al. [72] carried out a model-based study in the Pearl River delta and concluded that with the temperature increase of 1.5°C and 2.0°C condensed the duration of reproduction and physiological, leads to decrease rice yield for both the early and late maturing rice. Lv et al. [73] predicted that the grain-filling period will be shortened by 2–7 days and late maturing rice by 10–19 days in China during the period 2030s, 2050s, and 2070s. The single-crop rice in central China will be affected most in the future, but with CO_2 fertilization the yield reduction can be minimized. Further to cope up with the situation, there should be manipulation in planting date where early and single-crop rice should be plated early, whereas delayed planting is recommended for late rice to extend the grain-filling period and thus enhance rice productivity.

The CH_4 emission from flooded rice cultivation is considered to cause global warming. To minimize the CH_4 emission, varietal changes can be made. Further, avoidance of continuous water stagnation in the rice field, alternating dry–wet method of irrigation, and adoption of the system of rice intensification could minimize CH_4 emissions from rice fields. Integrated nutrient management and judicious use of N fertilizers can reduce NO_2 emissions from rice fields. In addition, primary tillage in rice cultivation, puddling, is involved with the burning of fossil fuels could be reduced by minimum tillage [33].

8.2.2 WHEAT

Change in global climatic parameters, namely, an elevated level of CO_2 in the atmosphere, the rise of temperature, variation in surface solar radiation, and quantity and distribution started to exert impact on food production [74, 75]. Wheat is highly sensitive to weather conditions. The crop has some critical stages to moisture sensitivity such as crown root initiation, tillering, flowering, and grain filling, and providing required irrigation improves productivity. Sufficient rainfall during these stages increases crop yield [76]. Temperature is another climatic factor and wheat shows sensitivity to temperature [77] The ideal temperature for germination is considered as 12–25 °C and at grain filling is 35.4 °C [78]. Increase in temperature beyond the threshold level is harmful in wheat productivity [79–81]. Higher temperature also enhances transpiration and accelerates crop phenology [82]. Heat stress during growth stage negatively impacts crop growth and assimilate production [83–86]. The elevated level of CO_2 is known to reduce N concentration in plant tissues of wheat reducing protein content in grain [86, 87]. In general, the projected impact of global warming and hot wave during maturity and erratic rain with dry spells can negatively influence wheat productivity in tropical and semi-arid regions [3, 88]

China is the foremost wheat producer in the world [89] and 50% of China's winter wheat is grown in the north China plain [90]. Another region of wheat production in the country is northern China, which belongs to the temperate climate and here spring wheat is grown [91]. In a multi-crop model simulation study, Xiao et al. [75] recorded that in the future increased temperature will alter crop phonological development and duration will be shortened. Enhanced rain will be helpful for wheat [76]. Studies revealed that change in solar radiation may not influence much [75], but the impact of CO_2 concentration is more on wheat productivity since wheat is a C_3 plant and highly sensitive to the concentration of CO_2 [74, 92]. The combined effect of temperature rise, more precipitation, and elevated CO_2 level and their interaction will increase wheat yield in China [93] as observed by Xiao et al. [75]. Adaptive measures suggested the need for the development of suitable varieties with more tolerance to high temperature and longer in duration [77, 94–97].

In India, Indo-Gangetic Plains is an imperative location for wheat cultivation globally and it accounts for 15% of global production and half of the area is heat stressed. Different studies showed that an upsurge of temperature by 1°C can reduce the wheat output of India by 4–5 mt [80, 81, 98]. Growth period is shortened due to an increase in temperature; even early senescence is caused leading to yield loss [99]. Under late sowing conditions, yield reduction is more due to the shortening of the growth period and increased temperature at the grain formation stage. There is a reduction of wheat production in irrigated areas by 6% in India [100]. A study clearly indicated that heat-wave negatively impacted wheat yield in India and 56% wheat area of the country is presently suffering from heat-wave [101]. Timely sowing, zero tillage sowing in the rice-based cropping system and adoption of heat-tolerant late varieties are suitable adaptation options to diminish the negative effects of global warming.

The climate of Iran is arid and semiarid, and wheat is a vital main food crop. The simulation-based studies indicated that as a consequence of climate change the yield of wheat in dry regions of Khorasan Razavi province in Iran will be highly affected [102, 103]. The country witnessed a momentous rise in temperature in the latest 60 years particularly in dry regions [104–107]. Ragab and Prudhomme [108] demonstrated a projection of climate variability from 2000 to 2015 in which an upsurge of temperature by 2.00–2.75°C and decrease of 20%–25% annual rain was predicted. As adaptation options, increased plant density (400 plants m^{-2}), choice of suitable varieties, the higher dose of N fertilizer, and manipulation in planting time may be considered [106].

Pakistan is another wheat-producing country, where it is treated as a staple food crop and productivity is only 2675 kg ha^{-1} [107]. The country is greatly exaggerated due to global warming and it is categorized as seventh in the Global Climate Risk Index from 1997 to 2016 [108]. Earlier there was an estimation of negative influences on wheat as a result of the changing climate [109] and a trend of reduction of wheat yield was already noted and reasons were prolonged winter associated with irregular rain [110]. However, no sufficient evidence on the hostile impressions due to the changing climate on food crops and adaptive actions are not available yet [111–115]. Different literature narrated some adaptation options such as manipulation of planting date, choice of varieties, nutrient management, and shifting to new crops [114, 116].

Ukraine is a leading wheat producer in the world and like other countries, the antagonistic effect of global warming is observed in Ukraine, due to increasing winter temperature and also reduction of rainfall pattern in the southern Steppe zone during 1961–2009 [117]. In general, climatic variability may cause yield variation of crops and the production risk [118]. But in Ukraine, the positive outcome is more prominent from climate change as the country will receive frost-free season for a longer time, increase in winter temperature, and elevated CO_2 level, which will facilitate wheat production [119, 120]. Out of total cereals area in the country, wheat occupies about 50% and it is the seventh-largest exporter of wheat [36]. There is an indication of an adverse impact on wheat productivity in southern Ukraine [121, 122], but northern Ukraine will be benefited from projected climate change in the future [123]. A study revealed that in Russia, temperature rise during the winter wheat sowing period might have a positive impact on yield, but warm autumns resulted in the reduction of yields [32]. Further, another study indicated a yield reduction of spring wheat due to the increase of temperature at the end of 21st century [124].

In temperate areas of European countries, the impact of climate change was noted. The elevated CO_2 and higher solar radiation increased wheat productivity in a study, but the increase in temperature reduced wheat yield [125]. In Northern Europe, low temperature and short crop growth period are the main issues of agriculture, but in Southern Europe, temperature rise and low precipitation are the concerns. The adversative effects of the changing climate will be prominent in countries like Hungary, Serbia, and Romania [126] due to the predicted enhancement of GHGs. Wheat yield may reduce by 6% for rising of a °C rise in temperature [74] along with drought [127, 128]. The Alentejo region of southern Portugal is the main wheat-growing area of the country, where typical Mediterranean climate has prevailed and rainfed wheat is grown [129]. During spring, water shortage is a common problem hampering flowering and grain filling [130, 131] and rise of temperature will cause yield loss of wheat [132–137] in south Portugal. A study predicted yield changes from -22% to 5% in 2021–2050 and -39% to -22% in 2051–2080 [138].

National Sustainable Agriculture Coalition [139] of the US mentioned that with warmer temperatures, the wheat yield will be increased in some regions of the USA because of enhancement in precipitation and C concentration. A study in Kansas, USA clearly mentioned and revealed in the short run that the rising temperature due to the changing climate will increase wheat yield, but in the long run it will adversely effect crop productivity. However, more rainfall will boost wheat yield in the near future, whereas in the long run, the impact will be negative [140]. In Kentucky, USA, wheat is cultivated for grain as well as silage and the impact of climate change is predicted on wheat yield. Future projections stated that in North America, CO_2 level may increase 550 and 700 ppm by 2050 and 2100, respectively, and temperature increase will be 5°F to 7°F by 2100 [141]. The possible effects of the increase in temperature are early tillering and more biomass production (by the rising temperature of 3.5°F), heat stress, and poor grain filling (above 86°F = 30°C). On the other hand, increased CO_2 level will ensure more assimilate production resulting in biomass yield and reduced flour quality and nutritional value. Further, there will be more weed, pest, and disease problems. Development of suitable cultivars with heat and frost tolerance and their adoption may be a suitable option to mitigate ill effects of climate change. But there will not be a major impact on total wheat production in Kentucky conditions [142]. For C_3 crops like wheat, an increase of CO_2 level and global temperature by 2.0°C would balance the adverse effect on yield [143], but water stress may reduce productivity. Predictions indicated that a rise of 2.0°C temperature and elevated CO_2 will not do any adverse effect on wheat yield in Canada [144].

The adverse effect of global warming is noticeably more prominent in wheat yield in African countries. There will be a reduction of 10% wheat yield in subtropical areas with an increased in temperature by 1 °C in Ethiopia [145, 146] and wheat cultivation is preferred in higher altitude. There will be a benefit of increased CO_2 concentration (confirmed by Aleminew and Abera [86]).

8.2.3 MAIZE

Maize is considered as the third vital cereals in the world and as a C_4, it is cultivating globally under diverse climatic conditions. Depending upon the variation in regional climatic diversity and management, yield differences are observed. Comparatively better productivity is observed in temperate regions of USA and Europe and China, Argentina, and South Africa. A simulation study indicated that under climate extremes due to global warming impact on maize yield will not be same in different countries [147]. In general, under the situation of global warming and projected temperature increase by 2°C, the production of maize cane be declined by 8% to 18% (53 million tons) in four major countries, namely, US, China, Brazil, and Argentina, which account for 68% of world maize output [148]. The current CO_2 concentration is at 400 ppm and is projected to increase from 794 to 1142 ppm by 2100 if not mitigation measures properly adopted [149] and precipitation will increase [150]. As maize is a C_4 plant, the effect of rise of CO_2 level will impact less. With an increase of CO_2 up to two times, an increase of biomass and yield was noted by 4% in maize [151]. Among different weather parameters, temperature and rainfall have an excessive impact on maize production. The range of temperature for maize during vegetative and reproductive phases are 8–38°C and 8–30°C [152, 153], while optimum temperature for growth is considered as 34°C [154] and above 35°C temperature is sensitive for pollen viability [155, 156]. If vapor pressure decreases, the pollination will be hampered [156].

In the US, major maize-growing states are Iowa, Illinois, Ohio, Indiana, Minnesota, Nebraska, South Dakota, and Wisconsin. In the region soil moisture stress is not the constraint as the region gets enough of precipitation (834.46 mm). Further increase of rainfall predicted due to climate change may create stress due to excess water and water logging as indicated in a study [147]. But the combined impact of temperature rise and decrease of precipitation may be fatal. Predictions revealed that an increase of temperature by 1.46°C and rainfall by 30% may reduce maize yield by 7.4% in 2030 in the Midwestern of the USA, the important maize growing tract of the USA. Actually, temperature and precipitation may influence a lot on maize productivity and Hatfield et al. [157] noted that increase of maximum temperature in July at above 32°C, the minimum temperature at more than 20°C, and precipitation during July and August is <150 mm may cause a disturbance in pollination (due to July high temperature), grain filling (August less minimum temperature), and water scarcity (high water requirement of the crop during the important growth stages and less rain during July to August). They found that there will be a yield drop by 50% in 2075 as predicted in the Southern part of the corn belt of the United States. Kukal and Irmak [158] studied the climate explained yield variability in the USA. Great Plains region and noted variability and dissimilarity among different states on the yield of maize due to differences of temperature and rainfall. They observed that the temperature alone caused yield erraticism in the southeast and northwest Nebraska and Iowa of the USA and accountable for up to 22% decline in productivity with a 1°C increase in temperature. In general, the choice of suitable cultivars and manipulation of sowing dates are the suitable adaptive measure to combat climate change [159].

Maize is the utmost significant cereal of Mexico and great diversity is noted in terms of climatic conditions and the growing environment as it is raised in arid hot and humid and temperate regions [160] with different cultivars [161]. The average productivity is less (2.9 t/ha) compared to the world average [162] In Mexico under changed climatic conditions in the future as predicted by Canadian climate center model in the earlier study, there will be a negative impact on rainfed maize production due to increase in temperature and less rainfall and fertilizer management was suggested as the adaptive measure [163]. Further, Bellon et al. [160] mentioned that by 2050 there will be less rain during May–October and increase in maximum (+1.4–3.5°C) and minimum (+1.4–2.6°C) temperatures. Recent studies also indicated that the problem is more crucial in rainfed areas in future, while the yield of maize will be stable in the irrigated field [164]. Varietal changes with local races as noted earlier with different crops [165–167], crop diversification [168, 169], and more equitable water distribution [161] could be ideal options for smallholders in rainfed conditions to combat the situation in the future, but under irrigated condition, more productive hybrids could be chosen [161].

The current scenario encompasses to predict changes in agriculture due to the changing climate in the Caribbean and Latin America. Possible changes include the rise of air and soil temperature, more CO_2 concentration, change in the hydrological cycle, and threat of natural calamity [168]. Jones and Thornton [29] predicted yield decline in eastern Brazil by 25% in 2055, but in Venezuela, there will be no impact though is not a major maize growing country of Latin America. In Colombia, maize yield will remain unchanged in 2055 as the study. However, in Argentina, Bolivia, Guatemala, Honduras, and Nicaragua yield reduction of maize was noted. Jones and Thornton [29] suggested that the development of drought-tolerant crop and cultivars and improved agronomic management may be the suitable options to manage the future threat due to climate change. In Brazil, Uruguay, and the Pampas region of Argentina, maize is an important cereal and there is a possibility to increase temperature and precipitation of CO_2 level. Yield reduction of maize by 5% per °C increase of temperature [169].

Maize is one of the most important cereals of China [170], and is responsible for the food security of the country as well as the world [171]. Middle China accounts for 38% [147] and Northeast China shares for 30% [171] of China's total maize production. Middle China is the "summer corn belt" where farmers choose early varieties [147]. In summer, if the crop is not adequately irrigated, maize yield may reduce and in Middle China maize cultivation is mostly rainfed. But the increase in precipitation along with temperature rise may yield more. Predictions mentioned that increase of temperature by 1.46°C and rainfall by 30% may yield 22.8% more and with the same temperature rise and 30%, less rain may decrease 10.7% more yield in 2030 than the productivity was in 2000 level [147]. Northeast China is vulnerable as a consequence of global warming with an increase of temperature by 0.38°C decade^{-1}, which leads to influence in the crop yield including maize also [172–174]. In the 2000s, maize yield is declined by 12.9% [175] and another study Lv et al. [174] also stated yield reduction of maize was due to the fluctuation temperature, rainfall, and sunshine hours during the crop growth stage. Introduction of temperature and water stress-tolerant cultivars may be the suitable options for sustaining maize yield in Northeast China against ill effects of climate change.

In European countries, climate change impacts on maize productivity were found minimum. A previous study mentioned that there will be a yield reduction of rainfed maize by 70% in 2050 and the maximum negative impact will be pronounced at latitudes 43–47°N, which is the important maize growing area of Europe if sowing dates and varieties remained unchanged [176, 177]. But in irrigated maize, there will not be many variations with either a slight increase or decrease of yield. However, early sowing can enhance the yield of both the irrigated and rainfed maize in 2050 as predicted by Parent et al. [177]. Further, there will be varietal change to cope up with the situation as observed in France in the recent past. The cultivars that were cultivated at latitudes of 44.8°–45.9° in 1996 were shifted to the latitude of 46.5°–47.5° in 2009 [177]. In Ireland, maize is cultivated mainly as forage and temperature rise will not influence maize productivity much in the country and an upsurge of temperature by 1.6°C and less rainfall will occur in 2055. The crop under the future predicted conditions may suffer from water stress, but the rising temperature will favor growth as the crop used to suffer due to low temperature [178]. The overall scenario of maize cultivation in Europe will not be affected much and by the adoption of adaptive measures productivity of corn can be sustained in the future.

Climate change influences agriculture and there is a great variation on the nature of the impact of climate change in different regions. The fluctuation of the climatic condition is more sensitive and fatal in tropical and subtropical Africa, because of temperature increase from 1°C to 2°C [179] and more frequency of climate-related calamities such as flood, drought, heat wave, and so on creating insufficient production, food and nutritional insecurity, and famine [180]. Maize is sensitive to a temperature beyond 30°C during the day and the majority of the predictions indicated that in Subsaharan African countries there will be an increase of temperature by 2°C [67]. Maize is an important cereal of different African countries, which already started to face problems due to climate change and in the future, the impacts will be fatal unless suitable adaptive measures are considered [181]. In Malawi, maize is grown by 97% of farmers, while increasing temperature and scarcity of rainfall already showed negative impacts in maize productivity. A crop yield modeling results indicated that in the mid- and end century, there will be a yield reduction of −0.73% to −14.33% and −13.19% to −31.86%, respectively, due to climatic variation. Several forecasts showed that there will be a rise of temperature by 1.26–2.20°C and 1.78–3.58°C in the mentioned periods, respectively, in African countries. For example, there is a chance of yield reduction of maize in Cameroon by 20.6%–69.6% in the 2080s due to climatic variation (mainly due to high-temperature stress) as evidenced by Tingem et al. [182]. In Zambia, due to the decline of water availability in 2100 by 13%, maize cultivation will be stress [183] and the decrease of yield by 25% [184]. There will be a significant reduction in maize yield in Swaziland due to delayed rainfall and hydrological disasters in future [185]. A crop model simulation study revealed that there will be yield reduction in rainfed maize in the southern and western highlands of Tanzania [186] due to heat stress and deficit rainfall [187, 188]. For most of the African countries, for sustaining maize productivity in the future change of cultivars, manipulation of sowing time, and efficient water management are the important adaptive measures.

Other than China, Indonesia and India are the leading maize-growing countries in Asia. Indonesia is an archipelagic country, where the temperature is by and large stable over the year [189]. The temperature may increase by 0.9–2.2°C and 1.1–3.2°C by the 2060s and 2100, respectively. There will be some changes in the quantity of rain and rainfall pattern. Rainfall may increase by 10% in 2050, but there will be a decrease of rain from July to September by 10%–25% [190]. The change in rainfall pattern may create drought, flood, and cyclone. Further, there is a chance of rising sea level and alteration in coastal agroecosystem. Ultimately, environmental factors may cause a yield reduction of maize by 40% in the future due to climate change in Indonesia [191]. In India, maize is mainly cultivated in the winter (October–March) and rainy (June–September) seasons. Due to the fluctuation of environmental factors, there will be a yield reduction of winter maize in the middle of Indo-Gangetic Plains (MIGP) and Southern Plateau (SP) regions as per predictions [41]. The yield of rainy season rainfed maize may also reduce by 35% in SP and 55% in MIGP [192], but there is less chance of yield decline in upper Indo-Gangetic Plains. Another study predicted yield loss of maize by 10%–21.4% in Punjab due to climate change [193, 194]. Singh et al. [195] mentioned that the winter crop of maize will be benefited by temperature rise by 2°C in upper Indo-Gangetic Plains. An increase in maximum temperature is a yield declining factors of maize in Telangana [196]. But proper adaptation practices such as sowing improved and stress-tolerant cultivars and more nitrogenous fertilizers can check the yield loss and even increase the yield of irrigated maize by 10 and 4 in 2050 and 2080 conditions [197].

8.2.4 PULSES

Rise of temperature was known to create hindrance in growth and decreased productivity of common beans (*Phaseolus vulgaris*) [198]. Stress due to high temperature caused floral and pod abortion, poor fertilization, pod set, and seed filling, causing fewer yields in food legumes [198]. High temperature also caused early senescence and reduced crop duration leading to a decrease in seed yield by 50% in different pulses [199]. The threshold temperatures of lentil (*Lens esculenta*) are >33°C and <15°C and beyond the temperatures terminal heat stress (THS) is pronounced. Due to THS, lentil showed declined biomass production, poor pod set, and inferior yields [200–202]. More than 35°C temperatures caused inferior pollen viability, pod set, small seed, reduced seed weight, and poor productivity in chickpea (*Cicer arietinum*) [202–204]. With 1°C increase of temperature beyond the threshold level can cause yield reduction in chickpea by 10%–15% [205–208]. Another study indicated that 0.1°C temperature increase with reduced rain can decline the yield of chickpea by 38.5 kg/ha [209–211]. Further, Vijaylaxmi [212] observed poor growth and yield along with inferior harvest index of pea (*Pisum sativum*) genotypes when the temperature was increased beyond 34.3°C. In pigeon pea (Cajanus cajan), 1°C rise in maximum temperature (32.3°C) was observed to reduce yield by 20.8% as stated by Mishra et al. [213]. But the rise in minimum temperature (26.3°C) showed a positive impact on yield and changes in rainfall due to climate change marginally influenced the productivity of pigeon pea in Gujarat, India. As adaptation measures, rainwater harvesting, creation of irrigation potential, and choice thermo-tolerant varieties were also suggested. Dubey et al. [214] studied in the Bundelkhand region of Uttar Pradesh, India and noted that reduced yield of chickpea, lentil, and pigeon pea because of the drop in rainfall and temperature rise. Chilling stress is also another temperature stress that reduces chickpea yield in India and Australia at a temperature below 15°C [215–217]. Chilling was noted to cause flower abortion, pollen sterility, infertility, declined seed size, and yield [218–221].

In India, the leading producer of pulses, these crops are mostly grown as rainfed or partially irrigated during monsoon season and residual soil moisture conditions in winter. The crops often face moisture stress and drought during the reproductive stage, which is further known as terminal drought [222]. The terminal drought is known to cause a disturbance in proper physiological and metabolic activities during the reproductive stage in different pulses. Further, drought stress reduces leaf area increase and accelerates senescence. The impact of drought stress was noted by different researchers from various countries. Drought during active flowering and podding stages reduced production of photosynthates and decreased pod formation, rate, and duration of seed growth with a decline of 30%–72% of yield in chickpea [223, 224]. The occurrence of drought during the reproductive stage caused a reduction in biomass production, pollen germination, pollen viability and podding duration, decreased pods and seeds plant^{-1}, seed yield, and duration of chickpea [225–228]. An enhancement of antioxidant enzymes activity was noted to cope up the drought stress; although, there was a yield reduction of up to 25% was recorded in lentil [229]. In lentil, reduction of leaf area, pod and seed number and enhanced seed abortion were with yield reduction of 70% was recorded by [230]. Lathyrus (*Lathyrus sativus*) also showed yield reduction due to terminal drought by expressing abortion of flower, ovule, and pod [231], declined stomatal conductance, assimilate production, pod, and seed size [232]. The development of climate-resilient varieties with traits like a deep root system, short duration, early flowering, and faster dry-matter production is a key consideration as an adaptation strategy [233]. Further, as an adaptation strategy, early sowing of pulses may also be considered to take advantage of sufficient moisture during the major growth period and to avoid terminal drought incidence [234].

In Myanmar, pulses are the second important crops next to rice. During recent times, climate change impacted on pulse yields. Rainfall in November (sowing time) and January (flowering) causes havoc yield losses of pulses [235]. In general, there was temperature rise in North America, Canada, and the United States by 0.7°C, 0.9°C, and 0.4°C compared to the 20th century [235] and its impact on the Northern Great Plains of North America is observed. Simulation studies indicated that soybean yield will be reduced under elevated CO_2 conditions. Under increased rainfall conditions, there will be more incidences of diseases in chickpea, lentil, and pea. Varietal development and manipulation of the growing season and climate-resilient agronomic management were suggested as adaptation options [235]. Impact of climate change on production of pulse cultivated in Bangladesh was also noted [236] and the most important pulse, green gram (*Vigna radiata*) varieties were developed ("BARI Mug-1 to BARI Mug-8," "BARI Sola-9," and "BARI Falon-1") with salt and heat tolerance [237] as an adaptation measure to cope up with the situation.

8.3 PROSPECTS OF FRONTIER AGRICULTURAL TECHNOLOGIES TO ENCOUNTER THE FOOD AND NUTRITIONAL SAFETY

Tremendous development has been accomplished throughout the world in the betterment of human lifestyle over the past century. With significant breakthroughs in technology, urbanization has been expanded and innovative production systems have been created that observed huge changes in societies. However, to achieve a world free from fear and need, this much progress is not sufficient. The world agricultural production has been increased almost three times between the period 1960 and 2015 as a consequence of the remarkable application of technologies, crop inputs in terms of fertilizer, water, and high-yielding varieties in the agricultural system [238]. Further, the simultaneous occurrence of industrialization and globalization of food and agriculture have lengthened the food supply chains increasing the corporeal detachment from farm to plate. The urbanization has improved the preference for prepared foods, processed, and packaged among the consumers. Although, the world has witnessed some significant developments in farmings, still persistent and widespread hunger and malnutrition are huge challenges in many parts of the world. The present rate of improvement in crop production will not be satisfactory to remove the starvation by 2030, and not even by 2050 [238].

8.3.1 FOOD SAFETY

Food safety can be well-defined as the condition "when most of the people all times have physical and socio-economic entree to sufficient, safe and nutritious foods for their hale and hearty life." From this definition, four major areas are to be considered namely physical accessibility of foodstuffs, actual and economic entrance to food, food utilization, and the stability of the previous three, to maintain food security for a longer duration. Simultaneous fulfillment of these four dimensions is a must need for realization of the purposes of food security. Indeed, we are producing more enough food than ever before with advanced technologies, which are adequate to feed the world. However, the poor distribution of the foods and different preferences for food among people are there and the income difference between different classes of populations seems to be the principal reason for these inequities [239].

8.3.2 FOOD SECURITY AND GROWTH OF POPULATION

The growth of population, climate fluctuation, and urbanization are three major realms of global challenge on food security [240]. The world's population is growing at an increasing rate every year. It is estimated to be 9.7 billion by 2050 and 10.9 billion in 2100 [241, 242]. The uncontrolled population increase has put the world at stake. The increasing population will boost the demand for food and nutrition. The availability of land resources has a greater correlation with crop production. Kendall and Pimentel [243] opined that only a third of the Earth's soil is appropriate for cultivation. Population increase requires more land, water, and air for survival, thus reduces the resource availability for producing food. The adverse effects of population explosion are many and all these ultimately change the climate.

8.3.3 ADVANCED AGRICULTURAL TECHNOLOGIES IN ATTAINING FOOD SECURITY

To meet the food demand of the increasing population the present agricultural productivity needs to be increased exponentially. Production can be increased both by a horizontal and vertical expansion of the cultivation area. However, under the present scenario of changing climate and population explosion, conventional agriculture has become a bane to itself as it adds to atmospheric GHGs that cause global warming. Increasing food production and economic growth during recent years have been achieved at a heavy cost to the natural environment [239].

Therefore, to safeguard food and nutritional safety for the present and the ever-growing population in the future, it is important to introduce the next-generation and climate-resilient agricultural technologies in the traditional farmings. With the modernization of agriculture use of machines has been introduced in the farms, which enhances the productivity of farms. Wide economic development and higher crop productivity might be fruitful to mitigate food and nutritional demand, and the former can reinforce the latter [244]. Now the present situation demands for an agricultural system that is more productive and less wasteful. The principal inputs for better agricultural production comprise of land, healthy soils, water, and plant genetic resources. An increase in the scarcity of these resources in many parts of the earth has urged for sustainable utilization and management of these. Integrated approaches and recycling the resources in crop production may be one of the best choices for the sustainable agricultural system. Increasing the productivity of existing agricultural lands and simultaneously ameliorating the degraded lands and problem soils through sustainable agricultural practices would reduce the pressure of removal of forests for the cultivation of crops. Judicious management of limited water in dryland areas through improved irrigation and water harvesting technologies along with the development of new drought-resistant cultivars can sustain the productivity of these areas. Adoption of conservation agriculture using resource conservation technologies can restore soil health as well as the biodiversity of the ecosystem. Several new concepts of farming have been put forward namely organic farming, bio-intensive farming, etc. for the restoration of ecological balance. Crop improvement through biotechnological tools has enabled the production of cultivars with resistant to numerous biotic and abiotic stresses and greater adaptability to changing climate. This can increase food production to ensure food security. Precision farming, hydroponics, and vertical farming have shown promising results in increasing agricultural production.

In the present situation, the optimum management of available resources along with the proper allocation of advanced technologies has become important in improving production and ecological health. A new approach, climate-smart agriculture (CSA) has been introduced by FAO that intentions to upsurge the crop production and incomes in a sustainable manner, to build flexibility to climate change and to reduce GHGs, where possible [245]. CSA is comprised of multidimensional site-specific concepts that identify the appropriate farming practices to withstand climatic abnormalities [246, 247]. Application of the integration of compatible technologies in agriculture can ensure food safety for the rising population simultaneously keeping the environment safe.

8.4 NEXT-GENERATION SEQUENCING AGRICULTURAL TECHNOLOGIES (NGSAT) AND THEIR APPLICATIONS

Significant improvements have been witnessed in all fields of science and technology in the past century. The success of transgenic technology in crops such as cotton, maize, and soybean has cemented the mode for biotechnological research in the field of agriculture. Urgency is there to enhance the yield and nutritive qualities of crop plants to safeguard food safety for the ever-growing population. Improvement in the quantitative and qualitative traits of crop plants has become possible with the advancement of highly productive omics technologies. The recent introduction of NGSAT further revolutionized the sequencing of the nucleic acid, which contributes to a new era in omics approaches [248].

8.4.1 NEXT-GENERATION SEQUENCING

The discovery of the double-helix structure of deoxyribonucleic acid (DNA) and the four DNA bases (A, T, C, and G) facilitate more knowledge regarding the genome sequencing and composition of DNA in an organism [249]. They have also pioneered the development of sequencing technologies. Before discovering the new era of sequencing technologies, this Sanger sequencing technology was commonly used by biotechnologists. The new era of NGSAT opened new perspectives for genomic studies. Roche 454 sequencer was the pioneer to develop these NGSAT [250], and it provides high sequencing rate at a very low cost. Thereafter, several other technologies of DNA sequencing have been developed by different companies (Illumina, Helicos, etc.). These technologies are also known as NGSAT or "high throughput sequencing technologies." Important features of this technology are summarized in Table 8.2.

TABLE 8.2   Important Features of the Next-Generation Sequencing Technologies in Omics Studies

Technology	Read length	Yield (Reads per run)	References
Roche 454	700	~700 thousand	Adopted from source [247]
IlluminaHiSeq	300	~300 billion	
SOLiD[a]	100	~200 billion	
Ion Torrent	200	~60 billion	
PacBio RS II	14,000	~47 thousand	

[a]SOLiD: Supported Oligonucleotide Ligation and Detection

8.4.2 AGRICULTURAL APPLICATION OF NEXT-GENERATION SEQUENCING TECHNOLOGIES

The development of NGSAT and accessibility to genetic information of model crop plants have helped in the discovery of new genes and study of gene expression profiles [251, 252]. The crop traits are generally controlled by multiple genes [253]. To identify key genes linked with desired phenotypic characters, massive sequencing and gene expression profiling are required [254]. The productivity and sustainability of crop plants have been enhanced with the application of biotechnological approaches in agriculture. The potential fields of application of NGSAT in agriculture include genomics, transcriptomics, proteomics, and metabolomics.

TABLE 8.3 Next-Generation Sequencing Predicted Genes Involve in Different Crops

Crop	The Haploid Number of Chromosome	Number of Genes Predicted	Reference
Rice	12	37,544	[255]
Soybean	20	46,430	[256]
Chickpea	8	28,269	[257]
Banana	11	36,542	[258]
Maize	10	32,540	[259]
Wheat	21	94,000	[260]
Pigeonpea	11	48,680	[261]
Cabbage	9	45,758	[262]
Mungbean	11	22,427	[263]
Pearlmillet	7	38,579	[257]

These NGSAT have been successfully exploited to study and manipulate the economic crop plants for enhancing yield and improving nutritional qualities. Therefore, the application of NGSAT in the above-mentioned fields along with bioinformatics tools is required to recognize the features, which affect the growth and yield of crops in the present situation of changing climate and to produce crops of greater adaptability. The first plant to be fully sequenced was *Arabidopsis thaliana* in 2000, which was annotated for identifying 25,000 functional genes and further used as a model crop for gene annotation of other crops [249]. NGSAT have helped in predicting new genes by comparative genomics studies that include a comparison of the newly sequenced genome with model crops. The number of genes predicted by next-generation sequencing technique in different crops is shown in Table 8.3.

8.4.3 GENOMICS IN AGRICULTURE

The techniques involved in genetics and molecular biology for gene mapping, DNA sequencing or complete genome study of target organisms, and subsequently arranging the results in a database and their application. It provides information regarding the genetical organization, gene mapping, total number of genes, and their functions [264]. Advances in genomic technologies with next-generation sequencing have enabled us to design agricultural crops with promising economical traits. The selective breeding procedure has become more targeted and about two to three times faster than before with genome sequence-based DNA tests. In certain cases, strains with new properties are created due to genetic modification that improving productivity and nutritional value of the crop. Massive sequences and the single nucleotide polymorphism (SNP) detection can be made by NGSAT [265]. These molecular markers help in the early detection of desired characters in the progeny. The number of markers detected in various crops using NGSAT are given in Table 8.4.

TABLE 8.4 Markers Discovered in Crop Plants using NGSAT

Crops	Number of Markers Identified	Sequencing Technique	References
Pigeonpea	309,052 SSR	Illumina	[257]
Maize	36,000 SNP	Roche GS20	[266]
Grapes	17 million SNP	Roche 454	[267]
Arabidopsis	Epigenetic modification	Roche 454	[268]
Chickpea	81,845 SSR	Illumina	[269]
Pearlmillet	88,256 SSR	Illumina	[270]

EST: Expressed Sequence Tag; SSR: Simple Sequence Repeat; SNP: Single Nucleotide Polymorphism.

Several crops with improved productivity and nutritional quality have already been created by using genomics studies. Genome sequencing and gene expression studies have helped in the identification of genes related to a particular trait, which could be utilized for the development of crops by addition of genes or masking the effect of a gene. Higher rice productivity, bananas with longer shelf life, tomatoes with longer shelf life, and drought-tolerant maize crop can be developed by using genomics technologies.

8.4.4 TRANSCRIPTOMICS IN AGRICULTURE

Transcriptomics is the knowledge of transcriptome that represents the comprehensive set of ribonucleic acid (RNA) transcripts formed by the genome under exact situations using high throughput procedures, such as microarray analysis. The transcriptome provides information of expressed gene sequences in a specific tissue at a specific time. Advances in NGSAT have helped in obtaining cost-effective transcriptome assemblies useful for gene annotation [271]. The transcriptome is dynamic as it is essentially a reflection of the genes that are actively expressed at any given time under various conditions. The patterns of changes in gene expression due to internal and external factors are determined by it. Functional genomics utilizes the genomic data and gene expression analysis together to identify putative genes linked with the specific biological trait. Transcriptomic techniques have enabled to study different traits of crop plants (Table 8.5) in detail.

TABLE 8.5 The Application of Transcriptomic Techniques on Various Plants

Crops	Stress Trait	Approach Applied	References
Chickpea	Drought tolerance	cDNA library	[272]
Pigeon pea	Fusarium wilt resistance	BSA	[272]
Soybean	Cyst nematode resistance	MAS	[273]
Pea	Stemphylium blight resistance	QTL	[274]
Cowpea	Strigartesistance	SCAR markers	[275]
Groundnut	Rust resistance	Back cross progeny	[276]
Alfalfa	Aluminum toxicity	Microarray	[277]

QTL, Quantitative Trait Analysis; MAS, Marker-Assisted Selection; BSA, Bulk Segregation Analysis; SCAR, Sequence Characterized Amplified Regions

8.4.5 PROTEOMICS IN AGRICULTURE

Proteomics is the wide range of study on proteome, which refers to a set of proteins generated in an organism, system, or biological context. The proteins involve in the plant process and also accelerate the stress-tolerant capacities have been discovered by using proteomics technology. Functional analyses of those proteins will contribute to developing stress-resistant/tolerant crops and artificially regulated crops. Two-dimensional electrophoresis gel and liquid chromatography-based analysis are two generally used techniques for proteome analysis and the latter is becoming increasingly common in many laboratories. The effective improvement of NGSAT is helpful for the documentation and explanation of proteins and their isoforms in a specific crop species have become easier.

The use of transgenic techniques instead of breeding methods is becoming more popular for rapid development of cultivars with desired characteristics. It is important to evaluate these genetically modified crops using proteomic techniques [278]. C_4 plants are more efficient than C_3 plants in conversing sunlight into starch owing to two types of chloroplasts in their cells. The substances that affect the sunlight conversion efficiency may be detected in C_4 plant with the help of comparative proteomic analyses [279]. The proteomic examination is anticipated to accompaniment traditional molecular genetics methods to study the mechanism of infection by the pathogen in cereal crops. The symbiotic relationship between crop and soil bacteria has also been detected by the use of proteomics [280].

8.4.6   METABOLOMICS IN AGRICULTURE

Metabolomics is the scientific study of all metabolites in a biological sample. It involves the identification and quantification of all the metabolites. With the advancement of modern techniques such as nuclear magnetic resonance (NMR), mass spectrometry (MS), and fourier-transform infrared spectroscopy, the analysis of metabolites has become easier. The metabolite concentration is the determinant of flavor, aroma, and storage capacity of crops [281]. The biochemical activities involve in plant growth, a defense mechanism against pathogens, and in stress tolerance ability can be detected by metabolomics techniques. In several crops, metabolic profiling has been done through MS and NMR techniques to ascertain metabolic responses to herbicides, investigate metabolic regulation, and metabolite changes to environmental conditions of light, temperature, humidity, soil type, salinity, fertilizers, pests and pesticides, and genetic perturbations. The detailed analysis of metabolites may become helpful to develop new pesticides, reduce pesticide use and increase in nutritive values or assist with other key traits [282].

8.5   NEXT GENERATION CLIMATE RESILIENT BREEDING AND BIOTECHNOLOGICAL APPROACHES

With the advance of time, agriculture production has been increased by many folds. The total production of wheat and rice has been increased three times since 1960 [283]. But population growth and climate change have been pressing challenges on food security. The global population is approaching to reach 9 billion by 2050, when the mean temperature may rise between 2°C and 5°C [21], among them one billion people are still going hungry every day [284]. However, it is almost impossible to calculate the consequences of climate change, which may include rising of temperature, CO_2 level [21, 48, 285, 286] frequency of absolute weather conditions like droughts and floods [21, 287, 288], new diseases, and pests infestation [286]. Moreover, these issues would not merely add up, their collective reinforcement and interaction would make the impact more complicated. So, food security in future will face fourfold challenges: increased demand, decreased supply, production have to be both resilient and sustainable. So, to combat with abiotic stresses through the development of resilient and stable cultivars, multi-disciplinary teamwork is needed where the frontline starting point is breeding.

In the 20th century, crop plants were improved through conventional breeding techniques and those techniques are still following in advance breeding or biotechnological strategies, where target genes are precisely monitored in their introduction and expression level. The main obstacle of conventional breeding is that it takes a long time, which has been shortened through different accelerated breeding strategy. However, the introduction of genes from unrelated species, silencing of the unwanted gene in special condition, expression of the desirable gene(s) in stress environment have also been possible through different advance approaches such as genetic transformation, mutagenesis, genome editing, and RNA mediated gene silencing. Here we will discuss next-generation breeding (with some modifications in conventional breeding also) and biotechnological approaches followed in agriculture to produce climate-resilient crop cultivars.

8.5.1 UTILIZATION OF GERMPLASM FOR SCREENING STRESS TOLERANCE

Elite gene pool produced over times through artificial selection has limited allelic diversity, which has to be broadened by mingling exotic stress-tolerant wild and landraces to tape-out novel genes. The international platforms such as CIMMYT, IRRI, ICARDA, ICRISAT, IITA, and CIAT maintain the huge number of accessions of different crops and so far only 5% of them has globally utilized for commercial breeding [289]. So, by screening that germplasm, researchers can develop different stress-tolerant trait-specific core sets, where genotypes number will be minimum and allelic diversity will be maximum [290, 291]. Phenotyping of those germplasms under specific stress condition such as drought, salinity, heat, cold, and waterlog will help to understand variability and develop core set, which will later use as source or donor population in both conventional or advanced breeding program. There are some stress-specific traits that should consider during phenotyping. For example, to produce a drought tolerance gene pool for maize leaf senescence, ASI and leaf rolling traits [292, 293] should consider; for heat tolerance in wheat higher stomatal conductance and higher leaf chlorophyll content [252] should consider; to exploit genetic diversity in rice against heat tolerance stay green [294], the higher number of spikelets per panicle under warm condition, flowers open in the early morning have to consider [295].

8.5.2   ACCELERATED LINE BREEDING

By screening germplasm, we can develop a gene pool with diversified genotypes where genes of different target traits remain either in heterozygous conditions or randomly distributed in different genotypes. So, the next and vital step is to fix those genotypes in homozygous condition to be used as potential parental lines. But it is a challenging task and time-consuming process as it may require 7–8 generations. By following different accelerated breeding approaches it can be done in a short period, which are stated in the following sections.

8.5.2.1   DOUBLED-HAPLOID

It is a technology to produce 100% homozygous pure or inbred lines immediately and thus it can save 6–7 generations. As it is carried on in tissue culture laboratories, the species that has an established tissue culture protocol to regenerate plants through the embryo, pollen, or organ culture can be subjected to double haploid program. It has been used for decades in rice breeding and several rice varieties have been developed through this process [296]. Anther culture followed by double haploid production has been stated in many cereals including rice [297], wheat [298], maize [299], etc. In maize breeding, it is a routine technology [300, 301], where haploid inducers are being utilized [302] followed by chromosome doubling. RWS [300] and sister line RWS-76 [301] are two haploid inducer lines, which have 8%–10% of induction rate in tropical maize [302] In wheat, flowers are pollinated with maize pollen followed by embryo development and after few days the embryo is rescued as it has a haploid number of chromosome and embryos are cultured at tissue culture medium where regeneration of haploid plants takes place. Those plants are treated with colchicine to produce double haploid homozygous wheat plants [303].

8.5.2.2 RAPID GENERATION ADVANCEMENT (RGA)

RGA, also known as single seed descent, is a rapider breeding technique, where crop growth conditions are manipulated. It reduces the life cycle of the crop so that several generation advancements per year can be achieved. The life cycle could be fastened by breaking seed dormancy, seed treatment, and temperature control, application of hormones, nutrient management, accelerated flowering, embryo rescue, and combinations thereof. As several generations advances in a single year, it can reduce time to produce homozygous lines. It has been practiced in many crops such as sorghum [304], chickpea [305], rice [306], and maize [307].

8.5.2.3 SHUTTLE BREEDING

Shuttle breeding is an accelerated breeding program, where crops are grown at different field locations to get an extra generation advancement. It improves the selection process as the same genotypes are screened in different environmental conditions [308]. This type of program is followed in CIMMYT and rice during 1982 [309].

8.5.3 PRECISION PHENOTYPING

Crop performance such as stress resilience, quality, and yield potentiality are mainly controlled by their genetic makeup, which can be measured through phenotyping. So, phenotyping is a process of calculating genotypic differences under stress condition. In fact, it is running through a purely statistical manner. High throughput precision phenotyping increases selection intensity, accuracy, repeatability, identification of new genetic variability, and ultimately speed up genetic gain. Currently, it is perceived as major limiting factors of breeding efficacy under stress conditions both in conventional and advanced breeding. It is considered as a bottleneck for further efficiency of the breeding program [310–312]. In stress breeding program, it is a challenging task to phenotype direct and indirect traits. For precision and high throughput phenotype, different phenotyping platforms have been introduced.

Visible light imaging [313, 314] or near-infrared imaging [315, 316] for the entire part of the plant, thermal imaging for leaf tissues or whole shoot [317], only for shoot 3D imaging [318], RGB data for measuring series of traits [319, 320], remotely-controlled unmanned aerial vehicles (UAVs) under open-field conditions, UAVs for target traits throughout the cropping period have been reported on phenotype.

8.5.4 DIFFERENT APPROACHES TO COMBAT STRESS IN PLANT

The choice of approaches and tactics for encouraging and refining resistance crop plant to stresses predominantly depends upon the accessibility of resistance sources. Once it is selected, the next step is to incorporate those traits through different breeding approaches into elite cultivars to produce stress-resilient varieties. The breeding tactics can be separated into traditional and contemporary procedures.

8.5.4.1 CONVENTIONAL BREEDING APPROACHES

The main drawback of conventional plant breeding methods is tardiness. It requires about 10–12 years in releasing a variety. However, different speed breeding techniques help to shorten this time. Furthermore, the main theme of breeding processes either in conventional or advanced is almost the same. The various conventional breeding methods used for stress breeding are discussed in following sections.

8.5.4.1.1 Hybridization and Cultivar Development

To combine required characters such as stress resistance and higher yield in a single genotype, elite high yielder commercial cultivar is hybridized with stress-tolerant or resistant genotypes and in their subsequent generation selection under stress condition is performed. In this way, high-yielding and stress-resistant cultivars for self-pollinated crops. The stem rust-resistant (Ug-99 race) was wheat cultivar Lasani-2008 was developed following this way [321]. If the trait is monogenic and expressed in heterozygous condition then hybridization between inbreds leads to stress resistance. A hybrid barley variety was developed by Brahim and Barrett [322], where various inbreds were hybridized following composite cross and the resulting hybrid barley was resistant to powdery mildew.

8.5.4.1.2   Backcross Breeding

The backcross is a proven successful breeding method to transfer single stress-resistant genes in an elite high yielding susceptible cultivar. In backcross, the elite cultivar is used as recipient parent and resistant genotype is used as a donor parent. The resultant hybrid is called backcross 1, where both parents have equal genetic content. In the subsequent years, the resistant genotype is crossed with the recipient elite cultivar and it continued until the subsequent progeny contain 99% genes of the elite cultivar. In every year, a subset of the progeny is screen out against target stress environment and the only tolerant or resistant progenies are proceed. Backcross was successfully practiced in the development of transgenic cotton, where *BtcryIA* gene was transferred to grow resistance against chewing insects [323]. Similarly, *Xanthomonas campestris* pv resistant *XAN-159* gene was transferred into "*Chase*" cultivar of bean and *Bt* toxin gene was introduced into wild *B. rapa* from *Bt B. napus* [324, 325].

8.5.4.1.3   Composite Crosses (Gene Pyramiding)

When a stress-tolerant or resistant trait is controlled by some minor genes or several traits are associated with a single stress tolerance that is distributed into several cultivars, the composite cross is helpful to combine them in a single cultivar. The number of composite crosses depends on how much parents contain those genes. At first, crosses are made between every two parents and the subsequent F1 hybrids are crisscrossed with each other and the resultant progeny contain the genetic components of four parents. The progenies have to cross again with each other and it is repeated until all the selected parents' genetic content are mingled in a single cultivar. Jackson et al. [326] produced scald disease resistance in barley populations using three composite crosses. Similarly, improved resistance for different diseases has also been reported in barley, for example, resistance for powdery mildew and blotch [327]. Gene pyramiding by pooled crosses delivers strong and durable cultivars against insect pests and pathogens [328].

8.5.4.1.4 Gene Pyramiding

It is also well-known as dirty crop approaches where a number of iso (homogenous) lines are involved to improve high-yielding stress-resistant cultivar. Here, the isolines are produced by backcrossing an elite cultivar with different target genes containing stress tolerant genes. Thus, the isolines are morphologically similar to that elite cultivar [329] but each line has different resistant genes. Then they are bulked to produce multiline. As the multiline possesses, a number of different resistant genes provides durable horizontal resistance to the crop. So, if the isolines contain the resistance gene of multiple stresses, then the resultant cultivar may provide resistance to multiple stresses. Such multiline has been introduced in rice for blast resistance [330–332] and it can also successfully replace susceptible cultivar [329].

8.5.4.2 ADVANCE BREEDING AND BIOTECHNOLOGICAL APPROACHES

Genetic drag, genetic erosion, non-specificity, reproductive obstacles, and tardiness are the major constraint of conventional breeding. So, there is an urgent need for novel techniques that can overcome those obstacles. Next-generation breeding and biotechnological approaches such as marker-assisted selection (MAS), marker-assisted backcrossing, and marker-assisted gene pyramiding can help to overcome those limitations. Moreover, advancement in plant genetic engineering (genetic transformation and genome editing) has made it possible to transfer genes from unrelated species and even from the nonplant organism. These methodologies are a pronounced opportunity to progress crop plants with momentous commercial possessions such as improved biotic and abiotic stress-resistant or tolerances, yield and nutritional quality.

8.5.4.2.1 Identification of the Genes Involved in Stress Response

For improvement of stress tolerance in crops, at first, we have to understand comprehensively their stress response mechanisms. The documentation of genes controlled in stress response is another important way, which is then tailed by purposeful categorization of the genes and thus decoding its mechanism. Finally, those genes have to introduce or incorporate through genetic engineering or molecular breeding to develop stress-tolerant crops. Some of those approaches are described briefly.

8.5.4.2.1.1   Advancing Genetics and the Candidate Gene Approach

So far researchers have identified many stress-responsive genes on a genome-wide scale following forward and reverse genetics. Using forward genetics salt-tolerant *SOS1* gene responsive for imperfect shoot growth under salinity has been recognized in rice, tomato, and poplar [333–335]. Heterologous overexpression of the *AtNHX1* inhibited the salt-sensitive phenotype in a yeast *nhx1* mutant and conferred salt tolerance to *Arabidopsis* as well as tomato and *Brassica* [16, 336–340].

8.5.4.2.1.2   Comparative Gene Expression Analysis

It is another widely adopted high throughput gene expression analysis stress-responsive genes are identified through technologies like microarray, articulated categorization tags, suppression subtractive hybridization, massively parallel signature sequencing, and serial analysis of gene expression. comparative gene appearance included analysis: (1) stressed and unstressed samples of the same species, (2) contrasting genotypes of the same species, (3) phenotypically contrasting organisms of different but related species, and (4) expression analysis of extremophiles [341]. Following these techniques, salinity and stress-responsive have been identified in many crop species [342–349].

8.5.4.2.1.3   Association Genetics

It has been used to discover, genotyping and mapping genetic marker, QTLs, and candidate genes. NGSAT-based association mapping has allowed to study hundreds of individuals at a time [350]. For identification of potential markers or alleles genome-wide association studies (GWAS) or whole-genome scan has proven useful. Huang et al. [351] have made a fine resolution genetic map of fourteen agronomic traits related to yield in stress tolerance following NGSAT and GWAS of 373 indica lines. Kumar et al. [352] have mapped three new QTLs in rice responsible for salinity tolerance on chromosome numbers 4, 6, and 7 [353].

8.5.4.2.1.4   Computational Tools and Databases

Stress tolerance is a very complicated process and many genes with primary and secondary effect influence the process. Moreover, different species may have similar stress-responsive genes. There are some web-based international database platforms for different organisms, where scientist or researchers share their genomic sequence-based findings that have become a resource for plant biologists [354]. By calculating these databases, one can identify stress response putative genes through sequence and structure homology in the same or distant species. But it is only a prediction that has to be validated later by biological assessment.

8.5.4.2.2   *Functional Characterization of Stress Tolerance Genes/ QTLs/Alleles*

After identifying stress-responsive QTLs, alleles, or genes, their functions have to be verified first in *vivo*. By comparing posttranscriptional or post-translational alteration of control and sensitive genotypes under specific stress condition one can classify or annotate stress-responsive QTLs/genes/allele. The next step is to transfer or introduce them into elite sensitive cultivars.

8.5.4.2.3   *Molecular Marker-Assisted Breeding*

It is a molecular particularly DNA marker-based novel breeding approach, where stress-responsive traits are edited or improved through the genotypic analysis of linkage maps or genome. It is a potential method of improving genetic make-up. It comprising of different approaches that are stated in the following sections.

8.5.4.2.3.1   *MAS*

It is an unintended selection method, where a morpho-biochemical or DNA/RNA distinction associated with stress tolerance trait is used as a marker, instead of the whole tolerance [355]. The connotation between markers and water stress tolerance and powdery mildew resistance in barley lines has been found [356]. In rice, QTLs have been found for dehydration tolerance, accumulation of abscisic acid, stomatal behavior, osmotic adjustment, root thickness and penetration, root number, and length [357]. In maize, the difference between pollen shedding and silk emergence days known as anthesis silking interval (ASI) can be a marker for drought tolerance and short ASI containing genotypes should be selected to get higher yield as it is little influence by drought [358].

8.5.4.2.3.2   *Marker-Assisted or Marker-Based Backcrossing (MABC)*

It is just as like the conventional backcrossing method and the only difference is rather than evaluating the phenotype, the markers linked to QTL or gene of stress tolerance are traced to confirm gene transfer. Researchers can confirm the transfer of gene or QTL at a very early stage as it requires a small portion of tissue (leaf or shoot) and thus it saves time, labor, space, and a huge number of the sample can be handled efficiently. There are some genes that have low heritability, difficult phenotypic expression, particular stress-responsive expression, recessive effect, phenotypes expressed in the later growth stage (such as flowers, fruits, and seeds) or gene pyramiding of different stress tolerances, can be easily confirmed or identified in early-stage through this technique [359]. Crop species including wheat, maize, rice, pearl millet, barley, tomato, soybean, etc. [360, 361] have been practiced through MABC to improve different traits that is drought tolerance, disease, and pest resistance.

8.5.4.2.3.3 *Quantitative Trait Analysis (QTLs)*

When two or more genes or QTLs are tried to combine in a single cultivar, it is known as gene pyramiding and for precise selection, if it is confirmed through markers linked to those genes or QTLs then it is called marker-assisted gene pyramiding. Genes or QTLs pyramids have been established in rice against blast and bacterial blight [362]; in barley for qualitative gene and QTLs for stripe rust, and also in wheat, barley, and soybean [363, 364].

8.5.4.2.3.4 *Marker-Assisted Recurrent Selection (MARS)*

It is a DNA marker-based recurrent selection, which enhances the efficiency and accelerates the progress of recurrent selection. Here genotypic selection and intermatingare carried out on the same cropping season for one cycle of selection. It is especially useful in introgression of multiple targeted genes or QTLs from numerous sources via recurring assortment based on a multiple-parental population [365]. It has been applied for many characters including grain yield and moisture [366], or stress tolerance [367] and numerous qualities are being beleaguered concurrently; in maize to develop the complex quantifiable characters like grain yield and stover quality [368] and drought tolerance [369]. The occurrence of desirable alleles of water deficit stress was improved by practicing MARS in maize [370].

8.5.4.2.3.5 *Genomic Selection (GS) or Genome-Wide Selection*

It is a cumulative method where all the marker loci distributed over whole-genome contribute to target trait expression [371] have to consider simultaneously [372, 373] to predict complicated traits precisely to grant the selection. In the NGSAT, SNPs are considered as variations and the cumulative effect of SNPs called as genomic estimated breeding value (GEBV) describes the expression of the trait. Then the predicted GEBV has to utilize for selecting desirable individuals in the selection program. It has enormous potentiality for accurate selection of complex traits, shortening breeding cycle for variety development time by increasing genetic gain [374]. In maize, GS method has been previously stated for drought tolerance. Beyene and colleagues [375] reported a 7.3% increased grain yield in maize through GS; Shikha et al. [376] have found the best model for drought-phenotyped genotypes; Vivek et al. [377] observed 10%–20% of GS than unadventurous phenotypic assortment under water deficit conditions.

8.5.4.2.4   *Mutation Breeding*

If there are no tolerance genes exist in using germplasm, then one way to familiarize heritable adjustments is a mutation in crops and select mutants responsive to particular stresses. There are two types of mutagens, that is physical (X-ray, gamma, and UV-irradiation) and chemical (MMS, EMS, BTH, and colchicine). For example, T-DNA insertion and transposable elements mutagenesis as biological mutagens have been already used success-fully by researchers [378]. It not only helps to create variations but also benefits to know the mechanism of resistance. While physical and chemical mutagenesis is a nonspecific arbitrary procedure and can occur anywhere in the whole genome, TILLING (Targeting Induced Local Lesions IN Genome) a high throughput and economical technique that can identify mutations (as SNPs) persuaded preciously utilizing chemical mutagens in the target genes [379]. TILLING is a reverse genetics tool that requires genomic sequence. Eco-TILLING is thoughtful of TILLING, which delivers the benefit of SNPs in usual mutants to monitor the plant populations for various stress condi-tions. For example, Mejlhede et al. [380] utilized eco-TILLING to learn the genetic dissimilarity in powdery mildew resistance in barley.

8.5.4.2.4.1   *Transgenic*

Plant genetic engineering has unlocked novel windows to adapt crops for particular requirements, such as improved yield, quality, and tolerant to various stress conditions [381, 382]. These technologies can incorporate external DNA into diverse plant cells for the development of transgenic plants with innovative anticipated characters [383–385]. Many earlier findings have shown the potentiality of influencing *CBF/DREB* genes to enhance adaptability against water deficit condition [386]. Similarly, genetic alteration with genes coding arginine decarboxylase upgraded the adaptability against abiotic stresses in numerous plants [387]. For example, the transformation of LEA proteins in other plant cells improved the adaptableness against drought and salinity in rice [388] and in wheat [389].

8.5.4.2.4.2 Cisgenesis

Another improved genetic renovation technique is "cisgenesis," which is first proposed by Schouten et al. [389]. It is one of the new plant breeding technology, which is recently recognized as an auspicious tool than the conventional transgenic method. Cisgenesis comprises all the genetic procedures of the T-DNA as introns, flanking regions, promoters, and terminators.

8.5.4.2.5 **Genome Editing**

Genome editing is the latest technology developed for precise and site-specific manipulation of the gene function. Genome editing method generally uses modified DNA cleavage chemicals and cellular DNA restoration mechanism [390]. The chemicals are typically concocted nucleases, which slash specific DNA at user-specified categorizations and these double-strand disruptions are renovated either by nonhomologous end joining or homologous recombination [391]. Through these methods, point mutation (deletion or insertion), activation or repression of genes, gene knockouts, and epigenetic alterations are conceivable [392] There are four kinds of engineered nucleases, which have been utilized for genome engineering in plants [391, 393–395]. A detail of these methods are discussed in the following sub-section.

8.5.4.2.5.1 Zinc Finger Nucleases (ZFNs)

ZFNs are targetable DNA cleavage components that have been espoused as gene-targeting implements. ZFNs is composed of two proteins that are connected artificially with peptide linker [396]. One protein targets specific DNA sequences and binds with it while the other, a bonded nuclease, generally FokI, cuts the target DNA in a nonprecise mode. They bind as monomers and each can recognizes 3 bpof DNA. Thus, three to six monomers can target DNA sequences of about 9–18-bases long and for more specificity longer

(24–36 bp) recognition sequences have to be edited, which can also reduce the off-site cleavage [397, 398]. ZFNs have been applied in apple and fig successfully through a targeted mutagenesis experiment [399].

8.5.4.2.5.2   *Transcription Activator-like Effector Nucleases (TALENs)*

Alike to the ZFNs, TALENs are also chimeric proteins containing customizing DNA-binding components bonded to endo-nuclease. Its DNA-binding dominions poised of 33–35 amino acids recurrences, each of which distinguishes a single DNA base pair. This represents an advantage in terms of design flexibility [400]. The target specifies of TALENs trusts on the appearance of two amino acids, occur at 12 and 13 positions are the exception, which is called as repeat-variable residues [401]. Each of the RVD recognizes a specific DNA base; for instance, NI repeats of RVDs bind to adenosine, HD to cytosine, NG to thymine, and NN to guanine or adenosine [402]. After the discovery of DNA recognition mechanism, it got attention instantly for its utilization in biotechnology [403]. It has been applied to get site-specific alterations in various plant species like *A. thaliana* [404], tobacco [405], and rice to developed heritable, disease-resistant lines.

8.5.4.2.5.3   *Clustered Regularly Interspaced Short Palindromic Repeats (CRISPR)/CRISPR-associated protein9 (Cas9)*

CRISPR is a bacterial DNA segment that contains repeats and Cas9 is a CRISPR-related protein. It is firstly exposed in 1987, to identify and break down the specific strands of DNA, complementary to the CRISPR sequence. In recent past years, the CRISPR/Cas9 genome editing approach has so far shown the greatest potential to develop crop cultivars, which are to grow under the hostile environment with desirable yield. The recent progress, prospects, and challenges of CRISPR/Cas-mediated genome editing approach for crop improvement against adversative environmental conditions [406, 407]. CRISPR–Cas9 genome editing method has become a revolution to increase yield along with improving stress tolerance that is diseases, drought, salinity, etc. [408].

8.6   CHALLENGES FOR APPLICATION OF NGSAT

In the field of agriculture, NGSAT have eased the crop improvement through the processes of genome sequencing, gene editing, gene mutation etc. and has substantially reduced the cost and time requirements than previously used sanger technology of sequencing [409]. Although with negligible hands-on procedures a lot of data are obtained, it is not very easy to do so. This huge amount of data through helpful in research, it is very hard to analyze those [410]. As NGSAT use in agriculture continues to increase, enormous challenges are there that hinder its development.

8.6.1 DIFFICULTY IN SEQUENCING

NGSAT are used in the sequencing of the whole genome within a few days, which includes massive parallel sequencing. The huge amount of data generated in this process requires the improvement of a rationalized, extremely automated pipeline to enable investigation [411].

8.6.2 DIFFICULTY IN THE SEQUENCE ALIGNMENT

After each run sequence alignment is done. However, it is a difficult task. There are chances of mapping a read to multiple locations as reading lengths are relatively short (36–250 bp). Sequencing error is likely to happen in next-generation sequencing because of the lack of quality of data generated through this technique, especially in repetitive zones [412].

8.6.3 PROBLEMS WHILE MAPPING MULTI-READS

Plant genomes are not unique or different. Numerous nucleotide arrangements are precisely like the other or with slight dissimilarities; between pseudo-genes, repetitive orders this matching makes the mapping of data to the genome difficult [413].

8.6.4 DIFFICULTY IN IDENTIFYING REDUNDANT SEQUENCES

It is very essential to recognize redundant sequences-duplicate reads, which are an outcome of polymerase chain reaction (PCR) amplification. Although, it is very problematic to identify arrangements that are changed by PCR [414].

8.6.5   LACK OF FOCUS AND INFRASTRUCTURE

Next-generation technology requires biotechnological laboratory equipped with all the instruments to carry out the sequencing procedure. In most of the developing countries, the requirement of this technology in the field of agriculture has not yet been identified and also not proper infrastructure facilities are there [415]. In the absence of identified national priorities, it is difficult to introduce new technologies.

8.6.6   POOR FUNDING FOR RESEARCH AND DEVELOPMENT

Although advancement in NGSAT have lowered the cost of genome sequencing to a great extent, the other instruments required in this process require heavy investment. In most of the developing countries, the budget for research and development is low, which makes it difficult to adopt new technologies [416].

8.6.7   INADEQUATE HUMAN RESOURCES AND EXPERTISE

NGSAT generate a huge amount of data in a short time. Highly skilled manpower or experts in this field are needed to handle and analyze such information [417]. However, in developing countries, there is a shortage of such persons.

8.6.8   FARM-LEVEL BARRIERS

The role of NGSAT in agriculture is the development of crops with improved traits [418]. However, the farmers (especially small and marginal) are reluctant to take the risk of using new cultivars in place of traditional varieties.

8.7   CONCLUSIONS AND FUTURE RESEARCH THRUSTS

It is projected that the global population is approaching 9.5 billion by 2050. One of the adverse effects as a result of the fluctuation of climate change is global warming, which means the rise of atmospheric temperature that influences the earth's ecosystem, particularly on the agroecosystem.

Anthropogenic changes occur within the short period of few decades causing harm to the environment and living beings, which are largely responsible for raising the ecological complications such as increasing climate change consequences decline the food and nutritional security and consequently mounting the food values. In current decades, global warming has to turn out to the extreme pressures to agricultural systems leads to global food and nutritional security. Since agriculture is amongst the greatly thoughtful systems prejudiced due to the change in ecological condition. These effects are more serious in developing countries due to their agricultural depend on the economy and also traditional farming. Therefore to eliminate hunger and poverty by 2050 and also reducing the GHGs emission, climate-smart agricultural technologies are essential in the traditional farming. The present chapter advocates next-generation climate-smart agricultural technologies in traditional agriculture for future food and environmental security.

KEYWORDS

- **agricultural technology**
- **traditional farming**
- **growing population**
- **food and nutritional security**

REFERENCES

1. Niles, M. T.; Ahuja, R.; Esquivel, J.; Mango, N.; Duncan, M.; Heller, M.; Tirado, C. Climate change and food systems: Assessing impacts and opportunities. Meridian Institute, **2017**. Available online: http://bit.ly/2oFucpe (accessed on 31 May 2020).
2. Loboguerrero, A.M.; Campbell, B.M.; Cooper, P.J.M.; Hansen, J.W.; Rosenstock, T., Wollenberg, E. Food and earth systems: Priorities for climate change adaptation and mitigation for agriculture and food systems. *Sustainability,* **2019**, *11*, 1372; doi:10.3390/su11051372.
3. IPCC. Intergovernmental Panel on Climate Change. Climate Change: Impacts, Adaptation and Vulnerability. In Contribution of Working Group II to the Fourth Assessment Report of the Intergovernmental Panel on Climate Change, **2007**, Cambridge University Press: Cambridge, UK.
4. Korea Meteorological Administration. (2008) Understanding of Climate Change and Scenario Application of climate change scenarios. Available online: http://www.kma.go.kr/download_01/2018english.pdf Accessed on 16 August 2020.

5. Aryal, J. P., Sapkota, T. B., Khurana, R., Khatri-Chhetri, A., Bahadur Rahut, D., et al. Climate change and agriculture in South Asia: adaptation options in smallholder production systems. *Environment, Development and Sustainability*, **2020**, *22*, 5045–5075. https://doi.org/10.1007/s10668-019-00414-4.

6. Gornall, J.; Betts, R., Burke, E. Implications of climate change for agricultural productivity in the early twenty-first century. *Philosophical Transactions of the Royal Society of London. Series B, Biological Sciences*, **2010**, *365*, 2973–2989.

7. Tun Oo, A.; Van Huylenbroeck, G.; Speelman, S. Measuring the economic impact of climate change on crop production in the dry zone of myanmar: a ricardian approach. *Climate*, **2020**, *8*, 9. doi:10.3390/cli8010009.

8. Iizumi, T. Emerging Adaptation to Climate Change in Agriculture. *In: Emerging Adaptation to Climate Change in Agriculture, Research and practices*, Iizumi T, Hirata R, Matsuda R (eds.), Springer Nature Singapore, **2019**, pp. 1–16. https://doi.org/10.1007/978-981-13-9235-1.

9. UNEP. The adaptation gap report, **2017**. UNEP, Nairobi. https://www.unenvironment.org/ resources/report/adaptation-gap-report-2017 (Accessed 25 April 2020)

10. Sugiura, T.; Sumida, H.; Yokoyama, S.; Ono, H. Overview of recent effects of global warming on agricultural production in Japan. *Japan Agricultural Research*, **2012**, *46*, 7–13. https://doi.org/10.6090/jarq.46.7

11. Nuttall, J. G.; O'Leary, G. J.; Panozzo, J. F.; Walker, C. K.; Barlow, K. M.; Fitzgerald, G. J. Models of grain quality in wheat–a review. *Field Crops Research*, **2017**, *202*, 136–145.

12. Ergon, Å.; Seddaiu, G.; Korhonenc, P.; Virkajärvic, P.; Bellocchid, G.; Jørgensene, M.; Østremf L.; Reheulg, D.; Volaire, F. How can forage production in Nordic and Mediterranean Europe adapt to the challenges and opportunities arising from climate change? *European Journal of Agronomy*, **2018**, *92*, 97–106. http://dx.doi.org/10.1016/j.eja.2017.09.016.

13. Ishimaru, T.; Hirabayashi, H.; Sasaki, K.; Ye, C.; Kobayashi, A. Breeding efforts to mitigate damage by heat stress to spikelet sterility and grain quality. *Plant Production Science*, **2016**, *19*, 12–21 doi.org/10.1080/1343943X.2015.1128113

14. Morita, S.; Wada, H.; Matsue, Y. Countermeasures for heat damage in rice grain quality under climate change. *Plant Production Science*, **2016**, *19*, 1–11.

15. Fujibe, F.; Yamazaki, N.; Kobayashi, K. Long-term changes of heavy precipitation and dry weather in Japan (1901–2004). *Journal of the Meteorological Society of Japan*, **2006**, *84*, 1033–1046.

16. Ishigooka, Y.; Fukui, S.; Hasegawa, T.; Kuwagata, T.; Nishimori, M.; Kondo, M. Large-scale evaluation of the effects of adaptation to climate change by shifting transplanting date on rice production and quality in Japan. *Journal of Agricultural Meteorology*, **2017**, *73*, 156–173.

17. Shimoda, S.; Kanno, H.; Hirota, T. Time series analysis of temperature and rainfall-based weather aggregation reveals significant correlations between climate turning points and potato (*Solanum tuberosum* L) yield trends in Japan. *Agricultural and Forest Meteorology*, **2018**, *263*, 147–155.

18. FAO. The state of food security and nutrition in the world. **2018**. Building climate resilience for food security and nutrition. FAO, Rome. http://www.fao.org/3/I9553EN/i9553en.pdf. Accessed 2 July2020.

19. IPCC, Summary for policymakers, Climate Change 2014: Impacts, Adaptation, andVulnerability. Part A: Global and Sectoral Aspects. Contribution of Working GroupII

to the Fifth Assessment Report of the Intergovernmental Panel on ClimateChange, Cambridge University Press, Cambridge, United Kingdom and New York, NY, USA, 2014.

20. Oyoshi, K.; Tomiyama, N.; Okumura, T.; Sobue, S.; Sato, J. Mapping rice-planted areas using time-series synthetic aperture radar data for the Asia-RiCE activity. *Paddy* and *Water Environment,* **2016**, *14*, 463–472. doi.org/10.1007/s10333–015-0515-x

21. Vrieling, A.; Meroni, M.; Mude, A. G.; Chantrat, S.; Ummenhofer, C. C.; de Bie, C. A. J. M. Early assessment of seasonal forage availability for mitigating the impact of drought on East African pastoralists. *Remote Sensing of Environment,* **2016**, *174*, 44–55.

22. Ohno, H.; Sasaki, K.; Ohara, G.; Nakazono, K. Development of grid square air temperature and precipitation data compiled from observed, forecasted, and climatic normal data. *Climate Biosphere,* **2016**, *16*, 71–79.

23. Hayashi, K.; Llorca, L.; Rustini, S.; Setyanto, P.; Zaini, Z. Reducing vulnerability of rainfed agriculture through seasonal climate predictions: a case study on the rainfed rice production in Southeast Asia. *Agricultural Systems,* **2018**, 162, 66–76.

24. Swaminathan, M. S.; Kesavan, P. C. Agricultural research in an era of climate change. *Agricultural Research,* **2012**, *1*, 3–11.

25. Ackerman, F. and Stanton, E. A. 2013. Climate Impacts on Agriculture: A Challenge to Complacency? *Global Development and Environment Institute Working.* Paper No. 13–01.

26. Islam, S.M.F.; Karim, Z. World's demand for food and water: The consequences of climate change, *In*: Desalination-Challenges and Opportunities. **2019**, https://doi.org/10.5772/intechopen.85919, Online First, Available online: https://www.intechopen.com (Accessed 16 June 2020).

27. Jones, P.G.; Thornton, P.K. The potential impacts of climate change on maize production in Africa and Latin America in 2055. *Global Environmental Change,* **2003**, *13*, 51–59.

28. Alcamo, J., Dronin, N., Endejan, M., Golubev, G. and Kirilenko, A. 2007. A new assessment of climate change impacts on food production shortfalls and water availability in Russia. *Global Environmental Change.* **2007**, 17(3–4), 429–444.

29. Belyaeva, M. and Bokusheva, R. 2017. Will climate change benefit or hurt Russian grain production? A statistical evidence from a panel approach, Discussion Paper, No. 161. Institute of Agricultural Development in Transition Economies (IAMO), Halle (Saale),

30. Nguyen, N.V. Global climate changes and rice food security. *IRC Rep,* **2012**, 24–31. Available online: http://www.fao.org/climatechange/15526–03ecb62366f779d1ed4528 7e698a44d2e.pdf (Accessed 01 July 2019).

31. Sage, T. L.; Bagha, S.; Lundsgaard-Nielsen, V.; Branch, H. A.; Sultmanis, S.; Sage, R. F. The effect of high temperature stress on male and female reproduction in plants. *Field Crops Research,* **2015**, *182*, 30–42. http://dx.doi.org/10.1016/j.fcr.2015.06.011.

32. Lesk, C.; Rowhani, P.; Ramankutty, N. Influence of extreme weather disasters on global crop production. *Nature,* **2016**, *529*, 84–87. http://dx.doi.org/10.1038/nature16467.

33. FAOSTAT. Crop production statistics. **2020**, Available online: http://www.fao.org/faostat/en/#data/QC; Accessed 21 June 2020.

34. Singh, K.; McCleanb, C. J.; Bükerc, P.; Hartleya, S. E.; Hill, J. K. Mapping regional risks from climate change for rainfed rice cultivation in India. *Agricultural Systems,* **2017**, *156*, 76–84. http://dx.doi.org/10.1016/j.agsy.2017.05.009

35. Yoshida, S. Climate environment and its influence. In: Fundamentals of Rice Crop Science. The International Rice Research Institute, Manila, **1981**, p. 269

Plant Abiotic Stress Physiology, Volume 1

36. Nguyen, D. N.; Lee, K. J.; Kim, D. I.; Anh, N. T.; Lee, B. W. Modeling and validation of high-temperature induced spikelet sterility in rice. *Field Crops Research,* **2014**, *156*, 293–302. http://dx.doi.org/10.1016/j.fcr.2013.11.009.

37. Van Oort, P.A.; Zhan, T.,; de Vries, M.E.; Heinemann, A.B., Meinke, H. Correlation between temperature and phenology prediction error in rice (*Oryza sativa* L.). *Agricultural and Forest Meteorology,* **2011**, *151*, 1545–1555. http://dx.doi.org/10.1016/j.agrformet.2011.06.012.

38. NAAS. Climate Resilient Agriculture in India. **2013**, Policy Paper No. 65, National Academy of Agricultural Sciences, New Delhi: 20 p.

39. Auffhammer, M., Ramanathan, V. and Vincent, J. R. Climate change, the monsoon, and rice yield in India. *Climatic Change,* **2012**, *111*(2), 411–424.

40. Soora, N.K., Aggarwal, P.K., Saxena, R., Rani, S., Jain, S. and Chauhan, N., 2013. An assessment of regional vulnerability of rice to climate change in India. *Climatic Change,* **2013**, *118*(3–4), 683–699.

41. World Bank. Overcoming drought: adaptation strategies for Andhra Pradesh, **2006**, Washington, DC, USA

42. Bhatt, D.; Sonkar, G.; Mall, R. K. Impact of Climate Variability on the Rice Yield in Uttar Pradesh: an Agro-Climatic Zone Based Study. *Environmental Processes,* **2019**, *6*, 135–153. https://doi.org/10.1007/s40710-019-00360-3

43. Hatfield, J.L., Prueger, J.H. Temperature extremes: effect on plant growth and development. *Weather and Climate Extremes,* **2015**, *10*, 4–10

44. Kelkar, S.M.; Kulkarni, A.; Koteswara Rao K. Impact of climate variability and change on crop production in Maharashtra, India. *Current Science,* **2020**, 118, 1235–1245.

45. Peng, S. B.; Huang, J. E.; Sheehy, J. E.; Laza, R. C.; Vispera, K, H.; Zhong, X. H.; Centeno, S.; Khush, G. S.; Cassman, K. G. Rice yields decline with higher night temperatures from global warming, *Proceedings of the National Academy of Sciences of the United States of America,* **2004**, *101*, 9971–9975.

46. Murdiyarso, D. Adaptation to climatic vulnerability and change: Asian perspectives on agriculture and food security. *Environmental Monitoring and Assessment,* **2000**, *61* (1):123–131.

47. Matsuda, M. Upland farming systems coping with uncertain rainfall in the central dry zone of Myanmar: how stable is indigenous multiple cropping under semi-arid conditions? *The Journal of Human Ecology,* **2013**, *41*, 927–936

48. Fraga, H., Guimarães, N., Santos, J. A. Future changes in rice bioclimatic growing conditions in Portugal. *Agronomy,* **2019**, *9*, 674. doi:10.3390/agronomy9110674

49. Hossain, M.S.; Qian, L.; Arshad, M.; Shahid, S.; Fahad, S.; Akhter, J. Climate change and crop farming in Bangladesh: an analysis of economic impacts. *International Journal of Climate Change Strategies and Management,* **2019**, *11*, 424–440. doi: 10.1108/IJCCSM-04-2018-0030.

50. Rahman, S.; Anik, A. R. Productivity and efficiency impact of climate change and agroecology on Bangladesh agriculture. *Land Use Policy,* **2020**, *94*, 104507. http://dx.doi.org/ 10.1016/j.landusepol.2020.104507

51. Shahid, S. Impact of climate change on irrigation water demand of dry season boro rice in northwest Bangladesh, *Climatic Change,* **2010**. doi: 10.1007/s10584-010-9895-5.

52. Schmutz, J.; Cannon, S. B.; Schlueter, J.; Ma, J.; Mitros, T.; Nelson, W.; Hyten, D. L. Genome sequence of the palaeopolyploid soybean. *Nature,* **2010**, *463*(7278), 178–183.

53. Shahid, S.; Harun, S. B.; Katimon, A. Changes in diurnal temperature range in Bangladesh during the time period 1961–2008. *Atmospheric Research*, **2012**, *118*, 260–270. doi: 10.1016/j. atmosres.2012.07.008.

54. Shahid, S.; Khairulmaini, O. S. Spatial and temporal variability of rainfall in Bangladesh. *Asia-Pacific The Journal of the Atmospheric Sciences*, **2009**, *45*, 375–389.

55. BBS (Bangladesh Bureau of Statistics), 2008. Compendium of Environment Statistics of Bangladesh. Government of Bangladesh, Dhaka, Bangladesh.

56. Asaduzzaman, M., Ringler, C., Thurlow, J. and Alam, S. 2010. Investing in Crop Agriculture in Bangladesh for Higher Growth and Productivity, and Adaptation to Climate Change. Bangladesh Food Security Investment Forum, Dhaka.

57. Sarker, M. A. R.; Alam, K.; Gow, J. Exploring the relationship between climate change and rice yield in Bangladesh: An analysis of time series data. *Agricultural Systems*, **2012**, *112*, 11–16. http://dx.doi.org/10.1016/j.agsy.2012.06.004

58. BBS (Bangladesh Bureau of Statistics), 2009. Yearbook of Agricultural Statistics of Bangladesh. Government of Bangladesh, Dhaka

59. Rahman, M. A.; Kang, S.; Nagabhatla, N.; Macnee, R. Impacts of temperature and rainfall variation on rice productivity in major ecosystems of Bangladesh. *Agriculture & Food Security*, **2017**, *6*, 10. http://dx.doi.org/10.1186/s40066-017-0089-5

60. Mamun, A.H.M.M.; Ghosh, B.C.; Rayhanul Islam, S.M. Climate Change and Rice Yield in Bangladesh: A Micro Regional Analysis of Time Series Data. *International Journal of Science and Research*, **2015**, *5*(2), 1–8

61. Ahmed, A. H. 2006. Bangladesh: Climate Change Impacts and Vulnerability. Climate change cell, Department of Environment, Comprehensive Disaster Management Programme, Bangladesh, https://www.preventionweb.net/files/574_10370.pdf Accessed 18 July 2020.

62. Krishnamurthy, P.K.; Lewis, K.; and Choularton, R.J. A methodological framework for rapidly assessing the impacts of climate risk on national-level food security through a vulnerability index. *Global Environmental Change*, **2014**, *25*, 121–132

63. Hossain, M.S.; Majumder, A.K. impact of climate change on agricultural production and food security: a review on coastal regions of Bangladesh. *International Journal of Agricultural Research, Innovation and Technology*, **2018**, *8*(1), 62–69.

64. Li, Z.; Liu, Z.; Anderson, W.; Yang, P.; Wu, W.; Tang, H.; You, L. Chinese rice production area adaptations to climate changes, 1949–2010. *Environmental Science & Technology*, **2015**, *49*, 2032–2037.

65. Liu, Z.; Li, Z.; Tang, P.; Li, Z.; Wu, W.; Yang, P.; You, L.; Tang, H. Change analysis of rice area and production in China during the past three decades. *Journal of Geographical Science*, **2013**, *23*, 1005–18.

66. Lobell, D.B.; Bänziger, M.; Magorokosho, C.; Vivek, B. Nonlinear heat effects on African maize as evidenced by historical yield trials. *Nature Climate Change*, **2011**, *1*, 42–45.

67. Hu, Y; Fan, L.; Liu, Z.; Yu, Q.; Liang, S.; Chen, S., You, L.; Wu, W.; Yang, P. Rice production and climate change in Northeast China: evidence of adaptation through land use shifts. *Environmental Research Letters*, **2019**, *14*, 024014, https://doi.org/10.1088/1748-9326/aafa55

68. Liu, L.; Wang, E.; Zhu, Y.; Tang, L. Contrasting effects of warming and autonomous breeding on single-rice productivity in China. *Agriculture, Ecosystems & Environment*, **2012**, *149*, 20–9.

69. Yohannes, H. A. Review on relationship between climate change and agriculture. *Journal of Earth Science and Climatic Change*, **2016**, *7*, 335. doi: 10.4172/2157-7617.1000335

70. Tao, F.; Zhang, Z.; Shi, W.; Liu, Y.; Xiao, D.; Zhang, S.; Zhu, Z.; Wang, M.; Liu, F. Single rice growth period was prolonged by cultivars shifts, but yield was damaged by climate change during 1981–2009 in China, and late rice was just opposite. *Global Change Biology*, **2013**, *19*, 3200–9.

71. Shi, Q-H.; Liu, J-G.; Wang, Z-H.; Tao, T-T.; Chen, F.; Chu, Q-Q. Change of rice yield gaps and influential climatic factors in southern China. *Acta Agronomica Sinica*, **2013**, *38*, 896–903.

72. Guo, Y.; Wu, W., Du, M., Liu, X.; Wang, J.; Bryant, C. R. Modeling Climate Change Impacts on Rice Growth and Yield under Global Warming of 1.5 and 2.0 °C in the Pearl River Delta, China. *Atmosphere,* **2019**, 10, 567. doi:10.3390/atmos10100567

73. Lv, Z.; Zhu, Y.; Liu, X.; Ye, H., Tian, Y.; Li, F. Climate change impacts on regional rice production in China. *Climatic Change,* **2018**, Available online: https://doi.org/10.1007/s10584-018-2151-0 (Accessed 25 June 2020)

74. Asseng, S., Foster, I. and Turner, N. The impact of temperature variability on wheat yields. *Global Change Biology*. **2011**, *17*, 997–1012.

75. Xiao, D.; Bai, H.; Liu, D. L. Impact of future climate change on wheat production: a simulated case for China's wheat system. *Sustainability*, **2018**, *10*, 1277; doi:10.3390/su10041277.

76. Mudasser, M.; Hussain, I.; Aslam, M. Constraints to Land-and Water Productivity of Wheat in India and Pakistan: A Comparative Analysis; International Water Management: Colombo, Sri Lanka, 2001.

77. Chen, Y.; Zhang, Z.; Tao, F. Impacts of climate change and climate extremes on major crops productivity in China at a global warming of 1.5 and 2.0 °C. *Earth System Dynamics*, **2018**, *9*, 543–562. https://doi.org/10.5194/esd-9-543-2018

78. Porter, J.R.; Gawith, M. Temperatures and the growth and development of wheat: A review. *The European Journal of Agronomy*, **1998**, *10*, 23–26.

79. Luo, Q. Temperature thresholds and crop production: A review. *Climatic Change*, **2011**, *109*, 583–598.

80. Zhao, C.; Liu, B.; Piao, S.; Wang, X.; Lobell, D.B.; Huang, Y.; Huang, M.; Yao, Y.; Simona, B.; Ciais, P.; Durand J-L, et al. Temperature increase reduces global yields of major crops in four independent estimates. *Proceedings of the National Academy of Sciences of the United States of America*, **2017**, *114*, 9326–9331.

81. Mukherjee, D. ed., Economic Integration in Asia: Key Prospects and Challenges with the Regional Comprehensive Economic Partnership, **2019,** Routledge.

82. Eduardo, D.O.; Helen, B.; Kadambot, S. H. M., Samuel, H., Jens, B., Jairo, P. A. Can elevated CO2 combined with high temperature ameliorate the effect of terminal drought in wheat? *Functional Plant Biology*. **2013**, *40*, 160–171.

83. Van Ittersum, M. K.; Howden, M. S.; Asseng, S. Sensitivity of productivity and deep drainage of wheat cropping systems in a Mediterranean environment to changes in CO_2, temperature and precipitation. *Agriculture Ecosystems & Environment.* **2003**, *97*, 255–273.

84. Fulco, L.; Senthold, A. Climate change impacts on wheat production in a Mediterranean environment in Western Australia. *Agricultural Systems*, **2006**, *90*, 159–179.

85. Lobell, D.B.; Gourdji, S.M. The influence of climate change on global crop productivity. *Plant Physiology*, **2012**, *160*, 1686–1697.

86. US Climate Change Science Program. The Effects of Climate Change on Agriculture, Land Resources, Water Resources, and Biodiversity in the United States. **2008,** 1717 Pennsylvania Avenue, Washington, D.C. 20006 USA, pp. 240

87. Aleminew, A.; Abera, M. Effect of climate change on the production and productivity of wheat crop in the highlands of Ethiopia: a review. *Agricultural Reviews.* **2020,** *41*(1), 34–42.

88. Hogy, P.; Wieser, H.; Kohler, P.; Schwadorf, K.; Breuer, J.; Erbs, M.; Weber, S.; Fangmeier, A. Does elevated atmospheric CO_2 allow for sufficient wheat grain quality in the future? *Journal of Applied Botany and Food Quality,* **2009,** *82,* 114–121.

89. Burney, J.; Ramanathan, V. Recent climate and air pollution impacts on Indian agriculture. *Proceedings of the National Academy of Sciences* U S A, **2014,** *111*(46), 16319–16324.

90. Tao, F.; Zhang, Z.; Xiao, D.; Zhang, S.; Rötter, R.; Shi, W.; Liu, Y.; Wang, M.; Liu, F.; Zhang, H. Responses of wheat growth and yield to climate change in different climate zones of China, 1981–2009. *Agricultural and Forest Meteorology,* **2014,** 189–190, 91–104.

91. Xiao, D.; Shen, Y.; Qi, Y.; Moiwo, J.; Min, L.; Zhang, Y.; Guo, Y.; Pei, H. Impact of alternative cropping systems on groundwater use and grain yields in the North China Plain Region. *Agricultural Systems,* **2017a,** *153,* 109–117.

92. Xiao, D.; Cao, J.; Bai, H.; Qi, Y.; Shen, Y. Assessing the impact of climate variables and sowing date on spring wheat yield in the Northern China. *International Journal of Agriculture and Biology,* **2017b,** *19,* 1551–1558.

93. Fu, C.; Wen, G. Variation of ecosystems over East Asia in association with seasonal, interannual and decadal monsoon climate variability. *Climate Change,* **2001,** *43,* 477–494.

94. Asseng, S., Ewert, F., Martre, P., Rötter, R.P., et al. Rising temperatures reduce global wheat production. *Nature Climate Change,* **2015,** *5,* 143–147.

95. Nonhebel, S. Effects of temperature rise and increase in CO_2 concentration on simulated wheat yields in Europe. *Climatic Change,* **1996,** *34,* 73–90.

96. Lin, E.; Xiong, W.; Ju, H.; Xu, Y.; Li, Y.; Bai, L.; Xie, L. Climate change impacts on crop yield and quality with $CO2$ fertilization in China. *Philosophical Transactions of the Royal Society B: Biological Sciences,* **2005,** *360,* 2149–2154.

97. Yang, Y.; Liu, D.; Anwar, M.; Zuo, H.; Yang, Y. Impact of future climate change on wheat production in relation to plant-available water capacity in a semiarid environment. *Theoretical and Applied Climatology,* **2014,** *115,* 391–410.

98. Wang, J.; Wang, E.; Yang, X.; Zhang, F.; Yin, H. Increased yield potential of wheat-maize cropping system in the North China Plain by climate change adaptation. *Climate Change,* **2012,** *113,* 825–840.

99. Sun, Q.; Kröbel, R.; Müller, T.; Römheld, V.; Cui, Z.; Zhang, F.; Chen, X. Optimization of yield and water-use of different cropping systems for sustainable groundwater use in North China Plain. *Agriculture Water Management,* **2011,** *98,* 808–814.

100. Xiao, D.; Tao, F. Contributions of cultivars, management and climate change to winter wheat yield in the North China Plain in the past three decades. *European Journal of Agronomy,* **2014,** *52,* 112–122.

101. Gupta, R.; Somanathan, E.; Dey, S. Global warming and local air pollution have reduced wheat yields in India. *Climatic Change,* **2016,** *140,* 593–604, DOI 10.1007/ s10584-016-1878-8

102. Jalota, S.K.; Kaur, H.; Kaur, S.; Vashisht, B.B. Impact of climate change scenarios on yield, water and nitrogen-balance and use efficiency of rice–wheat cropping system. *Agricultural Water Management,* **2013,** *116,* 29–38.

103. Danhassan, S. S.; Meena, A.; Usman, M. K.; Hussaini, A.; Abubakar, A.; Abubakar, M. J. Climate Change and the Fall of Agricultural Production in Some Selected Regions of India. *International Research Journal of Innovations in Engineering and Technology,* **2018**, *2*(9), 17- 23.

104. Chakraborty, D.; Sehgal, V. K.; Dhakar, R.; Ray, M.; Das, D. K. Spatio-temporal trend in heat waves over India and its impact assessment on wheat crop. *Theoretical and Applied Climatology,* **2019**. Available online https://doi.org/10.1007/s00704-019-02939-0 Accessed 06 July 2020.

105. Eyshi Rezaei, E.; Bannayan, M. Rainfed wheat yields under climate change in northeastern Iran. *Meteorological Applications,* **2011**, *19*, 346–354.

106. Bannayan, M., Paymard, P. and Ashraf, B. Vulnerability of maize production under future climate change: possible adaptation strategies. *Journal of the Science of Food and Agriculture.* **2016**, 96, 4465–4474.

107. Paymard, P.; Bannayan, M.; Haghighi, R. S. Analysis of the climate change efect on wheat production systems and investigate the potential of management strategies. *Natural Hazards,* **2018**. https://doi.org/10.1007/s11069-018-3180-8

108. Rahimzadeh, F.; Asgari, A.; Fattahi, E.; Variability of extreme temperature and precipitation in Iran during recent decades. *The International Journal of Climatology,* **2009**, *29*, 329–343.

109. Ragab, R.; Prudhomme, C. Climate change and water resources management in arid and semi-arid regions: prospective and challenges for the 21st century. *Biosys Engineering,* **2002**, *81*, 3–34.

110. Hochman, Z.; Holzworth, D.; Hunt, J.R. Potential to improve on-farm wheat yield and WUE in Australia. Crop and Pasture Science, **2009**, *60*, 708–716.

111. Khan, A. The Looming Food Security. In Economics and Bussiness Reviews; *The Daily Dawn,* **2011**, Karachi, Pakistan.

112. Kiani, A.; Iqbal, T. Climate change impact on wheat yield in Pakistan (An Application of ARDL Approach). NUST Journal of Social Sciences and Humanities, 2018, *4*(2), 240–262.

113. Siddique, K.H., Johansen, C., Turner, N.C., Jeuffroy, M.H., Hashem, A., Sakar, D., Gan, Y. and Alghamdi, S.S., Innovations in agronomy for food legumes. A review. *Agronomy for Sustainable Development,* **2012**, *32*(1), 45–64.

114. Ali, S., Liu, Y., Ishaq, M., Shah, T., Abdullah Ilyas, A. and Ud Din, I. Climate change and its impact on the yield of major food crops: evidence from Pakistan. *Foods.* **2017**, 6:39. doi:10.3390/foods6060039

115. Abid, M., Scheffran, J., Schneider, U. A. and Ashfaq, M. Farmers' perceptions of and adaptation strategies to climate change and their determinants: The case of Punjab Province, Pakistan. *Earth System Dynamics.* **2015**, *6*, 225–243. https://doi.org/10.5194/esd-6-225-2015.

116. Rauf, S.; Bakhsh, K.; Abbas, A.; Hassan, S.; Ali, A.; Kächele, H. How hard they hit? Perception, adaptation and public health implications of heat waves in urban and peri-urban Pakistan. Environmental Science and Pollution Research, **2017**, *24*(11), 10630–10639. DOI: https://doi.org/10.1007/ s11356-017-8756-4.

117. Stocker, T. F.; Dahe, Q.; Plattne, G. Climate Change 2013: The Physical Science Basis. Working Group I Contribution to the Fifth Assessment Report of the Intergovernmental Panel on Climate Change. **2013**; Summary for Policymakers. New York: IPCC.

118. Ali, A. and Erenstein, O. Assessing farmer use of climate change adaptation practices and impacts on food security and poverty in Pakistan. *Climate Risk Management.* **2017**, *16*, 183–194.

119. Abid, M., Schneider, U. A. and Scheffran, J. Adaptation to climate change and its impacts on food productivity and crop income: Perspectives of farmers in rural Pakistan. *Journal of Rural Studies*. **2016**, *47*, 254–266. DOI: https://doi.org/10.1016/j.jrurstud.2016.08.005

120. Morgounov, A.; Haun, S.; Lang, L.; Martynov, S.; Sonder, K. Climate change at winter wheat breeding sites in central Asia, eastern Europe, and USA, and implications for breeding. Euphytica, **2013**, *194*(2), 277–92.

121. Nikolayeva, L., Denisov, N., and Novikov, V. Climate change in Eastern Europe: Belarus, Moldova, Ukraine. In: Climate change in Eastern Europe: Belarus, Moldova, Ukraine, 2012, Environment and Security Initiative (ENVSEC), Zoï Environment Network (ZOI), Available online http://preventionweb.net/go/29358 Accessed 05 July 2020

122. Fischer, S.; Pluntke, T.; Pavlik, D.; Bernhofer, C. Hydrologic effects of climate change in a sub-basin of the Western Bug River, Western Ukraine. *Environmental Earth Sciences*, **2014**, *72*(12), 4727–44.

123. Lioubimtseva, E.; Beurs, K. M.; Henebry, G. M. Grain production trends in Russia, Ukraine, and Kazakhstan in the context of the global climate variability and change. In Grain Production Trends in Russia, Ukraine, and Kazakhstan in the Context of the Global Climate Variability and Change eds. T. Younos & C. A. Grady, 121–41. **2013**, Berlin: Springer.

124. Lioubimtseva, E.; Henebry, G. Grain production trends in Russia, Ukraine and Kazakhstan: New opportunities in an increasingly unstable world? *Frontiers in Earth Science*, **2012**, *6*(2), 157–66.

125. Supit, I.; van Diepen, C. A.; de Wit, A. J. W.; Wolf, J.; Kabat, P.; Baruth, B.; Ludwig, F. Assessing climate change effects on European crop yields using the Crop Growth Monitoring System and a weather generator. *Agricultural and Forest Meteorology*, **2012**, *164*, 96- 111.

126. Müller, D.; Jungandreas, A., Koch, F.; Schierhorn, F. Impact of Climate Change on Wheat Production in Ukraine. 2016, Institute for Economic Research and Policy Consulting, Kyiv Available online: https://www.apd-ukraine.de/images/APD_APR_05–2016_impact_on_wheat_eng_fin.pdf Accessed 07 July 2020

127. Pavlova, V.; Shkolnik, I.; Pikaleva, A.; Efimov, S.; Karachenkova, A.; Kattsov. V. Future changes in spring wheat yield in the European Russia as inferred from a large ensemble of high-resolution climate projections. *Environmental Research Letters*, **2019**, *14*, 034010. https://doi.org/10.1088/1748-9326/aaf8be

128. Gorst, A.; Dehlavi, A.; Groom, B. Crop productivity and adaptation to climate change in Pakistan. *Environment and Development Economics*, **2018**, *23*, 679–701. DOI: https://doi.org/10.1017/ S1355770X18000232

129. Wolf, J. Effects of climate change on wheat production potential in the European Community. *European Journal of Agronomy*, **1993**, *2*(4), 281–292.

130. Olesen, J.E.; Trnka, M.; Kersebaum, K.C.; Skjelvåg, A.; Seguin, B.; Peltonen-Sainio, P.; Rossi, F.; Kozyra, J.; Micale, F. Impacts and adaptation of European crop production systems to climate change. *The European Journal of Agronomy*, **2011**, *34*, 96–112.

131. Barnabás, B., Jäger, K. and Fehér, A. The effect of drought and heat stress on reproductive processes in cereals. *Plant Cell and Environment*. **2008**, *31*, 11–38.

132. Raza, A.; Razzaq, A.; Mehmood, S. S.; Zou, X.; Zhang, X.; Lv, Y.; Xu, J. Impact of Climate Change on Crops Adaptation and Strategies to Tackle Its Outcome: A Review. *Plants,* **2019**, *8*, 34. doi:10.3390/plants8020034

133. Valverde, P.; de Carvalho, M.; Serralheiro, R.; Maia, R.; Ramos, V.; Oliveira, B. Climate change impacts on rainfed agriculture in the Guadiana river basin (Portugal). *Agricultural Water Management*, **2015**, *150*, 35–45. https://doi. org/10.1016/j.agwat.2014.11.008

134. Costa, R.; Pinheiro, N.; Almeida, A.; Gomes, C.; Coutinho, J.; Coco, J.; Costa, A.; Maçãs, B. Effect of sowing date and seeding rate on bread wheat yield and test weight under mediterranean conditions". *Emirates Journal of Food and Agriculture*, **2017**, *25*(12), 951–956. doi:https://doi.org/10.9755/ejfa.v25i12.16731

135. Páscoa, P.; Gouveia, C. M.; Russo, A.; Trigo, R. M. The role of drought on wheat yield interannual variability in the Iberian Peninsula from 1929 to 2012. *The International Journal of Biometeorology*, **2017**, *61*, 439–451. https://doi.org/10.1007/s00484-016-1224-x

136. Dias, A.S.; Lidon, F. C. Evaluation of grain filling rate and duration in bread and durum wheat, under heat stress after anthesis. *Journal of Agronomy and Crop Science*, **2009**, *195*, 137–147. https://doi.org/10.1111/j.1439-037X.2008.00347.x

137. Scotti-Campos, P.; Semedo, J. N.; Pais, I.; Oliveira, M.; Passarinho, J.; Ramalho, J. C. Heat tolerance of Portuguese old bread wheat varieties. *Emirates Journal of Food and Agriculture*, **2014**, *26*, 170–179. https://doi.org/10.9755/ejfa.v26i2.16761

138. Rolim, J.; Teixeira, J. L.; Catalao, J.; Shahidian, S. The impacts of climate change on irrigated agriculture in Southern Portugal. *Irrigation Drainage*, **2017**, *66*, 3–18. https://doi.org/10.1002/ird.1996

139. Yang, C.; Fraga, H.; van Ieperen, W.; Trindade, H.; Santos, J. A. Effects of climate change and adaptation options on winter wheat yield under rainfed Mediterranean conditions in southern Portugal. *Climatic Change*, **2019**, *154*, 159–178. https://doi.org/10.1007/s10584-019-02419-4

140. National Sustainable Agriculture Coalition. Agriculture and Climate Change: Policy Imperatives and Opportunities to Help Producers Meet the Challenge. **2019**, 110 Maryland Avenue Washington, D.C. Available online: https://sustainableagriculture.net Accessed 07 July 2020

141. Howard, J.; Cakan, E.; Upadhyaya, K. P. Climate Change and Its Impact on Wheat Production in Kansas. International Journal of Food and Agricultural Economics, **2016**, *4*(2), 1–10.

142. Olmstead, A. L.; Rhode, P. W. Adapting North American wheat production to climatic challenges, 1839–2009. *Proceedings of the National Academy of Sciences of the United States of America*, **2011**, *108*(2), 480–485.

143. Russell, Kathleen.; Lee, Chad.; McCulley Rebecca, L.; Van Sanford, David.; "Impact of Climate Change on Wheat Production in Kentucky". *Plant Soil Sci Res Rep*. **2014**, 2. Available online: https://uknowledge.uky.edu/pss_reports/2 Accessed 07 July 2020

144. Ruane, A.C.; Antle, J.; Elliott, J.; Folberth, C.; Hoogenboom, G.; Mason-D'Croz, D.; Müller, C.; Porter, C. H.; Phillips, M.; Raymundo, R.; Sands, R.; Valdivia, R.; White, J.; Wiebe, K.; Rosenzweig, C. Biophysical and economic implications for agriculture of +1.5 °C and +2.0 °C global warming using AgMIP coordinated global and regional assessments, *Climatic Research*, **2018**, *76*, 17–39.

145. Qian, B.; Zhang, X.; Smith, W.; Grant, B.; Jing, Q.; Cannon, A. J.; Neilsen, D.; McConkey, B. Li G.; Bonsal, B.; Wan, H.; Xue, L.; Zhao, J.; Climate change impacts on Canadian yields of spring wheat, canola and maize for global warming levels of 1.5 °C, 2.0 °C, 2.5 °C and 3.0 °C. *Environmental Research Letters*, **2019**, *14*, 074005. https://doi.org/10.1088/1748-9326/ab17fb

146. Lobell, D.B. and Field, C.B., Global scale climate–crop yield relationships and the impacts of recent warming. *Environmental Research Letters*, **2007**, *2*(1), p.014002.

147. Adhikari, U., Nejadhashemi, A.P. and Woznicki, S.A., Climate change and eastern Africa: a review of impact on major crops. *Food and Energy Security*, **2015**, *4*(2), pp.110–132.

148. Li, X.; Takahashi, T.; Suzuki, N.; Kaiser, H.M. The impact of climate change on maize yields in the United States and China. *Agricultural Systems*, **2011**, *104*, 348–353 doi:10.1016/j.agsy.2010.12.006

149. Tigchelaara, M.; Battistia, D. S.; Naylorb, R. L.; Ray, D. K. Future warming increases probability of globally synchronized maize production shocks. **2018**. Available online: www.pnas.org/lookup/suppl/doi:10. 1073/pnas.1718031115/-/DCSupplemental. Accessed 08 July 2020

150. Collins, M.; Knutti, R.; Arblaster, J.; Dufresne, J-L.; Fichefet, T.; Friedlingstein, P.; Gao, X.; Gutowski, W.J.; Johns, T.; Krinner, G.; Shongwe, M.; Tebaldi, C.; Weaver, A.J.; Wehner, M. Longterm climate change: Projections, commitments and irreversibility. In: Stocker TF, Qin D, Plattner G-K, Tignor M, Allen SK, Boschung J, Nauels A, Xia Y, Bex V, Midgley PM, editors. Climate Change **2013**: The Physical Science Basis. Contribution of Working Group I to the Fifth Assessment Report of the Intergovernmental Panel on Climate Change. Cambridge, United Kingdom and New York, NY, USA: Cambridge University Press; 2013

151. Trenberth, K. E. Changes in precipitation with climate change. *Climate Research.* **2011**, *47*, 123–138.

152. Leakey, A.D.B.; Uribelarrea, M.; Ainsworth, E.A.; Naidu, S.L.; Rogers, A.; Ort, D.R.; Long, S.P. Photosynthesis, productivity, and yield of maize are not affected by open-air elevation of CO_2 concentration in the absence of drought. *Plant Physiology*, **2006**, *140*, 779–790

153. Badu-Apraku, B., Hunter, R. B. and Tollenaar, M. Effect of temperature during grain filling on whole plant and grain yield in maize (*Zea mays* L.). *Canadian Journal of Plant Science.* **1983**, *63*, 357–363.

154. Muchow, R.C.; Sinclair, T.R.; Bennett, J.M. Temperature and solar-radiation effects on potential maize yield across locations. *Agronomy Journal*, **1990**; *82*, 338–343

155. Kiniry, J.R.; Bonhomme, R. Predicting maize phenology. In: Hodges T, editor. Predicting Crop Phenology. Boca Raton, FL: CRC Press; **1991**, pp. 115–131.

156. Herrero, M.P.; Johnson, R.R. High temperature stress and pollen viability in maize. *Crop Science*, **1980**, *20*, 796–800.

157. Fonseca, A. E.; Westgate, M. E. Relationship between desiccation and viability of maize pollen. *Field Crops Research.* **2005**, *94*, 114–125.

158. Hatfield, J.L., Wright-Morton, L.; Hall, B. Vulnerability of grain crops and croplands in the Midwest to climatic variability and adaptation strategies. *Climatic Change*, **2018**, *146*, 263–275. DOI: 10.1007/s10584-017-1997-x

159. Kukal, M.S.; Irmak, S. Climate-Driven Crop Yield and Yield Variability and Climate Change Impacts on the U.S. Great Plains Agricultural Production. *Scientific Reports,* **2018**, *8*, 3450. DOI:10.1038/s41598-018-21848-2

160. Eulenstein, F., Lana, M., Schlindwein, S., Sheudzhen, A., Tauschke, M., Behrend, A., Guevara, E. and Meira, S., Trends of soybean yields under climate change scenarios. *Horticulturae*, **2017**, *3*(1), 10.

161. Bellon, M. R., Hodson, D. and Hellin, J. Assessing the vulnerability of traditional maize seed systems in Mexico to climate change. *Proceedings of the National Academy of*

Sciences of the United States of America. **2011**, *108*(33):13432–13437. www.pnas.org/cgi/doi/10.1073/pnas.1103373108.

162. Ureta, C.; González, E. J.; Espinosa, A.; Trueba, A.; Piñeyro-Nelson, A.; Álvarez-Buylla, E. R. Maize yield in Mexico under climate change. *Agricultural Systems*, **2020**, *177*, 102697. https://doi.org/10.1016/j.agsy.2019.102697.

163. Condel, C.; Liverman, D.; Flores, M.; Ferrer, R.; Araujo, R.; Betancourt, E.; Villarreal, G.; Gay, C. Vulnerability of rainfed maize crops in Mexico to climate change. *Climate Research*, **1997**, *9*, 17–23.

164. Thiele, G. Informal potato seed systems in the Andes: Why are they important and what should we do with them? *World Development*, **1999**, *27*, 83–99.

165. Nagarajan, L.; Smale, M.; Village seed systems and the biological diversity of millet crops in marginal environments of India. *Euphytica*, **2007**, *155*, 167–182.

166. Burke, M. B.; Lobell, D. B.; Guarino, L. Shifts in African crop climates by 2050, and the implications for crop improvements and genetic resources conservation. *Global Environmental Change*, **2009**, *19*, 317–325.

167. Morton, J.F. The impact of climate change on smallholder and subsistence agriculture. *Proceedings of the National Academy of Sciences of the United States of America*, **2007**, *104*, 19680–19685.

168. Morris, M.; Mekuria, M.; Gerpacio, R. Crop Variety Improvement and its effect on productivity: The impact of international agricultural research, eds. Evenson RE, Gollin D. **2003**, CABI Publishing, Wallingford, UK, pp. 135–158.

169. Vergara W,; Rios, A.R.; Trapido, P.; Malarín, H. Agriculture and Future Climate in Latin America and the Caribbean: Systemic Impacts and Potential Responses. Inter-American Development Bank Climate Change and Sustainability Division, **2014**, Discussion paper no. IDB-DP-329, available online: iadb.org/climatechange Accessed 09 July 2020

170. Travasso, M. I.; Magrin, G. O.; Baethgen, W. E.; Castaño, J. P.; Rodriguez, G. R.; Pires, J. L.; Gimenez, A.; Cunha, G.; Fernandes, M. Adaptation Measures for Maize and Soybean in Southeastern South America. **2006**, Assessments of Impacts and Adaptations to Climate Change (AIACC) Working Paper no. 28. Available: www.aiaccproject.org Accessed 08 July 2020.

171. Niu, X-K.; Xie, R-Z.; Liu, X.; Zhang, F-L.; Li, S-K.; Gao, S-J. Maize yield gains in northeast china in the last six decades. *Journal of Integrative Agriculture*, **2013**, *12*(4), 630–637. doi.org/10.1016/S2095-3119(13)60281-6.

172. Zhao, J.; Guo, J.; Mu, J.; Exploring the relationships between climatic variables and climate-induced yield of spring maize in Northeast China. *Agriculture, Ecosystems and Environment*, **2015**, *207*, 79–90. http://dx.doi.org/10.1016/j.agee.2015.04.006

173. Tao, F.; Zhang, Z. Adaptation of maize production to climate change in North China Plain: Quantify the relative contributions of adaptation options. *European Journal of Agronomy*, **2010**, *33*(2), 103–116.

174. Li, Z.; Yang, P.; Tang, H.; Yin, H.; Liu, Z., Zhang, L. Response of maize phenology to climate warming in Northeast China between 1990 and 2012. *Regional Environmental Change*, **2014**, *14*, 39–48. https://doi.org/10.1007/s10113-013-0503-x.

175. Lv, S.; Yang, X.G.; Lin, X.M.; Liu, Z.J.; Zhao, J.; Li, K.N.; Mu, C.Y.; Chen, X.C.; Chen, F.J.; Mi, G.H. Yield gap simulations using ten maize cultivars commonly planted in Northeast China during the past five decades. *Agricultural and Forest Meteorology*, **2015**, *205*, 1–10.

176. Chen, F.; Chen, Y.; Gao, Q.; Yuan, L.; Zhang, F.; Mi, G. Modern maize hybrids in Northeast China exhibit increased yield potential and resource use efficiency despite adverse climate change. *Global Change Biology*, **2013**, *19*(3), 923–936. https://doi.org/10.1111/gcb.12093.

177. Siebert, S. et al. Development and validation of the global map of irrigation areas. *Hydrology and Earth System Sciences,* **2005**, *9*, 535–547.

178. Parent, B., Leclere, M., Lacube, S., Semenov, M.A., Welcker, C., Martre, P. and Tardieu, F., Maize yields over Europe may increase in spite of climate change, with an appropriate use of the genetic variability of flowering time. *Proceedings of the National Academy of Sciences*, **2018**, *115*(42), 10642–10647.

179. Holden, N.M.; Brereton, A.J. Potential impacts of climate change on maize production and the introduction of soybean in Ireland. *The Irish Journal of Agricultural and Food Research*, 2003, *42*(1), 1–15.

180. Mendelsohn, R. The impact of climate change on agriculture in developing countries. *Journal of Natural Resources Policy Research*, **2008**, *1*, 5–19. doi:10.1080/1939045080 2495882.

181. Mulenga, B.P.; Wineman, A.; Sitko, N.J. Climate trends and farmers' perceptions of climate change in Zambia. *Environmental Management*, **2017**, *59*, 291–306.

182. Ngingi, S. N.; Climate change adaptation strategies: Water resources management options for smallholder farming systems in Sub-Saharan Africa. **2009**, The MDG Centre for East and Southern Africa, The Earth Institute at Columbia University. http://www.foresightfordevelopment.org/sobipro/55/197-climate-changeadaptation-strategies-water-resources-management-options-for-smallholder-farming-systems-in-subsaharan-africa. Accessed 03 July 2020.

183. Msowoya, K.; Madani, K., Davtalab, R., Mirchi, A.; Lund, J.R. Climate Change Impacts on Maize Production in the Warm Heart of Africa. *Water Resource Management*, **2016**, DOI 10.1007/s11269-016-1487-3.

184. Tingem, M.; Rivington, M.; Bellocchi, G.; Azam-Ali, S.; Colls, J. Effects of climate change on crop production in Cameroon. *Climate Research*. **2008**, *36*, 65–77. doi: 10.3354/cr00733.

185. Hamududu, B.H.; Ngoma, H. Impacts of climate change on water resources availability in Zambia: implications for irrigation development. *Environment, Development and Sustainability,* **2020**, *22*, 2817–2838. https://doi.org/10.1007/s10668-019-00320-9.

186. Mulungu, K.; Tembo, G.; Bett, H.; Ngoma, H. Climate Change and Crop Yields in Zambia: Correlative Historical Impacts and Future Projections. *Preprints* **2019**, 2019110249.

187. Oseni, T.O.; Masarirambi, M. T. Effect of Climate Change on Maize (*Zea mays*) Production and Food Security in Swaziland. *American-Eurasian Journal of Agricultural & Environmental Sciences,* **2011**, *11*(3), 385–391.

188. Luhunga, P.M. Assessment of the impacts of climate change on maize production in the southern and western highlands sub-agro ecological zones of Tanzania. *Frontiers in Environmental Science*, 2017, *5*, 51. doi: 10.3389/fenvs.2017.00051

189. Ahmed, S. A., Diffenbaugh, N. S., Hertel, T. W., Lobell, D. B., Ramankutty, N., Rios, A. R. and Rowhani, P. Climate volatility and poverty vulnerability in Tanzania. *Global Environmental Change*. **2011**, *21*, 46–55. doi: 10.1016/j.gloenvcha.2010.10.003

190. Luhunga, P.; Changa, L.; Djolov, G. Assessment of the impacts of climate change on maize production in the Wami Ruvu basin of Tanzania. *The Journal of Water and Climate Change*, **2016**, *5*, 142–164. doi: 10.2166/wcc.2016.055

Plant Abiotic Stress Physiology, Volume 1

191. Climate Change Profile: Indonesia. 2018. Ministry of Foreign Affairs of the Netherlands P.O. Box 20061, 2500 EB, The Hague, The Netherlands, Available at: https://reliefweb. int/report/indonesia/climate-change-profile-indonesia Accessed: 10 July 2020

192. Naylor, R.L.; Battisti, D.S.; Vimont, D.J.; Falcon, W.P.; Burke, M.B. Assessing risks of climate variability and climate change for Indonesian rice agriculture. *Proceedings of the National Academy of Sciences of the United States of America,* **2007**. http://www. pnas.org/content/104/19/7752.full.pdf

193. Restu Ananda, R.; Widodo, T. A. General Assessment of Climate Change—Loss of Agricultural Productivity in Indonesia. **2019**, Master and Doctoral Program, Economics Department, FEB Gadjad Mada University. Available at https://mpra.ub.uni-muenchen. de/91316/ MPRA Paper No. 91316 Accessed 10 July 2020

194. Byjesh, K.; Kumar, S. N.; Aggarwal, P. K. Simulating impacts, potential adaptation and vulnerability of maize to climate change in India. *Mitigation and Adaptation Strategies for Global Change,* **2010**, *15*(5), 413–431.

195. Danhassan, S. S.; Suleman, M.; Hussaini, A.; Abubakar, M. J.; Mukhtar, U. K.; Abubakar, A. Socio-economic impacts of climate change on selected crops and domesticated animal production in India. *International Journal of Novel Research in Interdisciplinary Studies,* **2019**, *6*(1), 1–8.

196. Singh, S.; Kattarkandi, B.; Deka, S.; Choudhar, R. Impact of climatic variability and climate change on maize productivity in north India. *The Current Advances in Agricultural Sciences,* **2010**, *2*(1), 5–9.

197. Guntukula, R.; Goyari, P. The impact of climate change on maize yields and its variability in Telangana, India: A panel approach study. *Journal of Public Affairs,* **2020**; e2088. DOI: 10.1002/pa.2088

198. Soora, N. K.; Singh.; Anil Kumar.; Agarwal, P. K.; Rao, V. U. M.; Venkateswarlu, B. Climate Change and Indian Agriculture: Impact, Adaptation and Vulnerability–Salient Achievements from ICAR Network Project, IARI Publication, New Delhi, India, **2012**, 32 p.

199. Rainey, K.; Griffiths, P. Evaluation of *Phaseolus acutifolius* A. Gray plant introductions under high temperatures in a controlled environment. *Genetic Resources and Crop Evolution,* **2005**, *52*, 117–120. doi: 10.1007/s10722-004-1811-2.

200. Sita, K.; Sehgal, A.; HanumanthaRao, B.; Nair, R. M.; Vara Prasad, P.; Kumar, S.; Gaur, P. M.; Farooq, M.; Siddique, K. H.; Varshney, R. K. Food legumes and rising temperatures: effects, adaptive functional mechanisms specific to reproductive growth stage and strategies to improve heat tolerance. *Frontiers in Plant Science,* **2017**, *8*, 1658.

201. Gowda, C. L.; Samineni, S.; Gaur, P. M.; Saxena, K. B. Enhancing the productivity and production of pulses in India. In: *Climate change and sustainable food security.* National Institute of Advanced Studies, Bangalore, **2013**, pp. 145–159.

202. Agrawal, S. K. 2017. Effects of heat stress on physiology and reproductive biology of chickpea and lentil. http://hdl.handle.net/20.500.11766/6324. Accessed on 19 June 2020.

203. Bhaduri, D.; Purakayastha, T. J.; Patra, A. K.; Singh, M.; Wilson, B. R. Biological indicators of soil quality in a long-term rice–wheat system on the Indo-Gangetic plain: combined effect of tillage–water–nutrient management. *Environment Earth Science,* **2017**, *76*, 202.

204. Delahunty, A. J.; Nuttall, J. G.; Brand, J. D. Improving lentil tolerance to heat stress. **2016**, Available: https://grdc.com.au/resources-and-publications/grdc-update-papers/

tab-content/grdc-updatepapers/2016/02/improving-lentil-tolerance-to-heat-stress Accessed 21 July 2020.

205. Basu, P. S., Ali, M. and Chaturvedi, S. K. 2009. Terminal heat stress adversely affects chickpea productivity in Northern India – Strategies to improve thermotolerance in the crop under climate change. In: *Impact of climate change on agriculture;* Panigrahy, S.; Shankar, S.; R. and Parihar, J. S. (Eds.) ISPRS Archives XXXVIII-8/W3 workshop proceedings: International Society for Photogrammetry and Remote Sensing, New Delhi, pp. 189–193.

206. Devasirvatham, V.; Tan, D. K. Y.; Trethowan, R. M.; Gaur, P. M.; Mallikarjuna, N. Impact of high temperature on the reproductive stage of chickpea. In: *Food Security from Sustainable Agriculture.* Dove, H. and Culvenor, R. A. (Eds.) Proceedings of the 15th Australian Society of Agronomy Conference. Lincoln, New Zealand, **2010**, pp. 15–18.

207. Kumar, M., Mishra, S., Dixit, V. et al. Synergistic effect of *Pseudomonas putida* and *Bacillus amyloliquefaciens* ameliorates drought stress in chickpea (*Cicer arietinum* L.). *Plant Signalling Behavior,* **2016**, 11, e1071004. https://doi.org/10.1080/15592324.201 5.1071004

208. Wang, J.; Gan, Y. T.; Clarke, F.; McDonald, C. L. Response of chickpea yield to high temperature stress during reproductive development. *Crop Science,* **2006**, *46,* 2171–2178.

209. Upadhaya, H.D.; Dronavalli, N.; Gowda, C.L.L.; Singh, S. Identification and evaluation of chickpea germplasam for tolerance to heat stress. *Crop Science,* **2011**, *51,* 2079–2094.

210. Kalra, N.; Chakraborty, D.; Sharma, A.; Rai, H.K.; Jolly, M.; Chander, S.; Kumar, P.R.; Bhadraray, S.; Barman, D.; Mittal, R.B.; et al. effect of temperature on yield of some winter crops in northwest India. *Currrent Science,* **2008**, *94,* 82–88.

211. Dubey, SK; Sah, U; Singh, SK. Impact of climate change on pulse productivity and adaptation strategies as practiced by the pulse growers of Bundelkhand region of Uttar Pradesh. *Journal of Food Legumes,* **2011**, *24*(3), 230–234

212. Devasirvatham, V.; Tan, D. K. Y. Impact of High Temperature and Drought Stresses on Chickpea Production. *Agronomy,* **2018**, *8,* 145. doi:10.3390/agronomy8080145

213. Vijaylaxmi. Effect of high temperature on growth, biomass and yield of field pea genotypes. *Legume Research,* **2013**, *36,* 250–254

214. Mishra, S.; Singh, R.; Kumar, R.; Kalia, A.; Panigrahy, S. R.Impact of climate change on pigeon pea. *Economic Affairs,* **2017**, *62*(3), 455–457. DOI: 10.5958/0976-4666.2017.00057.2

215. Dubey, S. K.; Sah, U.; Singh, S. K. Impact of climate change on pulse productivity and adaptation strategies as practiced by the pulse growers of Bundelkhand region of Uttar Pradesh. *Journal of Food Legumes,* **2011**, *24*(3), 230–234.

216. Saxena, M. C. Problems and potential of chickpea production in the nineties. *Chickpea Nineties,* **1990**, 13–25.

217. Kumar, S.; Thakur, P.; Kaushal, N.; Malik, J. A.; Gaur, P.; Nayyar, H. Effect of varying high temperatures during reproductive growth on reproductive function, oxidative stress and seed yield in chickpea genotypes differing in heat sensitivity. *Archives of Agronomy and Soil Science,* **2013**, *59,* 823–843. doi: 10.1080/ 03650340.2012.683424

218. Berger, J. D., Ali, M., Basu, P. S., Chaudhary, B. D., et al. Genotype by environment studies demonstrates the critical role of phenology in adaptation of chickpea (*Cicer arietinum* L.) to high and low yielding environments of India. *Field Crops Research.* **2006**, *98,* 230–244.

219. Nayyar, H.; Kaur, G.; Kumar, S.; Upadhyaya, H. D. Low temperature effects during seed filling on chickpea genotypes (*Cicer arietinum* L.): probing mechanisms affecting seed

reserves and yield. *Journal of Agronomy and Crop Science*, **2007**, *193*, 336–344. doi: 10.1111/j.1439–037X.2007.00269.x.

220. Kiran, A.; Kumar, S.; Nayyar, H.; Sharma, K. D. Low temperature induced aberrations in male and female reproductive organ development cause flower abortion in Chickpea. *Plant, Cell & Environment*, **2019**, *42*, 2075–2089. doi: 10.1111/ pce.13536.

221. Kumar, S.; Malik, J.; Thakur, P.; Kaistha, S.; Sharma, K. D.; Upadhyaya, H. D. et al. Growth and metabolic responses of contrasting chickpea (*Cicer arietinum* L.) genotypes to chilling stress at reproductive phase. *Acta Physiologiae Plantarum*, **2011**, *33*, 779–787. doi: 10.1007/s11738-010-0602-y

222. Rani, A.; Devi, P.; Jha, U. C.; Sharma, K. D.; Siddique, K. H. M.; Nayyar, H. Developing climate-resilient chickpea involving physiological and molecular approaches with a focus on temperature and drought stresses. *Frontiers in Plant Science*, **2020**, *10*, 1759. doi: 10.3389/fpls.2019.01759

223. Farooq, M.; Gogoi, N.; Barthakur, S.; Baroowa, B.; Bharadwaj, N.; Alghamdi, S. S.; Siddique, K. H. Drought stress in grain legumes during reproduction and grain filling. *Journal of Agronomy and Crop Science*, **2016**, *203*, 81–102.

224. Davies, S. L.; Turner, N. C.; Siddique, K. H.; Leport, L.; Plummer, J.A. Seed growth of desi and kabuli chickpea (*Cicer arietinum* L.) in a short-season Mediterranean-type environment. *Australian Journal of Experimental Agriculture*, **1999**, *39*, 181–188.

225. Behboudian, M. H., Ma, Q., Turner, N. C. and Palta, J. A. Reactions of chickpea to water stress: yield and seed composition. *Journal of the Science of Food and Agriculture*. **2001**, *81*, 1288–1291.

226. Leport, L.; Turner, N.C.; French, R.J.; Barr, M.D.; Duda, R.; Davies, S.L.; Tennant, D., Siddique, K.H. Physiological responses of chickpea genotypes to terminal drought in a Mediterranean-type environment. *The European Journal of Agronomy*, **1999**, *11*, 279–291

227. Fang, X.; Turner, NC.; Yan, G.; Li, F.; Siddique, K. H. Flower numbers, pod production, pollen viability, and pistil function are reduced and flower and pod abortion increased in chickpea (*Cicer arietinum* L.) under terminal drought. *Journal of Experimental Botany*, **2010**, *61*, 335–345.

228. Mafakheri, A.; Siosemardeh, A.; Bahramnejad, B.; Struik, P.C.; Sohrabi, Y. Effect of drought stress on yield, proline and chlorophyll contents in three chickpea cultivars. *Aust J Crop Sci*, **2010**, 4:580

229. Ghassemi-Golezani, K. A.; Mustafavi, S. H.; Shafagh-Kalvanagh, J. Field performance of chickpea cultivars in response to irrigation disruption at reproductive stages. *Research on Crops*, **2012**, *13*, 107–112.

230. Allahmoradi, P., Mansourifar, C., Saiedi, M. and Jalali H. S. Effect of different water deficiency levels on some antioxidants at different growth stages of lentil (*Lens culinaris* L.). *Advances in Environmental Biology*. **2013**, *7*, 535–543.

231. Shrestha, R.; Turner, N. C.; Siddique, K. H.; Turner, D. W.; Speijers, J. A water deficit during pod development in lentils reduces flower and pod numbers but not seed size. *Australian Journal of Agricultural Research*, **2006**, *57*, 427–438.

232. Gusmao, M.; Siddique, K.H.; Flower, K.; Nesbitt, H.; Veneklaas, E.J. Water deficit during the reproductive period of grass pea (*Lathyrus sativus* L.) reduced grain yield but maintained seed size. *Journal of Agronomy and Crop Science,* **2012**, 198, 430–441

233. Kong, H.; Palta, J.A.; Siddique, K.H.; Stefanova, K.; Xiong, Y.C.; Turner, N.C. Photosynthesis is reduced, and seeds fail to set and fill at similar soil water contents in grass pea (*Lathyrus sativus* L.) subjected to terminal drought. *Journal of Agronomy and Crop Science*, **2015**, 201, 241–252.

234. Basu, P. S.; Singh, U.; Kumar, A.; Praharaj, C. S.; Shivran, R. K. Climate change and its mitigation strategies in pulses production. *Indian Journal of Agronomy*. **2016**, *61*(4), 71–82.
235. Bahl, P.N. 2015. Climate change and pulses: Approaches to combat its impact. *Agricultural Research.* 4:103–108. DOI 10.1007/s40003-015-0163-9
236. Mar, S.; Nomura, H.; Takahashi, Y.; Ogata, K.; Yabe, M. Impact of Erratic Rainfall from Climate Change on Pulse Production Efficiency in Lower Myanmar. *Sustainability*, **2018**, *10*, 402. doi:10.3390/su10020402
237. Cutforth HW, McGinn SM, McPhee KE, and Miller PR., Adaptation of pulse crops to the changing climate of the Northern Great Plains. *Agron. J.,* **2007**, 99, 1684–1699. doi:10.2134/agronj2006.0310s
238. Ali, M. Y; Hossain, M. E. Profiling climate smart agriculture for southern coastal region of bangladesh and its impact on productivity, adaptation and mitigation. *EC Agriculture*. **2019**, *5*(9):530–544.
239. Rahaman, M. A.; Rahman, M. M.; Hossain, M. S. Climate-Resilient Agricultural Practices in Different Agro-ecological Zones of Bangladesh W. Leal Filho (Ed.), Handbook of Climate Change Resilience, **2019**, Springer Nature Switzerland AG, doi:10.1007/978-3-319-71025-9_42-1
240. FAO. The future of food and agriculture – Trends and challenges. Rome, **2017**.
241. Hazell, P.; Wood, S. Drivers of change in global agriculture. *Philosophical Transactions of the Royal Society B: Biological Sciences*, **2008**, *363*(1491), 495–515.
242. Havas, K.; Salman, M. Food security: its components and challenges. *International Journal of Food Safety, Nutrition and Public Health*, **2011**, *4(1)*, 4–11.
243. UN. 2019. United Nations. World Population Prospects 2019: Highlights. *Statistical Papers-United Nations (Ser. A), Population and Vital Statistics Report*.
244. Kendall, H; Pimentel, D. Constraints on the expansion of the global food supply. *Ambio*, **1994**, *23(3)*, 198–205.
245. HLPE. Sustainable agricultural development for food security and nutrition: what roles for livestock? A report by the High Level Panel of Experts on Food Security and Nutrition of the Committee on World Food Security, Rome, **2016**.
246. FAO. Climate smart agriculture-Source book. Rome, **2013**.
247. Mondal, M.; Garai, S.; Banerjee, H. Smart Practices and Adaptive Technologies for Climate Resilient Agriculture. In: *Advanced Agriculture*. Maitra, S. and Pramanick, B. (Eds.). New Delhi Publishers; New Delhi; 2020, pp. 3–35.
248. Garai, S.; Mondal, M.; Mukherjee, S. Resource Conservation Technologies for Achieving Sustenance in Agricultural Production System. In: *Advanced Agriculture*. Maitra, S. and Pramanick, B. (Eds.). New Delhi Publishers; New Delhi; **2020**, pp. 325–349.
249. Shokralla, S.; Spall, J. L.; Gibson, J. F.; Hajibabaei, M. Next-generation sequencing technologies for environmental DNA research. *Molecular Ecology*, **2012**, *21*, 1794–1805.
250. Sanger, F.; Nicklen, S.; Coulson, A. R. DNA sequencing with chain terminating inhibitors. *PNAS*, **1977**, *74(12)*, 5463–5467.
251. Kchouk, M.; Gibrat, J. F.; Elloumi, M. Generations of sequencing technologies: from first to next generation. *Biology and Medicine,* **2017**, *9(3)*, 395. doi: 10.4172/0974-8369.1000395.
252. Vlk, D.; Repkova, J. Application of next generation technology in plant breeding. *Czech Journal of Genetics and Plant Breeding,* **2017**, *53(3),* 89–96. doi: 10.17221/192/2016-CJGPB
253. Mochida, K.; Kawaura, K.; Shimosaka, E.; Kawakami, N.; Shin-I, T.; Kohara, Y. Tissue expression map of a large number of expressed sequence tags and its application to

Plant Abiotic Stress Physiology, Volume 1

in silico screening of stress response genes in common wheat. *Molecular Genetics & Genomic,* **2006,** *276,* 304–312.

254. Yu, J.; Hu, S.; Wang, J.; Wong, K. S.; Li, S.; Liu, B.; Deng, Y. A draft sequence of the rice genome (*Oryza sativa* L.). *Science,* **2002,** *296*(5565), 79–92.

255. Varshney, R. K.; Song, C.; Saxena, R. K.; Azam, S.; Yu, S.; Sharpe, A. G.; Cannon, S. Draft genome sequence of chickpea (*Cicer arietinum*) provides are source for trait improvement. *Nature Biotechnology,* **2013,** *31(3),* 240–248.

256. D' Hont, A.; Denoeud, F.; Aury, J.M.; Baurens, F.C.; Carreel, F.; Garsmeur, O.; Noel, B. The banana (*Musa acuminata*) genome and the evolution of monocotyledonous plants. *Nature,* **488,** *7410,* 213–217.

257. Schnable, P. S.; Ware, D.; Fulton, R. S.; Stein, J. C.; Wei, F.; Pasternak, S.; Liang, C. The B73 maize genome: Complexity, diversity and dynamics. *Science,* **2009,** *326*(5956), 1112–1115.

258. Brenchley, R.; Spannag, M.; Pfeifer, M.; Barker, G. L. A.; D'Amore, R.; Allen, A. M.; McKenzie, N. Analysis of the bread wheat genome using whole genomes hot gun sequencing. *Nature,* **2012,** *491*(7426), 705–710.

259. Varshney, R. K.; Chen, W.; Li, Y.; Bharti, A. K.; Saxena, R. K.; Schlueter, J. A. Draft genome sequence of pigeonpea (*Cajanus cajan*), an orphan legume crop of resource-poor farmers. *Nature Biotechnology,* **2011,** *30(1),* 83–90.

260. Liu, S.; Liu, Y.; Yang, X.; Tong, C.; Edwards, D.; Parkin, I. A.; Zhao, M. The *Brassica oleracea* genome reveals the asymmetrical evolution of polyploidy genomes. *Nature Communications,* **2014,** *5,* 3930.

261. Kang, Y. J.; Kim, S. K.; Kim, M. Y.; Lestari, P.; Kim, K. H.; Ha, B. K.; Jun, T. H. Genome sequence of mung bean and insights into evolution within *Vigna* species. *Nature Communications,* **2014,** *5,* 5443.

262. Varshney, R. K., Shi, C., Thudi, M., Mariac, C., Wallace, J. Pearl millet genome sequence provides a resource to improve agronomic traits in arid environments. *Nature Biotechnology,* **2017,** *35,* 969–976.doi:10.1038/nbt.3943.

263. Shalini, S.; Singla, A.; Goyal, M.; Kaur, V.; Kumar, P. Omics in agriculture: Applications, challenges and future perspectives. In: *Crop Improvement for Sustainability*; Kumar, P.; Kumar, S.; Kumar, S. and Yadav, R. C. (Eds.), Astral Publications; New Delhi; **2018,** pp. 343–360.

264. Barbazuk, W. B., Emrich, S. J., Chen, H. D., Li, L. and Schnable, P.S. SNP discovery via 454 transcriptome sequencing. *Plant Journal. 2007,* 51, 910–918.

265. Novaes, E.; Drost, D. R.; Farmerie, W. G.; Pappas, G. J.; Grattapaglia, D.; Sederoff, R. R.; Kirst, M. High throughput gene and SNP discovery in *Eucalyptus grandis,* an uncharacterized genome. *BMC Genomics,* **2008,** *9,* 312.

266. Lister, R.; O'Malley, R.C.; Tonti-Filippini, J.; Gregory, B. D.; Berry, C. C.; Millar, A. H.; Ecker, J. R. Highly integrated single-base resolution maps of the epigenome in *Arabidopsis. Cell,* **2008,** *133,* 395–397.

267. Salgotra, R. K.; Gupta, B. B.; Stewart, C.N. From genomics to functional markers in the era of next-generation sequencing. *Biotechnology Letters,* **2014,** *36(3),* 417–426.

268. Mochida, K.; Shinozaki, K. Genomics and bioinformatics resources for crop improvement. *Plant and Cell Physiology,* **2010,** *51(4),* 497–523.

269. Kotresh, H.; Fakrudin, B.; Punnuri, S.; Rajkumar, B.; Thudi, M.; Paramesh, H.; Lohithswa H.; Kuruvinashetti, M. S. Identification of two RAPD markers genetically linked to a recessive allele of a Fusarium wilt resistance gene in pigeonpea (*Cajanus cajan* L.). *Euphytica,* **2006,** *149,* 113–120.

270. Cahill, D. J.; Schmidt, D. H. Use of marker assisted selection in a product development breeding program. In: *Proceedings of the 4th International Crop Science Congress*, Brisbane, Australia, **2004**.

271. Saha, G. C.; Sarker, A.; Chen, W.; Vandemark, G. J.; Muehlbauer, F. J. Inheritance and linkage map positions of genes conferring resistance to stem phylium blight in lentil. *Crop Science*, **2010**, *50*, 1831–1839.

272. Omoigui, L. O.; Kamara, A. Y.; Ishiyaku, M. F.; Boukar, O. Comparative responses of cowpea breeding lines to Striga and Alectra in the dry savannah of northeast Nigeria. *African Journal of Agricultural Research*, **2012**, *7*, 747–754.

273. Varshney, R. K.; Pandey, M. K.; Janila, P.; Nigam, S. N.; Sudini, H.; Gowda, M. V. C. Marker-assisted introgression of a QTL region to improve rust resistance in three elite and popular varieties of peanut (*Arachis hypogaea* L.). *Theoretical and Applied Genetics*, **2014**, *127(8)*, 1771–1781.

274. Shalini, S., Singla, A., Goyal, M., Kaur, V. and Kumar, P., Omics in agriculture: applications, challenges and future perspectives. In Crop Improvement for Sustainability. 2018, pp. 332–360.

275. Gong, C. Y.; Wang, T. Proteomic evaluation of genetically modified crops: current status and challenges. *Frontier in Plant Science*, **2013**, *4*, 41.

276. Zhao, Q.; Chen, S.; Dai, S. C_4 photosynthetic machinery: insights from maize chloroplast proteomics. *Frontier in Plant Science*, **2013**, *4*, 85.

277. Shrestha, K. M.; Tamot, B.; Pratt, E. P. S.; Saitie, S.; Bräutigam, A.; Weber, A. P. M. et al. Comparative proteomics of chloroplasts envelopes from bundle sheath and mesophyll chloroplasts reveals novel membrane proteins with a possible role in C4-related metabolite fluxes and development. *Frontier in Plant Science*, **2013**, *4*, 65.

278. Salavati, A.; Shafeinia, A.; Klubicova, K.; Bushehri, A. A. S.; Komatsu, S. Proteomic insights into intra- and intercellular plant–bacteria symbiotic association during root nodule formation. *Frontier in Plant Science*, **2013**, *4*, 28.

279. Memelink, J. Tailoring the plant metabolome without a loose stitch. *Trends in Plant Science*, **2005,** *10*, 305–307.

280. Van Emon, J. M. The geomics revolution in agricultural research. *Journal of Agricultural and Food Chemistry*, **2016**, *64(1)*, 36–44.

281. FAOSTAT, http://faostat.fao.org (**2018**).

282. Rijsberman F. CGIAR: a global research partnership for a food secure future. Retrieved, 2012, 2/5/13, from http://www.cgiar.org/consortiumnews/cgiar-global-research-partnership-for-a-food-secure-future/

283. Wassmann, R., Jagadish, S.V.K., Heuer, S., Ismail, A., Redona, E., Serraj, R., Singh, R.K., Howell, G., Pathak, H. and Sumfleth, K., Climate change affecting rice production: the physiological and agronomic basis for possible adaptation strategies. *Advances in Agronomy*, **2009**, *101*, 59–122.

284. G.C. Nelson, M.W. Rosegrant, J. Koo, R. Robertson, T. Sulser, T. Zhu, C. Ringler, S. Msangi, A. Palazzo, M. Batka, Climate Change: Impact on Agriculture and Costs of Adaptation, International Food Policy Research Institute (IFPRI), Washington, D.C., USA, 2009.

285. Wagena, M.B.; Easton, Z.M., Agricultural conservation practices can help mitigate the impact of climate change. *Science of The Total Environment*, **2018**, *635*, 132–143.

286. Hay J.E., D. Easterling, K.L. Ebi, M. Parry, Introduction to the special issue: observed and projected changes in weather and climate extremes, *Weather and Climate Extreme*, **2016**, *11*, 1–3. https://doi.org/10.1016/J.WACE.2015.08.006.

287. Mirza, M.M.Q., Climate change, flooding in South Asia and implications, *Regional Environmental Change*, **2011**, *11*, 95–107, https://doi.org/10.1007/s10113-010-0184-7.

288. Hoisington, D.; Khairallah, M.; Reeves, T.; Ribaut, J.M.; Skovmand, B.; Taba, S.; Warburton, M., Plant genetic resources: What can they contribute toward increased crop productivity?. *Proceedings of the National Academy of Sciences*, **1999**, 96(11), 5937–5943.

289. Wang, S.; Kang, S.; Zhang, L.; Li, F., Modelling hydrological response to different land-use and climate change scenarios in the Zamu River basin of northwest China. *Hydrological Processes: An International Journal*, **2008**, *22*(14), 2502–2510

290. Heisey, P.W. and Edmeades, G.O., Maize production in droughtstressed environment. In World Maize Facts and Trends 1997 per 1998. *CIMMYT, Mexico*, **1999.**

291. Monneveux, P., Sanchez, C., Beck, D. and Edmeades, G.O., Drought tolerance improvement in tropical maize source populations: evidence of progress. *Crop Science*, **2006**, *46*(1), 180–191.

292. Dupuis I, Dumas C. Influence of temperature stress on *in vitro* fertilization and heatshock protein synthesis in maize (*Zea mays* L.) reproductive tissues. *Plant Physiology*, 1990, 94, 665–670.

293. Reynolds MP, Ortiz-Monasterio JI, McNab A, editors. Application of Physiology in Wheat Breeding. Mexico, DF: CIMMYT; 2001, pp. 124–135

294. Gonzalez Fontes A, editor. Abiotic Stress Adaptation in Plants; Houston: Studium PressLlc. 2010. pp. 387–415

295. Prakash, V., Kumar, S., Dwivedi, S.K., Rao, K.K. and Mishra, J.S., Impact, Adaptation Strategies and Vulnerability of Indian Agriculture Towards the Climate Change. In *Conservation Agriculture* (pp. 437–457). Springer, Singapore, **2016**.

296. Misoo S, Hirabayashi T, Kamijima O, and Sawano M. Efficient induction of diploidizedplants in anther culture of rice by colchicine pretreatment of cold-preserved spikes. *Plant Tissue Culture Letter*, **1991**, 8, 82–86.

297. Grauda D, Lepse N, Strazdiņa V, Kokina I, Lapiņa L, Miķelsone A, Ļubinskis L and Rashal I. Obtaining of doubled haploid lines by anther culture method for the Latvian wheat breeding. *Agronomy Research*, **2010,** 8, 545–552.

298. Eder J, Chalyk S. In vitro haploid induction in maize. *Theoretical and Applied Genetics*, **2000**, 104, 703–708.

299. Chang, M.T.; Coe, E.H., Doubled haploids. In *Molecular genetic approaches to maize improvement* (pp. 127–142). Springer, Berlin, Heidelberg, **2009.**

300. Geiger, H.H. and Gordillo, G.A., Doubled haploids in hybrid maize breeding. *Maydica*, **2009**, *54*(4), 485–499.

301. De La Fuente, G.N., Frei, U.K., Trampe, B., Ren, J., Bohn, M., Yana, N., Verzegnazzi, A., Murray, S.C. and Lübberstedt, T., A diallel analysis of a maize donor population response to in vivo maternal haploid induction: II. Haploid male fertility. *Crop Science*, **2020**, *60*(2), 873–882.

302. Hussain B., Muhammad Ahsan Khan, Qurban Ali and ShadabShaukat. Double haploid production in wheat through microspore culture and wheat X maize crossing system: an overview. *IJAVMS*, **2012**, 6(5), 332–344 DOI:10.5455/ijavms.168

303. Rizal, G., Karki, S., Alcasid, M., Montecillo, F., Acebron, K., Larazo, N., et al. Shortening the breeding cycle of sorghum, a model crop for research. *Crop Science*. **2014**, *54*, 520–529. doi: 10.2135/cropsci2013.07.0471.

304. Gaur, P.M.; Srinivasan, S.; Gowda, C. L. L.; Rao, B. V. Rapid generationadvancement in chickpea. *Journal of Semi-Arid Tropical Agricultural Research*. 2007, *3*, 1–3.

305. Tanaka, J.; Hayashi, T.; Iwata, H.. A practical, rapid generationadvancementsystem for rice breeding using simplified biotron breeding system. *Breeding Science.* **2016**, *66*, 542–551. doi:10.1270/jsbbs.15038

306. Nepolean, T.; Kaul, J.; Mukri, G.; Mittal, S. Genomics-enablednext-generation breedingapproaches for developingsystem-specific drought toleranthybrids in maize. *Frontiers in Plant Science.* **2018**, 9, 361.doi: 10.3389/fpls.2018.00361

307. Ortiz, R., Trethowan, R., Ferrara, G.O., Iwanaga, M., Dodds, J.H., Crouch, J.H., Crossa, J. and Braun, H.J., High yield potential, shuttle breeding, genetic diversity, and a new international wheat improvement strategy. *Euphytica*, **2007**, *157*(3), 365–384.

308. D.J. Mackill, B.C.Y. Collard, G.N. Atlin, A.M. Ismail, S. Sarkarung, Overview of andhistorical perspectives on the EIRLSBN, in: B.C.Y. Collard, A.M. Ismail, B. Hardy(Eds.), EIRLSBN Twenty Years Achiev. Rice Breed, International Rice ResearchInstitute, 2013, pp. 1–6.

309. Abera Desta, Z. and Ortiz, R. Genomic selection: genome-wide prediction in plant improvement. *Trends in Plant Science.* **2014**, 19, 592–601.

310. Araus, J. L. and Cairns, J. E. Field high-through put phenotyping: the new crop breeding frontier. *Trends in Plant Science.* **2014**, *19*, 52–61.

311. Araus, J. L., Kefauver, S. C., Zaman-Allah, M. A., Olsen, M. S. and Cairns, J. E. Translating high-throughput phenotyping into genetic gain. *Trends in Plant Science.* **2018**, *23*(5), 451–466. https://doi.org/10.1016/j.tplants.2018.02.001.

312. Grift, T.E., Novais, J. and Bohn, M., 2011. High-throughput phenotyping technology for maize roots. *Biosystems Engineering*, **2011**, *110*(1), 40–48.

313. Nagel, K.A., Putz, A., Gilmer, F., Heinz, K., Fischbach, A., Pfeifer, J., Faget, M., Blossfeld, S., Ernst, M., Dimaki, C. and Kastenholz, B., GROWSCREEN-Rhizo is a novel phenotyping robot enabling simultaneous measurements of root and shoot growth for plants grown in soil-filled rhizotrons. *Functional Plant Biology*, **2012**, *39*(11), 891–904.

314. Spielbauer, G., Armstrong, P., Baier, J.W., Allen, W.B., Richardson, K., Shen, B. and Settles, A.M., High-throughput near-infrared reflectance spectroscopy for predicting quantitative and qualitative composition phenotypes of individual maize kernels. *Cereal Chemistry*, **2009**, *86*(5), pp.556–564.

315. Hong, G., Lee, J.C., Robinson, J.T., Raaz, U., Xie, L., Huang, N.F., Cooke, J.P. and Dai, H., Multifunctional in vivo vascular imaging using near-infrared II fluorescence. *Nature Medicine*, **2012**, *18*(12), 1841–1846.

316. Zhang, Y. and Zhang, N., Imaging technologies for plant high-throughput phenotyping: A review. *Frontiers of Agricultural Science and Engineering*, **2018**, *5*, 406–419.

317. Klose, R.; Penlington, J.; Ruckelshausen, A., Usability study of 3D time-of-flight cameras for automatic plant phenotyping. *Bornimer Agrartechnische Berichte*, **2009**, *69*, 93–105.

318. Frank, Eulenstein.; Marcos Alberto Lana.; Sandro, Luis Schlindwein.; Askhad, Khasrethovich Sheudzhen.; Marion, Tauscke.; Axel, Behrendt.; Edgardo, Guevara.; Santiago, Meira. Regionalization of maize responses to climate change scenarios, n use efficiency and adaptation strategies. *Horticulturae,* **2015**, *3*, 9. doi:10.3390/horticulturae3010009

319. Zaman-Allah, M., Vergara, O., Araus, J.L., Tarekegne, A., Magorokosho, C., Zarco-Tejada, P.J., Hornero, A., Albà, A.H., Das, B., Craufurd, P. and Olsen, M., Unmanned aerial platform-based multi-spectral imaging for field phenotyping of maize. *Plant Methods*, **2015**, *11*(1), 1–10.

320. Singh, R.P., Hodson, D.P., Huerta-Espino, J., Jin, Y., Bhavani, S., Njau, P., Herrera-Foessel, S., Singh, P.K., Singh, S., Govindan, V. Theemergence of Ug99 races of the stem

rust fungus is a threat toworld wheat production. *The Annual Review of Phytopathology*, **2011**, *49*, 465–481.

321. Brahim KM, Barrett JA. Evolution of mildew resistance in ahybrid bulk population of barley. *Heredity,* **1991**, *67*, 247–256.

322. Zhang, B.H.; Guo, T.-L.; Wang, Q.-L. Inheritance and segregationof exogenous genes in transgenic cotton. *Journal of Genetics*, **2000**, *79*, 71–75.

323. Mutlu, N.; Miklas, P.; Reiser, J.; Coyne, D. Backcross breeding forimproved resistance to common bacterial blight in pinto bean(*Phaseolus vulgaris* L.). *Plant Breed,* **2005**, *124*, 282–287.

324. Zhu, B.; Lawrence, J.R.; Warwick, S.I.; Mason, P.; Braun, L.; Halfhill, M.D.; Stewart, C.N. Stable *Bacillus thuringiensis* (Bt) toxincontent in interspecific F1 and backcross populations of wild*Brassicarapa*afterBt gene transfer. *Molecular Ecology,* **2004**, *13*, 237–241.

325. Jackson, L.F.; Webster, R.K.; Allard, R.W.; Kahler, A.L. Geneticanalysis of changes in scald resistance in barley compositecross V. *Phytopathology,* **1982**, *72*, 1069–1072.

326. Maroof MAS, Webster RK, Allard RW. Evolution of resistanceto scald, powdery mildew, and net blotch in barley compositecross II populations. *Theoretical and Applied Genetics*, **1983**, *66*, 279–283.

327. Finckh, M.; Gacek, E.; Goyeau, H.; Lannou, C.; Merz, U. Cerealvariety and species mixtures in practice, with emphasis ondisease resistance. *Agronomie EDP Science,* **2000**, *20*, 813–837.

328. Keneni, G.; Bekele, E.; Imtiaz, M.; Dagne, K. Genetic vulnerabilityof modern crop cultivars: causes, mechanism and remedies. International Journal of Plant Research, **2012**, 2, 69–79.

329. Ashizawa, T., Zenbayashi, K. and Koizumi, S. Development ofa simulation model for forecasting rice blast epidemics inmultiline. *Japanese Journal of Phytopathology,* **2001**, *67*, 194–197.

330. Ishizaki K, Hoshi T, Abe S, Sasaki Y, Kobayashi K, KasaneyamaH, Matsui T, Azuma S. Breeding of blast resistantlines in rice variety "Koshihikari" and evaluation of theircharacters. Breed Sci **2005**, *55*, 371–377.

331. Sattari A, Fakheri B, Hassan FSC, Noroozi M. Blast resistancein rice: a review of breeding and biotechnology. International Journal of Agriculture and Crop Sciences, **2014**, *7*, 329–333.

332. Martínez-Atienza, J.; Jiang, X.; Garciadeblas, B.; Mendoza, I.;Zhu, J.-K.; Pardo, J. M.; Quintero, F. J. Conservation of the saltoverly sensitive pathway in rice. *Plant Physiology,* **2007**, *143*, 1001–12.

333. Tang, R. J.; Liu, H.; Bao, Y.; Lv, Q. D.; Yang, L.; Zhang, H. X.The woody plant poplar has a functionally conserved salt overlysensitive pathway in response to salinity stress. *Plant Molecular Biology, 2010, 74*, 367–380.

334. Olías, R.; Eljakaoui, Z.; Pardo, J. M.; Belver, A. The Na+/H+exchanger SOS1 controls extrusion and distribution of Na+ intotomato plants under salinity conditions. *Plant Signaling & Behavior,* **2009**, *4*, 973–976.

335. Gaxiola, R. A.; Rao, R.; Sherman, A.; Grisafi, P.; Alper, S. L.; Fink, G. R. The *Arabidopsis thaliana* proton transporters, *AtNhx1*and*Avp1*, can function in cation detoxification in yeast. *Proceedings of the National Academy of Sciences of the United States of America,* **1999**, *96*, 1480–1485.

336. Quintero, F. J.; Blatt, M. R.; Pardo, J. M. Functional conservation between yeast and plant endosomal Na+/H+ antiporters. *FEBS Letter,* **2000**, *471*, 224–228.

337. Apse, M. P. and Blumwald, E. Engineering salt tolerance in plants. *Current Opinion in Biotechnology*, **2002**, *13*, 146–150.
338. Kumar, S., Kalita, A., Srivastava, R. and Sahoo, L., Co-expression of Arabidopsis NHX1 and bar improves the tolerance to salinity, oxidative stress, and herbicide in transgenic mungbean. *Frontiers in Plant Science*, **2017**, *8*, 1896. https://doi.org/10.3389/fpls.2017.01896
339. Zhang, H. X.; Blumwald, E. Transgenic salt-tolerant tomato plants accumulate salt in foliage but not in fruit. *Nature Biotechnology*, **2001**, *19*, 765–768.
340. Pandey GK. Elucidation of Abiotic Stress Signaling in Plants: Functional Genomics Perspective, Vol.1.; Springer: New York, 2015.
341. Matsumura, H.; Nirasawa, S. Transcript profiling in rice (Oryzasativa L.) seedlings using serial analysis of gene expression(SAGE). *Plant Journal*, **1999**, *20*, 719–726.
342. Kawasaki, S.; Borchert, C.; Deyholos, M.; Wang, H.; Brazille, S.;Kawai, K.; Galbraith, D.; Bohnert, H. J. Gene expression profilesduring the initial phase of salt stress in rice. *Plant Cell*, **2001**, *13*, 889–905.
343. Kawaura, K.; Mochida, K.; Yamazaki, Y.; Ogihara, Y.Transcriptome analysis of salinity stress responses in commonwheat using a 22k oligo-DNA microarray. Functional & Integrative Genomics, **2006**, *6*, 132–42.
344. Houde, M.; Belcaid, M.; Ouellet, F.; Danyluk, J.; Monroy, A. F.;Dryanova, A.; Gulick, P.; Bergeron, A.; Laroche, A.; Links, M. G.;MacCarthy, L.; Crosby, W. L.; Sarhan, F. Wheat EST resources forfunctional genomics of abiotic stress. *BMC Genomics*, **2006**, *7*, 149–170.
345. Ouyang, B.; Yang, T.; Li, H.; Zhang, L.; Zhang, Y.; Zhang, J.; Fei, Z.; Ye, Z. Identification of early salt stress response genes intomato root by suppression subtractive hybridization andmicroarray analysis. *The Journal of Experimental Botany*, **2007**, *58*, 507–20.
346. Kawaura, K.; Mochida, K.; Ogihara, Y. Genome-wide analysis foridentification of salt-responsive genes in common wheat. *Functional & Integrative Genomics*, **2008**, *8*, 277–86.
347. Mustafiz, A.; Singh, A. K.; Pareek, A.; Sopory, S. K.; Singla-Pareek, S. L. Genome-wide analysis of rice and Arabidopsisidentifies two glyoxalase genes that are highly expressed in abioticstresses. *Functional & Integrative Genomics*, **2011**, *11*, 293–305.
348. Li, L.; Wang, W.; Wu, C.; Han, T.; Hou, W. Construction of two suppression subtractive hybridization libraries and identificationof salt-induced genes in soybean. *Journal of Integrative Agriculture*, **2012**, *11*, 1075–1085.
349. Davey, J. W.; Hohenlohe, P. A; Etter, P. D.; Boone, J. Q.; Catchen, J. M.; Blaxter, M. L. Genome-wide genetic marker discovery andgenotyping using next-generation sequencing. *Nature Reviews Genetics*, **2011**, *12*, 499–510.
350. Huang, X.; Wei, X.; Sang, T.; Zhao, Q.; Feng, Q.; Zhao, Y.; Li, C.;Zhu, C.; Lu, T.; Zhang, Z.; Li, M.; Fan, D.; Guo, Y.; Wang, A.;Wang, L.; Deng, L.; Li, W.; Lu, Y.; Weng, Q.; Liu, K.; Huang, T.;Zhou, T.; Jing, Y.; Li, W.; Lin, Z.; Buckler, E. S.; Qian, Q.; Zhang, Q.-F.; Li, J.; Han, B. Genome-wide association studies of 14agronomic traits in rice landraces. *Nature Genetics*, **2010**, *42*, 961–7.
351. Kumar, V.; Singh, A.; Mithra, S. V. A.; Krishnamurthy, S. L.;Parida, S. K.; Jain, S.; Tiwari, K. K.; Kumar, P.; Rao, A. R.;Sharma, S. K.; Khurana, J. P.; Singh, N. K.; Mohapatra, T.Genome-wide association mapping of salinity tolerance in rice (*Oryza sativa*). *DNA Research*, 2015, 1–13.
352. Ma, Y.; Qin, F.; Tran, L.-S. P. Contribution of genomics to genediscovery in plant abiotic stress responses. *Molecular Plant*, **2012**, *5*, 1176–8.

353. Rosyara UR. Requirement of robust molecular marker technology for plant breeding applications. *Journal of Plant Breeding Group.* **2006**; 1, 67–72.
354. Ivandic V, Thomas WTB, Nevo E, Zhang Z, Forster BP. Association of SSRs with quantitative traitvariation including biotic and abiotic stress tolerance in *Hordeumspontaneum. Plant Breeding,* **2003***, 122*, 300–304.
355. Zhang J, Chandra Babu R, Pantuwan G, Kamoshita A, Blum A, Wade L, Sarkarung S, O'Toole, JC, Nguyen NT (1999) Molecular dissection of drought tolerance in rice: from physio- morphologicaltraits to field performance. In: Ito, O., O'Toole, J. and Hardy, B. (eds) *Genetic Improvement of Rice for Water-limited Environments.* International Rice Research Institute (IRRI), Manila, Philippines, pp. 331–343.
356. Khan H, S.H. Wani, A.M. Iqbal. 2014. Molecular Approaches to Enhance Abiotic Stress Tolerance. In: *Innovations in Plant Sciences and Biotechnology.* Shabir H Wani, C.P. Malik, Amandeep Hora, Ritesh Kaur (Eds.), Agrobios, 1st Edition, India, 2014, pp. 151–170.
357. Jiang GL. Plant marker-assisted breeding and conventional breeding: challenges and perspectives. *Advances in Crop Science and Technology***2013**, *1*, e106.
358. Collard BCY, Jahufer MZZ, Brouwer JB, Pang ECK. An introduction to markers, quantitative trait loci (QTL) mapping and marker-assisted selection for crop improvement: the basic concepts. *Euphytica.* **2005**, 142(1–2), 169–196.
359. Dwivedi SL, Crouch JH, Mackill DJ, Xu Y, Blair MW, Ragot M, et al. The molecularization of public sector crop breeding: progress, problems, and prospects. *Advances in Agronomy.* **2007**, *95*, 163–318.
360. Luo Y, Yin Z. Marker-assisted breeding of Thai fragrance rice for semi-dwarf phenotype, submergence tolerance and disease resistance to rice blast and bacterial blight. *Molecular Breeding.* **2013**, *32*(3), 709–721.
361. Castro AJ, Capettini F, Corey AE, Filichkina T, Hayes PM, Kleinhofs A *et al.* Mapping and pyramiding of qualitative and quantitative resistance to stripe rust in barley. *Theoretical and Applied Genetics,* **2003**, *107*(5), 922–930.
362. Li, R.; Fan, W.; Tian, G.; Zhu, H.; He, L. et al. The sequence and de novo assembly of the giant panda genome. *Nature,* **2010**, *463(7279),* 311–317.
363. Castro, A.J., Chen, X., Hayes, P.M., Knapp, S.J., Line, R.F., Toojinda, T. and Vivar, H., Coincident QTL which determine seedling and adult plant resistance to stripe rust in barley. *Crop Science,* **2002**, *42*(5), 1701–1708.
364. Lloyd A, Plaisier CL, Carroll D, Drews GN. Targeted mutagenesis using zinc-finger nucleases in Arabidopsis. *Proceedings of the National Academy of Sciences of the United States of America.* **2005**, *102*(6), 2232–2237.
365. Eathington SR. Practical applications of molecular technology in the development of commercial maize hybrids. Proc. of the 60th Annual Corn and Sorghum Seed Res. Conf., Chicago [CD-ROM]. 7–9 December 2005. American Seed Trade Association, Washington, DC.
366. Ragot M, Gay G, Muller JP, Durovray J (2000) Efficient selection for the adaptation to the environmentthrough QTL mapping and manipulation in maize. In: Ribaut, J. M. and Poland, D. (eds) *MolecularApproaches for the Genetic Improvement of Cereals for Stable Production in Water-limited Environments*, Centro Internacional de Mejoramiento de Maiz y Trigo (CIMMYT), Mexico, DF, pp.128–130.
367. Massman, J. M., Jung, H. J. G., and Bernardo, R.. Genomewideselection versus marker-assisted recurrent selection to improve grain yieldand stover-quality traits for cellulosic ethanol in maize. *Crop Science.* **2013**, 53, 58–66. doi: 10.2135/cropsci2012.02.0112

368. Beyene, Y., Semagn, K., Mugo, S., et al. Genetic gains in grain yield through genomic selection in eight biparentalmaize populations under drought stress. *Crop Science.* **2015**, *55*, 154–163. doi: 10.2135/cropsci2014.07.0460.

369. Bankole, F., Menkir, A., Olaoye, G., Crossa, J., Hearne, S., Unachukwu, N., et al. Genetic gains in yield and yield related traits under drought stressand favorable environments in a maize population improved using markerassisted recurrent selection. *Frontiers in Plant Science,* **2017**, *8*, 808. doi: 10.3389/fpls.2017.00808

370. Meuwissen, T. H.; Hayes, B. J.; anGoddard, M. E. Prediction oftotal genetic value using genome-wide dense marker maps. *Genetics,* **2001**, *157*, 1819–1829.

371. Heffner, E. L.; Sorrells, M. E.; and Jannink, J. Genomic selection for cropimprovement. *Crop Science.* **2009**, *49*, 1–12. doi: 10.2135/cropsci2008.08.0512.

372. Guo, Z., Tucker, D. M., Lu, J., Kishore, V., and Gay, G. Evaluationof genome-wide selection efficiency in maize nested association mappingpopulations. *Theoretical and Applied Genetics,* **2012**, *124*, 261–275. doi: 10.1007/s00122-011-1702-9

373. Heffner, E.L., Lorenz, A.J., Jannink, J.L., Sorrells, M.E., Plant breeding with Genomic selection: gain per unit time and cost, *Crop Science.* **2010**, *50*, 1681–1690, https://doi.org/10.2135/cropsci2009.11.0662.

374. Beyene Y., Semagn K., Mugo S., Tarekegne A., Babu R., Meisel B., Sehabiague P., Makumbi D., Magorokosho C., Oikeh S., Gakunga J. Genetic gains in grain yield through genomic selection in eight bi-parental maize populations under drought stress. *Crop Science.* **2015**, *55*(1), 154–63.

375. Shikha, M., Kanika, A., Rao, A. R., Mallikarjuna, M.G., Gupta, H. S., and Nepolean, T. Genomic selection for drought tolerance using genome-wide SNPs inmaize. *Frontiers in Plant Science.* **2017**, *8*, 550. doi: 10.3389/fpls.2017.00550.

376. Vivek, B. S., Krishna, G. K., Vengadessan, V., Babu, R., Zaidi, P. H., Kha, L. Q., et al.. Use of genomic estimated breeding values results inrapid genetic gains for drought tolerance in maize. *Plant Genome,* **2017**, 10, 1–8. doi: 10.3835/plantgenome2016.07.0070.

377. Alonso, J. M. and Joseph, R. Moving forward in reverse genetictechnologies to enable genome-wide phenomic screens in Arabidopsis. Nature Reviews Genetics. **2006**, *7*, 524–536 doi:10.1038/nrg1893.

378. Colbert, T.; Till, B. J.; Tompa, R.; Reynolds, S.; Steine, M. N.;Yeung, A. T.; Mccallum, C. M.; Comai, L.; Henikoff, S.; Division, B. S.; Hutchinson, F.; Washington, T. C. High-ThroughputScreening for Induced Point Mutations. *Plant Physiology,* **2001**, *98109*, 480–484.

379. Mejlhede, N., Kyjovska, Z., Backes, G., Burhenne, K., Rasmussen, S.K., Jahoor, A. Eco-TILLING for the identification of allelicvariation in the powdery mildew resistance genes Mlo and Mlaof barley. *Plant Breed,* **2006**, 125, 461–467.

380. Rao AQ, Bakhsh A, Kiani S, Shahzad K, Shahid AA, Husnain T, et al. The myth of plant transformation. *Biotechnology Advances.* **2009**, 27, 753–763.

381. Fagoaga C, Tadeo FR, Iglesias DJ, Huerta L, Lliso I, Vidal AM *et al.* Engineering of gibberellin levels in citrus by sense and antisense overexpression of a GA 20-oxidase gene modifies plant architecture. *Journal of Experimental Botany.* **2007**, *58*(6):1407–1420.

382. Newell CA. Plant transformation technology; developments and applications. *Molecular Biotechnology.* **2000**, *16*, 53–65.

383. Arbona, V. and Gómez-Cadenas, A. Hormonal modulation ofcitrus responses to fooding. *Journal of Plant Growth Regulation.* **2008**, *27*(3), 241–250.

384. Xiong, Y. and S. Z. Fei, Functional and phylogenetic analysisof a DREB/CBF-like gene in perennial ryegrass (*Loliumperenne* L.), *Planta,* **2006**, *224, 4*, 878–888.

385. Wang, J., P. P. Sun, C. L. Chen, Y. Wang, X. Z. Fu, and J. H.Liu, An arginine decarboxylase gene *PtADC*from *Poncirustrifoliata*confers abiotic stress tolerance and promotes primaryroot growth in *Arabidopsis, Journal of Experimental Botany*, **2011**, *62*(8), 2899–2914.

386. Rohila, J. S., R. K. Jain, and R. Wu, Genetic improvementof Basmati rice for salt and drought tolerance by regulatedexpression of a barley Hva1 cDNA, *Plant Science*, **2002**, *163*(3), 525–532.

387. Amudha, J. and Balasubramani, G. Recent molecular advancesto combat abiotic stress tolerance in crop plants. *Biotechnologyand Molecular Biology Review*. **2011**, *6*(2), 31–58.

388. Schouten HJ, Krens FA, Jacobsen E. Do cisgenic plants warrant less stringent oversight? *Nature Biotechnology*, **2006**, 24, 753.

389. Orellana, S.; Yañez, M.; Espinoza, A.; Verdugo, I.; González, E.;Ruiz-Lara, S.; Casaretto, J. a The transcription factor SlAREB1confers drought, salt stress tolerance and regulates biotic andabiotic stress-related genes in tomato. *Plant, Cell & Environment*, **2010**, *33*, 2191–2208.

390. Carroll, D. Genome engineering with targetable nucleases. *Annu.Rev. Biochem.*, **2014**, *83*, 409–39.

391. Kamburova, V. S., Nikitina, E. V., Shermatov, S. E., Buriev, Z. T., Kumpatla, S. P., Emani, C., et al. Genome editing in plants: an overview of tools andapplications. *International Journal of Agronomy*, **2017**, 15. doi: 10.1155/2017/7315351

392. Weinthal, D.; Tovkach, A.; Zeevi, V.; Tzfira, T. Genome editing inplant cells by zinc finger nucleases. *Trends in Plant Science*, **2010**, *15*, 308–21.

393. Silva, G.; Poirot, L.; Galetto, R.; Smith, J.; Montoya, G.;Duchateau, P.; Pâques, F. Mega-nucleases and other tools fortargeted genome engineering: perspectives and challenges for genetherapy. *Current Gene Therapy*, **2011**, *11*, 11–27.

394. Curtin, S. J.; Voytas, D. F.; Stupar, R. M. Genome engineeringof crops with designer nucleases. *Plant Genome Journal*, **2012**, *5*, 42.

395. Hartung, F., and Schiemann, J. Precise plant breeding using new genomeediting techniques: opportunities, safety and regulation in the EU. *Plant Journal*, **2014**, *78*, 742–752. doi: 10.1111/tpj.12413

396. Miller, J. C., Holmes, M. C., Wang, J., Guschin, D. Y., Lee, Y. L., Rupniewski, I., et al. An improved zinc-finger nuclease architecturefor highly specific genome editing. *Nature Biotechnology* **2007**, 25, 778–785. doi: 10.1038/nbt1319.

397. Petolino, J. F.. Genome editing in plants via designed zinc finger nucleases in vitro cell. *Developmental Biology*, **2015**, *51*, 1. doi: 10.1007/s11627-015-9663-3

398. Peer, R., Rivlin, G., Golobovitch, S., Lapidot, M., Gal-On, A., Vainstein, A., et al. Targeted mutagenesis using zinc-finger nucleases in perennial fruittrees. *Planta* **2015**, *241*, 941–951. doi: 10.1007/s00425-014-2224-x

399. Gaj, T., Gersbach, C. A., and Barbas, C. F. ZFN, TALEN, and CRISPR/Casbasedmethods for genome engineering. *Trends Biotechnology*, **2013**, *31*, 397–405.doi: 10.1016/j.tibtech.2013.04.004

400. Deng, D., Yan, C., Pan, X., Mahfouz, M., Wang, J., Zhu, J. K., et al. Structuralbasis for sequence-specific recognition of DNA by TAL effectors. *Science* **2012**, *335*, 720–723. doi: 10.1126/science.1215670

401. Nakayama TJ, Borém A, Chiari L, Molinari HBC and Nepomuceno AL. Precision genetic engineering. In: *Omics in Plant Breeding*, Aluízio Borém, Roberto Fritsche-Neto (Eds.). John Wiley & Sons, Inc., River Street Hoboken, NJ, USA, **2014**, 187–205.

402. Bogdanove AJ, Schornack S, Lahaye T. TAL effectors: finding plant genes for disease and defense. *Current Opinion in Plant Biology.* **2010,** *13*(4), 394–401.

403. Cermak T, Doyle EL, Christian M, Wang L, Zhang Y, Schmidt C *et al.* Efficient design and assembly of custom Talen and other TAL effector-based constructs for DNA targeting. *Nucleic Acids Research.* **2011,** *39*(12), 82–82.

404. Mahfouz MM, Li L, Piatek M, Fang X, Mansour H, Bangarusamy DK, et al. Targeted transcriptional repression using a chimeric TALE-SRDX repressor protein. *Plant Molecular Biology.* **2012,** 78(3), 311–321.

405. Song G, Jia M, Chen K, Kong X, Khattak K, Xie C *et al.* CRISPR/Cas9: A powerful tool for crop genome editing. *The Crop Journal,* **2016,** *4*(2), 75–82.

406. Paul JW, Qi Y. CRISPR/Cas9 for plant genome editing: accomplishments, problems and prospects. *Plant Cell Reports.* **2016,** *35,* 1417.

407. Patidar OP, Gautam C, Tantuway G, Kumar S, Yadav A, Meena DS, Nagar A. RNA-guided genome editing tool CRISPR-Cas9: its applications and achievements in model and crop plants. *Journal of Pure and Applied Microbiology.* **2016,** *10*(4), 3035–3042.

408. Tiwari, M.; Sharma, D.; Trivedi, P. K. Artificial microRNA mediated gene silencing in plants: progress and perspectives. Plant Molecular Biology, **2014,** *86,* 1–18.

409. WMD3-Web MicroRNA Designer. http://wmd3.weigelworld.org/cgi-bin/webapp.cgi (Accessed April 5, 2015).

410. Aversano, R., Ercolano, M. R., Caruso, I., Fasano, C., Rosellini, D. and Carputo, D. Molecular tools for exploring polyploid genomes in plants. *International Journal of Molecular Sciences.* **2012,** *13*(8), 10316–10335.

411. Wei, L.; Xiao, M.; Hayward, A.; Fu, D. Applications and challenges of next-generation sequencing in Brassica species. *Planta,* **2013,** *238,* 1005–102. DOI 10.1007/s00425-013-1961-6.

412. Snowdon, R. J.; Luy, F. L. I. Potential to improve oilseed rape and canola breeding in the genomics era. *Plant Breeding,* **2012,** *131(3),* 351–360.

413. Varshney, R. K.; Nayak, S. N.; May, G. D.; Jackson, S. A. Next-generation sequencing technologies and their implications for crop genetics and breeding. *Trends in Biotechnology,* **2009,** *27(9),* 522–530.

414. Mundry, M.; Bornberg-Bauer, E.; Sammeth, M.; Feulner, P. G. Evaluating characteristics of de novo assembly software on 454 transcriptome data: a simulation approach. *PLoS One,* **2012,** *7(2),* e31410.

415. Edwards, D.; Batley, J.; Snowdon, R. J. Accessing complex crop genomes with next-generation sequencing. *Theoretical and Applied Genetics,* **2013,** *126(1),* 1–11.

416. Xu, M.; Dong, Y.; Zhang, Q.; Zhang, A.; Luo, Y.; Sun, J.; Fan, Y.; Wang, L. Identification of miRNAs and their targets from *Brassica napus* by high-throughput sequencing and degradome analysis. *BMC Genomics,* **2012,** *13,* 42.

417. Liu, L., Li, Y., Li, S., Hu, N., He, Y., Pong, R., Lin, D., Lu, L. and Law, M.. Comparison of next-generation sequencing systems. *BioMed Research International,* 2012, Article ID 251364, 11pagesdoi:10.1155/2012/25136

CHAPTER 9

CRISPR/Cas-Mediated Genome Editing Technologies in Plants

DEEPU PANDITA

Government Department of School Education, Jammu,
Jammu and Kashmir, India; E-mail: deepupandita@gmail.com

ABSTRACT

Clustered regularly interspaced short palindromic repeats/CRISPR-associated (Cas) proteins (CRISPR/Cas) is an adaptive immunity system in bacterial and archaeal groups for protection against invading RNA/DNA genetic material and bacteriophages. The CRISPR/Cas system is the most adaptable, instrumental, and revolutionary genome editing approach in history of molecular biology with diverse applications in modification of plant genomes and exploring the understanding of biological phenomenon, gene functional research, and improvement of vital agronomic traits. CRISPR/Cas is now a keystone in genome engineering and greatly expedited the progress of gene editing from the concept to practice with unparalleled ease, accuracy, robustness, simplicity, cost effectiveness, flexibility precision, and high efficiency. CRISPR/Cas may be categorized into Class I and Class II, with six types and 34 different subtypes. Out of these, CRISPR/Cas9 (Csn1), CRISPR/Cas12 (Cpf1), and recently developed novel endoribonuclease CRISPR/Cas13 for RNA interference have been optimized with broad applications across plant species as genome engineering platforms. The Class II type VI CRISPR/Cas13 systems guided by gRNA bind and cleave single-stranded RNA rather than DNA as in case of CRISPR/Cas9 and CRISPR/Cas12 and, therefore, offer novel prospectives for transcriptome engineering and fighting RNA viruses. In this chapter, several CRISPR/Cas-based approaches with successful examples to engineer the plants with desirable traits related to yield, quality, plant architecture, abiotic, and biotic stress have been discussed.

9.1 INTRODUCTION

Directed breeding of crops enables increase in yield, nutritional value, and consumer-valued characteristics of the produce. However, limited genetic diversity limits the volume of crop improvement that can be accomplished by conventional breeding approaches [1]. Unparalleled social and political resistance gridlocked the wide-ranging propagation of genetically modified (GM) crops. Because of that, only few traits are successfully introduced in the world market. The precision, ease, and economical genome editing (GE) technologies decrease the technological and economic drawbacks of genetically modified organisms (GMOs), but public approval cannot be assured [2, 3]. Conventional breeding [multiple genes transfer along with gene of interest (GOI)], GM crops (only GOI is inserted) and GE mechanisms (GOI is edited) differ in efficiency, quality, and number of genes transferred (see Figure 9.1). GE tools have revolutionized the precise single-base-pair DNA modifications and manipulations in eukaryotes, through insertions and/or deletions (indels), substitutions, and conversions of nucleotide bases [4–6]. The current five most common GE tools for targeted GE include oligonucleotide-directed mutagenesis [7, 8], which originated in the early 1980s with use in plant science ~15 years ago and four kinds of engineered nucleases, which includes zinc finger nucleases (ZFNs), mega nucleases, transcription activator-like effector nucleases (TALENs) and clustered regularly interspersed short palindromic repeat/CRISPR-associated protein (CRISPR/Cas) systems [9–13], which revolutionized area of genome manipulation. The arena of GE with site-specific nucleases initiated with ZFNs in 2002, when *Drosophila melanogaster* was subjected to mutagenesis and targeted chromosomal cleavage [14]. ZFNs are fusion proteins with synthetic zinc finger DNA binding domain, which is specific to three base pairs at the target site and DNA cleavage domain [15]. The chimeric proteins of two ZFNs with *FokI* monomer nuclease act as a restriction enzyme and create a spacer of 5–6 bp, dimerization and targeted double-stranded breaks (DSBs) in DNA [14, 16, 17]. This stimulates cellular DNA repair machinery. ZFNs find wide use in various plant species, for instance, *Arabidopsis thaliana* [18–20], *Nicotiana tabacum* [21–23], *Zea mays* [24], and *Oryza sativa* [25]. ZFNs are difficult to engineer, suffer from target site availability, function inconsistently across different genetic loci, and may cause occasional toxicity [26–29]. TALENs initially characterized in *Xanthomonas* are second class of sequence-specific nucleases for site-directed mutagenesis, highly specific, broader target range, easy to engineer, and more mutagenic than ZFNs [30–33]. TALENs consist of transcription activator-like effectors joined to the *FokI* endonuclease catalytic domain and target specifically one base pair of the target site [34, 35]. TALENs have various success stories in flowering plants and bryophytes [36–38], including *A. thaliana* [39], *Hordeum vulgare* [40], *O. sativa* [36], *Brachypodium distachyon* [36], *Zea mays* [41], *Solanum tuberosum* [42], *Solanum lycopersicum* [43], *Glycine max* [44], and *Triticum aestivum* [45]. However, constraints in engineering of DNA-binding domain proteins, inefficiency in process of transfection, complications in designing, authentication, and limitations on multiplexed mutations hinder widespread application of ZFNs and TALENs [46].

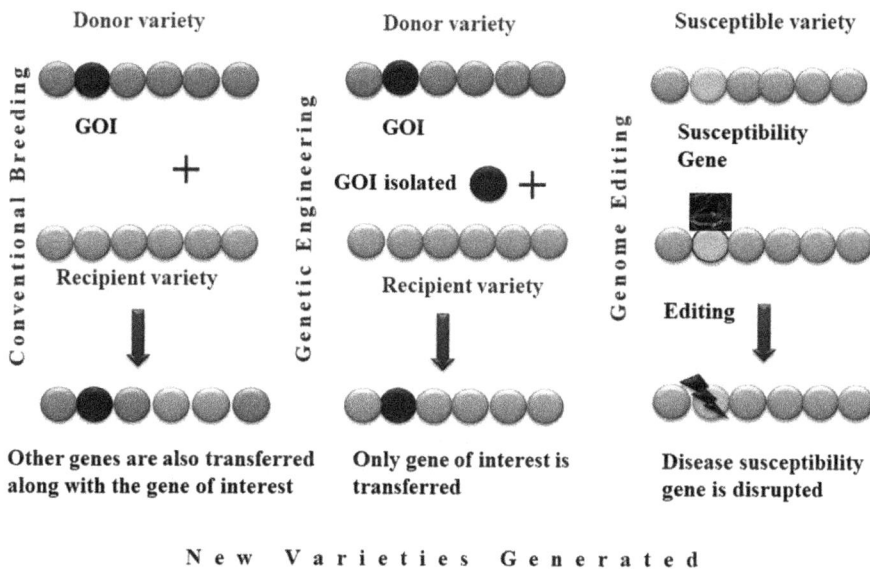

FIGURE 9.1   Conventional breeding, genetically modified crops, and genome editing.

CRISPR/Cas systems are present in bacterial (46%) and archaeal (90%) genomes, to defend against RNA and DNA viruses; however, distribution and classification differ prominently within and transversely in phylogenetic clades [47]. GE by CRISPR/Cas as the most potent extensive use due to accuracy, design, efficiency, versatility, simplicity, unprecedented ease, cost-effectiveness, highly accurate, specificity and precision, multiplexing, broad application range and flexibility, dominant, and game-changing tool in history of modern molecular biology around the world has not only gained acceptance as trait development tool but also achieved legal/regulatory approvals for development of the products in many nations [11, 48, 49]. Numerous variants of CRISPR/Cas systems are available, but the drawback of off-target effects exists [13]. Various plant base editors (BEs) have been recognized by fusion of catalytically impaired complex of nucleobase deaminases with Cas9, Cas12a, or Cas13 proteins and guide RNA (gRNA). Adenine BEs show efficient conversion of adenine (A) to guanine (G), while cytosine BEs show conversion of cytosine (C) to thymine (T) with high efficiency in target regions. The RNA BEs of Cas13 editing tools substitute adenine (A) to inosine (I) or specific exchange of cytosine (C) to uracil (U) known as RESCUE in RNA [50]. Several types of simple and precise plant BE systems have been discovered [51]. BE3 [46], BE4 [52], targeted AID [53], and dCpf1-BE [54] base editing systems find application in most important crops. These base editing systems use Cas9 or Cas12a for engaging cytidine deaminases to create specific C–T changes via DNA mismatch repair pathways [46, 52, 53, 54].

9.2   CRISPR/CAS SYSTEM

9.2.1   DISCOVERY

CRISPR/Cas was discovered as an acquired immunity system in genomes of bacteria and archaea to guard from invading mobile genetic elements, plasmid, and/or bacteriophage DNA or RNA [55–62]. The name CRISPR originated from the detection of islands known as clusters of regularly interspaced repeats [55]. Hitherto, only some type I CRISPR/Cas systems have been characterized, best remarkable being the model type I-E CRISPR/Cas system from *Escherichia coli*, which was, in reality, first identified CRISPR locus in bacteria in 1987 [55]; however, their function was hidden till 2005 [63, 64], and in 2007, CRISPRs along with Cas proteins proved to afford immunity to bacteria [65]. In 2012, GE by the CRISPR/Cas system was reported for the first time in a cell-free system [66], and afterward, five independent research groups used this tool for editing genes in animals [67–70]. The first report of the most advanced third-generation CRISPR/Cas9-based GE approach was in *A. thaliana*, *Nicotiana benthamina* [71, 72], and *O. sativa* [73]. The historical timeline of CRISPR/Cas systems is highlighted in Figure 9.2.

9.2.2   CLASSIFICATION OF THE CRISPR/CAS SYSTEM

There is notable diversity in organization and function of CRISPR/Cas in prokaryotic systems of bacteria and archaea to defend against invasive

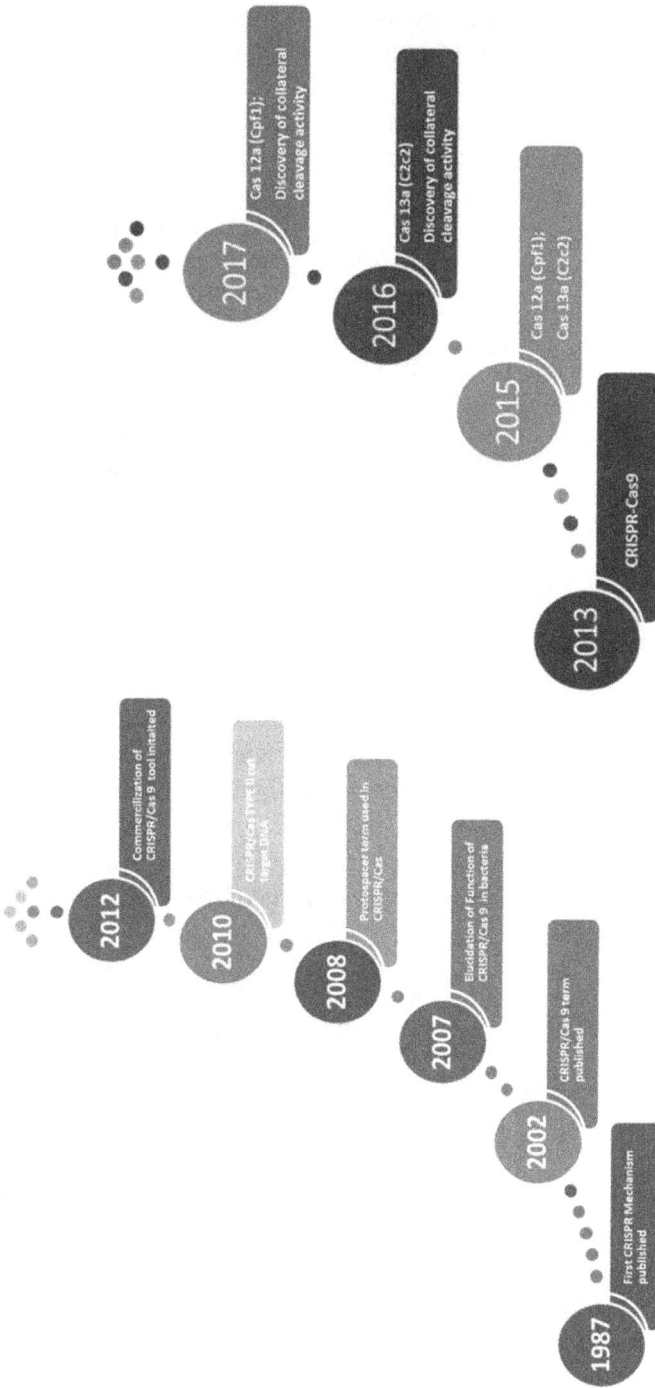

FIGURE 9.2 Historical timetable of the CRISPR/Cas system.

viruses and mobile genetic elements [56, 60, 61]. The CRISPR/Cas system is cataloged into two major classes on the basis of modular composition, assortment of Cas genes, and their crRNP effector protein complexes and structure. CRISPR/Cas systems with two classes (1 and 2) have been described, which are further categorized into six types (I–VI) [56, 74] and 34 subtypes [75]. CRISPR/Cas Class 1 includes three distinct types I, III, and IV, which possess a set of CRISPR-associated multiprotein effector complexes of four to seven signature Cas nuclease protein components, few in close-fitting association with the CRISPR RNA (crRNA) for antiviral defense (Cascade). In contrast, the CRISPR/Cas class 2 system comprises II, V, and VI types relying on crRNPs with a single- and multidomain effector nucleases, such as Cas9, Cas12, or Cas13 [56, 75, 76]. CRISPR/Cas I, II, and V (and likely IV) target DNA and VI (Cas13) system exclusively targets RNA molecules [51, 77, 78]. Subtype III (Csm/Cmr) systems recognize and support cleavage of DNA as well as RNA [79]. DNA targeting CRISPR/Cas systems have evolved to depend on recognition of a 2–5 base pair sequence known as protospacer adjacent motif (PAM) for degradation, which remains located nearby target sequence of invading DNA. PAM is absent nearby identical sequence inside CRISPR array, due to flanked repeats, which avoids self-targeting [80]. Type III and VI target RNA depend on consensus PAMs. Nevertheless, activities of several type III [81–84] and type VI effector crRNPs [85–87] were reported to be under regulation of protospacer flanking sequences (PFS) that flank the protospacers of target RNAs. Because of simple design in the structure of effector complexes, CRISPR/Cas class 2 has become an outstanding choice to develop a state-of-the-art generation of GE tools. The outline of classification of CRISPR/Cas systems is highlighted in Figure 9.3.

9.2.3 MECHANISM OF CRISPR/CAS SYSTEMS

The CRISPR/Cas system has characteristically dual portions: (1) Cas proteins function for defense and acquisition of invader DNA/RNA and (2) CRISPR array, which comprises alternating conserved direct repeat domains and invader-derived (Spacer) variable DNA sequences with identical size flanked by a leader region. CRISPR array remembers invader sequence [88]. The CRISPR/Cas system mechanism has specified-sequence way using recognition and cleavage of foreign nucleic acids. As a guard mechanism in prokaryotes, CRISPR/Cas has four distinctive phases.

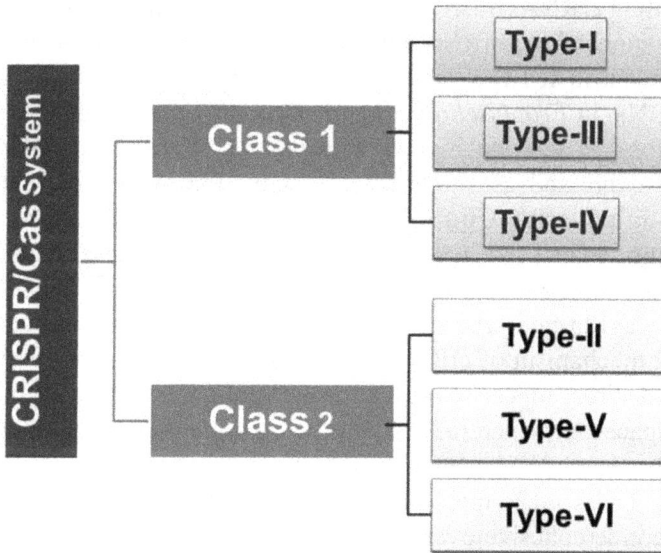

FIGURE 9.3   Classification of the CRISPR/Cas system.

i)   *Adaptation (spacer acquisition):* Adaptation is the process of integrating invading nucleotide fragments into CRISPR locus after capture by the host organism with the help of Cas1 and Cas2 [89].

ii) *Precursor crRNA (pre-crRNA) expression:* The expression process involves the production of large pre-crRNA/gRNA through transcription of CRISPR array [90, 91].

iii) *Maturation (crRNA biogenesis):* The maturation of transcribed guide CRISPR pre-crRNA includes specific recognition and processing by cleavage through ribonuclease (RNase) to produce a small, mature crRNAs [91].

iv) *Interference (RNA targeting):* The mature crRNAs associate with or guide monomeric or multimeric Cas effector protein complexes to produce numerous crRNA-Cas (crRNP) effector complexes [91], which scan and capture the cellular nucleic acids for presence of target sequence complementary to crRNAs known as PAM and degrade invading nucleic acids either through the ribonucleoprotein (RNP) complex itself or by recruitment of an supplementary factor exhibiting the nuclease activity [92, 93] (see Figure 9.4a).

Cleavage by PAM recognition avoids attack to bacterial CRISPR locus [94]. In contrast to CRISPR/Cas9 systems, type V systems do not require transactivating crRNA (tracrRNA) and RNase III for maturation of crRNAs. The maturation of pre-crRNA into mature crRNA (42 to 44 nt length) is mediated by intrinsic RNase activities of Cas12a domains. The biogenesis of mature crRNA in *Francisella novicida* starts with the recognition of 27–32 base-pair-long spacer positioned adjacent to 36 base-pair-long repeats, which are expressed as single transcript by FnCas12a [95]. The repeat sequences in pre-crRNA transcript form a pseudoknot structure, which is readily identified by Cas12a [96–98]. Pseudoknot binding to divalent cations such as Mg^{2+} or Ca^{2+} augments binding of crRNA to Cas12a. The WED domain of Cas12a catalyzes the processing of 5′ end of crRNA, but the 3′ end processing mechanism of crRNA is still obscure. A mature crRNA consists of 19–20 nt direct repeat sequence (5′ pseudoknot structure) and 20–24 nt guide or spacer sequence [53, 98]. In *A. thaliana* with high AT content in genome, the novel PAM sequence prerequisite increases targetable genomic sequences. The Cas12a mechanism generates staggered DSBs with 5′-OH 5 bp nucleotide cohesive overhangs in distal end of the target cleaved site, which increase efficiency and application of promote site-directed integration of new DNA GE with compatible sticky ends, AT-rich regions (Like untranslated and promoter regions) by the NHEJ mechanism [95, 99–101]. Cas9 generates blunt ends [95, 101]. Cas12a requires "T" rich (5′-TTTN-3′) PAM sequences, which increase repertoire of protospacers and the number of possible plant genetic manipulations [95, 102–104]. PAM of Cas12a is relatively longer and more specific than Cas9. However, a long PAM sequence requirement in plant genomes is also a limitation for Cas12a GE [105]. The target sites without T-rich PAM sequence motifs are difficult to be identified and edited by Cas12a. The improved and modified Cas12a variants (impLb-Cas12a) have been engineered, which recognize altered PAM specificities [100, 105, 106], and two Cas12a variants (ceCas12a and beCas12a) with a more stringent PAM site requirement in order to minimize off-target events were identified [107]. The mechanism of adaptive immunity by CRISPR/Cas systems and mode of action is demonstrated in Figure 9.4.

9.2.4   IMPORTANCE AND APPLICATIONS OF THE CRISPR/CAS SYSTEM

ZFNs and TALENs were the first-generation GE tools that came into existence some 20 years ago providing the recognition of target sequences. The CRISPR/Cas system has revolutionized the arena of GE and outshined the ZFN and TALEN systems in the area of genome modification [66, 108]. The CRISPR/Cas system is a robust, rapid, simple, economical, and accurate

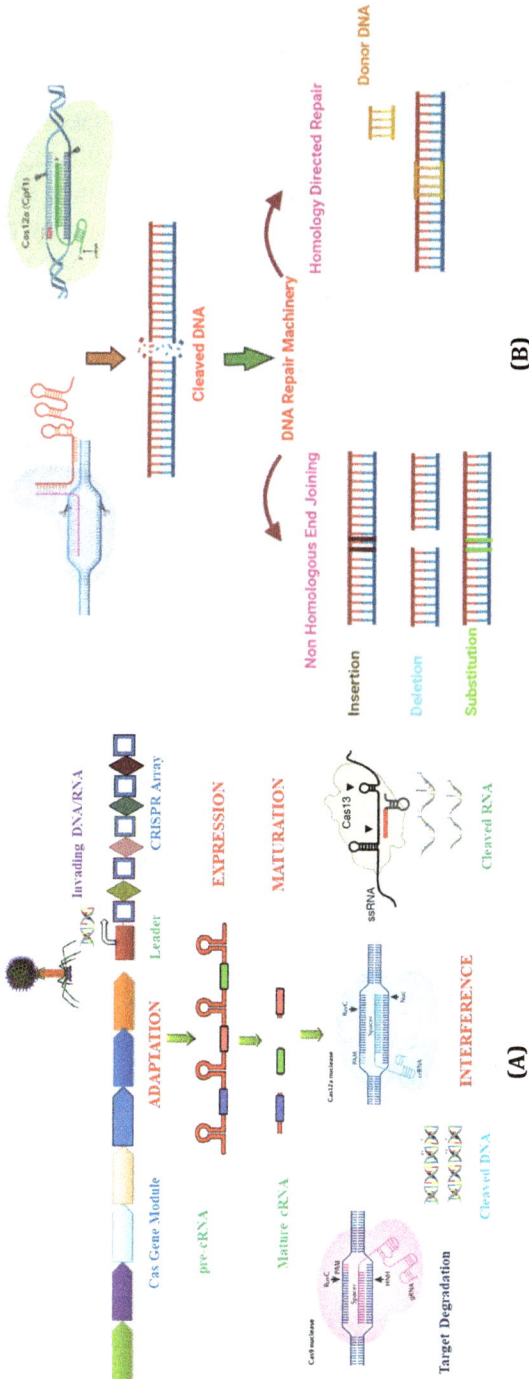

FIGURE 9.4 (a) Overview of adaptive immunity by CRISPR/Cas systems, including adaptation, expression, maturation, and interference stages. (b) Mechanism of target cleavage by CRISPR/Cas9 and CRISPR/Cas12 systems.

technique of genome manipulation to produce both cisgenics and transgenics [109], out of which the cisgenesis technique is ethically unbiased [110]. GMOs face moral, regulatory, and biosafety hazards to anthropological health and environment [111], while GE by CRISPR/Cas systems faces (via the NHEJ repair mechanism), unpredicted threats of off-target editing and potential exploitation but are comparatively less risky and morally more acceptable [93, 112–114]. Genome-edited crops have lower threat to security of man [114, 115]. The agriculture and food science has extensive importance in utilizing the CRISPR/Cas toolbox to raise the quantity and quality of food [116]. The initial agricultural investigation of the CRISPR/Cas9 system signposted its greater prospective in GE of *O. sativa* [117]. The CRISPR/Cas system and tissue culture in plants introduces genetic modifications directly into relevant lines with improved swiftness and specificity [118]. The upgraded single-guide RNA (sgRNA) designs improved GE efficiency of CRISPR/Cas in *Zea mays* [45], *T. aestivum* [119], *O. sativa* [120, 121], *Glycine max* [122, 123], *Citrus sinensis* [124], *S. lycopersicum* [125], *Brassica napus* subsp. napus [126], *Lolium perenne* [127], and *Physcomitrella patens* [128] for improvement of imperative genetic traits, such as abiotic and biotic stress tolerance, nutrient uptake, nutritional value, harvest improvement, methylome editing, and architecture (see Figure 9.5). CRISPR/Cas-mediated editing of methylome in *A. thaliana* utilized a modified SunTag approach and earlier characterized catalytic domain Domains Rearranged Methyltransferase of *N. tabacum* to serve as methylation effector for modification of FWA and SUPERMAN genes [129]. CRISPR/Cas9-based base editing was done in cotton by single-base modification. Base editing systems generate precise alterations in amino acids by change in particular single base or development of gene knockout through introduction of a premature stop codon [130]. Base editing of ALS and FTIP1e in rice resulted in resistance to two herbicides [131] and further can improve 14 agronomic traits in *O. sativa* and flavor and fruit mass in *S. lycopersicum* [132]. The base editing system finds effective use in *O. sativa*, *T. aestivum, Zea mays, S. tuberosum,* and *S. lycopersicum* to reduce off-target editing [130, 131, 133–136] and delivers an opportunity to squeeze the vital single-base-pair mutation-based agronomic traits.

9.3   TYPE I CRISPR/CAS SYSTEM

Type I CRISPR/Cas systems and signature Cas3 helicase establish the most copious and prevalent CRISPR/Cas system in genera of bacteria and archaea [47, 137] and find use in in situ editing in bacteria [137].

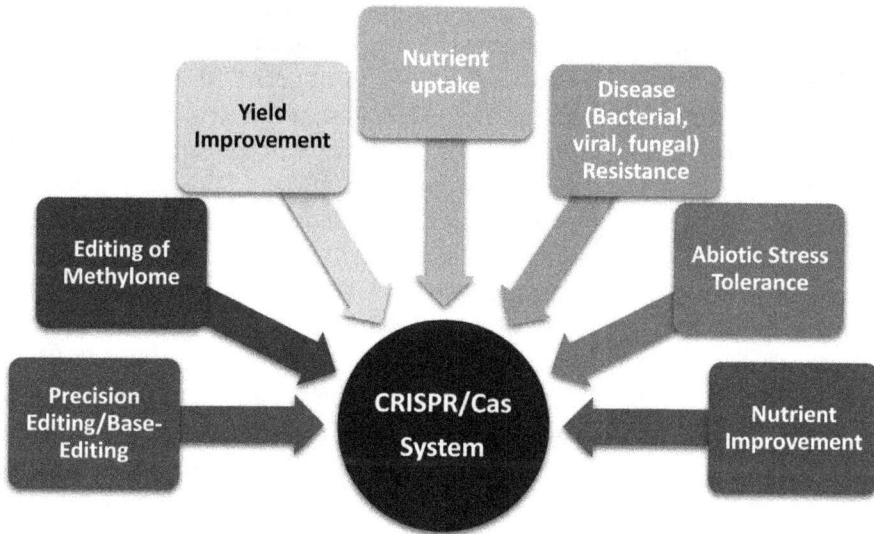

FIGURE 9.5 Applications of CRISPR/Cas systems.

9.4 TYPE II CRISPR/CAS SYSTEM (CRISPR/CAS9)

The type II CRISPR/Cas system from Class 2 is subdivided into II-A, II-B, and II-C subtypes. Moreover, CRISPR array and Cas proteins encode extra tracrRNA, which mediate crRNAs and Cas9 interaction [51]. Cas9 is the maximum investigated and utmost recurrent DNA-specific multifunctional and multiple-domain endonuclease. Cas9 are present in *Streptococcus pyogenes, S. thermophilus, S. aureus, Francisella novocida,* and *Brevibacillus laterosporus. S. pyogenes* Cas9 (SpCas9) has most applicability [138]. CRISPR/Cas9 is a prokaryotic adaptive immune system used to protect bacteria and archaea against attack by foreign DNA and specific phages [46, 139]. CRISPR/Cas9 machinery acquires short fragments of such foreign DNA within an array of CRISPRs [55, 140–142]. CRISPR/Cas9 has two principal parts of CRISPR-associated protein 9 (Cas9) and noncoding crRNAs, including tracrRNA and pre-crRNA [143]. The pre-crRNAs generate 23–47 base pairs repeat or spacer sequences. The tracrRNA has sequence complementary to a nucleotide of pre-crRNA for the formation of two-component sgRNA duplex requisite for the maturation of crRNA and crRNA-mediated specific DNA cleavage [143–145]. Cas9 is crRNA-dependent endonuclease having two distinctive nuclease domains, that is, HNH and RuvC-like domain. HNH and RuvC nuclease domains cleave target DNA strands and nontarget DNA strands respectively [76, 143]. The CRISPR/Cas9 system introduces DSBs at DNA target sites. Homologous base pairing of sgRNA with target sequence regulates specificity binding and DNA Cleavage by Cas9 at selected genomic locus adjacent to 5'-NGG-3' triplet PAM by generating a DSB at the target-specific site [145–147]. The 5'-NGG-3' PAM sequence exists around 5–10 times in every 100 bp in model plant species [138]. Thus, PAM prerequisite is presently a bottleneck for target sites of Cas9. For solving this problem, many Cas9 variants and Cas9 orthologs with numerous PAM preferences are practiced to accomplish the same outcomes as the wild-type CRISPR/ Cas9 system. A cell naturally utilizes dual mechanisms of nonhomologous end joining (NHEJ) and homology-directed repair (HDR) pathways to repair DSB (see Figure 9.4b). Imprecise NHEJ repair pathway creates indels (insertions and deletions) that result into loss-of-function mutations. NHEJ can be used to generate gene knockouts. Another mechanism of precise HDR pathway repairs pre-existing mutations by introduction and insertion of template donor DNA sequence to repair the break [148]. HDR can be used for precise editing of DNA sequences. Therefore, NHEJ is extra efficient than HDR across eukaryotic cells mainly in nondividing cells. This has made precise gene editing challenging and restricted its application in gene therapy. Moreover, precision gene editing necessitates presence of a homologous template containing the desired change for HDR. The synchronized transfer of this template to the DSB in the target cells has additionally limited applications of the gene editing [149]. Cas9 can be altered into nickase with potential of generating single-strand cleavage via mutation in either HNH domain or RuvC-like domain [138] or Nuclease-dead Cas9 (CRISPR/dCas9) DNA binding protein mutant of Cas9 by mutation in Asp10→Ala, His840→Ala domains [138]. Solitary or multiple genes undergo repression or activation by enzymatically inactive and nonfunctional dead, deactivated Cas9 endonuclease re-engineering tool paired with an activator, for instance, transcriptional activator domain VP64 or by a repressor protein. Transcriptional activation has usage in immunity of plants [150, 151], revelation of regulatory networks, and improved plant metabolite production [152, 153]. CRISPR/dCas9 applications consist of activation and repression of genes, editing of epigenome, live-cell chromatin imaging, modulation of chromatin topology, and DNA-free genetic modification [154].

9.4.1 APPLICATIONS OF THE CRISPR/CAS9 SYSTEM IN GENOME EDITING OF PLANTS

Promising engineering tool of CRISPR/Cas9 has been utilized for efficacious GE of more than 25 species and 100 genes to generate various necessary qualities in major crops [155, 156]. The abiotic and biotic stress tolerance along with nutritional value of plants can be improved by the CRISPR/Cas9 toolbox. CRISPR/Cas9 machinery joined with RNPs defends *O. sativa* against blast fungus *Magnaporthe oryzae* [157] and increased nutrient value of *S. tuberosum* by knock down of granule-bound starch synthase gene [158]. The pYL CRISPR/Cas9 multiplex vector system in *S. lycopersicum* allowed manipulation of metabolic pathway for higher amounts of c-amino-butyric acid (GABA) [159]. CRISPR/Cas9 has been applied to remove the immunogenic gluten epitopes (gliadings) in *T. aestivum* L., which cause coeliac disease in some individually [160]. Integrated multiplexed CRISPR/Cas9-based high-throughput targeted mutagenesis with high-throughput gene editing can effectively target 743 candidate genes related to important agronomical and nutritional traits [161]. Some applications of the CRISPR/Cas9 system in plant GE are given in Table 9.1.

9.5 TYPE III CRISPR/CAS SYSTEM

Type III CRISPR/Cas systems belong to Class 1 and target both invasive invading RNA and DNA by two endoribonucleases and one DNase that match their crRNAs. In the absence of invaders, Type III crRNPs and three nuclease actions retain inactive state. Interaction between recently expressed foreign target RNA and CRISPRRNA of effector crRNPs activate RNase and DNase causing cotranscriptional cleavage of nontemplate strand of invader DNA and target RNA for both Type III-A [188, 189, 190–196] and Type III-B systems [81, 197–201]. Relatively large type Csm1 (III-A) and Cmr2 (III-B) (of Cas10 superfamily) systems are two effector crRNPs effector complexes [56], which harbor HD domains having HD DNA nuclease active site for providing DNase action of Type III CRISPR/Cas on active transcription of target DNA [79, 81, 195, 197, 198, 200, 202–207] and are regulated by short PFS located just 3' of the target RNA matching to 5' CRISPR repeat-derived tag sequence [83, 189, 204]. The 3' flanking sequences beside target RNA help in activating Type III-B (Cmr) DNA cleavage in *Pyrococcus furious* [81]. A complex of crRNA and six Cas proteins (Cmr1, Cmr2, Cmr3, Cmr4, Cmr5, and Cmr6) identify and cleave RNAs complementary to CRISPRRNA

TABLE 9.1 Applications of the CRISPR/Cas9 System in Plant Genome Editing

Genes Targeted	Effects	References
Four EBEs of OsSWEET14 promoter	Resistance against bacterial blight in Super Basmati rice	[162]
Badh2	Controls rice fragrance	[163, 164]
C3H, C4H, 4CL, CCR and IRX	Lignocellulose biosynthesis in *Dendrobium officinale*	[165]
eIF4G	Candidate rice tungro disease resistance gene	[166]
ENGase	N-glycans modification in grains of *Hordeum vulgare* cv. "Golden Promise"	[167]
EPFL9	Stomatal developmental	[168]
Fatty acid desaturase2 (FAD2)	Enhancement of seed oil composition of *Camelina sativa*	[90]
GhMYB25 like-sgRNA1 and sgRNA2	Allotetraploid cotton resistant to wilt caused by Verticillium	[169]
Gn1a, DEP1	Grain yield	[170]
Gn1a, GS3	Grain yield	[171]
Granule-bound starch synthase	Starch quality (amylopectin) in *S. tuberosum*	[172]
Heat Shock Protein 90 (HSP90)	Thermotolerance in *Tetraselmis suecica*	[173]
HvPM19 BolC.GA4.a	Pod shatter and control of dormancy in *Brassica oleracea* and *Hordeum vulgare*	[174]
Improved expression of ARGOS8	Enhanced drought tolerance in *Zea mays*	[175]
OsALS	Confers herbicide resistance	[131]
OsERF922	Responsible for rice blast resistance	[132]
OsNramp5	Metal transporter gene	[177]
OsRR22 and OsNAC041	Improvement in salinity tolerance	[178–180]
OsSWEET11	Rice caryopsis development	[181]

TABLE 9.1 (Continued)

Genes Targeted	Effects	References
Plastific large subunit of ADP-glucose pyrophosphorylase (OsAGPL4)	Starch synthesis pathway in pollen grains of O. sativa	[182]
PYL1, PYL4, PYL6	Control plant growth and stress responses	[183]
RIN gene (ripening inhibitor)	Inhibition of fruit ripening in S. lycopersicum	[184]
SBEI and SBEIIb	Increased amylose and resistant starch levels in high-amylose O. sativa for improved nutritional value	[185]
SlIAA9	Generation of parthenocarpic S. lycopersicum plants	[186]
TaMLO-A1 and TaMLO-B1	Bread wheat immune to powdery mildew	[119]
TIFY1a, TIFY1b	Plant adaption to cold	[187]

guides. Type III-A (Csm) shows a similar mechanism of RNA target cleavage [79, 192, 193, 204, 206]. GGDD motif of Csm1/Cmr2 conserved palm domain stimulated functions in synthesis of short cyclic or linear oligoadenylate molecules that bind conserved N-terminal CARF domains of Csm6 (III-A) or Csx1 (III-B) and activate their RNase activities [208–212]. Target and cleavage of RNA was initially discovered in Type III-B (Cmr) of *P. furiosus* [213]. Type III-associated Csm6 and Csx1 endoribonuclease systems share N-terminal CARF and RNase R-X4-6-H active catalytic site motifs inside C terminal higher eukaryotes and prokaryotes nucleotide (HEPN) binding domains [195, 214–219] and do not form stable physical associations with associated effector crRNP complexes [191, 193, 212, 220]. The crRNA pairing to target RNA complementary sequence allows III-A and III-B effector crRNPs to identify and cleave target RNA by Csm3 or Cmr4 RNases located along crRNP backbone of effector complexes [83, 189, 204]. Cas protein endoribonuclease Csm3 in III-A and Cmr4 in III-B of Cas7 superfamily occur in four or extra replicas along backbone of complex [56, 192, 193, 199, 221–224] and cleave target RNA at six nucleotide intervals. Purified recombinant Csm6 [195, 209, 210, 215–219] and Csx1 [81, 208, 211] cleave various RNA molecules in laboratory conditions and demonstrates specificity for cleavage after few bases.

9.6   TYPE IV CRISPR/CAS SYSTEM

The type IV CRISPR/Cas system mainly encoded by bacterial plasmids or rarely by prophages [225] lack greatly conserved adaptation module and effector nuclease so are functionally defective but may play a defensive role in prokaryotes [226]. Type IV is subdivided into IV-A and IV-B possessing effector module proteins and extremely diverged Cas7 (Csf2), Cas5 (Csf3), and a smaller version of Cas8 (Csf1)1. IV-A encodes DinG family helicase (Csf4), a type IV-specific Cas6-like protein (Csf5) with a role in crRNAs maturation and successive formation of Cascade-like crRNA-guided effector complex of Csf1, Csf3, Csf5, and multiple copies of Csf2 [227], while IV-B lacks dinG, csf5, but encodes a putative (Cas11) [228, 229].

9.7   TYPE V CRISPR/CAS SYSTEM

CRISPR/Cas12a (previously CRISPR-Cpf1) from *Prevotella* and *Francisella* is single RNA-guided endonuclease of class II, type V CRISPR effector that serve as simple and attractive tool as DNA interference GE system, which lacks HNH domain and possess RuvC-like domain [95, 102, 103]. Cas12a/ Cpf1 has three RuvC domains and a matchless fold domain because of absence of HNH domain resembling RuvC domain. The positively charged central channel of nuclease (NUC) lobe determines cleavage of target sequence after catalytic residue mutations in Cas12a RuvC domain of *Acidaminococcus* sp. prevent cleavage in target and nontarget sequences [96]. Cas12b is deficient in HNH domain and has NUC domain determining cleavage of target resembling CRISPR/Cas12a. NUC domain of Cas12b is noticeably dissimilar from Cas12a NUC domain. Cas12a and Cas12b utilize RuvClike domain for cleavage of target DNA subsequent to substantial changes in conformation related to initial cleaving of nontarget strand. NUC domain participates in guide target binding [230]. Cpf1-mediated GE involves only shorter crRNA with protospacer functioning as gRNA and does not require tracrRNA, while Cas9 needs both crRNA and tracerRNA [101, 104, 106, 231, 232]. The short crRNAs can lead to secondary structures in RNA and Cpf1 edited lines require specific genomic evaluation due to Cpf1 genomic reorganizations in locations adjacent to target sites [233]. Cas12b is smaller in size than Cas12a. Cpf1 is guided by sgRNAs of 42 bp as against 100 bp for Cas9 to a target site, where it recognizes a thymidine-rich PAM sequence (TTTN) PAM on opposition of gRNA-targeted strand in a target location [99, 231]. Cas12a targets both double-stranded DNA (dsDNA) and single-stranded DNA (ssDNA), which enhanced scope of CRISPR/Cas12a-based GE [234]. Cas12a is temperature sensitive, which limits its effectiveness in GE of the plants [235, 236], but engineered variants with enhanced activities have been generated [237]. Cas12a GE systems AsCas12a, FnCas12a, and LbCas12a with varying efficiency have been demonstrated in plants [238]. Cpf1 editing has been used in rice [107, 239, 240, 241], tomato [242], tobacco [243], Arabidopsis [103, 237], and more recalcitrant genomes, including the allotetraploid cotton [244], citrus [245], soybean [231], wheat [246], crop domestication [127, 247], and gene stacking [248] and offers an unlimited substitute to Cas9 and extensive variety of targetable genes other than the ones provided by Cas9. Transcriptional repression of miR159b by deactivated nuclease domains of dAsCas12a (D908A) and dLbCas12a (D832A) in *A. thaliana* has been reported [103]. Research studies in numerous plant species have recommended lesser editing efficiencies linked with Cas12a than Cas9 [161, 249, 250]. Cas9 and Cas12a was used to target same loci, and in one case, a higher efficiency of mutation with Cas9 was observed [249], and in another, Cas12a was more efficient [241].

9.7.1 APPLICATIONS OF THE CRISPR/CAS12 SYSTEM IN GE OF PLANTS

CRISPR/Cas12/Cpf1systems are progressively applied in major crops and have been successfully used in cotton [249], maize [249], and rice [240, 251]. Some applications of the CRISPR/Cas12 system in plant GE are given in Table 9.2.

CRISPR/Cas14 systems with 38 variants divided into eight families (Cas14a to Cas14h) possess RuvC nuclease domain and occur exclusively within DPANN superphylum of archaea. Cas14 is another new promising tool, which encodes small Cas protein (40–70 kDa) and is one-third the size of the Cas9 protein of class 2 CRISPR-Cas systems. It cleaves ssDNA effectively without a PAM restriction but not dsDNA, so finds application only as single-nucleotide polymorphism genotyping diagnostic toolkit/Cas14-DETECTR of the dsDNA, ssDNA and RNA [257].

9.8 TYPE VI CRISPR/CAS SYSTEM (CAS13)

CRISPR/Cas type VI (Cas13) is the only prokaryotic CRISPR/Cas immune system reported to selectively target RNA. Cas13 shows promise as a tool for the detection and manipulation of RNA [258]. VI effector crRNPs comprise single Cas13 nuclease along with a crRNA. Cas13 has two conserved HEPN domains, which jointly compose a complex RNase active center for cleavage of the target RNA [58, 216, 205, 259]. Cas13 endonucleases have distinct RNase activity for processing and maturation of the precursor crRNA into mature crRNA [216, 217, 218, 259]. Four distinctive subtypes VI-A (Cas13a/C2c2 variant), VI-B (Cas13b/C2c6), VI-C (Cas13c/C2c7), and VI-D (Cas13d) are reported based on Cas13 phylogeny [58, 86, 87] possessing two HEPN domains positioned nearby terminal ends of Cas13 for RNA-targeted nucleolytic activity [74]. The other distinguishing features of VI subtypes include difference in size and sequence of Cas13 and co-occurrence of supplementary Cas genes (e.g., Csx27 or Csx28 in VI-B) and WYL domain proteins (in IV-D) encoding accessory proteins for modulation of Cas13crRNP RNA cleavage in a positive or inhibitory method [84, 87]. Cas13a (VI-A), Cas13b (VI-B), Cas13c (VI-C), and Cas13d (VI-D) distinguish from Cas12/Cas9 due to lack in a DNAse domain and presence of two external HEPN domains. Target specificity of Cas13 is based on a 28–30-nt spacer and needs gRNA of ~64 nt [82, 260]. Cas13a and poly-crRNAs expressing genome edited plants can target diverse RNA viruses

TABLE 9.2 Applications of the CRISPR/Cas12 System in Plant Genome Editing

Genes Targeted	Effects	CRISPR/Cas12 Type	References
CsPDS	Carotenoid biosynthetic pathway	LbCas12a	[245]
Gl2	Epicuticular wax formation Regulation of fatty acid elongase pathway	LbCas12a	[249]
OsALS	Herbicide resistance	LbCas12a	[54]
OsDEP1, OsPDS, and OsEPFL9	Regulating carbon nitrogen metabolism Yield; Carotenoid biosynthetic pathway; Abiotic stress tolerance	FnCas12a	[238]
OsDEP1, OsROC5	Regulating carbon nitrogen metabolism Yield; leaf rolling controlling; Negatively modulates bulliform cells	LbCas12a	[252]
OsDL, OsALS, OsNCED1, OsAO1	Herbicide resistance; Abscisic acid regulation-stress tolerance; caroteniod catabolism	FnCas12a	[253]
OsEPFL9	Regulates stomatal density in leaves	CRISPR-Cas12a	[254]
OsPDS	Carotenoid biosynthetic pathway	AsCas12a, LbCas12a	[241]
OsPDS and OsSBEIIb	Encodes phytoene desaturase and starch branching enzyme	CRISPR-Cas12a	[255]
OsPDS, OsGS3, OsALS, OsNAL	Carotenoid biosynthetic pathway; Grain length-yield; Herbicide resistance	LbCas12a	[256]
OsROC5 and OsDEP1	Leaf rolling control; Negatively modulates bulliform cells; Regulating carbon, nitrogen metabolism, Yield	FnCas12a, AsCas12a, LbCas1	[250]
SlHKT1;2	Salt tolerance, Abiotic stress tolerant	LbCas12a	[242]
TaWaxy and TaMTL	Haploid induction gene; Waxy- starch synthase gene involved in flour quality	LbCas12a	[246]

and bestow plants with constant immunity against viruses. Virus resistant plants can defend themselves on exposure to mixed infections of widespread viruses [261]. Cas13a and Cas13b have potential for RNA knockdowns, while Cas13d orthologs can control endogenous transcript splicing with the probability of in vivo transport owing to its trivial size (~930 Å) among the class 2 CRISPR effectors [262].

9.8.1   APPLICATIONS OF THE CRISPR/CAS13 SYSTEM IN GE OF PLANTS

CRISPR/Cas13 finds application as an effective tool for enhancement of the plant immunity against viruses. CRISPR/Cas13 can be utilized in RNA targeting [263–265], RNA tracking, and RNA editing [266]. CRISPR/Cas13-engineered *S. lycopersicum* and *Nicotiana benthamiana* are resistant to Tomato Yellow Leaf Curl Virus [267]. CRISPR/Cas13a in *N. benthamiana* degrades single-stranded RNA of Turnip Mosaic Virus at diverse sites [268]. crRNAs demonstrated most effective RNA interference and cleavage of viral genome after targeting the gene sequences encoding proteinase silencing suppressor (HC-Pro). Besides LshpCas13a, recently, several well-characterized variants of Cas13 proteins such as LwaCas13a (*Leptotrichia wadei*), BzCas13b (*Bergeyella zoohelcum*), PspCas13b (Prevotella sp. P5-125), and CasRx13d (*Rumino-coccus flavefaciens* XPD3002) have been tested to find out better variant against the plant viruses [265]. Some applications of the CRISPR/Cas13 system in plant GE are given in Table 9.3.

9.9   LIMITATIONS OF CRISPR/CAS SYSTEM IN PLANTS

Several research investigations have shown off-target effects, which may cause rearrangements in chromosomes and other mutations [145, 269], through CRISPR/Cas9 in animal [270–272], human, mouse, and rat cell lines [273–275], and functional plant genome studies [255, 276]. RNA-guided base CRISPR/Cas editing systems show off-targets in the mammalian system as well [277–279]. Cas13 has some off-target effects in plants [280] and shows collateral RNase activity after binding to a target transcript, resulting in degradation of nontarget RNAs as well [217].

Very scarce off-target mutations in *Gossypium* edited by CRISPR/Cas9 were reported indicating that off-target effects of CRISPR/Cas rare with a few low-frequency off-target mutations [281, 282]. Transgenic soybean showed that off-target editing creates fewer genetic variations compared to mutations from radiations [276]. Some of the major limitations of CRISPR/Cas9 systems include the restriction of edits to regions of high GC content because of a "G"-rich PAM sequence requirement and its relatively large size and lower efficiency of GE [233]. Although efforts are underway to engineer near "PAMless" Cas9 variants [283], this work has not been replicated in plant systems. As recently shown in maize, Cas9 was used in combination with FLP recombinase to engineer gene stacks, thus greatly facilitating breeding efforts to stack traits of interest [284]. Both Cas9 and Cas12a have unique advantages and disadvantages for genome engineering, as off-target mutations occur at low frequencies with both enzymes [285].

TABLE 9.3   Applications of the CRISPR/Cas13 System in Plant Genome Editing

Genes Targeted	**Virus-Free Plants Generated**	**CRISPR/ Cas13**	**References**
5-Enolpyruvylshikimate-3-phosphate synthase (EPSP), hydroxycinnamoyl-CoA shikimate/quinate hydroxycinnamoyl transferase (HCT), phytoene desaturase (PDS)	*O. sativa*	LwaCas13a	[86]
Turnip mosaic virus genome (TuMV-GFP)	*N. Benthamiana*	LshpCas13a	[263]
TuMV virus	*thaliana*	LshpCas13a	[264]
Plant viruses		LwaCas13a, PspCas13b, CasRx13d	[265]

9.10   CONCLUSION

CRISPR/Cas-mediated GE platforms are new age of robust, accurate, simple, economical, flexible, popular, widely used, and formidable genetic and biotechnological tools for investigating functional genomics and improvement of plant traits. The CRISPR/Cas GE toolbox includes CRISPR/Cas9, CRISPR/Cas12a, CRISPR/Cas13a, CRISPR/Cas13b, CRISPR/Cas13d, and CRISPR/Cas14 and well-developed CRISPR/Cas-derived BEs with distinguishable pros and cons. The off-target effects and HDR inefficiency may be serious prerequisites to acquire desirable traits. The improvements in plant traits can be accomplished through discovery of genes/traits and ingression with GE tools. The "DNA designing" nature of CRISPR technology may guarantee designing of biotic or abiotic stress-resistant plants and improved traits and characteristics of plants (see Tables 9.1–9.3).

KEYWORDS

- genome editing
- zinc finger nucleases
- transcription activator-like effector nucleases
- CRISPR/Cas
- CRISPR/Cas9
- CRISPR/Cas12
- CRISPR/Cas13
- engineered plants

REFERENCES

1. Li, Q., Sapkota, M., and van der Knaap, E. (2020). Perspectives of CRISPR/Cas-mediated cis engineering in horticulture: Unlocking the neglected potential for crop improvement. *Horticulture Res.* 7, 36. https://doi.org/10.1038/s41438-020-0258-8.

2. Smart, R. D., Blum, M., and Wesseler, J. (2017). Trends in approval times for genetically engineered crops in the United States and the European Union. *J. Agric. Econ.* 68, 182–198, doi:10.1111/1477-9552.12171.

3. Callaway, E. (2018). CRISPR plants now subject to tough GM laws in European Union. Nature 560, 16, doi: 10.1038/d41586-018-05814-6.

4. Carroll, D. (2017). Focus: Genome editing: Genome editing: Past, present, and future. *Yale J. Biol. Med.* 90, 653.

5. Bak, R. O., Gomez-Ospina, N., and Porteus, M. H. (2018). Gene editing on center stage. *Trends Genet.* 34(8): 600–611, doi:10.1016/j.tig.2018.05.004.

6. Naso, G. and Petrova, A. (2019). CRISPR/Cas9 gene editing for genodermatoses: Progress and perspectives. *Emerg. Top. Life. Sci.* 3, 313–326.

7. Wallace, R. B., Schold, M., Johnson, M. J., Dembek, P., and Itakura, K. (1981). Oligonucleotide directed mutagenesis of the human β-globin gene: A general method for producing specific point mutations in cloned DNA. *Nucleic Acids Res.* 9, 3647–3656, doi:10.1093/nar/9.15.3647.

8. Sauer, N. J. et al. (2016) Oligonucleotide-directed mutagenesis for precision gene editing. *Plant Biotechnol. J.* 14, 496–502.

9. Carroll, D. (2014). Genome engineering with targetable nucleases. *Annu. Rev. Biochem.* 83, 409–439.

10. Petolino, J. F. (2015) Genome editing in plants via designed zinc finger nucleases. *Vitro Cell Dev. Biol. Plant* 51, 1–8.

11. Barrangou, R. and Doudna, J. A. (2016). Applications of CRISPR technologies in research and beyond. *Nat. Biotechnol.* 34(9): 933–941.

12. Metje-Sprink, J., Menz, J., Modrzejewski, D., and Sprink, T. (2019). DNA-free genome editing: Past, present and future. *Front. Plant Sci.* 9: 1957, doi: 10.3389/fpls.2018.01957.

13. Manghwar, H., Li, B., Ding, X., Hussain, A., Lindsey, K., Zhang, X., and Jin, S. (2020). CRISPR/Cas systems in genome editing: Methodologies and tools for sgRNA design, off-target evaluation, and strategies to mitigate off-target effects. *Adv. Sci.* 7, 1902312.

14. Bibikova, M., Golic, M., Golic, K. G., and Carroll, D. (2002). Targeted chromosomal cleavage and mutagenesis in Drosophila using zinc-finger nucleases. *Genetics* 161, 1169–1175.

15. Rai, K. M., Ghose, K., Rai, A., Singh, H., Srivastava, and R., Mendu, V. (2019). Genome engineering tools in plant synthetic biology. In *Current Developments in Biotechnology and Bioengineering*; Elsevier: Amsterdam, The Netherlands, pp. 47–73.

16. Bibikova, M., Beumer, K., Trautman, J. K., and Carroll, D. (2003). Enhancing gene targeting with designed zinc finger nucleases. *Science* 300, 764.

17. Kim, Y. G. Cha, J., and Chandrasegaran, S. (1996). Hybrid restriction enzymes: Zinc finger fusions to Fok I cleavage domain. *Proc. Natl. Acad. Sci. USA* 93, 1156–1160.

18. Lloyd, A., Plaisier, C. L., Carroll, D., and Drews, G. N. (2005). Targeted mutagenesis using zinc-finger nucleases in Arabidopsis. *Proc. Natl. Acad. Sci. USA* 102, 2232–2237.

19. Osakabe, K., Osakabe, Y., and Toki, S. (2010). Site-directed mutagenesis in Arabidopsis using custom-designed zinc finger nucleases. *Proc. Natl. Acad. Sci. USA* 107, 12034–12039. 15.

20. Gallego-Bartolome et al. (2019). Co-targeting RNA polymerases IV and V promotes efficient de novo DNA methylation in Arabidopsis. *Cell.* 176, 1068–1082, doi: 10.1016/j.cell.2019.01.029.

21. Wright, D. A., Townsend, J. A., Winfrey, R. J., Jr., Irwin, P. A., Rajagopal, J., Lonosky, P. M., Hall, B. D., Jondle, M. D., and Voytas, D. F. (2005). High-frequency homologous recombination in plants mediated by zinc-finger nucleases. *Plant J.* 44, 693–705.

22. Townsend, J. A., Wright, D. A., Winfrey, R. J., Fu, F., Maeder, M. L., Joung, J. K., and Voytas, D. F. (2009). High-frequency modification of plant genes using engineered zinc-finger nucleases. *Nature* 459, 442.

23. Petolino, J. F., Worden, A., Curlee, K., Connell, J., Moynahan, T. L. S., Larsen, C., and Russell, S. (2010). Zinc finger nuclease-mediated transgene deletion. *Plant Mol. Biol.* 73, 617–628.

24. Shukla, V. K., Doyon, Y., Miller, J. C., DeKelver, R. C., Moehle, E. A., Worden, S. E., Mitchell, J. C., Arnold, N. L., Gopalan, S., and Meng, X. (2009). Precise genome modification in the crop species *Zea mays* using zinc-finger nucleases. *Nature* 459, 437–441.

25. Ainley, W. M et al. (2013). Trait stacking via targeted genome editing. *Plant Biotechnol. J.* 11, 1126–1134, doi: 10.1111/pbi.12107.

26. Carroll, D. (2011). Genome engineering with zinc-finger nucleases. *Genetics* 188, 773–782.

27. Sander, J. D., Dahlborg, E. J., Goodwin, M. J., Cade, L., Zhang, F., Cifuentes, D., Curtin, S. J., Blackburn, J. S., Thibodeau-Begany, S., Qi, Y., Pierick, C. J., Hoffman, E., Maeder, M. L., Khayter, C., Reyon, D., Dobbs, D., Langenau, D. M., Stupar, R. M., Giraldez, A. J., Voytas, D. F., Peterson, R. T., Yeh, J. R., and Joung, J. K. (2011). Selection-free zinc-finger-nuclease engineering by context-dependent assembly (CoDA). *Nat. Methods* 8, 67–69.

28. Reyon, D., Kirkpatrick, J. R., Sander, J. D., Zhang, F., Voytas, D. F., Joung, J. K., Dobbs, D., and Coffman, C. R. (2011). ZFNGenome: A comprehensive resource for locating zinc finger nuclease target sites in model organisms. *BMC Genomics* 12, 83.

29. Hafez, M. and Hausner. G. (2012). Homing endonucleases: DNA scissors on a mission. *Genome* 55: 553–569.

30. Bogdanove, A. J. and Voytas, D. F. (2011). TAL effectors: Customizable proteins for DNA targeting. *Science* 333, 1843–1846.

31. Doyle, E. L., Booher, N. J., Standage, D. S., Voytas, D. F., Brendel, V. P., Vandyk, J. K., and Bogdanove. A. J. (2012). TAL effector-nucleotide targeter (TALE-NT) 2.0: Tools for TAL effector design and target prediction. *Nucl. Acids Res.* 40, W117–W122.

32. Chen, S., Oikonomou, G., Chiu, C. N., Niles, B. J., Liu, J., Lee, D. A., Antoshechkin, I., and Prober, D. A. (2013). A large-scale in vivo analysis reveals that TALENs are significantly more mutagenic than ZFNs generated using context-dependent assembly. *Nucl. Acids Res.* 41, 2769–2778.

33. Juillerat, A., Dubois, G., Valton, J., Thomas, S., Stella, S., Marechal, A., Langevin, S., Benomari, N., Bertonati, C., Silva, G. H., Daboussi, F., Epinat, J. C., Montoya, G., Duclert, A., and Duchateau, P. (2014). Comprehensive analysis of the specificity of transcription activator-like effector nucleases. *Nucl. Acids Res.* 42, 5390–5402.

34. Boch, J., Scholze, H., Schornack, S., Landgraf, A., Hahn, S., Kay, S., Lahaye, T., Nickstadt, A., and Bonas, U. (2009). Breaking the code of DNA binding specificity of TAL-type III effectors. *Science* 326, 1509–1512.

35. Christian, M., Cermak, T., Doyle, E. L., Schmidt, C., Zhang, F., Hummel, A., Bogdanove, A. J., and Voytas, D. F. (2010). Targeting DNA double-strand breaks with TAL effector nucleases. *Genetics* 186, 757–761.

36. Shan, Q., Wang, Y., Chen, K., Liang, Z., Li, J., Zhang, Y., Zhang, K., Liu, J., Voytas, D. F., and Zheng, X. (2013). Rapid and efficient gene modification in rice and Brachypodium using TALENs. *Mol. Plant* 6, 1365–1368.

37. Li, T., Liu, B., Spalding, M. H., Weeks, D. P., and Yang, B. (2012) High-efficiency TALEN-based gene editing produces disease-resistant rice. *Nat. Biotechnol.* 30, 390.

38. Kopischke, S., Schüßler, E., Althoff, F., and Zachgo, S. (2017). TALEN-mediated genome-editing approaches in the liverwort *Marchantia polymorpha* yield high efficiencies for targeted mutagenesis. *Plant Methods* 13, 20.

39. Christian, M., Qi, Y., Zhang, Y., and Voytas, D. F. (2013). Targeted mutagenesis of Arabidopsis thaliana using engineered TAL effector nucleases. *G3. (Bethesda)* 3, 1697–1705.

40. Budhagatapalli, N., Rutten, T., Gurushidze, M., Kumlehn, J., and Hensel, G. (2015). Targeted modification of gene function exploiting homology-directed repair of TALEN-mediated double-strand breaks in barley. *G3. (Bethesda)* 5, 1857–1863.

41. Char, S. N. et al. (2015). Heritable site-specific mutagenesis using TALENs in maize. *Plant Biotechnol. J.* 13, 1002–1010.

42. Clasen, B. M. et al. (2016). Improving cold storage and processing traits in potato through targeted gene knockout. *Plant Biotechnol. J.* 14, 169–176.

43. Lor, V. S., Starker, C. G., Voytas, D. F., Weiss, D., and Olszewski, N. E. (2014). Targeted mutagenesis of the tomato PROCERA; Gene using transcription activator-like effector nucleases. *Plant Physiol.* 166, 1228, doi: 10.1126/science.aad5725.

44. D., H., Zeng et al. (2016). Efficient targeted mutagenesis in soybean by TALENs and CRISPR/Cas9. *J. Biotechnol.* 217, 90–97, doi: 10.1016/j.jbiotec.2015.11.005.

45. Liang, Z., Zhang, K., Chen, K., and Gao, C. (2014). Targeted mutagenesis in Zea mays using TALENs and the CRISPR/Cas system. *J. Genet. Genom.* 41, 63 –68.

46. Doudna, J. A. and Charpentier, E. (2014). The new frontier of genome engineering with CRISPR-Cas9. *Science* 346, 1258096. https://doi.org/10.1126/science.

47. Crawley, A. B., Henriksen, J. R., and Barrangou, R. (2018). CRISPRdisco: An automated pipeline for the discovery and analysis of CRISPR-Cas systems. *CRISPR J.* 1, 171–181.

48. Ledford, H. (2015). CRISPR, the disruptor. *Nat. News* 522, 20.

49. Wang, M., Wang, S., Liang, Z., Shi, W., Gao, C., and Xia, G. (2017). From genetic stock to genome editing: Gene exploitation in wheat. *Trends Biotechnol.* 36, 160–172, doi: 10.1016/j.tibtech.2017.10.002.

50. Monsur, M. B., Shao, G., Lv, Y., Ahmad, S., Wei, X., Hu, P., and Tang, S. (2020). Base editing: The ever expanding clustered regularly interspaced short palindromic repeats (CRISPR) tool kit for precise genome editing in plants. *Genes* 11, 466, doi: 10.3390/genes11040466.

51. Burmistrz, M., Krakowski, K., and Krawczyk-Balska, A. (2020). RNA-targeting CRISPR–CAS systems and their applications. *Int. J. Mol. Sci.* 21, 1122, doi: 10.3390/ijms21031122.

52. Komor, A. C., Kim, Y. B., Packer, M. S., Zuris, J. A., and Liu, D. R. (2016). Programmable editing of a target base in genomic DNA without double-stranded DNA cleavage. *Nature* 533, 420–424.

53. Safari, F., Zare, K., Negahdaripour, M., Barekati-Mowahed, M., and Ghasemi, Y. (2019) CRISPR Cpf1 proteins: Structure, function and implications for genome editing. *Cell Biosci.* 9(36): 1–21, doi: 10.1186/s13578-019-0298-7.

54. Li, S., Zhang et al. (2018). Expanding the scope of CRISPR/Cpf1—Mediated genome editing in rice. *Mol. Plant.* 11, 995–998, doi: 10.1016/j.molp.2018.03.009.

55. Ishino, Y., Shinagawa, H., Makino, K., Amemura, M., and Nakata, A. (1987). Nucleotide sequence of the Iap gene, responsible for alkaline phosphatase isozyme conversion in Escherichia coli, and identification of the gene product. *J. Bacteriol.* 169, 5429–5433. https://doi.org/10.1128/jb.169.12. 5429-5433.1987

56. Makarova, K. S et al. (2015). An updated evolutionary classification of CRISPR-Cas systems. *Nat. Reviews Microbiol.* 13, 722–736.

57. Makarova, K. S., Grishin, N. V., Shabalina, S. A., Wolf, Y. I., and Koonin, E. V. (2006). A putative RNA-interference-based immune system in prokaryotes: Computational analysis of the predicted enzymatic machinery, functional analogies with eukaryotic RNAi, and hypothetical mechanisms of action. *Biol. Direct* 1, 7.

58. Shmakov, S. et al. (2015). Discovery and functional characterization of diverse class 2 CRISPR-Cas systems. *Mol. Cell* 60 (3): 385–397. https://doi.org/10.1016/j.molcel.2015.10.008

59. Sovov, T., Kerins, G., Demnerov K., and Ovesn, J. (2016). Genome editing with engineered nucleases in economically important animals and plants: State of the art in the research pipeline. *Curr. Issues Mol. Biol.* 21, 41 –62.

60. Jackson, R. N., van Erp, P. B., Sternberg, S. H., and Wiedenheft, B. (2017). Conformational regulation of CRISPR-associated nucleases. *Curr. Opin. Microbiol.* 37, 110–119.

61. Hille, F., Richter, H., Wong, S. P., Bratovic, M., Ressel, S., and Charpentier, E. (2018). The biology of CRISPR-Cas: Backward and forward. *Cell* 172, 1239–1259.

62. Albitar, A., Rohani, B., Will, B., Yan, A., and Gallicano, G. I. (2018). The application of CRISPR/Cas technology to efficiently model complex cancer genomes in stem cells. *J. Cell. Biochem.* 119, 134–140.

63. Mojica, F. J., García-Martínez, J., and Soria, E. (2005). Intervening sequences of regularly spaced prokaryotic repeats derive from foreign genetic elements. *J. Mol. Evol.* 60, 174–182.

64. Bolotin, A., Quinquis, B., Sorokin, A., and Ehrlich, S. D. (2005). Clustered regularly interspaced short palindrome repeats (CRISPRs) have spacers of extrachromosomal origin. *Microbiology* 151, 2551–2561.

65. Barrangou, R., Fremaux, C., Deveau, H., Richards, M., Boyaval, P., Moineau, S., Romero, D. A., and Horvath, P. (2007). CRISPR provides acquired resistance against viruses in prokaryotes. *Science* 315, 1709–1712.

66. Jinek, M., Chylinski, K., Fonfara, I., Hauer, M., Doudna, J. A., and Charpentier, E. J. S. (2012). A programmable dual-RNA-guided DNA endonuclease in adaptive bacterial immunity. *PLoS One*, 337, 816–821.

67. Jinek, M., East, A., Cheng, A., Lin, S., Ma, and E., Doudna, J. (2013). RNA-programmed genome editing in human cells. *Elife* 2, e00471.

68. Cho, S. W., Kim, S., Kim, J. M., and Kim, J. S. (2013). Targeted genome engineering in human cells with the Cas9 RNA-guided endonuclease. *Nat. Biotechnol.* 31, 230.

69. Hwang, W. Y., Fu, Y., Reyon, D., Maeder, M. L., Tsai, S. Q., Sander, J. D., Peterson, R. T., Yeh, J. J., and Joung, J. K. (2013). Efficient genome editing in zebrafish using a CRISPR-Cas system. *Nat. Biotechnol.* 31, 227.

70. Mali, P., Yang, L., Esvelt, K. M., Aach, J., Guell, M., DiCarlo, J. E., Norville, J. E., and Church, G. M. (2013). RNA-guided human genome engineering via Cas9. *Science* 339, 823–826.

71. Li, J.-F., Norville, J. E., Aach, J., McCormack, M., Zhang, D., Bush, J., Church, G. M., and Sheen, J. (2013). Multiplex and homologous recombination-mediated genome editing in *Arabidopsis* and *Nicotiana benthamiana* using guide RNA and Cas9. *Nat. Biotechnol.* 31, 688–691.

72. Nekrasov, V., Staskawicz, B., Weigel, D., Jones, J. D., and Kamoun, S. (2013). Targeted mutagenesis in the model plant *Nicotiana benthamiana* using Cas9 RNA-guided endonuclease. *Nat. Biotechnol.* 31, 691–693.

73. Shan, Q., Wang, Y., Li, J., Zhang, Y., Chen, K., Liang, Z., Zhang, K., Liu, J., Xi, J. J., and Qiu, J. (2013). Targeted genome modification of crop plants using a CRISPR-Cas system. *Nat. Biotechnol.* 31, 686–688.

74. Shmakov, S. et al. (2017). Diversity and evolution of class 2 CRISPR-Cas systems. *Nat. Rev. Microbiol.* 15(3): 169–182. https://doi.org/10.1038/nrmicro.2016.184

75. Makarova, K. S., Wolf, Y. I., and Koonin, E. V. (2018). Classification and nomenclature of CRISPR-Cas systems: Where from here? *CRISPR J.* 1, 325–336.

76. Ishino, Y., Krupovic, M., and Forterre, P. (2018). History of CRISPR-Cas from encounter with a mysterious repeated sequence to genome editing technology. *J. Bacteriol.* 200, e00580-17.

77. Koonin, E. V., Makarova, K. S., and Zhang, F. (2017). Diversity, classification and evolution of CRISPR-Cas systems. *Curr. Opin. Microbiol.* 37, 67–78.

78. Gasiunas, G., Barrangou, R., Horvath, P., and Siksnys, V. (2012). Cas9–crRNA ribonucleoprotein complex mediates specific DNA cleavage for adaptive immunity in bacteria. *Proc. Natl. Acad. Sci.* 109, E2579–E2586.

79. Samai, P., Pyenson, N., Jiang, W., Goldberg, G. W., Hatoum-Aslan, A., and Marraffini, L. A. (2015). Co-transcriptional DNA and RNA cleavage during type III CRISPR-Cas immunity. *Cell*, 161, 1164–1174.

80. Leenay, R. T., Maksimchuk, K. R., Slotkowski, R. A., Agrawal, R.N., Gomaa, A. A., Briner, A. E., Barrangou, R., and Beisel, C. L. (2016). Identifying and visualizing functional PAM diversity across CRISPR-Cas systems. *Mol. Cell* 62, 137–147.

81. Elmore, J. R., Sheppard, N. F., Ramia, N., Deighan, T., Li, H., Terns, R. M., and Terns, M. P. (2016). Bipartite recognition of target RNAs activates DNA cleavage by the Type III-B CRISPR-Cas system. *Genes Develop.* 30, 447–459.

82. Gootenberg, J. S. et al. (2017). Nucleic acid detection with CRISPR-Cas13a/C2c2. *Science* 356, 438–442. https://doi.org/10.1126 /science.aam9321

83. Pyenson, N. C., Gayvert, K., Varble, A., Elemento, O., and Marraffini, L. A. (2017). Broad targeting specificity during bacterial type III CRISPR-Cas immunity constrains viral escape. *Cell Host Microbe* 22, 343–353.

84. Yan, W. X., Chong, S., Zhang, H., Makarova, K. S., Koonin, E. V., Cheng, D. R., and Scott, D. A. (2018). Cas13d is a compact RNA-targeting type VI CRISPR effector positively modulated by a WYL-domain containing accessory protein. *Mol. Cell* 70, 327–339.

85. Abudayyeh, O. O. et al. (2016). C2c2 is a single-component programmable RNA-guided RNA-targeting CRISPR effector. *Science* 353, aaf5573.

86. Abudayyeh, O. O. et al. (2017). RNA targeting with CRISPR-Cas13. *Nature* 550(7675): 280–284. https://doi.org/10.1038/nature24049

87. Smargon, A. A. et al. (2017). Cas13b is a type VI-B CRISPR-associated RNA-guided RNase differentially regulated by accessory proteins Csx27 and Csx28. *Mol. Cell* 65, 618–630.

88. Wang, F. et al. (2019). Advances in CRISPR-Cas systems for RNA targeting, tracking and editing. *Biotechnol. Adv.* 37, 708–729.

89. Sternberg, S. H., Richter, H., Charpentier, E., and Qimron, U. (2016). Adaptation in CRISPR-Cas systems. *Mol. Cell* 61, 797–808.

90. Jiang, W. Z., Henry, I. M., Lynagh, P. G., Comai, L., Cahoon, E. B., and Weeks, D. P. (2017). Significant enhancement of fatty acid composition in seeds of the allohexaploid, Camelina sativa, using CRISPR/Cas9 gene editing. *Plant Biotechnol. J.* 15, 648–657. https://doi.org/10.1111/pbi.12663

91. Terns, M. P. (2018). CRISPR-based technologies: Impact of RNA-targeting systems. *Mol. Cell* 72(3): 404–412, doi: 10.1016/j.molcel.2018.09.018.

92. Hille, F. and Charpentier, E. (2016). CRISPR-Cas: Biology, mechanisms and relevance. *Phil. Trans. R. Soc. B.* 371, 20150496.

93. Wang, C., Zhai, X., Zhang, X., Li, L., Wang, J., and Liu, D. P. (2019). Gene-edited babies: Chinese Academy of Medical Sciences' response and action. *Lancet*, 393, 25–26.

94. Gleditzsch, D. Pausch, P., Müller-Esparza, H., Özcan, A Guo, X., Bange, G., and Randau, L. (2019). PAM identification by CRISPR-Cas effector complexes: Diversified mechanisms and structures. *RNA Biol.* 16: 4, 504–517, doi: 10.1080/15476286.2018.1504546.

95. Zetsche, B., Gootenberg, J. S., Abudayyeh, O. O., Slaymaker, I. M., Makarova, K. S., Essletzbichler, P., Volz, S. E., Joung, J., Van Der Oost, J., and Regev, A. (2015). Cpf1 is a single RNA-guided endonuclease of a class 2 CRISPR-Cas system. *Cell* 163, 759–771, doi: 10.1016/j.cell.2015.09.038.

96. Yamano, T. et al. (2016). Crystal structure of Cpf1 in complex with guide RNA and target DNA. *Cell* 165, 949–962. https://doi. org/10.1016/J.CELL.2016.04.003

97. Dong, D. et al. (2016). The crystal structure of Cpf1 in complex with CRISPR RNA. *Nature* 532, 522–526, doi: 10.1038/nature17944.

98. Swarts, D. C., and Jinek, M. (2018). Cas9 versus Cas12a/Cpf1: Structure–function comparisons and implications for genome editing. *Wiley Interdiscip. Rev. RNA* 9, 1–19, doi: 10.1002/wrna.1481.

99. Paul, J. W. and Qi, Y. (2016). CRISPR/Cas9 for plant genome editing: Accomplishments, problems and prospects. *Plant Cell Rep.* 35, 1417–1427, doi: 10.1007/s00299-016-1985-z.

100. Gao, L. et al. (2017). Engineered Cpf1 variants with altered PAM specificities increase genome targeting range. *Nat. Biotechnol.* 35, 789–792, doi: 10.1038/nbt.3900.

101. Vanegas, K. G., Jarczynska, Z. D., Strucko, T., and Mortensen, U. H. (2019). Cpf1 enables fast and efficient genome editing in Aspergilli. *Fungal Biol. Biotechnol.* 6, 6.

102. Mahfouz, M. M. (2017) Genome editing: The efficient tool CRISPR–Cpf1. *Nat. Plants*, 3, 17028.

103. Tang, X., Lowder et al. (2017). A CRISPR Cpf1 system for efficient genome editing and transcriptional repression in plants. *Nat. Plants* 3, 1–5, doi: 10.1038/nplants.2017.18.

104. Alok, A. et al. (2020). The rise of the CRISPR/Cpf1 system for efficient genome editing in plants. *Front. Plant Sci.* 11, 264, doi: 10.3389/fpls.2020.00264.

105. Tóth, E. et al. (2020). Improved LbCas12a variants with altered PAM specificities further broaden the genome targeting range of Cas12a nucleases. *Nucl. Acids Res.* 48, 3722–3733, doi: 10.1093/nar/gkaa110.

106. Kleinstiver, B. P. et al. (2016). Genome-wide specificities of CRISPR-Cas Cpf1 nucleases in human cells. *Nat. Biotechnol.* 34, 869–874.

107. Chen, P. et al. (2020). A Cas12a ortholog with stringent PAM recognition followed by low off-target editing rates for genome editing. *Genome Biol.* 21, 1–13, doi: 10.1186/s13059-020-01989-2.

108. Waryah, C. B., Moses, C., Arooj, M., and Blancafort, P. (2018). Zinc fingers, TALEs, and CRISPR systems: A comparison of tools for epigenome editing. In *Epigenome Editing*, (Jeltsch, A. and Rots, M. G., Eds.), New York, NY, USA: Humana Press, pp. 19–63.

109. Palmgren, M. G. et al. (2015). Are we ready for back-to-nature crop breeding? *Trends Plant Sci.* 20, 155–164.

110. Weigel, D. (2017). Ethische Fragen spielen keine Rolle—Einsatz Chancen und Risiken von CRISPR/Cas bei Pflanzen. Forschung & Lehre. https://www.forsc hung-und-lehre. de/ethische-fragen-spielen-keine-rolle-1292/(27 June 2019, last assessed).

111. Globus, R. and Qimron, U. (2018). A technological and regulatory outlook on CRISPR crop editing. *J. Cell. Biochem.* 119, 1291–1298.

112. Steinbrecher, R. A. (2015). Genetic engineering in plants and the "New Breeding Techniques (NBTs)." Inherent risks and the need to regulate. Econexus Briefing, pp. 1–8.

113. Bartkowski, B., Theesfeld, I., Pirscher, F., and Timaeus, J. (2018). Snipping around for food: Economic, ethical and policy implications of CRISPR/Cas genome editing. *Geoforum*, 96, 172–180.

114. Li, G., Liu, Y. G., and Chen, Y. (2019). Genome-editing technologies: The gap between application and policy. *Sci. China Life Sci.* 62 (11): 1534–1538.

115. Schulman, A. H., Oksman-Caldentey, K. M., and Teeri, T. H. (2019). European Court of Justice delivers no justice to Europe on genome-edited crops. *Plant Biotechnol. J.* 18, 8–10.

116. Es, I., Gavahian, et al. (2019). The application of the CRISPRCas9 genome editing machinery in food and agricultural science: Current status, future perspectives, and associated challenges. *Biotechnol. Adv.* 37, 410–421.

117. Feng, Z. et al. (2013). Efficient genome editing in plants using a CRISPR/Cas system. *Cell Res.* 23, 1229–1232.

118. Adli, M. (2018). The CRISPR tool kit for genome editing and beyond. *Nat. Commun.* 9, 1911.

119. Wang, Y., Cheng, X., Shan, Q., Zhang, Y., Liu, J., Gao, C., and Qiu, J. L. (2014). Simultaneous editing of three homoeoalleles in hexaploid bread wheat confers heritable resistance to powdery mildew. *Nat. Biotechnol.* 32, 947–951.

120. Li, C., Li, W., Zhou, Z., Chen, H., Xie, C., and Lin, Y. (2020). A new rice breeding method: CRISPR/Cas9 system editing of the Xa13 promoter to cultivate transgene-free bacterial blight-resistant rice. *Plant Biotechnol. J.* 18, 313–315.

121. Miao, J. et al. (2013). Targeted mutagenesis in rice using CRISPR-Cas system. *Cell Res.* 23, 1233–1236.

122. Bai, M. et al. (2020). Generation of a multiplex mutagenesis population via pooled CRISPR-Cas9 in soya bean. *Plant Biotechnol. J.* 18, 721–731.

123. Chilcoat, D., Liu, Z.-B., and Sander, J. (2017). Use of CRISPR/Cas9 for crop improvement in maize and soybean. In *Progress in Molecular Biology and Translational Science*, vol. 149, (Teplow, D. B., ed), Cambridge, MA, USA: Academic Press, pp. 27–46.

124. Jia, H. and Wang, N. (2014). Targeted genome editing of sweet orange using Cas9/sgRNA. *PLoS One*, 9, e93806.

125. Brooks, C., Nekrasov, V., Lippman, Z. B., and Van Eck, J. (2014). Efficient gene editing in tomato in the first generation using the clustered regularly interspaced short palindromic repeats/CRISPR-associated9 system. *Plant Physiol.* 166, 1292–1297.

126. Zheng, M. et al. (2020). Knockout of two Bna MAX 1 homologs by CRISPR/Cas9-targeted mutagenesis improves plant architecture and increases yield in rapeseed (*Brassica napus* L.). *Plant Biotechnol. J.* 18, 644–654.

127. Zhang, Y., Pribil, M., Palmgren, M., and Gao, C. (2020). A CRISPR way for accelerating improvement of food crops. *Nat. Food* 1, 200–205, doi: 10.1038/s43016-020-0051-8.

128. Yi, P. and Goshima, G. (2019). Transient cotransformation of CRISPR/Cas9 and oligonucleotide templates enables efficient editing of target loci in Physcomitrella patens. *Plant Biotechnol. J.* 18, 599–601.

129. Papikian, A., Liu, W., Gallego-Bartolomé, J., and Jacobsen, S. (2019). Site-specific manipulation of Arabidopsis loci using CRISPR-Cas9 SunTag systems. *Nat. Commun.* 10, 729.

130. Qin, L., Li, J., Wang, Q., Xu, Z., Sun, L., Alariqi, M., Manghwar, H., Wang, G., Li, B., and Ding, X. (2020). High efficient and precise base editing of C•G to T•A in the allotetraploid cotton (*Gossypium hirsutum*) genome using a modified CRISPR/Cas9 system. *Plant Biotechnol. J.* 18, 45–56.

131. Shimatani, Z. et al. (2017). Targeted base editing in rice and tomato using a CRISPR-Cas9 cytidine deaminase fusion. *Nat. Biotechnol.* 35, 441–443, doi: 10.1038/nbt.3833.

132. Tieman, D., Zhu, G., Resende, M. F., Lin, T., Nguyen, C., Bies, D., Rambla, J. L., Beltran, K. S. O., Taylor, M., and Zhang, B. J. S. (2017). A chemical genetic roadmap to improved tomato flavor. *Science* 355, 391–394.

133. Ren, B. et al. (2018). Improved base editor for efficiently inducing genetic variations in rice with CRISPR/Cas9-guided hyperactive hAID mutant. *Mol. Plant* 11, 623–626.

134. Zong, Y., Wang, Y., Li, C., Zhang, R., Chen, K., Ran, Y., Qiu, J. L., Wang, D., and Gao, C. (2017). Precise base editing in rice, wheat and maize with a Cas9-cytidine deaminase fusion. *Nat. Biotechnol.* 35, 438–440.

135. Veillet, F., Perrot, L., Chauvin, L., Kermarrec, M.-P., Guyon-Debast, A., Chauvin, J.-E., Nogué, F., and Mazier, M. (2019). Transgene-free genome editing in tomato and potato plants using agrobacterium-mediated delivery of a CRISPR/Cas9 cytidine base editor. *Int. J. Mol. Sci.* 20, 402.

136. Lu, Y. and Zhu, J.-K. (2017). Precise editing of a target base in the rice genome using a modified CRISPR/Cas9 system. *Mol. Plant* 10, 523–525.
137. Sinkunas, T., Gasiunas, G., Fremaux, C., Barrangou, R., Horvath, P., and Siksnys, V. (2011). Cas3 is a single-stranded DNA nuclease and ATP-dependent helicase in the CRISPR/Cas immune system. *EMBO J.*, 30(7): 1335–1342. https://doi.org/10.1038/emboj.2011.41
138. Xie, K., Zhang, J., and Yang, Y. (2014). Genome-wide prediction of highly specific guide RNA spacers for CRISPR–Cas9-mediated genome editing in model plants and major crops. *Mol. Plant* 7, 923–926.
139. Hsu, P. D., Lander, E. S., and Zhang, F. (2014). Development and applications of CRISPR-Cas9 for genome engineering. *Cell* 157, 1262–1278. https://doi.org/10.1016/j.cell.2014.05.010
140. Barrangou, R. (2013). CRISPR-Cas systems and RNA-guided interference. *Wiley Interdiscip. Rev. RNA* 4, 267–278. https://doi.org/10.1002/wrna.1159
141. Jansen, R., van Embden, J. D. A., Gaastra, W., and Schouls, L. M. (2002). Identification of genes that are associated with DNA repeats in prokaryotes. *Mol. Microbiol.* 43, 1565–1575. https://doi.org/10.1046/j.1365-2958.2002.02839.x
142. Mojica, F. J. M., Diez-Villasenor, C., Soria, E., and Juez, G. (2000). Biological significance of a family of regularly spaced repeats in the genomes of Archaea, Bacteria and mitochondria. *Mol. Microbiol.* 36, 244–246. https://doi.org/10.1046/j.1365-2958.2000.01838.x
143. Bhaya, D., Davison, and M., Barrangou, R. (2011). CRISPR-Cas systems in bacteria and archaea: Versatile small RNAs for adaptive defense and regulation. *Annu. Rev. Genet.* 45, 273–297.
144. Horvath, P. and Barrangou, R. (2010). CRISPR/Cas, the immune system of bacteria and archaea. *Science* 327(5962): 167–170.
145. Cong, L. et al. (2013). Multiplex genome engineering using CRISPR/Cas systems. *Science* 339(6121): 819–823, doi: 10.1126/science.1231143.
146. Braff, J. L., Yaung, S. J., Esvelt, K. M., and Church, G. M. (2016). Characterization of Cas9–guide RNA orthologs. *Cold Spring Harb. Protoc.* 2016. https://doi. org/10.1101/pdb.top086793
147. Chneiweiss, H. et al. (2017). Fostering responsible research with genome editing technologies: A European perspective. *Transgenic Res.* 26, 709–713.
148. Roy, B. et al. (2018). CRISPR/Cascade 9-mediated genome editing-challenges and opportunities. *Front. Genet.* 9, 240.
149. Sternberg, S. H. and Doudna, J. A. (2015). Expanding the biologist's toolkit with CRISPR-Cas9. *Mol. Cell* 58, 568–574. https://doi.org/10.1016/j.molcel. 2015.02.032
150. Stuiver, M. H. and Custers, J. H. (2001). Engineering disease resistance in plants. *Nature* 411, 865–868. https://doi.org/10.1038/35081200
151. Wally, O. and Punja, Z. K. (2010). Genetic engineering for increasing fungal and bacterial disease resistance in crop plants. *GM Crops* 1, 199–206. https://doi.org/10.4161/gmcr.1.4.13225
152. Butelli, E., Titta, L., Giorgio, M., Mock, H.-P., Matros, A., Peterek, S., Schijlen, E. G. W. M., Hall, R. D., Bovy, A. G., Luo, J., and Martin, C. (2008). Enrichment of tomato fruit with health-promoting anthocyanins by expression of select transcription factors. *Nat. Biotechnol.* 26, 1301–1308. https://doi.org/10.1038/nbt.1506
153. Zorrilla-Lo´pez, U., Masip, G., Arjo, ´G., Bai, C., Banakar, R., Bassie, L., Berman, J., Farre, ´G., Miralpeix, B., Pe´rez-Massot, E., Sabalza, M., Sanahuja, G., Vamvaka, E., Twyman, R. M., Christou, P., Zhu, C., and Capell, T. (2013). Engineering metabolic

pathways in plants by multigene transformation. *Int. J. Dev. Biol.* 57, 565–576. https://doi.org/10.1387/ijdb.130162pc

154. Moradpour, M. and Abdulah, S. N. A. (2019). CRISPR/dCas9 platforms in plants: Strategies and applications beyond genome editing. *Plant Biotechnol. J.*, 18, 32–44. https://doi.org/10.1111/pbi.13232

155. Zhou, X., Jacobs, T. B., Xue, L. J., Harding, S. A., and Tsai, C. J. (2015). Exploiting SNPs for biallelic CRISPR mutations in the outcrossing woody perennial Populus reveals 4-coumarate: CoA ligase specificity and redundancy. *New Phytol.* 208, 298–301. http://dx.doi.org/10.1111/nph.13470

156. Malzahn, A., Lowder, L. and Qi, Y. (2017). Plant genome editing with TALEN and CRISPR. *Cell Biosci.* 7, 21. https://doi.org/10.1186/s13578-017-0148-4

157. Foster, K., Kalter, J., Woodside, W., Terns, R. M., and Terns, M. P. (2019). The ribonuclease activity of Csm6 is required for anti-plasmid immunity by Type III-A CRISPR-Cas systems. *RNA Biol.* 16, 449–460.

158. Andersson, M., Turesson, H., Nicolia, A., Falt, A. S., Samuelsson, M., and Hofvander, P. (2017). Efficient targeted multiallelic mutagenesis in tetraploid potato (Solanum tuberosum) by transient CRISPR-Cas9 expression in protoplasts. *Plant Cell Rep.* 36, 117–128.

159. Li, R. et al. (2018). Multiplexed CRISPR/Cas9-mediated metabolic engineering of c-aminobutyric acid levels in Solanum lycopersicum. *Plant Biotechnol. J.* 16, 415–427.

160. Jouanin, A., Gilissen, L. J. W. J., Schaart, J. G., Leigh, F. J., Cockram, J., Wallington, E. J., Boyd, L. A., van den Broeck, H. C., van der Meer, I. M., America, A. H. P., Visser, R. G. F., and Smulders, M. J. M. (2020). CRISPR/Cas9 gene editing of gluten in wheat to reduce gluten content and exposure—Reviewing methods to screen for coeliac safety. *Front. Nutr.* 7, 51, doi: 10.3389/fnut.2020.00051.

161. Liu, H. J. et al. (2020). High-throughput CRISPR/Cas9 mutagenesis streamlines trait gene identification in maize. *Plant Cell* 32, 1397–1413.

162. Zafar, K., Khan, M. Z., Amin, I., Mukhtar, Z., Yasmin, S., Arif, M., Ejaz, K., and Mansoor, S. (2020). Precise CRISPR-Cas9 mediated genome editing in super basmati rice for resistance against bacterial blight by targeting the major susceptibility gene. *Front. Plant Sci.* 11, 575, doi: 10.3389/fpls.2020.00575.

163. Shan, Q., Zhang, Y., Chen, K., Zhang, K., and Gao, C. (2015). Creation of fragrant rice by targeted knockout of the Os BADH 2 gene using TALEN technology. *Plant Biotechnol. J.* 13, 791–800, doi: 10.1111/pbi.12312.

164. Shao, G. et al. (2017). CAS9-mediated editing of the fragrant gene Badh2 in rice. *Chin. J. Rice Sci.* 31, 216–222, doi: 10.16819/j.1001-7216.2017.6098.

165. Kui, L. et al. (2017). Building a genetic manipulation tool box for orchid biology: Identification of constitutive promoters and application of CRISPR/Cas9 in the orchid, Dendrobium officinale. *Front. Plant Sci.* 7, 2036. https://doi.org/10.3389/fpls.2016.02036

166. Macovei, A. et al. (2018). Novel alleles of rice eIF4G generated by CRISPR/Cas9-targeted mutagenesis confer resistance to rice tungro spherical virus. *Plant Biotechnol. J.* 16, 1918–1927, doi: 10.1111/pbi.12927.

167. Kapusi, E., Corcuera-Gómez, M., Melnik, S., and Stoger, E. (2017). Heritable genomic fragment deletions and small indels in the putative ENGase gene induced by CRISPR/Cas9 in barley. *Front. Plant Sci.* 8, 540. https://doi.org/10.3389/fpls.2017.00540

168. Yin, X. et al. (2017). CRISPR-Cas9 and CRISPR-Cpf1 mediated targeting of a stomatal developmental gene EPFL9 in rice. *Plant Cell Rep.* 36, 745–757. https://doi.org/10.1007/s00299-017-2118-z

                                              Plant Abiotic Stress Physiology, Volume 1

169. Li, C., Unver, T. and Zhang, B. (2017). A high-efficiency CRISPR/Cas9 system for targeted mutagenesis in cotton (Gossypium hirsutum L.). *Sci. Rep.* 7, 43902.

170. Huang, L. et al. (2018). Developing superior alleles of yield genes in rice by artificial mutagenesis using the CRISPR/Cas9 system. *Crop J.* 6, 475–481, doi: 10.1016/j.cj.2018.05.005.

171. Shen, L. et al. (2018). QTL editing confers opposing yield performance in different rice varieties. *J. Integr. Plant Biol.* 60, 89–93, doi: 10.1111/jipb.12501.

172. Andersson, M., Turesson, H., Nicolia, A., Fält, A.-S., Samuelsson, M., and Hofvander, P. (2016). Efficient targeted multiallelic mutagenesis in tetraploid potato (Solanum tuberosum) by transient CRISPR-Cas9 expression in protoplasts. *Plant Cell Rep.* 36, 117–128. https://doi.org/10.1007/s00299-016-2062-3

173. Xu, J., Soni, V., Chopra, M., and Chan, O. (2020). Genetic modification of the HSP90 gene using CRISPR-Cas9 to enhance thermotolerance in T. Suecica. *URNCST J.* 21, 4(4). https://doi.org/10.26685/urncst.178

174. Lawrenson, T. et al. (2015). Induction of targeted, heritable mutations in barley and Brassica oleracea using RNA-guided Cas9 nuclease. *Genome Biol.* 16, 258. https://doi.org/10.1186/s13059-015-0826-7

175. Shi, J., Gao, H., Wang, H., Lafitte, H. R., Archibald, R. L., Yang, M., Hakimi, S. M., Mo, H., and Habben, J. E. (2017). ARGOS 8 variants generated by CRISPR-Cas9 improve maize grain yield under field drought stress conditions. *Plant Biotechnol. J.* 15, 207–216.

176. Wang, F. et al. (2016). Enhanced rice blast resistance by CRISPR/Cas9-targeted mutagenesis of the ERF transcription factor gene OsERF922. *PLoS One* 11, e0154027, doi: 10.1371/journal.pone.0154027.

177. Tang, L. et al. (2017). Knockout of OsNramp5 using the CRISPR/Cas9 system produces low Cd accumulating indica rice without compromising yield. *Sci. Rep.* 7, 14438, doi: 10.1038/s41598-017-14832-9.

178. Farhat, S. et al. (2019). CRISPR-Cas9 directed genome engineering for enhancing salt stress tolerance in rice. *Semin. Cell Dev. Biol.* 96, 91–99, doi: 10.1016/j.semcdb.2019.05.003.

179. Zhang, A. et al. (2019). Enhanced rice salinity tolerance via CRISPR/Cas9-targeted mutagenesis of the OsRR22 gene. *Mol. Breed.* 39, 47, doi: 10.1007/s11032-019-0954-y.

180. Bo, W., Zhaohui, Z., Huanhuan, Z., Xia, W., Binglin, L., Lijia, Y., Xiangyan, H., Deshui, Y., Xuelian, Z., and Chunguo, W. (2019). Targeted mutagenesis of NAC transcription factor gene, OsNAC041, leading to salt sensitivity in rice. *Rice Sci.* 26, 98–108.

181. Ma, L., Zhang, D., Miao, Q., Yang, J., Xuan, Y., and Hu, Y. (2017). Essential role of sugar transporter OsSWEET11 during the early stage of rice grain filling. *Plant Cell Physiol.* 58, 863–873. https://doi.org/10.1093/pcp/pcx040

182. Lee, S. K. et al. (2016). Plastidic phosphoglucomutase and ADP-glucose pyrophosphorylase mutants impair starch synthesis in rice pollen grains and cause male sterility. *J. Exp. Bot.* 67, 5557–5569. https://doi.org/10.1093/jxb/erw324

183. Miao, C. et al. (2018). Mutations in a subfamily of abscisic acid receptor genes promote rice growth and productivity. *Proc. Natl. Acad. Sci. USA* 115, 6058–6063, doi: 10.1073/pnas.1804774115.

184. Ito, Y., Nishizawa-Yokoi, A., Endo, M., Mikami, M., and Toki, S. (2015). CRISPR/Cas9-mediated mutagenesis of the RIN locus that regulates tomato fruit ripening. *Biochem. Biophys. Res. Commun.* 467, 76–82. https://doi.org/10.1016/j.bbrc.2015.09.117

185. Sun, Y. et al. (2017). Generation of high-amylose rice through CRISPR/Cas9-mediated targeted mutagenesis of starch branching enzymes. *Front. Plant Sci.* 8, 298. https://doi.org/10.3389/fpls.2017.00298

186. Ueta, R. et al. (2017). Rapid breeding of parthenocarpic tomato plants using CRISPR/Cas9. *Sci. Rep.* 7, 507. https://doi.org/10.1038/s41598-017-00501-4

187. Huang, X., Zeng, X., Li, J., and Zhao, D. (2017). Construction and analysis of tify1a and tify1b mutants in rice (*Oryza sativa*) based on CRISPR/Cas9 technology. *J. Agric. Biotech.* 25, 1003–1012.

188. Marraffini, L. A. and Sontheimer, E. J. (2008). CRISPR interference limits horizontal gene transfer in staphylococci by targeting DNA. *Science* 322, 1843–1845.

189. Marraffini, L. A. and Sontheimer, E. J. (2010). Self versus non-self discrimination during CRISPR RNAdirected immunity. *Nature* 463, 568–571.

190. Millen, A. M., Horvath, P., Boyaval, P., and Romero, D. A. (2012). Mobile CRISPR/Cas-mediated bacteriophage resistance in Lactococcus lactis. *PLoS One* 7, e51663.

191. Hatoum-Aslan, A., Maniv, I., Samai, P., and Marraffini, L. A. (2014). Genetic characterization of antiplasmid immunity through a type III-A CRISPR-Cas system. *J. Bacterial.* 196, 310–317.

192. Tamulaitis, G., Kazlauskiene, M., Manakova, E., Venclovas, C., Nwokeoji, A. O., Dickman, M. J., Horvath, P., and Siksnys, V. (2014). Programmable RNA shredding by the type III-A CRISPR-Cas system of Streptococcus thermophilus. *Mol. Cell* 56, 506–517.

193. Staals, R. H. et al. (2014). RNA targeting by the type III-A CRISPR-Cas Csm complex of Thermus thermophilus. *Mol. Cell* 56, 518–530.

194. Cao, L., Gao, C. H., Zhu, J., Zhao, L., Wu, Q., Li, M., and Sun, B. (2016). Identification and functional study of type III-A CRISPR-Cas systems in clinical isolates of Staphylococcus aureus. *Int. J. Med. Microbiol.* 306, 686–696.

195. Jiang, W., Samai, P., and Marraffini, L. A. (2016). Degradation of phage transcripts by CRISPR-associated RNases enables type III CRISPR-Cas immunity. *Cell* 164, 710–721.

196. Ichikawa, H. T., Cooper, J. C., Lo, L., Potter, J., Terns, R. M., and Terns, M. P. (2017). Programmable type III-A CRISPR-Cas DNA targeting modules. *PLoS One* 12, e0176221.

197. Deng, L., Garrett, R. A., Shah, S. A., Peng, X., and She, Q. (2013). A novel interference mechanism by a type IIIB CRISPR-Cmr module in Sulfolobus. *Mol. Microbiol.* 87, 1088–1099.

198. Estrella, M. A., Kuo, F. T., and Bailey, S. (2016). RNA-activated DNA cleavage by the type III-B CRISPRCas effector complex. *Genes Develop.* 30, 460–470.

199. Hale, C. R., Cocozaki, A., Li, H., Terns, R. M., and Terns, M. P. (2014). Target RNA capture and cleavage by the Cmr type III-B CRISPR-Cas effector complex. *Genes Develop.* 28, 2432–2443.

200. Han, W. et al. (2016). A type III-B CRISPR-Cas effector complex mediating massive target DNA destruction. *Nucl. Acids Res.* 45, 1983–1993.

201. Zhang, J. et al. (2012). Structure and mechanism of the CMR complex for CRISPR-mediated antiviral immunity. *Mol. Cell* 45, 303–313.

202. Ramia, N. F., Tang, L., Cocozaki, A. I., and Li, H. (2014). Staphylococcus epidermidis Csm1 is a 3–5′ exonuclease. *Nucl. Acids Res.* 42, 1129–1138.

203. Jung, T. Y., An, Y., Park, K. H., Lee, M. H., Oh, B. H., and Woo, E. (2015). Crystal Structure of the Csm1 Subunit of the Csm complex and its single-stranded DNA-specific nuclease activity. *Structure* 23, 782–790.

204. Kazlauskiene, M., Tamulaitis, G., Kostiuk, G., Venclovas, C., and Siksnys, V. (2016). Spatiotemporal control of type III-A CRISPR-Cas immunity: Coupling DNA degradation with the target RNA recognition. *Mol. Cell* 62, 295–306.

205. Liu, L., Li, X., Ma, J., Li, Z., You, L., Wang, J., Wang, M., Zhang, X., and Wang, Y. (2017). The molecular architecture for RNA-guided RNA cleavage by Cas13a. *Cell.* 170, 714–726.

206. Liu, T. Y., Iavarone, A. T., and Doudna, J. A. (2017). RNA and DNA Targeting by a reconstituted thermus thermophilus Type III-A CRISPR-Cas system. *PLoS One* 12, e0170552.

207. Park, K. H. et al. (2017). RNA activation-independent DNA targeting of the Type III CRISPR-Cas system by a Csm complex. *EMBO Rep.* 18, 826–840.

208. Han, W., Pan, S., Lopez-Mendez, B., Montoya, G., and She, Q. (2017). Allosteric regulation of Csx1, a type IIIB-associated CARF domain ribonuclease by RNAs carrying a tetraadenylate tail. *Nucl. Acids Res.* 45, 10740–10750.

209. Kazlauskiene, M., Kostiuk, G., Venclovas, C., Tamulaitis, G., and Siksnys, V. (2017). A cyclic oligonucleotide signaling pathway in type III CRISPR-Cas systems. *Science* 357, 605–609.

210. Niewoehner, O., Garcia-Doval, C., Rostol, J. T., Berk, C., Schwede, F., Bigler, L., Hall, J., Marraffini, L. A., and Jinek, M. (2017). Type III CRISPR-Cas systems produce cyclic oligoadenylate second messengers. *Nature* 548, 543–548.

211. Rouillon, C., Athukoralage, J. S., Graham, S., Gruschow, S., and White, M. F. (2018). Control of cyclic oligoadenylate synthesis in a type III CRISPR system. *ELife* 7, e36734.

212. Hale, C. R., Zhao, P., Olson, S., Duff, M. O., Graveley, B. R., Wells, L., Terns, R. M., and Terns, M. P. (2009). RNAguided RNA cleavage by a CRISPR RNA-Cas protein complex. *Cell* 139, 945–956.

213. Anantharaman, V., Makarova, K. S., Burroughs, A. M., Koonin, E. V., and Aravind, L. (2013). Comprehensive analysis of the HEPN superfamily: Identification of novel roles in intra-genomic conflicts, defense, pathogenesis and RNA processing. *Biol. Direct* 8, 15.

214. Sheppard, N. F., Glover, III, C. V., Terns. R., and Terns, M. P. (2016). The CRISPR-associated Csx1 protein of Pyrococcus furiosus is an adenosine-specific endoribonuclease. *RNA* 22, 216–224.

215. Niewoehner, O., and Jinek, M. (2016). Structural basis for the endoribonuclease activity of the type III-A CRISPR-associated protein Csm6. *RNA.* 22, 318–329.

216. East-Seletsky, A., O'Connell, M. R., Burstein, D., Knott, G. J., and Doudna, J. A. (2017). RNA targeting by functionally orthogonal type VI-A CRISPR-Cas enzymes. *Mol. Cell* 66, 373–383.

217. East-Seletsky, A., O'Connell, M. R., Knight, S. C., Burstein, D., Cate, J. H., Tjian, R., and Doudna, J. A. (2016). Two distinct RNase activities of CRISPR-C2c2 enable guide-RNA processing and RNA detection. *Nature* 538(7624): 270–273. https://doi.org/10.1038/nature19802

218. Knott, G. J., East-Seletsky, A., Cofsky, J. C., Holton, J. M., Charles, E., O'Connell, M. R., and Doudna, J. A. (2017). Guide-bound structures of an RNA-targeting A-cleaving CRISPR-Cas13a enzyme. *Nat. Struct. Mol. Biol.* 24, 825–833.

219. Foster, A. J., Martin-Urdiroz, M., Yan, X., Wright, H. S., Soanes, D. M., and Talbot, N. J. (2018). CRISPR-Cas9 ribonucleoprotein-mediated co-editing and counter selection in the rice blast fungus. *Sci. Rep.* 8, 14355.

220. Staals, R. H. et al. (2013). Structure and activity of the RNA-targeting Type III-B CRISPR-Cas complex of Thermus thermophilus. *Mol. Cell* 52, 135–145.

221. Hatoum-Aslan, A., Samai, P., Maniv, I., Jiang, W., and Marraffini, L. A. (2013). A ruler protein in a complex for antiviral defense determines the length of small interfering CRISPR RNAs. *J. Biol. Chem.* 288, 27888–27897.

222. Benda, C., Ebert, J., Scheltema, R. A., Schiller, H. B., Baumgartner, M., Bonneau, F., Mann, M., and Conti, E. (2014). Structural model of a CRISPR RNA-silencing complex reveals the RNA-target cleavage activity in Cmr4. *Mol. Cell* 56, 43–54.

223. Ramia, N. F. et al. (2014). Essential structural and functional roles of the Cmr4 subunit in RNA cleavage by the Cmr CRISPR-Cas complex. *Cell Rep.* 9, 1610–1617.

224. Zhu, X. and Ye, K. (2015). Cmr4 is the slicer in the RNA-targeting Cmr CRISPR complex. *Nucl. Acids Res.* 43, 1257–1267.

225. Koonin, E. V. and Makarova, K. S. (2017). Mobile genetic elements and evolution of CRISPR-Cas systems: All the way there and back. *Genome Biol. Evol.* 9, 2812–2825

226. Koonin, E. V. and Makarova, K. S. (2019). Origins and evolution of CRISPR-Cas systems. *Philos. Trans. R. Soc. Lond. B Biol. Sci.* 374, 20180087.

227. Özcan, A. et al. (2019). Type IV CRISPR RNA processing and effector complex formation in Aromatoleum aromaticum. *Nat. Microbiol.* 4(1): 89–96, doi: 10.1038/s41564-018-0274-8.

228. Faure, G. et al. (2019). CRISPR–Cas in mobile genetic elements: Counter-defense and beyond. *Nat. Rev. Microbiol.* 17 (8): 513–525, doi: 10.1038/s41579-019-0204-7.

229. Shmakov, S. A., Makarova, K. S., Wolf, Y. I., Severinov, K. V., and Koonin, E. V. (2018). Systematic prediction of genes functionally linked to CRISPR-Cas systems by gene neighborhood analysis. *Proc. Natl. Acad. Sci. USA* 115, E5307–E5316.

230. Swarts, D. C., van der, Oost, J., and Jinek, M. (2017). Structural basis for guide RNA processing and seed-dependent DNA targeting by CRISPR-Cas12a. *Mol. Cell* 66, 221–233. https://doi.org/10.1016/j.molcel.2017.03.016

231. Kim, H., Kim, S. T., Ryu, J., Kang, B. C., Kim, J. S., and Kim, S. G. (2017). CRISPR/Cpf1-mediated DNA-free plant genome editing. *Nat. Commun.* 8, 1–7, doi: 10.1038/ncomms14406

232. Stella, S., Alcón, P., and Montoya, G. J. N. (2017). Structure of the Cpf1 endonuclease R-loop complex after target DNA cleavage. *Nature* 546, 559.

233. Bernabé-Orts, J. M., Casas-Rodrigo, I., Minguet, E. G., Landolfi, V., Garcia-Carpintero, V., Gianoglio, S., Vázquez-Vilar, M., Granell, A., and Orzaez, D. (2019). Assessment of Cas12a-mediated gene editing efficiency in plants. *Plant Biotechnol. J.* 17, 1971–1984, doi: 10.1111/pbi.13113

234. Chen, J. S. et al. (2018). CRISPR-Cas12a target binding unleashes indiscriminate singlestranded DNase activity. *Science* 360, 436–439, doi: 10.1126/science.aar6245.

235. Swarts, D. C. (2019). Making the cut(s): How Cas12a cleaves target and non-target DNA. *Biochem. Soc. Trans.* 47, 1499–1510, doi: 10.1042/BST20190564.

236. Wang, J., Zhang, C., and Feng, B. (2020). The rapidly advancing class 2 CRISPR-Cas technologies: A customizable toolbox for molecular manipulations. *J. Cell. Mol. Med.* 24, 3256–3270, doi: 10.1111/jcmm.15039.

237. Schindele, P. and Puchta, H. (2020). Engineering CRISPR/LbCas12a for highly efficient, temperature-tolerant plant gene editing. *Plant Biotechnol. J.* 18, 1118–1120, doi: 10.1111/pbi.13275.

238. Zhong, Z. et al. (2018). Plant genome editing using FnCpf1 and LbCpf1 nucleases at redefined and altered PAM sites. *Mol. Plant* 11, 999–1002, doi: 10.1016/j.molp.2018.03.008.

239. Wang, M., Mao, Y., Lu, Y., Tao, X., and Zhu, J. K. (2017). Multiplex gene editing in rice using the CRISPR-Cpf1 system. *Mol. Plant* 10, 1011–1013.

240. Tang, X., Liu, G., Zhou, J., Ren, Q., You, Q., Tian, L., Xin, X., Zhong, Z., Liu, B., and Zheng, X. (2018). A large-scale whole-genome sequencing analysis reveals highly

specific genome editing by both Cas9 and Cpf1 (Cas12a) nucleases in rice. *Genome Boil.* 19, 84.

241. Banakar, R., Schubert, M., Collingwood, M., Vakulskas, C., Eggenberger, A. L., and Wang, K. (2020). Comparison of CRISPR-Cas9/Cas12a Ribonucleoprotein complexes for genome editing efficiency in the rice phytoene desaturase (OsPDS) gene. *Rice* 13, 4, doi: 10.1186/s12284-0190365-z.

242. Vu, T. Van et al. (2020). Highly efficient homology-directed repair using CRISPR/Cpf1 geminiviral replicon in tomato. 18, 2133–2143, doi: 10.1111/pbi.13373.

243. Endo, A. and Toki, S. (2019). Targeted mutagenesis using FnCpf1 in tobacco. *Meth. Mol. Biol.* 1917, 269–281, doi: 10.1007/978-1-49398991-1_20.

244. Li, B., Rui et al. (2019). Robust CRISPR/Cpf1 (Cas12a)-mediated genome editing in allotetraploid cotton (*Gossypium hirsutum*). *Plant Biotechnol. J.* 17, 1862–1864, doi: 10.1111/pbi.13147.

245. Jia, H., Orbović, V., and Wang, N. (2019). CRISPRLbCas12a-mediated modification of citrus. *Plant Biotechnol. J.* 17, 1928–1937, doi: 10.1111/pbi.13109.

246. Liu, H. et al. (2020). Efficient induction of haploid plants in wheat by editing of TaMTL using an optimized Agrobacterium-mediated CRISPR system. *J. Exp. Bot.* 71, 1337–1349, doi: 10.1093/jxb/erz529.

247. Van Tassel et al. (2020). New food crop domestication in the age of gene editing: Genetic, agronomic and cultural change remain co-evolutionarily entangled. *Front. Plant Sci.* 11, 1–16, doi: 10.3389/fpls.2020.00789.

248. Razzaq, A. et al. (2019). Modern trends in plant genome editing: An inclusive review of the CRISPR/Cas9 toolbox. *Int. J. Mol. Sci.* 20, 4045, doi: 10.3390/ijms20164045.

249. Lee, K. et al. (2019). Activities and specificities of CRISPR/Cas9 and Cas12a nucleases for targeted mutagenesis in maize. *Plant Biotechnol. J.* 17, 362–372, doi: 10.1111/pbi.12982.

250. Malzahn, A. A. et al. (2019). Application of CRISPR-Cas12a temperature sensitivity for improved genome editing in rice, maize, and Arabidopsis. *BMC Biol.* 17, 1–14, doi: 10.1186/s12915-019-0629-5.

251. Xu, R., Qin, R., Li, H., Li, J., Yang, J., and Wei, P. (2018). Enhanced genome editing in rice using single transcript unit CRISPR-LbCpf1 systems. *Plant Biotechnol. J.* 17, 553–555.

252. Tang, X. et al. (2019). Single transcript unit CRISPR 2.0 systems for robust Cas9 and Cas12a mediated plant genome editing. *Plant Biotechnol. J.* 17, 1431–1445, doi: 10.1111/pbi.13068.

253. Endo, A., Masafumi, M., Kaya, H., and Toki, S. (2016). Efficient targeted mutagenesis of rice and tobacco genomes using Cpf1 from *Francisella novicida*. *Sci. Rep.* 6, 1–9, doi: 10.1038/srep38169.

254. Yin, X., Anand, A., Quick, P., and Bandyopadhyay, A. (2019). Editing a stomatal developmental gene in rice with CRISPR/Cpf1. In *Plant Genome Editing with CRISPR Systems. Methods in Molecular Biology*; Vol. 1917, Y. Qi, Ed. New York, NY: Humana Press.

255. Li, C. et al. (2018). Expanded base editing in rice and wheat using a Cas9-adenosine deaminase fusion. *Genome Biol.* 19, 59, doi: 10.1186/s13059-018-1443-z

256. Xu, R., Qin, R., Li, H., Li, J., Yang, J., and Wei, P. (2019). Enhanced genome editing in rice using single transcript unit CRISPR-LbCpf1 systems. *Plant Biotechnol. J.* 17, 553–555, doi: 10.1111/pbi.13028.

257. Harrington, L. B. et al. (2018). Programmed DNA destruction by miniature CRISPR-Cas14 enzymes. *Science* 362, 839–842, doi: 10.1126/science.aav4294

258. O'Connell, M. (2018). Molecular mechanisms of RNA-targeting by Cas13-containing type VI CRISPR-Cas systems. *J. Mol. Biol.* 431, 66–87.

259. Liu, L., Li, X., Wang, J., Wang, M., Chen, P., Yin, M., Li, J., Sheng, G., and Wang, Y. (2017). Two distant catalytic sites are responsible for C2c2 RNase activities. *Cell* 168, 121–134.

260. Gootenberg, J. S., Abudayyeh, O. O., Kellner, M. J., Joung, J., Collins, J. J., and Zhang, F. (2018). Multiplexed and portable nucleic acid detection platform with Cas13, Cas12a, and Csm6. *Science.* 360, 439–444, https://doi.org/10.1126/science.aaq0179.

261. Clough, S. J., Bent, A. F. (1998). Floral dip: A simplified method for agrobacterium-mediated transformation of arabidopsis thaliana. *Plant J.* 16, 735–743.

262. Konermann, S., Lotfy, P., Brideau, N. J., Oki, J., Shokhirev, M. N., and Hsu, P. D. (2018). Transcriptome engineering with RNA-targeting type VI-D CRISPR effectors. *Cell,* 173(3): 665–676. https://doi. org/10.1016/j.cell.2018.02.033

263. Aman, R. et al. (2018). RNA virus interference via CRISPR/Cas13a system in plants. *Genome Biol.* 19(1): 1–9. https://doi.org/10.1186/s13059-017-1381-1

264. Aman, R., Mahas, A., Butt, H., Aljedaani, F., and Mahfouz, M. (2018). Engineering RNA virus Interference via the CRISPR/Cas13 machinery in *Arabidopsis. Viruses* 10 (12): 732. https://doi.org/10.3390/v10120732

265. Mahas, A., Aman, R., and Mahfouz, M. (2019). CRISPR-Cas13d mediates robust RNA virus interference in plants. *Genome Biol.* 20(1): 263. https://doi.org/10.1186/s13059-019-1881-2

266. Abudayyeh, O. O., Gootenberg, J. S., Kellner, M. J., and Zhang, F. (2019). Nucleic acid detection of plant genes using CRISPR-Cas13. *CRISPR J.* 2(3): 165–171. doi: 10.1089/crispr.2019.0011

267. Tashkandi, M., Ali, Z., Aljedaani, F., Shami, A., and Mahfouz, M. M. (2018). Engineering resistance against tomato yellow leaf curl virus via the CRISPR/Cas9 system in tomato. *Plant Signal. Behav.* 13, e1525996.

268. Aman, R. et al. (2018). RNA virus interference via CRISPR/Cas13a system in plants. *Genome Biol.* 19, 1.

269. Ghosh, D., Venkataramani, P., Nandi, S., and Bhattacharjee, S. (2019). CRISPR–Cas9 a boon or bane: The bumpy road ahead to cancer therapeutics. *Cancer Cell Int.* 19, 12.

270. Ma, X., Zhu, Q., Chen, Y., and Liu, Y. G. (2016). CRISPR/Cas9 platforms for genome editing in plants: Developments and applications. *Mol. Plant* 9, 961–974.

271. Saha, S. K., Saikot, F. K., Rahman, M. S., Jamal, M. A. H., Rahman, S. K., Islam, S. R., and Kim, K. H. (2018). Programmable molecular scissors: Applications of a new tool for genome editing in biotech. *Mol. Ther-Nucl. Acids,* 14, 212–238.

272. Zhang, X. H., Tee, L. Y., Wang, X. G., Huang, Q. S., and Yang, S. H. (2015). Off target effects in CRISPR/Cas9-mediated genome engineering. *Mol. Ther. Nucl. Acids,* 4, e264.

273. Anderson, K. R. et al. (2018). CRISPR off-target analysis in genetically engineered rats and mice. *Nat. Methods,* 15, 512.

274. Aryal, N. K., Wasylishen, A. R., and Lozano, G. (2018). CRISPR/Cas9 can mediate high-efficiency off-target mutations in mice in vivo. *Cell Death Dis.* 9, 1099.

275. Fu, Y., Foden, J. A., Khayter, C., Maeder, M. L., Reyon, D., Joung, J. K., and Sander, J. D. (2013). High-frequency off-target mutagenesis induced by CRISPR-Cas nucleases in human cells. *Nat. Biotechnol.* 31, 822–826.

276. Anderson, J. E., Michno, J.-M., Kono, T. J., Stec, A. O., Campbell, B. W., Curtin, S. J., and Stupar, R. M. (2016). Genomic variation and DNA repair associated with soybean transgenesis: A comparison to cultivars and mutagenized plants. *BMC Biotechnol.* 16, 41.

277. Grunewald, J., Zhou, R., Garcia, S. P., Iyer, S., Lareau, C. A., Aryee, M. J., and Joung, J. K. J. N. (2019). Transcriptome-wide off-target RNA editing induced by CRISPR-guided DNA base editors. *Nature* 569, 433–437.

278. Kim, D., Kim, D.-E., Lee, G., Cho, S. I. and Kim, J. S. (2019). Genome-wide target specificity of CRISPR RNA-guided adenine base editors. *Nat. Biotechnol.* 37, 430.

279. Zuo, E. et al. (2019). Cytosine base editor generates substantial off-target single-nucleotide variants in mouse embryos. *Science* 19, 289–292.

280. Wang, W., Hou, J., Zheng, N., Wang, X., and Zhang, J. (2019). Keeping our eyes on CRISPR: The "Atlas" of gene editing. *Cell Biol. Toxicol.* 35, 285–288. https://doi.org/10.1007/s10565-019-09480-w

281. Feng, C. et al. (2018). High efficiency genome editing using a dmc1 promoter-controlled CRISPR/Cas9 system in maize. *Plant Biotechnol. J.* 16, 1848–1857.

282. Li, J. et al. (2019). Whole genome sequencing reveals rare off-target mutations and considerable inherent genetic or/and somaclonal variations in CRISPR/Cas9edited cotton plants. *Plant Biotechnol. J.* 17, 858–868.

283. Walton, R. T., Christie, K. A., Whittaker, M. N., and Kleinstiver, B. P. (2020). Unconstrained genome targeting with near-PAMless engineered CRISPRCas9 variants. *Science* 80 (368): 290–296, doi: 10.1126/science. aba8853.

284. Gao, H. et al. (2020). Complex trait loci in maize enabled by CRISPR-Cas9 mediated gene insertion. *Front. Plant Sci.* 11, 1–14, doi: 10.3389/fpls.2020.00535.

285. Nathaniel, G. et al. (2020). Plant genome editing and the relevance of off-target changes. *Plant Physiol.* 183, 1453–1471, doi: 10.1104/pp.19.01194.

CHAPTER 10

Use of Ornamental Plants for the Phytoremediation of Metal-Contaminated Soils

AISHA ABDUL WARIS, KHURRAM NAVEED, MUHAMMAD UMAIR, MUHAMMAD ASHAR AYUB, MUHAMMAD ZIA UR REHMAN*, UMAIR RIAZ, and TALHA SALEEM

Institute of Soil and Environmental Sciences, Faculty of Agriculture, University of Agriculture, Faisalabad 38000, Pakistan

Corresponding author. E-mail: ziasindhu1399@gmail.com

ABSTRACT

Soil heavy metal pollution is a substantial aspect of plant abiotic stresses and needs some remediation in contaminated fields to ensure food security. Phytoremediation is a cost-effective technology to remediate metal-contaminated soils by decreasing the total amount of heavy metals in soils. Numbers of hyperaccumulator plants have been practiced for the remediation of metal-contaminated soils. However, ornamental plants are grabbing the attention of the scientific community by showing their tolerance against pollution stress without losing their economic value. Ornamental plants have shown significant pollutant extractability in literature with wide end-use options. They are capable of accumulating toxins in their tissues and can be used for phytoremediation as being nonfood crops and less interaction with the human food chain. They grow in wild and evolved to grow in a wide range of terrains, including contaminated soil, which can also strengthen their option for phytoremediation. Phytoextractability of these plants can be enhanced with the application of different amendments. These amendments include different fertilizers, acidifying materials, plant-growth-promoting rhizobacterias, and chelating agents, which either enhance the biomass of plants or increase the bioavailability of contaminants in the soils. The safe utilization of these hyperaccumulating plants is ever since an issue that needs to be addressed comprehensively. This chapter summarizes the significance of ornamental plants as a phytoremediation tool for contaminated soils. The compiled draft will help to understand the potential for growing ornamental plants in contaminated soil, option for enhancement in their growth and extractability, and their safe disposal/utilization.

10.1 INTRODUCTION

Pollution of soil environment with heavy metals (HMs) is a matter of global concern as they have a toxic role in the ecosystem [1]. Accumulation of these toxic metal ions in the agricultural soils may lead to their entry into the food chain and ultimately in humans, resulting in severe health issues [2, 3]. They have a variable range of short-to-long-term influences on the health of living beings including humans. Major exposure pathways of these HMs include the ingestion of contaminated food (animals and plants), gases or dust, skin contact, and consuming contaminated water [4]. HMs such as Cd, Cr, Pb, Hg, and As are toxic for humans and animals even in trace amounts [5]. Total contents of certain HM ions such as Cd, Pb, Cr, Zn, Ni, Cu, Co, Hg, Tl, As, Mn, and Mo have been raised to toxic levels in agricultural soils, which are hazardous for plant growth, yield, and food quality [6]. Several natural and anthropogenic activities are associated with the accumulation of HMs in environmental matrices [7]. Natural sources of HMs in the environment include volcanic eruptions, mineral weathering, and erosion [8, 9]. Anthropogenic activities involved in the contamination of the environment with HMs are mining, industrial effluents, urbanization, fossil fuel burning, agricultural materials (fertilizers, pesticides, sludges, etc.), smelting of ores, and technogenic materials [9–12]. The prevalence of these HMs in the food chain is the root cause of gastrointestinal cancer [13, 14]. Prolonged ingestion of these HMs via food products can lead to their build-up in body organs (kidney, liver, etc.) and disrupts several metabolic processes.

Several physical and chemical approaches are present for the removal of these toxic metals from agricultural soils to check their entrance into the food chain, but their economic viability is not feasible to clean up the soils [15] and demand for alternative strategies. Phytoremediation has gained attention from past some decades as it uses green technology to remediate the soils to render them innocuous [16]. In phytoremediation, plants mostly use the phytoextraction process to decontaminate the soil and concentrate the toxic HMs in their aerial portions [17]. Therefore, phytoextraction can be an appropriate option owing to its economical and ecoviability. The use of ornamental plants for phytoextraction of HMs is a novel option rather than edible crop plants as they beautify the environment, generate by-products, are economically more beneficial, and are more importantly not linked with the food chain [19–23]. Phytoremediation ability of ornamental plants (OPs) for HMs involves antioxidative responses, deposition of metal ions in cells or inactive tissues, and sequestration of metal ions in vacuoles by making chelates [25, 26]. This ability of phytoextraction can be improved further by using certain amendments and inoculants such as the application of chelating agents, fertilizers, and microbes, which usually enhance the bioavailability, uptake, tolerance, and accumulation potential of HMs.

10.2 CONCENTRATION AND SOURCES OF HEAVY METALS IN ENVIRONMENT

HMs are present in all types of soil, but their concentrations vary extensively depending on their parent material and contamination level. Some metal ions may be present in traces or below the detection limit, while some may be present in anomalously high amounts. The concentration of metal ions in soils is divided into two forms, including "total concentration" and "available concentration." Total concentration counts for all the forms of metal ions present in a certain soil including ions present in the crystal structure of the minerals and solid-state organic matter, adsorbed on the surfaces of the oxides, clays, and carbonates, free and soluble organic or inorganic complexes, and labile forms [27–29]. Available concentration represents the fraction of metal ions that are free for the plant's uptake. The HMs contents present in certain agricultural soil are either derived from parent material or sourced from a wide range of anthropogenic activities [35]. Natural sources of HMs in soils are indeed from the weathering of the rocks or unconsolidated lithogenic material form which the soils develop, while other sources include atmospheric deposition, wastewaters, agrochemicals, sewage sludges, composts, livestock wastes, fossil fuel burning, erosion, and technogenic materials [30–37]. The net or total concentration of HMs present in a certain soil is the difference of addition and losses of HMs in the soil. The losses include crop removal, leaching, volatilization, and soil erosion. Despite the losses, there has been the significant accumulation of HMs in the soils since the past few decades due to a greater fraction of input sources [35–37]. The agricultural soils are of serious concern in this matter as they are going to contaminate the food chain leading to health risks. Total contents of certain HM ions such as Cd, Pb, Cr, Zn, Ni, Cu, Co, Hg, Tl, As, Mn, and Mo has raised to serious levels in agricultural soils, which are hazardous for plant growth, yield, and food quality. The reported average concentrations of these HMs in world soils are 1.1, 25, 42, 62, 18, 14, 6.9, 0.1, 0.6, 4.7, 418, and 1.8 mg kg^{-1} for Cd, Pb, Cr, Zn, Ni, Cu, Co, Hg, Tl, As, Mn, and Mo, respectively [38].

10.2.1 NATURAL OCCURRENCE

The natural (lithogenic) origins are the major sources that determine the concentration of HMs in the soils. Earth crust holds HMs in traces, and usually, they are present up to 1000 mg kg^{-1} of soil except for ore minerals, in which individual concentration is even higher. Parent materials undergo chemical and physical weathering and release their constituents in forms available for the plants. HM ions incorporated in the crystals of the primary minerals, which took part in the rock formation, also release during weathering [35–38]. These ions entered the crystal structure by the substitution of the constituent ions such as silicon, iron, and aluminum. Ferromagnesian minerals such as augite, olivine, and hornblende have large amounts of trace metals including several essential micronutrients of plants besides HM ions [35–37]. Owing to their high elemental contents and fast weathering rate, the soils developed from basalts and other rocks incorporating these minerals are relatively more fertile. Oil shales and/or black shales also have high concentrations of toxic HM ions (Cd, Mo, As, Cu, U, Zn, and V), while sandstones have relatively lower contents of HMs in them [36–38].

Concentration and type of HMs present in a certain rock vary significantly depending on the nature of the minerals. These rocks/minerals include black shales, limestones, phosphorites, ultramafic rocks, sedimentary ironstones, and metallogenic minerals. They can be a source of elevated concentrations of several metalloids in the soils developed on them, such as As, Ba, Cr, Hg, Ni, Sb, Se, Tl, V, Zn, Ag, Au, Cd, Cu, Mo, Pb, Th, W, and U. Black shales are the source of a substantial concentration of Cd in soils. Several sites have been reported for elevated concentrations of Cd; 24 mg kg^{-1} Cd in Carboniferous shales, Derbyshire [39], and 60 mg kg^{-1} Cd in carbonaceous siltstone and black shales in Wushan Country [40]. These soils also contain high concentrations of other HMs such as Ni, Zn, Mo, and Sb up to 388, 962, 99, and 15 mg kg^{-1}, respectively. Food crops grown on these soils (black shales) have been reported to concentrate Cd and Mo up to 76.5 mg kg^{-1}

in maize and 5.4 mg kg^{-1} in beans, respectively, and have a potential health risk for consumers [40]. Similarly, the soils developed on limestones and had been mineralized previously from sulfides by sedimentary processes and hydrothermal fluids anomalously contain a high concentration of different HMs. These soils contain HMs (mg kg^{-1}) in the ranges of 3–47 (Cd), 20–50 (Co), 62–410 (Ni), 107–10180 (Pb), 47–55 (Tl), and 420–3820 (Zn), which are very high for agriculture use [41, 42].

Phosphate minerals containing phosphorites are enriched with several HMs even more than the black shales. They mainly consist of apatite accompanied by some other sedimentary strata such as shales, sandstones, limestone, and cherts. A study reported that maximum concentrations of Cd, Cr, Cu, Ni, Pb, U, and Zn in phosphate rocks around the world were found to be 62.5, 490, 110, 180, 500, 150, and 1850 mg kg^{-1}, respectively [43]. Very high concentrations of Cd have been reported in soils developed on the phosphorites in Jamaica. Garrett et al. [44] found an anomalously high concentration of Cd, Zn, and U up to 6200, 12,300, and 166 mg kg^{-1} in Hope phosphorite rock present in Manchester Parish, Jamaica. Another similar deposit of Spitzbergen phosphorite in Jamaica found containing Cd and Zn up to 16,540 and 6103 mg kg^{-1}, respectively. Due to their anomalously high concentration of Cd in phosphorites, they are also known as "cadmiferous." The soils developed from these rocks or use of these rocks for phosphatic fertilizers has serious concerns for soil contamination risks. Soils developed on ultramafic rocks also contain some HMs such as Co, Cu, Ni, and Cr [45–49]. Iron-oxide-rich soils developed on ironstones containing more than 30% of iron oxides have been found generally containing a high concentration of As, Cu, Mo, Pb, Zn, Ba, Mn, Ni, and V [50, 51]. The dispersion of these soils by natural transportation processes resulted in the contamination of the undisturbed areas and referred to as the natural pollution of soils. The major processes involved in the transport of HMs from source to agricultural soils are wind currents, moving water, topography, and anthropogenic activities.

10.2.2 ANTHROPOGENIC SOURCES

The contamination of soil occurs by anthropogenic activities, which include various sources such as urbanization, agricultural inputs, mining activities, industrial wastes, and shooting ranges [53]. The comparative studies between anthropogenically contaminated soils and naturally affected soils showed higher concentrations of HMs in the former. The maximum values of some

HMs were up to 32 (Cd) times higher in anthropogenic contaminated soil profiles than naturally affected and 10 times with Pb and Cu. Agricultural practices have been found to add substantial amounts of HMs in the soils such as the application of wastewater irrigation, livestock manures, sewage sludge, inorganic fertilizers, and pesticides. Nanos and Rodríguez Martín [54] described agricultural practices as a major source of ingress of HMs in anthropogenically contaminated soils. Atmospheric deposition and irrigation water are also an important source of HMs contamination in soils [55, 56].

The application of livestock manures adds a significant amount of HMs in agricultural soils. A large amount of Zn and Cu are used in the feed of livestock as growth promoters, which lead to elevated concentrations of these metals in the manures [57]. Luo et al. [58] described high concentrations of Cu and Zn in farmyard manure. Similarly, sewage sludge is recycled to agricultural soils due to its beneficial fertilizer values of nitrogen, phosphorus, and organic matter present in it. However, these sewage sludges also contain a large number of HMs in them. The HMs present in the sewage sludge are usually sourced from the sewage waters of households, industrial effluents, and runoff from roads and roofs. The world ranges of some HMs in sewage sludges have been reported by literature compiled by Kabata-Pendias [38], including As (<26), Mn (<39,000), V (<400), Ba (<4000), and Sn (<700). Fertilizers containing macronutrients are also an important source of HM contamination in agricultural soils as they are applied at higher rates. The inorganic compounds or the rocks used for the formation of fertilizers contain substantial amounts of toxic metals in them. A considerable amount of these HMs pass on to the fertilizer products and ultimately to the soil when applied. Among them, phosphatic fertilizers are of serious concern as they contain the highest amounts of several toxic HMs, including Cd, U, As, Zn, and Th. Frequent application of phosphorite rock-based phosphatic fertilizers in soils leads to the accumulation of Cd ($30–60$ g ha^{-1}) in soils [59].

Urban soils are found to accumulate more contents of HMs than rural or agricultural soils. The group HMs found in higher concentrations in the urban environment include Pb, Cd, Sb, Cu, Hg, Zn, Ba, Ti, As, and Th [60, 61]. However, they may change concerning the location and some others may be dominant in other areas, but Pb is found to be common in several areas. The main sources of HMs in urban soils include deposition, fuel combustions, corrosions of metallic structures, technogenic materials, paints, and fires. These contents also go to make the part of agricultural soils transported by wind depositions and runoffs by rain or floods.

10.3 SIGNIFICANCE OF ORNAMENTAL PLANTS IN THE CONTAMINATED ENVIRONMENT

Due to an increase in polluted lands all over the world, their rehabilitation, and safety management in the future have become a concern for the scientific community [62, 63]. The use of chemical, physical, and biological remedial technologies poses limitations for HM soil remediation, such as alteration in pH of the soil, imbalance in nutrient availability, and even sometimes increase the mobility of the metals as in case of organic amendments [64]. Therefore, a complementary method is required for metal removal from contaminated soils. The use of ornamental plants for phytoextraction of HMs can be a novel option rather than conventional techniques as they beautify the environment, generate by-products, are economically more beneficial, and are more importantly not linked with the food chain [18–23]. Ornamental and medicinal plants such as Rosemary (*Salvia rosmarinus*), Vetiver (*Chrysopogon zizanioides*), Citronella (*Cymbopogon nardus*), Chamomile (*Matricaria chamomilla*), Lavender (*Lavandula*), Sage (*Salvia officinalis*), Palmarosa (*Cymbopogon martini*), Thyme (*Thymus vulgaris*), Basil (*Ocimum basilicum*), Mint (*Mentha*), Tulsi (*Ocimum tenuiflorum*), Geranium (*Pelargonium*), and Lemon Grass (*Cymbopogon*) Plants, when grown in the metal-contaminated soils, showed the ability for soil rehabilitation and effective use [62, 63, 65]. Most of the ornamental plants accumulate HMs in their roots including *Antirrhinum majus*, *Quamoclit pinnata* [21], *Nerium oleander* [66], *Erica australis* and *Erica andevalensis* [67], *Tagetes patula* [68, 69], and *Calendula officinalis* [70]. Vetiver grass was found as the best phytoremediator of HMs as it has significant efficacy of As removal from contaminated sites [72]. Cadmium accumulation efficacy of some native plants was tested in a study. It was found that *Tagetes patula* has significant Cd bioaccumulation potential [70, 71]. *T. patula* has also been described as a good phytoremediation of Fe, Pb, and Cu along with Cd [24, 68, 73] and some organic pollutants. *Amaranthus caudatus*, *Chlorophytum comosum*, and *A. hypochondriacus* had significant remediation potential for Cd contaminated soils [74–76]. *Gynura pseudochina* has found to accumulate a substantial amount of Cd, Zn [77], and Cr from contaminated soils [78]. *Cyperus rotundus* and *Panicum maximum* found very beneficial for covering the Cd contaminated lands [79, 80]. Lemon geraniums (Pelargonium sp.) accumulated Ni, Cd, Cu, and Pb to much extent from soil [81]. Hence, ornamental and medicinal plants can remediate the metal-contaminated soils, beautify the surroundings, and could be sold for revenue.

It has also been observed that the by-products of these plants, such as oils extracted from them, were free from HM accumulation, which means that they can be safely grown without deteriorating the quality and utility of these plants [63]. They can successfully prevent food chain contamination and improve the quality of soil; therefore, they are extremely suitable candidates for phytoremediation [65]. These plants can accumulate metals in their different parts, such as roots, grains, shoot, and fruit, irrespective of the amount present in the soil, and they can be safely grown in cadmium, lead, mercury, zinc, copper, and other metal-contaminated soils without being seriously affected in terms of their quality and output.

10.4 PHYSIOLOGICAL CHARACTERISTICS AND TOLERANCE BEHAVIOR OF ORNAMENTAL PLANTS GROWN IN METAL-CONTAMINATED SOIL

Ornamental plants have developed different mechanisms that are quite helpful in fighting against HMs stress. Phytoremediation ability of ornamental plants for HMs involves antioxidative responses, deposition of metal ions in cells or inactive tissues, and sequestration of metal ions in vacuoles by making chelates [24, 82]. Being a sustainable approach, the use of ornamental plants is highly recommended as they have a strong defense system against environmental stress, that is, HM toxicity [83]. There are some families of ornamental plants that are most promising and tolerant against stressed conditions. They act as hyperaccumulating, metallophytes, and photostabilizers [65]. They fight against HM stress in such a way to develop certain mechanisms. The HMs stress involve the production of reactive oxygen species such as hydroxyl radical, superoxides, and hydrogen peroxide; similarly, many variations occur at the cellular and molecular levels [84]. A lot of secondary metabolites are produced in OPs that perform different cellular functions that are necessary for physiological processes and to combat against these stress conditions [70]. The tools and techniques of molecular biology are much helpful in understanding the process of signaling and pathways which are involved in the production of secondary metabolites [85].

Plants tolerate metal toxicity by accumulating them in different parts of their body. Phytoextraction of ornamental plants consists of compartmentalization of HMs in the cell walls, metabolically inactive tissues, and vacuoles [17]. These plants have the instinctive ability to detoxify HMs utilizing numerous ways like sequestration, exclusion, and chelation [86]. It is found that HM stress boosts the essential oil production of certain aromatic plants [65]. The tolerance mechanism of ornamental plants against abiotic stress can be expedited through different ways likewise to enhance the phenolic and antioxidant capacity of these plants with the treatment of certain growth regulators. Similarly, the combination of a polyamine with cadmium makes it possible to decrease the disturbances caused by metal stress, such as the recovery of weight, and the content of photosynthetic pigments as well as the stimulation of antioxidant enzymes [83]. The use of ornamental plants reduced the adverse impact of Cd on soil enzyme activity and microbial communities and improved the ecological landscape of contaminated soil [17, 83].

10.5 PHYTOREMEDIATION RESPONSE OF ORNAMENTAL PLANTS IN METAL-CONTAMINATED SOILS

Heavily contaminated soils are unsuitable for agriculture, and cultivation of ornamental plants is a good alternative for the decontamination of such fields [62, 63, 65]. Generally, phytoremediation is a wide term, which includes removal of all type of metals from the soil; however, its potential benefits vary from one case to another [12]. There are several limitations associated with this technique; therefore, the idea of its use requires selection of plant species that are hyperaccumulators in nature and have high economic yield so that profound benefits could be derived. It is believed that specific species of ornamental plants can be grown in such soils, along with the estimation of the extent to which they could be used, and their value addition to the problem soil can help us convert this loss into profit [63]. Ornamental plants can extract metals from the soil and sediments insoluble form and further deposit them into roots and shoots. As far as ornamental plants are concerned, it is documented that their above-ground parts accumulate a high level of metals, but the extracted essential oil is always free from HM contamination. Therefore, they cannot be used as food, but under strict regulations and standards, or through the steam distillation process [65], their oil can be accepted in the market with the least risk of HM contamination [63]. HM phytoextraction of some ornamental plants is listed in Table 10.1. Moreover, in recent times, the term "phyto-managemnet" is derived from the concept of phytoremediation, which means that extensive research and development need to be done to extract all benefits of phytoremediation [65].

TABLE 10.1 Phytoremediation Ability of Ornamental Plants in Heavy Metal Soils

Sr. No.	Plant Species	Heavy Metal	Concentration (mg kg−1)	Remarks	Reference
1	*Gladiolus grandiflorus* L.	Cd	0–100	High Cd amount recorded in leaves of the plant. This plant has a high tolerance for Cd without showing negative growth symptoms.	[87]
	Chrysanthemum indicum L.	Cd	0–100	High Cd contents accumulated in leaves of the plants.	[88]
2	*Lonicera japonica*	Cd	5–200	The plant has high accumulation and tolerance capacity for Cd and high concentration of Cd did not affect plant parameters compared to control.	
3	*Calendula officinalis* L.	Cd	0–80	Plant accumulated a high amount of Cd in root and shoot regions.	[89]
		Pb	0–300	The plant can accumulate high Pb but with a significant reduction in growth.	
5	*Althaea rosea*	Cd	0–100	This plant accumulated high Cd concentration in shoots compared to roots with a slight decrease in height.	[71]
6	*Antirrhinum majus* L.	Pb	0–5000	This plant accumulated high Pb concentration in shoot without decreasing shoot biomass.	[21]
	Celosia cristata pyramidalis	Pb	0–5000		
6	*Chlorophytum comosum*	Pb	0–2000	This plant accumulated high Pb concentration in the root region and proved tolerant of Pb stress with a mild decrease in growth parameters at high concentration.	[91]
7	*Chrysanthemum indicum* L.	Pb	0–50	Plant accumulated the highest Pb concentration in roots with a significant reduction in growth parameters.	[92]
8	*Calendula officinalis* L.	Cu	0–400	High Cu amount accumulated in shoot and root tissues without showing toxicity symptoms.	[93]

TABLE 10.1 *(Continued)*

Sr. No.	Plant Species	Heavy Metal	Concentration (mg kg−1)	Remarks	Reference
9	*Mirabilis jalapa* L	Cr	0–102.5	This plant showed a high phytoextraction ability for Cr.	[94]
10	*Nymphaea spontanea landon*	Cr	0–10	The highest accumulation of Cr was recorded in the roots of the plant with a significant reduction in chlorophyll, sugar, and protein contents.	[95]

10.6 USE OF DIFFERENT AMENDMENTS TO ENHANCE THE PHYTOEXTRACTION OF ORNAMENTAL PLANTS IN SOIL

The efficiency of phytoextraction is governed by the ability of plants to concentrate the HMs in their aerial parts [96, 97]. Regardless of the fact that the type of plant being used counts for the phytoextraction potential, the bioavailability of HM ions in the soil is the key factor controlling the efficiency of the phytoextraction process. Therefore, it depends upon both the mobility of HMs in soil and the uptake potential of plants, which can be enhanced by using different amendments [96–100]. Metal accumulation potential of high biomass varieties of ornamental plants can also be used to enhance by using different agents, which can increase the mobility of HMs in soil without diminishing the yield of plants. Several studies are presently describing the role of chelating agents, fertilizers, and microbes in improved phytoextraction of HMs in ornamental plants [101–103].

10.6.1 ADDITION OF CHELATING AGENTS

Chelating agents can boost the phytoextraction of ornamental plants by forming complexes of metal ions and enhancing their solubility and translocation [74]. Some chelating agents can be helpful in this respect including diethylene triamine penta-acetic acid, sodium dodecyl sulfate (SDS), nitrilotriacetic acid, ethylene glycol tetra-acetic acid (EGTA), ethylene diamine succinate, ethylene diamine tetra-acetic acid (EDTA), and low-molecular-weight organic acids [73, 104–106]. However, excessive use of chelating agents may also lead to environmental contamination owing to a higher mobilization rate of HMs in soil, which needs to be considered while applying chelates. Nevertheless, ornamental plants showed improved growth with the application of chelates in concentration equimolar to metals concentration [21, 90].

Wang and Liu [106] reported significant translocation ability of EGTA when applied to *Mirabilis jalapa*. EGTA was found better and effective regarding the lesser risk of metal leaching than EDTA. Moreover, Cd translocation was higher with EGTA application than EDTA. Calendula officinalis and *Althaea rosea* can be used as potential Cd hyperaccumulators with the application of EGTA and SDS as these chelating agents significantly improved growth and accumulation [70]. Metal ions form chelates with these chemical agents outside the roots and then translocate from roots to shoot after uptake in the form of the metal–chelate complex. Wenzel et al. [107] reported that EDTA forms a complex with metals after entering into the root cells and then helps in translocation of metal ions to the aerial parts. In another study, it was reported that the highly stable EDTA metal complexes may get dissociate totally (iron) or partially (lead) after entering into the plants (*P. vulgaris*) depending upon the nature of the metals ions [108]. Likewise, the chelating agent's application in soil has been proposed in many studies for increasing HMs uptake by ornamental plants [73, 104–106]. Turgut et al. [101] reported increased accumulation of Ni, Cd, and Cr in *Helianthus annuus* by application of EDTA and citric acid. Similarly, *Mirabilis jalapa* L. showed increased growth and phytoextraction of Cd, Cu, Zn, and Pb by applying citric acid and EDTA. However, EDTA showed better extraction potential than citric acid [20]. Farid et al. [109] also explained substantially improved growth and Cr accumulation in sunflower tissues with a citric acid application. Lai and Chen [110] reported 5-mmol kg^{-1} soil EDTA for enhanced Zn, Cd, and Pb accumulation by rainbow pink indicating the chelating potential of EDTA. Chandra Sekhar et al. [111] concluded that hydroxyl ethylene diamine tetraacetic acid and EDTA enhanced the Pb uptake from the soil in the Indian sarsaparilla (*Hemidesmus indicus*), a hyperaccumulating plant. It can be summarized that these chelates make the phytoextraction process efficient either by increasing the tolerance of ornamentals plants or by escalating the mobility of HM ions.

10.6.2 APPLICATION OF NITROGEN, PHOSPHORUS, AND POTASSIUM (NPK) FERTILIZERS

Application of macronutrients (NPK) for phytoremediation of HM-contaminated soils is suggested in many studies. They can enhance the growth and tolerance of the plants by the provision of nutrition [75]. An increase in the biomass of the plants is the key factor in determining the efficiency of the remediation process. They also affect the bioavailability of HMs by altering the surface charge and pH of the soil. Phosphorus being an essential nutrient plays an important role in growth, while its low availability due to high fixation and low solubility is a limitation for plant growth. Yu and Zhou [112] described that with the addition of phosphorus (20–100 mg kg^{-1}) at various levels of Cd contamination (10–100 mg kg^{-1}), the growth of *Mirabilis jalapa* L. increased significantly. While the concentration of Cd in the aerial parts reduced with an increase in the application rate of phosphorus, the bioaccumulation factor also reduced with an increase in the P concentration that might be due to precipitation of phosphorus with Cd ions as Cd-phosphate. However, the translocation factor found maximum at the phosphorus concentration level of 100 mg kg^{-1}. Hence, it can be concluded that the application of phosphorus at a suitable rate might be helpful for better growth of plants and remediation of soils. Cao et al. [113] reported enhanced accumulation (265%) of As in *Pteris vittata* by application of phosphorus. Contrary to this, Thomas and Omueti [126] reported fair elevation in soil HMs (Cd, Zn, Cu, and Pb) concentration after application of phosphatic fertilizers and a judicious accumulation of HMs in tissues of *A. caudatus* in control (without P fertilizers). Nitrogen and potassium also take part in the phytoextraction of HMs by improving the growth of plants. Nitrogen has a major role in the production of biomass of plants, which is directly related to the accumulation of metal ions in tissues. Similarly, potassium application may also enhance the extraction potential by regulating the plant homeostasis and metabolic processes. Some metal ions have antagonistic effects with potassium which may counterpart the accumulation process.

10.6.3   INOCULATION OF PLANT-GROWTH-PROMOTING RHIZOBACTERIA (PGPRS)

Soil bacteria directly affect the metal ions bioavailability and accumulation by crop plants [3]. It is controlled by several bacterial activities including acidification, methylation, chelation, biosorption, complexation, and redox reactions [3, 102, 114]. They can transform, detoxify, or mobilize the HM ions. Among them, PGPR can improve the phytoextraction process of ornamental plants by enhancing their growth, tolerance to metal ions, and bioavailability of nutrients to them [117]. Several substances are produced by PGPRs, which can enhance the tolerance ability of ornamental plants and uptake of HMs by altering their bioavailability, including ACC deaminase, siderophores, IAA production, nitrogen fixation, and phosphate solubilizers [116]. For example, sunflower inoculated with Cr-tolerant rhizobacteria, including SS1, SS3, and SS6, showed enhanced accumulation and tolerance of Cr. They significantly improved the morpho-physiological characteristics of sunflower compared to uninoculated control. SS6 was found to be the best strain among them regarding growth and accumulation [117]. The efficiency of the phytoextraction process is based on the optimum growth of the plants, so better growth is the key factor in determining it [30, 112]. *Microbacterium* sp. and *Klebsiella* sp. are found to increase the Pb phytoextraction potential of *Mesembryanthemum criniflorrum* and *Pelargonium hortorum. P. hortorum* showed 1.9 folds more extraction of Pb than *Mesembryanthemum criniflorium* with *Microbacterium* sp. followed by *Klebsiella* sp. (1.8 folds). This study reveals that Pb resistant PGPRs are effective for enhancing the phytoextraction capability of ornamental plants [102]. Similarly, Dimkpa et al. [118] explained better growth and accumulation of Cd and Fe by *Helianthus annuus* with inoculation of siderophore producing strain *Streptomyces tendae* 4. It showed higher Fe uptake even than EDTA from the rhizosphere, indicating the potential use of PGPRs in phytoextraction. In another study, inoculation of Mirabilis jalapa with *Pseudomonas fluorescence* increased uptake of Cd, Ni, Cr, Pb, Zn, and Cu [119]. These studies indicate the influences of PGPRs on the bioavailability of HMs and the potential increase in phytoextraction of ornamental plants in metal soils.

10.7 UTILIZATION OF ORNAMENTAL PLANTS GROWN IN METAL-CONTAMINATED SOILS

Cultivation of ornamental plants on HM-contaminated soils is considered as more feasible and profitable techniques than the old phytoremediation methods [65]. Nowadays, this technique has acquired increased momentum around the world. It has vast applications in industries such as biofuel, cosmetic, and perfumery industrial sectors [83, 120]. Similarly, production of essential oils from different crops of ornamental plants such as industrial hemp (*Cannabis sativa*), tulsi (*Ocimum tenuiflorum*), peppermint (*Mentha balsamea*), and lemongrass (*Cymbopogon*) having sustainable efficiency of accumulating HMs and the extraction of oil from these plants occur through a process of evaporation and condensation, which makes oils risk free from metal contamination [121].

There are some families of aromatic medicinal plants that are most promising for HM remediation of contaminated sites, that is, *Geraniaceae, Lamiaceae*, and *Poaceae*. They act as hyperaccumulative, metallophytes, and stabilizers; these all being valuable economic crops can be grown on metal-contaminated soils instead of food crops [65]. These plants are also much resistant against harsh climatic conditions as climate change is a very burning issue these days, so the commercial cultivation of such crops is a very viable option for human economic growth [16, 17, 25, 86, 94]. It is estimated that among the world population, 80% is reliant on herbal medicine and 60% of which is based on medicinal and aromatic plants available in markets. Most of the drugs of antimicrobial, immunosuppressive, and anticancer are extracted from ornamental and medicinal plants [122]. At the end of 2050, the global demand for essential oils produced by plants will be increased up to five trillion US $ [123]. Therefore, there is a need to grow essential oil-producing grasses to combat the expected global demand. It has been described that the mode of obtaining these oils through different industrial processes makes them risk free from HM contamination [124]. After the extraction of oil, the remaining residues can be used for energy production in the form of biomass by direct burning or by obtaining biogas through it. This unified approach not only reduces the prices of petroleum oil, but also curtails the metal load from the soil and mitigates many other environmental stresses such as reduction in greenhouse gases and pollution alleviation [121]. Similarly, this integrated approach has so many advantages over traditional phytoremediation techniques particularly due to cost effectiveness, acceptability by the people, and ease in the reclamation of contaminated soil [125].

10.8   CONCLUSION

It is recommended that perennial or annual ornamental plants containing high biomass should be used, from which multiple time harvesting can be obtained in a year. As these plants are nonfood crops, they are a suitable candidate for phytoremediation of metal-contaminated soils. These plants can play a major role in protecting humans from environmental pollution. In the current scenario of food chain contamination, a new regime of ornamental plants has been emerging at the horizon of phytoremediation providing two-way solution, that is, metal-free soil and upgraded green economy.

KEYWORDS

- **abiotic stress**
- **metal contamination**
- **ornamental plants**
- **phytoremediation**

REFERENCES

1. Carré, F.; Caudeville, J.; Bonnard, R.; Bert, V.; Boucard, P.; Ramel, M. Soil Contamination and Human Health: A Major Challenge for Global Soil Security. In *Progress in*

Soil Science; Springer International Publishing, **2017**; pp. 275–295. https://doi.org/10.1007/978-3-319-43394-3_25.

2. Fahimirad, S.; Hatami, M. Heavy Metal-Mediated Changes in Growth and Phytochemicals of Edible and Medicinal Plants. In *Medicinal Plants and Environmental Challenges*; Springer, **2017**; pp. 189–214. https://doi.org/10.1007/978-3-319-68717-9_11.

3. Sarwar, N.; Imran, M.; Shaheen, M. R.; Ishaque, W.; Kamran, M. A.; Matloob, A.; Rehim, A.; Hussain, S. Phytoremediation Strategies for Soils Contaminated with Heavy Metals: Modifications and Future Perspectives. *Chemosphere* **2017**, *171*, 710–721.

4. Steffan, J. J.; Brevik, E. C.; Burgess, L. C.; Cerdà, A. The Effect of Soil on Human Health: An Overview. *Eur. J. Soil Sci.* **2018**, *69* (1), 159–171.

5. Azeh Engwa, G.; Udoka Ferdinand, P.; Nweke Nwalo, F.; N. Unachukwu, M. Mechanism and Health Effects of Heavy Metal Toxicity in Humans. *Poisoning in the Modern World-New Tricks for an Old Dog?*, IntechOpen; **2019**. https://doi.org/10.5772/intechopen.82511.

6. Chibuike, G. U.; Obiora, S. C. Heavy Metal Polluted Soils: Effect on Plants and Bioremediation Methods. *Appl. Environ. Soil Sci.* **2014**, *2014*, 1–12.

7. Asgari Lajayer, B.; Najafi, N.; Moghiseh, E.; Mosaferi, M.; Hadian, J. Removal of Heavy Metals (Cu2+ and Cd2+) from Effluent Using Gamma Irradiation, Titanium Dioxide Nanoparticles and Methanol. *J. Nanostructure Chem.* **2018**, *8* (4), 483–496.

8. Antoniadis, V.; Levizou, E.; Shaheen, S. M.; Ok, Y. S.; Sebastian, A.; Baum, C.; Prasad, M. N. V; Wenzel, W. W.; Rinklebe, J. Trace Elements in the Soil-Plant Interface: Phytoavailability, Translocation, and Phytoremediation—A Review. *Earth-Science Rev.* **2017**, *171*, 621–645.

9. Asgari Lajayer, B.; Ghorbanpour, M.; Nikabadi, S. Heavy Metals in Contaminated Environment: Destiny of Secondary Metabolite Biosynthesis, Oxidative Status and Phytoextraction in Medicinal Plants. *Ecotoxicol. Environ. Saf.* **2017**, *145*, 377–390.

10. Tauqeer, H. M.; Ali, S.; Rizwan, M.; Ali, Q.; Saeed, R.; Iftikhar, U.; Ahmad, R.; Farid, M.; Abbasi, G. H. Phytoremediation of Heavy Metals by Alternanthera Bettzickiana: Growth and Physiological Response. *Ecotoxicol. Environ. Saf.* **2016**, *126*, 138–146.

11. Feng, N. X.; Yu, J.; Zhao, H. M.; Cheng, Y. T.; Mo, C. H.; Cai, Q. Y.; Li, Y. W.; Li, H.; Wong, M. H. Efficient Phytoremediation of Organic Contaminants in Soils Using Plant–Endophyte Partnerships. *Sci. Total Environ.* **2017**, *583*, 352–368.

12. Mahar, A.; Wang, P.; Ali, A.; Awasthi, M. K.; Lahori, A. H.; Wang, Q.; Li, R.; Zhang, Z. Challenges and Opportunities in the Phytoremediation of Heavy Metals Contaminated Soils: A Review. *Ecotoxicol. Environ. Saf.* **2016**, *126*, 111–121.

13. Yuan, W.; Yang, N.; Li, X. Advances in Understanding How Heavy Metal Pollution Triggers Gastric Cancer. *Biomed Res. Int.* **2016**, *2016*, 1–10.

14. Türkdoğan, M. K.; Kilicel, F.; Kara, K.; Tuncer, I.; Uygan, I. Heavy Metals in Soil, Vegetables and Fruits in the Endemic Upper Gastrointestinal Cancer Region of Turkey. *Environ. Toxicol. Pharmacol.* **2003**, *13* (3), 175–179.

15. Xu, C.; Yang, W.; Zhu, L.; Juhasz, A. L.; Ma, L. Q.; Wang, J.; Lin, A. Remediation of Polluted Soil in China: Policy and Technology Bottlenecks. *Environ. Sci. Technol.* **2017**, *51* (24), 14027–14029.

16. Liu, J.; Xin, X.; Zhou, Q. Phytoremediation of Contaminated Soils Using Ornamental Plants. *Environ. Rev.* **2018**, *26* (1), 43–54.

17. Asgari Lajayer, B.; Khadem Moghadam, N.; Maghsoodi, M. R.; Ghorbanpour, M.; Kariman, K. Phytoextraction of Heavy Metals from Contaminated Soil, Water and

Atmosphere Using Ornamental Plants: Mechanisms and Efficiency Improvement Strategies. *Environ. Sci. Pollut. Res.* **2019**, *26* (9), 8468–8484.

18. Han, Y. L.; Yuan, H. Y.; Huang, S. Z.; Guo, Z.; Xia, B.; Gu, J. Cadmium Tolerance and Accumulation by Two Species of Iris. *Ecotoxicology* **2007**, *16* (8), 557–563.

19. Wang, Y.; Yan, A.; Dai, J.; Wang, N.; Wu, D. Accumulation and Tolerance Characteristics of Cadmium in *Chlorophytum Comosum*: A Popular Ornamental Plant and Potential Cd Hyperaccumulator. *Environ. Monit. Assess.* **2012**, *184* (2), 929–937.

20. Sun, Y. B.; Sun, G. H.; Zhou, Q. X.; Xu, Y. M.; Wang, L.; Liang, X. F.; Sun, Y.; Qing, X. Induced-Phytoextraction of Heavy Metals from Contaminated Soil Irrigated by Industrial Wastewater with Marvel of Peru (*Mirabilis jalapa* L.). *Plant, Soil Environ.* **2011a**, *57* (8), 364–371.

21. Cui, S.; Zhang, T.; Zhao, S.; Li, P.; Zhou, Q.; Zhang, Q.; Han, Q. Evaluation of Three Ornamental Plants for Phytoremediation of Pb-Contamined Soil. *Int. J. Phytoremediation* **2013**, *15* (4), 299–306.

22. Ramana, S.; Biswas, A. K.; Singh, A. B.; Ajay, A.; Ahirwar, N. K.; Subba Rao, A. Tolerance of Ornamental Succulent Plant Crown of Thorns (*Euphorbia Milli*) to Chromium and Its Remediation. *Int. J. Phytoremediation* **2015**, *17* (4), 363–368.

23. Selamat, S. N.; Abdullah, S. R. S.; Idris, M. Phytoremediation of Lead (Pb) and Arsenic (As) by *Melastoma malabathricum* L. from Contaminated Soil in Separate Exposure. *Int. J. Phytoremediation* **2014**, *16* (7–8), 694–703.

24. Sun, Y.; Zhou, Q.; Xu, Y.; Wang, L.; Liang, X. Phytoremediation for Co-Contaminated Soils of Benzo[a]Pyrene (B[a]P) and Heavy Metals Using Ornamental Plant *Tagetes Patula*. *J. Hazard. Mater.* **2011b**, *186* (2–3), 2075–2082.

25. Liu, J. N.; Zhou, Q. X.; Sun, T.; Wang, X. F. Feasibility of Applying Ornamental Plants in Contaminated Soil Remediation. *Chinese J. Appl. Ecol.* **2007**, *18* (7), 1617–1623.

26. Liu, R.; Jadeja, R. N.; Zhou, Q.; Liu, Z. Treatment and Remediation of Petroleum-Contaminated Soils Using Selective Ornamental Plants. *Environ. Eng. Sci.* **2012**, *29* (6), 494–501.

27. Mahanta, M. J.; Bhattacharyya, K. G. Total Concentrations, Fractionation and Mobility of Heavy Metals in Soils of Urban Area of Guwahati, India. *Environ. Monit. Assess.* **2011**, *173* (1–4), 221–240.

28. Marschner, H. *Mineral Nutrition of Higher Plants*. London: Academic Press; **1995**.

29. Palmer, C. M.; Guerinot, M. Lou. Facing the Challenges of Cu, Fe and Zn Homeostasis in Plants. *Nat. Chem. Biol.* **2009**, *5* (5), 333–340.

30. Rehman, M. Z. ur; Rizwan, M.; Ghafoor, A.; Naeem, A.; Ali, S.; Sabir, M.; Qayyum, M. F. Effect of Inorganic Amendments for in Situ Stabilization of Cadmium in Contaminated Soils and Its Phyto-Availability to Wheat and Rice under Rotation. *Environ. Sci. Pollut. Res.* **2015**, *22* (21), 16897–16906.

31. Hou, D.; O'Connor, D.; Nathanail, P.; Tian, L.; Ma, Y. Integrated GIS and Multivariate Statistical Analysis for Regional Scale Assessment of Heavy Metal Soil Contamination: A Critical Review. *Environ. Pollut.* **2017**, *231*, 1188–1200.

32. Ali, B.; Wang, B.; Ali, S.; Ghani, M. A.; Hayat, M. T.; Yang, C.; Xu, L.; Zhou, W. J. 5-Aminolevulinic Acid Ameliorates the Growth, Photosynthetic Gas Exchange Capacity, and Ultrastructural Changes Under Cadmium Stress in *Brassica napus* L. *J. Plant Growth Regul.* **2013**, *32* (3), 604–614.

33. Murtaza, G.; Javed, W.; Hussain, A.; Wahid, A.; Murtaza, B.; Owens, G. Metal Uptake via Phosphate Fertilizer and City Sewage in Cereal and Legume Crops in Pakistan. *Environ. Sci. Pollut. Res.* **2015**, *22* (12), 9136–9147.

34. Abbas, T.; Rizwan, M.; Ali, S.; Zia-ur-Rehman, M.; Farooq Qayyum, M.; Abbas, F.; Hannan, F.; Rinklebe, J.; Sik Ok, Y. Effect of Biochar on Cadmium Bioavailability and Uptake in Wheat (*Triticum aestivum* L.) Grown in a Soil with Aged Contamination. *Ecotoxicol. Environ. Saf.* **2017**, *140*, 37–47.

35. Nagajyoti, P. C.; Lee, K. D.; Sreekanth, T. V. M. Heavy Metals, Occurrence and Toxicity for Plants: A Review. *Environ. Chem. Lett.* **2010**, *8* (3), 199–216.

36. Alloway, B. J. Heavy Metals in Soils. *Heavy Met. Soils* **1990**, *15* (2), VIII–IX.

37. Alloway, B. J. Sources of Heavy Metals and Metalloids in Soils. *Environmental Pollution.* Springer Netherlands **2013**, pp. 11–50. https://doi.org/10.1007/978–94-007–4470-7_2.

38. Kabata-Pendias, A. *Trace Elements in Soils and Plants.* Springer, **2010**. https://doi.org/10.1201/b10158.

39. Marples, A.; Thornton, I. The Distribution of Cd Derived from Geochemical and Industrial Sources in Agricultural Soil and Pasture Herbage in Parts of Britain. In *'Cadmium 79' Conference*; **1980**.

40. Tang, J.; Xiao, T.; Wang, S.; Lei, J.; Zhang, M.; Gong, Y.; Li, H.; Ning, Z.; He, L. High Cadmium Concentrations in Areas with Endemic Fluorosis: A Serious Hidden Toxin? *Chemosphere* **2009**, *76* (3), 300–305.

41. Baize, D.; Chrétien, J. Les Couvertures Pédologiques de La Plate-Forme Sinémurienne En Bourgogne: Particularités Morphologiques et Pédo-Géochimiques (in French, with English Summary). *Étude Gest. des Sols* **1994**, *1* (2), 7–27.

42. Tremel, A.; Masson, P.; Sterckeman, T.; Baize, D.; Mench, M. Thallium in French Agrosystems-I. Thallium Contents in Arable Soils. *Environ. Pollut.* **1997**, *95* (3), 293–302.

43. Silva, E. F. da; Mlayah, A.; Gomes, C.; Noronha, F.; Charef, A.; Sequeira, C.; Esteves, V.; Marques, A. R. F. Heavy Elements in the Phosphorite from Kalaat Khasba Mine (North-Western Tunisia): Potential Implications on the Environment and Human Health. *J. Hazard. Mater.* **2010**, *182* (1–3), 232–245.

44. Garrett, R. G.; Porter, A. R. D.; Hunt, P. A.; Lalor, G. C. The Presence of Anomalous Trace Element Levels in Present Day Jamaican Soils and the Geochemistry of Late-Miocene or Pliocene Phosphorites. *Appl. Geochem.* **2008**, *23* (4), 822–834.

45. Goldhaber, M. B.; Morrison, J. M.; Holloway, J. A. M.; Wanty, R. B.; Helsel, D. R.; Smith, D. B. A Regional Soil and Sediment Geochemical Study in Northern California. *Appl. Geochem.* **2009**, *24* (8), 1482–1499.

46. Miranda, M.; Benedito, J. L.; Blanco-Penedo, I.; López-Lamas, C.; Merino, A.; López-Alonso, M. Metal Accumulation in Cattle Raised in a Serpentine-Soil Area: Relationship between Metal Concentrations in Soil, Forage and Animal Tissues. *J. Trace Elem. Med. Biol.* **2009**, *23* (3), 231–238.

47. Oze, C.; Fendorf, S.; Bird, D. K.; Coleman, R. G. Chromium Geochemistry in Serpentinized Ultramafic Rocks and Serpentine Soils from the Franciscan Complex of California. *Am. J. Sci.* **2004**, *304* (1), 67–101.

48. Proctor, J.; Baker, A. J. M. The Importance of Nickel for Plant Growth in Ultramafic (Serpentine) Soils. In: *Toxic Metals in Soil–Plant Systems,* Ross, S.M. (ed.), John Wiley & Sons Ltd **1994**, pp. 417–432.

49. Ross, S. M; Wood, M. D.; Copplestone, D.; Warriner, M.; Crook, P. UK Soil and Herbage Pollutant Survey Environmental Concentrations of Heavy Metals in UK Soil and Herbage. *Environ. Agency, Bristol* **2007**, *7*, 21.

50. Robinson, G. R.; Larkins, P.; Boughton, C. J.; Reed, B. W.; Sibrell, P. L. Assessment of Contamination from Arsenical Pesticide Use on Orchards in the Great Valley Region, Virginia and West Virginia, USA. *J. Environ. Qual.* **2007**, *36* (3), 654–663.

51. Breward, N. Arsenic and Presumed Resistate Trace Element Geochemistry of the Lincolnshire (UK) Sedimentary Ironstones, as Revealed by a Regional Geochemical Survey Using Soil, Water and Stream Sediment Sampling. *Appl. Geochem.* **2007**, *22* (9), 1970–1993.

52. Burt, R.; Wilson, M. A.; Mays, M. D.; Lee, C. W. Major and Trace Elements of Selected Pedons in the USA. *J. Environ. Qual.* **2003**, *32* (6), 2109–2121.

53. Bai, L. Y.; Zeng, X. B.; Su, S. M.; Duan, R.; Wang, Y. N.; Gao, X. Heavy Metal Accumulation and Source Analysis in Greenhouse Soils of Wuwei District, Gansu Province, China. *Environ. Sci. Pollut. Res.* **2015**, *22* (7), 5359–5369.

54. Nanos, N.; Rodríguez Martín, J. A. Multiscale Analysis of Heavy Metal Contents in Soils: Spatial Variability in the Duero River Basin (Spain). *Geoderma* **2012**, *189–190*, 554–562.

55. Hou, Q.; Yang, Z.; Ji, J.; Yu, T.; Chen, G.; Li, J.; Xia, X.; Zhang, M.; Yuan, X. Annual Net Input Fluxes of Heavy Metals of the Agro-Ecosystem in the Yangtze River Delta, China. *J. Geochemical Explor.* **2014**, *139*, 68–84.

56. Xia, X.; Yang, Z.; Cui, Y.; Li, Y.; Hou, Q.; Yu, T. Soil Heavy Metal Concentrations and Their Typical Input and Output Fluxes on the Southern Song-Nen Plain, Heilongjiang Province, China. *J. Geochemical Explor.* **2014**, *139*, 85–96.

57. Eckel, H.; Roth, U.; Döhler, H.; Schultheiß, U. *Concerted Action Aromis Assessment and Reduction of Heavy Metal Input into Agro-Ecosystems*; Kuratorium für Technik und Bauwesen in der Landwirtschaft eV (KTBL), **2005**.

58. Luo, L.; Ma, Y.; Zhang, S.; Wei, D.; Zhu, Y. G. An Inventory of Trace Element Inputs to Agricultural Soils in China. *J. Environ. Manage.* **2009**, *90* (8), 2524–2530.

59. McLaughlin, M. J.; Palmer, L. T.; Tiller, K. G.; Beech, T. A.; Smart, M. K. Increased Soil Salinity Causes Elevated Cadmium Concentrations in Field-Grown Potato Tubers. *J. Environ. Qual.* **1994**, *23* (5), 1013–1018.

60. Kabata-Pendias, A.; Mukherjee, A. B. Trace Elements from Soil to Human. *Trace Elements from Soil to Human.* Springer Berlin Heidelberg **2007**, pp. 1–550. https://doi.org/10.1007/978-3-540-32714-1.

61. Chen, T.; Liu, X.; Zhu, M.; Zhao, K.; Wu, J.; Xu, J.; Huang, P. Identification of Trace Element Sources and Associated Risk Assessment in Vegetable Soils of the Urban-Rural Transitional Area of Hangzhou, China. *Environ. Pollut.* **2008**, *151* (1), 67–78.

62. Maiti, S. K.; Kumar, A. Energy Plantations, Medicinal and Aromatic Plants on Contaminated Soil. *Bioremediation and Bioeconomy.* Elsevier **2016**, pp 29–47. https://doi.org/10.1016/B978-0-12-802830-8.00002-2.

63. Lydakis-Simantiris, N.; Fabian, M.; Skoula, M. Cultivation of Medicinal and Aromatic Plants in Heavy Metal-Contaminated Soils. *Glob. Nest J.* **2016**, *18* (3), 630–642.

64. Kim, H. S.; Seo, B. H.; Kuppusamy, S.; Lee, Y. B.; Lee, J. H.; Yang, J. E.; Owens, G.; Kim, K. R. A DOC Coagulant, Gypsum Treatment Can Simultaneously Reduce As, Cd and Pb Uptake by Medicinal Plants Grown in Contaminated Soil. *Ecotoxicol. Environ. Saf.* **2018**, *148*, 615–619.

65. Pandey, J.; Verma, R. K.; Singh, S. Suitability of Aromatic Plants for Phytoremediation of Heavy Metal Contaminated Areas: A Review. *Int. J. Phytoremediation* **2019**, *21* (5), 405–418.

66. Trigueros, D.; Mingorance, M. D.; Rossini Oliva, S. Evaluation of the Ability of *Nerium oleander* L. to Remediate Pb-Contaminated Soils. *J. Geochemical Explor.* **2012**, *114*, 126–133.

67. Pérez-López, R.; Márquez-García, B.; Abreu, M. M.; Nieto, J. M.; Córdoba, F. *Erica Andevalensis* and *Erica Australis* Growing in the Same Extreme Environments: Phytostabilization Potential of Mining Areas. *Geoderma* **2014**, *230–231*, 194–203.

68. Chaturvedi, N.; Ahmed, M. J.; Dhal, N. K. Effects of Iron Ore Tailings on Growth and Physiological Activities of *Tagetes patula* L. *J. Soils Sediments* **2014**, *14* (4), 721–730.

69. Chintakovid, W.; Visoottiviseth, P.; Khokiattiwong, S.; Lauengsuchonkul, S. Potential of the Hybrid Marigolds for Arsenic Phytoremediation and Income Generation of Remediators in Ron Phibun District, Thailand. *Chemosphere* **2008**, *70* (8), 1532–1537.

70. Liu, J. N.; Zhou, Q. X.; Sun, T.; Ma, L. Q.; Wang, S. Identification and Chemical Enhancement of Two Ornamental Plants for Phytoremediation. *Bull. Environ. Contam. Toxicol.* **2008a**, *80* (3), 260–265.

71. Liu, J. nv; Zhou, Q. xing; Sun, T.; Ma, L. Q.; Wang, S. Growth Responses of Three Ornamental Plants to Cd and Cd-Pb Stress and Their Metal Accumulation Characteristics. *J. Hazard. Mater.* **2008b**, *151* (1), 261–267.

72. Datta, R.; Quispe, M. A.; Sarkar, D. Greenhouse Study on the Phytoremediation Potential of Vetiver Grass, *Chrysopogon Zizanioides* L., in Arsenic-Contaminated Soils. *Bull. Environ. Contam. Toxicol.* **2011**, *86* (1), 124–128.

73. Wei, J. L.; Lai, H. Y.; Chen, Z. S. Chelator Effects on Bioconcentration and Translocation of Cadmium by Hyperaccumulators, *Tagetes Patula* and *Impatiens Walleriana*. *Ecotoxicol. Environ. Saf.* **2012**, *84*, 173–178.

74. Wang, K.; Liu, Y.; Song, Z.; Wang, D.; Qiu, W. Chelator Complexes Enhanced *Amaranthus hypochondriacus* L. Phytoremediation Efficiency in Cd-Contaminated Soils. *Chemosphere* **2019**, *237*, 124480.

75. Li, N.; Li, Z.; Fu, Q.; Zhuang, P.; Guo, B.; Li, H. Agricultural Technologies for Enhancing the Phytoremediation of Cadmium-Contaminated Soil by *Amaranthus hypochondriacus* L. *Water. Air. Soil Pollut.* **2013**, *224* (9).

76. Bosiacki, M.; Kleiber, T.; Kaczmarek, J. Evaluation of Suitability of *Amaranthus caudatus* L. and *Ricinus communis* L. in Phytoextraction of Cadmium and Lead from Contaminated Substrates. *Arch. Environ. Prot.* **2013**, *39* (3), 47–59.

77. Panitlertumpai, N.; Nakbanpote, W.; Sangdee, A.; Thumanu, K.; Nakai, I.; Hokura, A. Zinc and/or Cadmium Accumulation in *Gynura pseudochina* (L.) DC. Studied in Vitro and the Effect on Crude Protein. *J. Mol. Struct.* **2013**, *1036*, 279–291.

78. Mongkhonsin, B.; Nakbanpote, W.; Nakai, I.; Hokura, A.; Jearanaikoon, N. Distribution and Speciation of Chromium Accumulated in *Gynura pseudochina* (L.) DC. *Environ. Exp. Bot.* **2011**, *74* (1), 56–64.

79. Rungruang, N.; Babela, S.; Parkpian, P. Screening of Potential Hyperaccumulator for Cadmium from Contaminated Soil. *Desalin. Water Treat.* **2011**, *32* (1–3), 19–26.

80. Sao, V.; Nakbanpote, W.; Thiravetyan, P. Cadmium Accumulation by *Axonopus Compressus* (Sw.) P. Beauv and *Cyperus Rotundas* Linn Growing in Cadmium Solution and Cadmium-Zinc Contaminated Soil. *Songklanakarin J. Sci. Technol.* **2007**, *29* (3), 881–892.

81. Dan, T. V.; KrishnaRaj, S.; Saxena, P. K. Metal Tolerance of Scented Geranium (*Pelargonium Sp. 'Frensham'*): Effects of Cadmium and Nickel on Chlorophyll Fluorescence Kinetics. *Int. J. Phytoremediation* **2000**, *2* (1), 91–104.

82. Liu, Y. T.; Chen, Z. S.; Hong, C. Y. Cadmium-Induced Physiological Response and Antioxidant Enzyme Changes in the Novel Cadmium Accumulator, *Tagetes Patula*. *J. Hazard. Mater.* **2011**, *189* (3), 724–731.

                                               Plant Abiotic Stress Physiology, Volume 1

83. Pirzadah, T. B.; Malik, B.; Dar, F. A. Phytoremediation Potential of Aromatic and Medicinal Plants: A Way Forward for Green Economy. *J. Stress Physiol. Biochem.* **2019**, *15* (3).

84. Emamverdian, A.; Ding, Y.; Mokhberdoran, F.; Xie, Y. Heavy Metal Stress and Some Mechanisms of Plant Defense Response. *Sci. World J.* **2015**, *2015*, 1–18.

85. Isah, T. Stress and Defense Responses in Plant Secondary Metabolites Production. *Biol. Res.* **2019**, *52* (1), 39.

86. Zeng, P.; Guo, Z.; Cao, X.; Xiao, X.; Liu, Y.; Shi, L. Phytostabilization Potential of Ornamental Plants Grown in Soil Contaminated with Cadmium. *Int. J. Phytoremediation* **2018**, *20* (4), 311–320.

87. Lal, K.; Minhas, P. S.; Shipra; Chaturvedi, R. K.; Yadav, R. K. Extraction of Cadmium and Tolerance of Three Annual Cut Flowers on Cd-Contaminated Soils. *Bioresour. Technol.* **2008**, *99* (5), 1006–1011.

88. Liu, Z.; He, X.; Chen, W.; Yuan, F.; Yan, K.; Tao, D. Accumulation and Tolerance Characteristics of Cadmium in a Potential Hyperaccumulator-*Lonicera Japonica* Thunb. *J. Hazard. Mater.* **2009**, *169* (1–3), 170–175.

89. Tabrizi, L.; Mohammadi, S.; Delshad, M.; Moteshare Zadeh, B. Effect of Arbuscular Mycorrhizal Fungi On Yield and Phytoremediation Performance of Pot Marigold (*Calendula officinalis* L.) Under Heavy Metals Stress. *Int. J. Phytoremediation* **2015**, *17* (12), 1244–1252.

90. Cui, S.; Zhou, Q. X.; Wei, S. H.; Zhang, W.; Cao, L.; Ren, L. P. Effects of Exogenous Chelators on Phytoavailability and Toxicity of Pb in *Zinnia Elegans* Jacq. *J. Hazard. Mater.* **2007**, *146* (1–2), 347–355.

91. Wang, Y.; Tao, J.; Dai, J. Lead Tolerance and Detoxification Mechanism of *Chlorophytum Comosum*. *African J. Biotechnol.* **2011**, *10* (65), 14516–14521.

92. Mani, D.; Kumar, C.; Patel, N. K.; Sivakumar, D. Enhanced Clean-up of Lead-Contaminated Alluvial Soil through *Chrysanthemum indicum* L. *Int. J. Environ. Sci. Technol.* **2015**, *12* (4), 1211–1222.

93. Goswami, S.; Das, S. Copper Phytoremediation Potential of *Calandula officinalis* L. and the Role of Antioxidant Enzymes in Metal Tolerance. *Ecotoxicol. Environ. Saf.* **2016**, *126*, 211–218.

94. Miao, Q.; Yan, J. Comparison of Three Ornamental Plants for Phytoextraction Potential of Chromium Removal from Tannery Sludge. *J. Mater. Cycles Waste Manag.* **2013**, *15* (1), 98–105.

95. Choo, T. P.; Lee, C. K.; Low, K. S.; Hishamuddin, O. Accumulation of Chromium (VI) from Aqueous Solutions Using Water Lilies (*Nymphaea Spontanea*). *Chemosphere* **2006**, *62* (6), 961–967.

96. Rehman, M. Z. ur; Rizwan, M.; Sohail, M. I.; Ali, S.; Waris, A. A.; Khalid, H.; Naeem, A.; Ahmad, H. R.; Rauf, A. Opportunities and Challenges in the Remediation of Metal-Contaminated Soils by Using Tobacco (*Nicotiana tabacum* L.): A Critical Review. *Environ. Sci. Pollut. Res.* **2019**, *26* (18), 18053–18070.

97. Chen, H.; Cutright, T. EDTA and HEDTA Effects on Cd, Cr, and Ni Uptake by *Helianthus Annuus*. *Chemosphere* **2001**, *45* (1), 21–28.

98. Rizwan, M.; Ali, S.; Qayyum, M. F.; Ok, Y. S.; Zia-ur-Rehman, M.; Abbas, Z.; Hannan, F. Use of Maize (*Zea mays* L.) for Phytomanagement of Cd-Contaminated Soils: A Critical Review. *Environ. Geochem. Health* **2017**, *39* (2), 259–277.

99. Rizwan, M.; Ali, S.; Zia ur Rehman, M.; Rinklebe, J.; Tsang, D. C. W.; Bashir, A.; Maqbool, A.; Tack, F. M. G.; Ok, Y. S. Cadmium Phytoremediation Potential of Brassica Crop Species: A Review. *Sci. Total Environ.* **2018**, *631–632*, 1175–1191.

100. Khalid, H.; Zia-ur-Rehman, M.; Naeem, A.; Khalid, M. U.; Rizwan, M.; Ali, S.; Umair, M.; Sohail, M. I. Solanum Nigrum L.: A Novel Hyperaccumulator for the Phyto-Management of Cadmium Contaminated Soils. In *Cadmium Toxicity and Tolerance in Plants: From Physiology to Remediation*; Elsevier, **2018**; pp. 451–477.

101. Turgut, C.; Katie Pepe, M.; Cutright, T. J. The Effect of EDTA and Citric Acid on Phytoremediation of Cd, Cr, and Ni from Soil Using *Helianthus Annuus*. *Environ. Pollut.* **2004**, *131* (1), 147–154.

102. Manzoor, M.; Gul, I.; Ahmed, I.; Zeeshan, M.; Hashmi, I.; Amin, B. A. Z.; Kallerhoff, J.; Arshad, M. Metal Tolerant Bacteria Enhanced Phytoextraction of Lead by Two Accumulator Ornamental Species. *Chemosphere* **2019**, *227*, 561–569.

103. Li, C.; Zhou, K.; Qin, W.; Tian, C.; Qi, M.; Yan, X.; Han, W. A Review on Heavy Metals Contamination in Soil: Effects, Sources, and Remediation Techniques. *Soil Sediment Contam.* **2019**, *28* (4), 380–394.

104. Tapia, Y.; Eymar, E.; Gárate, A.; Masaguer, A. Effect of Citric Acid on Metals Mobility in Pruning Wastes and Biosolids Compost and Metals Uptake in *Atriplex Halimus* and *Rosmarinus Officinalis*. *Environ. Monit. Assess.* **2013**, *185* (5), 4221–4229.

105. Tahmasbian, I.; Safari Sinegani, A. A. Chelate-Assisted Phytoextraction of Cadmium from a Mine Soil by Negatively Charged Sunflower. *Int. J. Environ. Sci. Technol.* **2014**, *11* (3), 695–702.

106. Wang, S.; Liu, J. The Effectiveness and Risk Comparison of EDTA with EGTA in Enhancing Cd Phytoextraction by *Mirabilis jalapa* L. *Environ. Monit. Assess.* **2014**, *186* (2), 751–759.

107. Wenzel, W. W.; Unterbrunner, R.; Sommer, P.; Sacco, P. Chelate-Assisted Phytoextraction Using Canola (*Brassica napus* L.) in Outdoors Pot and Lysimeter Experiments. *Plant Soil* **2003**, *249* (1), 83–96.

108. Sarret, G.; Vangronsveld, J.; Manceau, A.; Musso, M.; D'Haen, J.; Menthonnex, J. J.; Hazemann, J. L. Accumulation Forms of Zn and Pb in *Phaseolus Vulgaris* in the Presence and Absence of EDTA. *Environ. Sci. Technol.* **2001**, *35* (13), 2854–2859.

109. Farid, M.; Ali, S.; Rizwan, M.; Ali, Q.; Abbas, F.; Bukhari, S. A. H.; Saeed, R.; Wu, L. Citric Acid Assisted Phytoextraction of Chromium by Sunflower; Morpho-Physiological and Biochemical Alterations in Plants. *Ecotoxicol. Environ. Saf.* **2017**, *145*, 90–102.

110. Lai, H. Y.; Chen, Z. S. The EDTA Effect on Phytoextraction of Single and Combined Metals-Contaminated Soils Using Rainbow Pink (*Dianthus Chinensis*). *Chemosphere* **2005**, *60* (8), 1062–1071.

111. Chandra Sekhar, K.; Kamala, C. T.; Chary, N. S.; Balaram, V.; Garcia, G. Potential of Hemidesmus Indicus for Phytoextraction of Lead from Industrially Contaminated Soils. *Chemosphere* **2005**, *58* (4), 507–514.

112. Yu, Z.; Zhou, Q. Growth Responses and Cadmium Accumulation of *Mirabilis jalapa* L. under Interaction between Cadmium and Phosphorus. *J. Hazard. Mater.* **2009**, *167* (1–3), 38–43.

113. Cao, X.; Ma, L. Q.; Shiralipour, A. Effects of Compost and Phosphate Amendments on Arsenic Mobility in Soils and Arsenic Uptake by the Hyperaccumulator, *Pteris vittata* L. *Environ. Pollut.* **2003**, *126* (2), 157–167.

114. Pardo, T.; Bernal, P.; Clemente, R. The Use of Olive Mill Waste to Promote Phytoremediation. In *Olive Mill Waste: Recent Advances for Sustainable Management*; Elsevier, **2017**; pp. 183–204.

115. Meena, V. S.; Meena, S. K.; Verma, J. P.; Kumar, A.; Aeron, A.; Mishra, P. K.; Bisht, J. K.; Pattanayak, A.; Naveed, M.; Dotaniya, M. L. Plant Beneficial Rhizospheric

Microorganism (PBRM) Strategies to Improve Nutrients Use Efficiency: A Review. *Ecol. Eng.* **2017**, *107*, 8–32.

116. Wu, G.; Kang, H.; Zhang, X.; Shao, H.; Chu, L.; Ruan, C. A Critical Review on the Bio-Removal of Hazardous Heavy Metals from Contaminated Soils: Issues, Progress, Eco-Environmental Concerns and Opportunities. *J. Hazard. Mater.* **2010**, *174* (1–3), 1–8.

117. Bahadur, A.; Ahmad, R.; Afzal, A.; Feng, H.; Suthar, V.; Batool, A.; Khan, A.; Mahmood-ul-Hassan, M. The Influences of Cr-Tolerant Rhizobacteria in Phytoremediation and Attenuation of Cr (VI) Stress in Agronomic Sunflower (*Helianthus annuus* L.). *Chemosphere* **2017**, *179*, 112–119.

118. Dimkpa, C. O.; Merten, D.; Svatoš, A.; Büchel, G.; Kothe, E. Siderophores Mediate Reduced and Increased Uptake of Cadmium by *Streptomyces Tendae* F4 and Sunflower (*Helianthus Annuus*), Respectively. *J. Appl. Microbiol.* **2009**, *107* (5), 1687–1696.

119. Petriccione, M.; Di Patre, D.; Ferrante, P.; Papa, S.; Bartoli, G.; Fioretto, A.; Scortichini, M. Effects of Pseudomonas Fluorescens Seed Bioinoculation on Heavy Metal Accumulation for *Mirabilis Jalapa* Phytoextraction in Smelter-Contaminated Soil. *Water, Air, Soil Pollut.* **2013**, *224* (8), 1645.

120. Nakbanpote, W.; Meesungnoen, O.; Prasad, M. N. V. Potential of Ornamental Plants for Phytoremediation of Heavy Metals and Income Generation. *Bioremediation and Bioeconomy*. Elsevier **2016**, pp. 179–217.

121. Jisha, C. K.; Bauddh, K.; Shukla, S. K. Phytoremediation and Bioenergy Production Efficiency of Medicinal and Aromatic Plants. *Phytoremediation Potential of Bioenergy Plants*. Springer Singapore; **2017**, pp. 287–304.

122. Ekor, M. The Growing Use of Herbal Medicines: Issues Relating to Adverse Reactions and Challenges in Monitoring Safety. *Front. Neurol.* **2014**, *4*, 177.

123. Verma, S. K.; Singh, K.; Gupta, A. K.; Pandey, V. C.; Trivedi, P.; Verma, R. K.; Patra, D. D. Aromatic Grasses for Phytomanagement of Coal Fly Ash Hazards. *Ecol. Eng.* **2014**, *73*, 425–428.

124. Zinicovscaia, I.; Gundorina, S.; Vergel, K.; Grozdov, D.; Ciocarlan, A.; Aricu, A.; Dragalin, I.; Ciocarlan, N. Elemental Analysis of Lamiaceae Medicinal and Aromatic Plants Growing in the Republic of Moldova Using Neutron Activation Analysis. *Phytochem. Lett.* **2020**, *35*, 119–127.

125. Suman, J.; Uhlik, O.; Viktorova, J.; Macek, T. Phytoextraction of Heavy Metals: A Promising Tool for Clean-up of Polluted Environment? *Front. Plant Sci.* **2018**, *871*.

126. Thomas, E.; Omueti, J. The Effect of Phosphate Fertilizer on Heavy Metal in Soils and *Amaranthus Caudatus*. *Agric. Biol. J. North Am.* **2012**, *3* (4), 145–149.

CHAPTER 11

Role of Electromagnetic Radiation in Abiotic Stress Tolerance

ABDELGHAFAR M. ABU-ELSAOUD[1], AWATIF M. ABDULMAJEED[2], HAIFA ABDULAZIZ S. ALHAITHLOUL[3], and MONA H. SOLIMAN[4*]

[1]*Botany Department, Faculty of Science, Suez Canal University, Ismailia, Egypt*

[2]*Biology Department, Faculty of Science, University of Tabuk, Umluj 46429, Saudi Arabia*

[3]*Biology Department, College of Science, Jouf University, Sakaka 2014, Kingdom of Saudi Arabia*

[4]*Botany and Microbiology Department, Faculty of Science, Cairo University, Giza 12613, Egypt*

Corresponding author. E-mail: monahsh1@gmail.com.

ABSTRACT

Plants and specially crop plants are exposed to numerous abiotic stresses that restrict germination, growth, and development. Chilling, salinity, radiations, water deficit, and heat abiotic stresses have a deleterious impact on crop plants. In the current climate change period, a great drop in crop productivity is compromising the efforts/strategies used for sustainable agricultural practices. Accordingly, plant physiologists and photo biologists are trying to enhance the tolerance against the deleterious effects of abiotic stresses especially during the current era of climate change. In the current chapter, an assessment has been made in the introduction section to firstly the nature and role of light in plant life, with a great concern to various environmental electromagnetic radiations (EMRs). The effect of EMRs on plants cell physiology and metabolic processes were also assessed in the current chapter.

EMRs have a noticeable effect on plants' physiology especially water uptake and cell turgor, membranes, minerals and nutrients, cell signals, energy and cell bioenergetics, reactive oxygen species (ROS), and oxidative stresses. In the second part of this chapter, we attempt to summarize the role of various EMRs (e.g., laser, polarized light, light-emitting diodes (LEDs), visible light, etc.) in enhancing germination, plant growth, and abiotic stress tolerance. This chapter and assessment may contribute to the development of various strategies for protecting crop plants against abiotic stress enhancing their tolerance.

11.1   INTRODUCTION TO LIGHT AND LIFE

For most life on earth, light is considered as a primary energy source. Various living species can absorb and use light energy for example plants/autotrophs can scavenge and photosynthesize light energy. Nevertheless, light is more than catalyst of energy-requiring metabolism; it reflects on the status of atmosphere by its efficiency, strength, and association with other environmental influences [1]. It is therefore significant for understanding the nature of organisms and their functioning [2].

Photomorphogenesis is the organism's developmental response to light, either quality, quantity, and direction or the relative day/night lengths (photo period). Light perception is carried out by photoreceptors, molecules that scavenge light leading to biological reactions [1]. Photostimulation process is involved in activating/enhancing biological processes by light wavelengths and the light used is the photostimulator.

11.1.1   *LIGHT AS A PART OF ELECTROMAGNETIC SPECTRUM*

Solar energy is a result of nuclear fusion in the sun, which is considered as the essential nonnuclear energy source on the earth. Total yearly solar energy captured by earth is approximately 5.62×1024 J, of which photosynthesis uses 3.16×1021 J year^{-1}. Visible light is a portion of EMR spectra ranges from γ- and X-rays to radio waves. Light is characterized by both "wave" and "particle." Light move as individual "photons" that travel in waves. The light wavelength (λ) is generally expressed in nanometers (nm). Visible light spectrum ranges from wavelength of 380–760 nm as shown in Figure 11.1. Equation (11.1) represents the relation between light wavelength, frequency, and the speed of light [1].

$$c = \nu\lambda \qquad (11.1)$$

$$E = hc/\lambda \qquad (11.2)$$

where c = light speed ($\sim 300 \times 10^6$ m s^{-1}) and h = Planck's constant (4.14×10^{-15} eV s), E = 1240 eV nm.

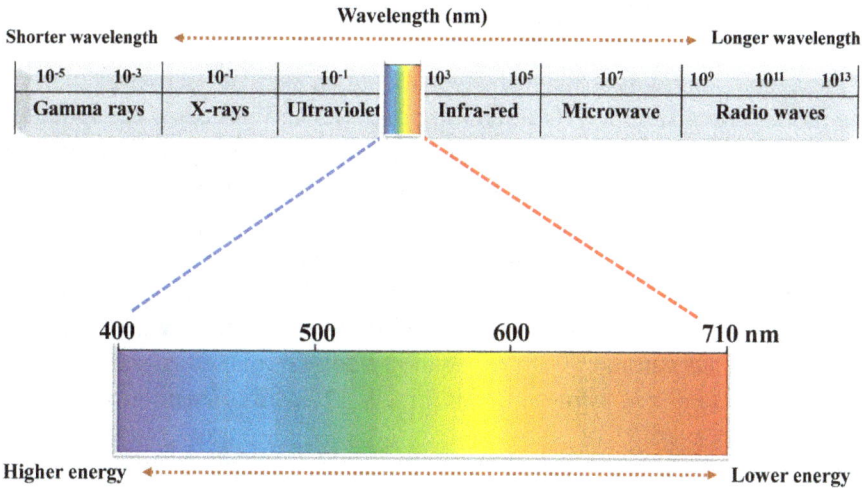

FIGURE 11.1   The electromagnetic spectrum along with visible light, energy is presented in J mol^{-1} [1].

11.1.2 PHOTOBIOLOGY

Photobiology is the science of interrelationship between life and light through photoreceptors, molecules that scavenge light leading to biological responses [1]. Photostimulation involves light activation of biological process and the type of light used is the photostimulators. Photoreceptors enable the organism to track various environmental fluctuations and rhythms and to appropriately adjust metabolism and physiology.

Plants and microorganisms contain various photoreceptors including phytochromes, phototropins, cryptochromes, and photosynthetic pigments, each possess a specific absorption spectrum, which stimulates specific responses. Graphing specific physiological outcome versus the wavelengths, it creates action spectrum, which identify photoreceptor for that response.

EMR including ultraviolet (UV), visible light, microwave, and laser can promote bioactive metabolites production from actinomycetes. In a study, a volcanic cave strain of actinomycetes strains in western Canada in exposure to UV light produced a new antimicrobial compounds versus six pathogens of multidrug resistant [3]. Four ways were hypothesized by which EM radiations are fundamental for life, these include mutagenesis, photosynthesis, thermal effects, and photomorphogenesis.

11.1.3   LIGHT POLARIZATION

Light is considered as a transverse wave, that is, oscillate perpendicularly to the propagation direction. No vibrating particles in light case, but change in magnetic and electric fields, both are perpendicular to propagation direction and to one another. Polarized light is composed of a beam with electric field component that are parallel [2]. The polarization plane is the plane comprising both the propagation line and the electric field direction. Light can be plane polarized when two beams propagate in similar direction having the same phase. Moreover, light can be circularly polarized, where electrical field is directed spirally with propagation line. Circular polarization can be left- and right handed as presented in Figure 11.2. Circular-polarized light is assumed as a product of similarly two strong right-angled plane-polarized compartments and 90 Å phase change [2].

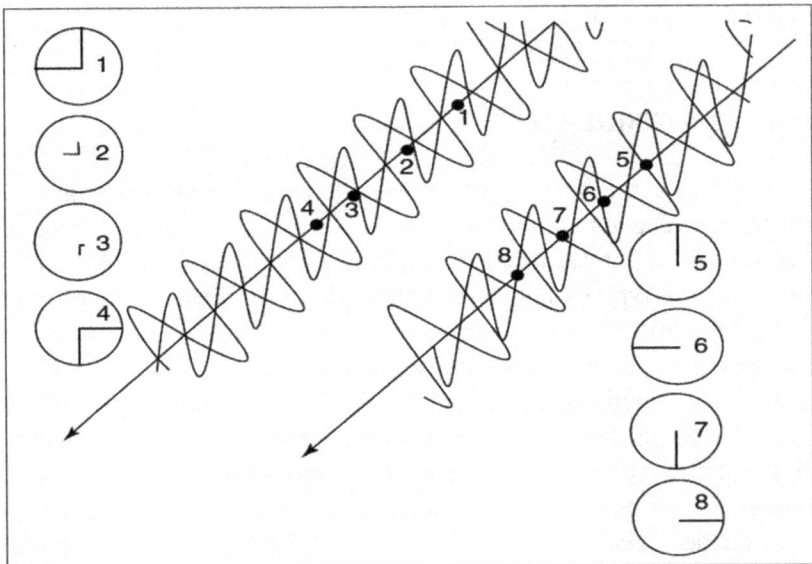

FIGURE 11.2 Plane-polarized wave, consisting of vertical and horizontally polarized portion, in the upper left part of the diagram. However, a circularly polarized beam in the lower right section.

Natural light is mostly unpolarized, that is, a spontaneous combination of every probable polarization. The light becomes partly plane polarized after reflected in water. Sunlight is a combination of circular and plane polarized, that is, elliptically polarized. Human cannot receive and recognize the light polarization directly; however, insects could and regularly apply polarization as a tool in positioning. In certain cases, plants respond differently to plane-polarized light according to their polarization axis. This holds in seed and nonseed plants, algae for chloroplast orientation, and for the development of gametophytes.

11.2 LIGHT GENERATION FOR PHOTOBIOLOGICAL RESEARCH

Nearly, all-natural light on earths comes from the sun, which in its totality, radiates like glowing blackbody with a temperature of ~6000 K. Certain sunlight wavelength components were presented here to be used in photobiological research including UV, IR, etc.

11.2.1 UV RADIATIONS

Solar light is important to life, nevertheless UV portion can have a deleterious effect on living organisms. UV radiations show higher quantum features than both visible and infrared (IR). Due to its theoretical impact, it is randomly recognized into UV-A (315–400 nm), UV-B (280–315 nm), and UV-C (100–280 nm). Owing to the absorption of stratospheric ozone, earth's atmosphere is mostly shielded from the most harmful UV-B wavelengths. UV-A is the least hazardous and is most widely found since it has the least energy. It is also referred to as blacklight, which is known for its apparent harmlessness. UV-A lamps are found in most phototherapies and tanning booths [4–6]. Although, UV-A radiation is less harmful than UV-B, it has various effects on air and water quality, troposphere, and can cause various human immune suppression [7].

UV-B is the most damaging UV wavelengths, since it has sufficient energy to destroy tissues, but not sufficient for the environment to consume it entirely. UV-B is thought to cause skin cancer. Since much UV-B radiation is absorbed by the atmosphere, a slight shift in ozone layer may significantly cause increased skin cancer. UV radiations particularly UV-B could have a positive regulation for plant response against plants' pests and pathogens [6]. UV radiation can either have a negative effect on microorganisms, UV-A, and UV-B. UV-C is absorbed entirely in the air, as its photons from interact with oxygen helping in ozone formation. UV-C lamps can be used to sterilized air and water, as it can kill bacteria.

11.2.2   LEDS

LEDs have various applications that do not require very strong light. Several types of spectral emission LEDs are currently available including red, green, blue, yellow, IR, and UV (A, B, and C). LEDs sources are not monochromatic sources, they represent a broadband spectrum. For certain forms, the distribution of pollution varies with the current activity. LEDs have the advantage that can be operated by a source of low volt (1.5–5 V) using a resistor that restricts current [8].

Traditional LEDs are composed of inorganic semiconductor, for example, GaAsP, SiC, InGaN, and GaN. Several labs and corporations have recently initiated an organic-based LEDs (OLEDs). OLEDs would possibly expand the spectra of available LEDs. A number of infrared emitting diodes with peak emission wavelengths of more than >4.5 μm and a 210 nm LED are being developed [9]. A significant recent invention is the design of an LED capable of processing a single photon at a time [8].

11.2.3   LASERS

Laser is an abbreviation for Light Amplification by Stimulated Emission of Radiation. Stimulated emission happens when a photon stimulates a molecule in an excited state to emits a second photon and decay to ground state. No special equipment is required for the stimulated emission as such. It happens occasionally as molecules are excited by photons with the correct energy encounter. In general, however, excited molecules are very unusual relative to ground-state molecules. For a laser to operate, an appropriate optical setup must also reduce photon errors, often requiring mirrors. Laser light has peculiar characteristics including coherence, collimated and monochromatic, and laser could be plane polarized.

Even the low-power lasers, for example, helium–neon (He–Ne) laser, should be handled with some precaution as it has a damaging effect on eye and retina. The low-power He–Ne laser is the most frequently used with wavelength of 632.8 nm. Yttrium aluminum garnet (YAG) laser is also used in photobiological purposes. They can emit IR radiations of 1060 nm wavelength. For photobiology research, YAG laser can be associated with frequency doubler consisting of crystals of potassium phosphate, generating a green light with wavelength of 530 nm (Table 11.1).

11.2.4 IR RADIATION

IR radiation is part of the electromagnetic spectrum, where *infra* means "below,": and thus spectrum lay just below the region of red color [15]. This form of EMR comprises approximately half of the sun's energy. Although, it cannot be physically observed or felt, IR radiation transfers rapidly as heat from one body to the other. The spectrum of IR radiation is divided into near-IR, mid-IR, and far-IR that corresponds to 760–1400 nm, 1400–3000 nm, and 3000 nm–1 mm, respectively [16]. Mid-IR is called the "growth ray" or "resonant frequency" as it is absorbed efficiently by living tissues and helps promote their growth, while far-IR is called "biogenetic radiation" [17].

11.3 EMR AFFECTS PLANT CELLS PHYSIOLOGICAL AND METABOLIC RESPONSES

Plants are essential components of a successful ecosystem and have a key role in the survival of living producers; therefore, the plants' biochemical and physiological response to specific magnetic fields at molecular and whole levels are crucial and beneficial. The electromagnetic bioeffect facilitates the learning experience as a new and dynamic knowledge of research involving both synthetic biology and electromagnetic frequencies (EMF) sciences. In this particular respect, in vitro and in vivo experiments should be seen as environmental signs that contribute to magnetic fields [18]. In addition, electromagnetic irradiation has led to many alterations in enzyme activity [19] and gene expression [20, 21].

11.3.1 WATER UPTAKE AND CELL TURGOR

Water permeation rate is the initial important phase of germination. Daily oxygen supply, optimum temperature, and water status are the crucial phases for germination and seedling development in natural conditions. Stressful environmental conditions, such as salinity, significantly affecting water quality thereby reducing the seed-imbibing water availability and decreasing

TABLE 11.1 Photostimulation by EMR

EM Source	Wavelength (λ; nm)	Photo Stimulation	Species	References
Nd-YAG laser	500–600	Enhanced cell division, synthesis of protein and cholesterol deprivation	*Actinomycetes: Streptomyces fradiae*	[10]
R-Polarized laser and rLED (red-LED)	632.8	Enhanced Ag-NPs biosynthesis	*Fungi: Trichoderma viridae Chaetomium globosum*	[11]
UV-A, UV-B	320–400 280–320	Changed fungal structure, enhanced production of UV-absorbing compounds, especially MAAs	*Fungi: Drechslera* sp. *Paecilomyces* sp.,	(unpublished data) [12]
He–Ne laser	400–500, 600	Enhanced cell division and synthesis of protein	*Streptomyces fradiae*	
He–Ne laser		Enhanced phenol hydroxylase enzyme activity	*Candida tropicalis*	[13]
He–Ne laser, UV		Increased production/antioxidant activities of endo-polysaccharides	*Phellinus igniarius*	[14]

root initiation [22]. The exact mechanisms for germination and growth were still elusive under influence of MF exposure. Walleczek [23] showed critical physiological effects, with a special relationship to cation levels such as Ca^{2+} ions on the cell membrane.

This induced changes and imbalances in ion strength that influenced the absorption of water through the structure of the plasma membrane [24, 25]. Lower turgidity, for example, leaf growth, root penetration in the soil and stomatal conductivity contributes to a significant reduction in carbon dioxide (CO_2) inflows and photosynthetic activity, as well as water permeability, which contributes to an obstacle to plant growth and development [26].

11.3.2 THE MEMBRANE PHASE RESPONSES

Several experiments have shown that the plasma cellular membrane region is the target area of the RF–EMF [27, 28]. The energy is required to produce the regular power source, which is considered to be the mediator of the entire plant's injury and radiation dose response [29, 30]. Increasing the activity of the plasma membrane, enzyme has been shown to be the target point of contact with electromagnetic waves. In addition, EM-induced genetic information has already been stated in cell-free formulations [31–33]. As with conventional contact between a magnetic field and any mobile charge, it can interact with the moving charge (i.e., load flow) of a cell with a similar propensity to the mobile interactions [34]. If load flow is connected to a biofunction, such as the enzyme activity, the procedure involves changes in reaction rate as seen in modified Na, K-ATPase, or cytochrome oxidase responses, in proportion to load release [34–37].

11.3.3 MINERALS AS ESSENTIAL NUTRIENTS AND CELL SIGNALS

The enzyme activity and the reinforcement of defense mechanisms, K^+ and Ca^{2+} ions exhibit a significant function in permanent stability and cellular membranes function [38–40]. Ca^{2+} is highly modulated by EMR exposure and can be observed through the orientated activation of a signal sensor pathway [41, 42] via the interface between the membrane and cytoplasm and directly on the ion itself. This factor is important in the existence of biochemical and gene expression mechanisms and in ion equilibrium regimes [32, 43]. Moreover, the combination of tobacco (Nicotiana tabacum) and soybean protoplast was studied with the use of ferromagnetic shelters for the effects

Plant Abiotic Stress Physiology, Volume 1

of thin, alternate MF, which were modified with the cyclotron frequency Ca^{2+} and K^+ ions. Protoplastic fusion was observed to increase its frequency by two and three occasions under these conditions [44, 45], with Ca^{2+} ions involved in protoplastic fusion induction. Available evidence by Shine et al. [46] recommended that Ca^{2+} have a vital role as cellular signal ions and is influenced by magnetic field impacts in improving the rate of mitotic division, cell growth, and all cell conductivity [47]. Esitken and Turan [48] stated that MF could worsen the nutrient intake for major macronutrients at identified potentials. The electrical phase includes changes to the exchange of ions. The impact of large municipal MF diffusion rate of ions on spatial dispersion across the roots, cytosol flowing, and cell growth mechanisms associated with intracellular weight and load transmission has been established [49].

11.3.4   EMR AND ENERGY RESPONSES OF THE CELL

Plants that are sessile and unable to flee the ecosystem need to try extensive modifications and should continuously track their behavior [21, 50–52]. Apparently, plants can identify electromagnetic waves [51]. The potential internal energy of the plant cells is significantly affected by the production of ATP, and AEC levels and plant cells require a sufficient concentration of AEC for functional maintenance and variability in this range leading to changes in cell energy metabolism [53]. Loss in ATP and AEC is mainly due to the modulation of cellular processes under stressful conditions such as water shortage and hypoxia and, as a result, profoundly damaging cell metabolism [54]. The partial deficiency of ATP is most probably due to its utilization during transcriptional regulation of energy-dependent mechanisms. Partial depletion of ATP is highly likely to result from its intake through energy-dependent transcription and translation. The usual RF–EMF-evoked mRNA accumulating was totally eliminated, indicating a straight linkage concerning energy potential of cells and stress response genesis, when proton gradient was interrupted by the usage of protonophore CCPC along-with subsequent decrease in ATP formation [55].

11.3.5   EMR SEEKS TO PROMOTE PHOTOSYNTHETIC ACTIVITY AND BIOMASS PRODUCTION

Photosynthesis is a biochemical process that begins with the absorption of light energy from proteins called reaction centers, transforming light into potential stored energy and associated biomass output. Electromagnetic waves in terms of quality and quantity of frequency range are the motivating factors for beneficial effects in stimulating and enhancing physiological plant reactions [56]. Various growth parameters have been strongly promoted in the sense of longer exposures to EMR for 30 min, such as mitotic activity [57] and superoxide dismutase (SOD) and catalase (CAT) activity [58], all cumulatively improve photosynthetic efficiency. From the applied perspective, light efficacy has been stated to take a marked consequence on metabolic pathways of exopolysaccharides (mainly EPS) in cyanobacteria [59] and algae [60]. The paramagnetic nature of certain molecules in plant cells and pigments, that is, chloroplast, may be used as a probability to describe the beneficial impact of magnetic field. Activation and increased performance of cellular components are explained by the magnetic molecular structures and the ability to accumulate greater energy and convert it into biological energy (ATP) [61, 62]. Results show that the improvement in potato tuber biomass production can be explained partially by microwaves under selected strengths and wavelengths [63].

Besides these, more stimulating results on bean, and other plant species have been recorded [64–66]. Significant improvement in plant morphology, such as plant height; deeper roots enable absorbing water and nutrient; and more biomass would have a greater photosynthetic rate and increase the stress tolerance of plant water shortages [67]. In contrast to the control of sunflower seedlings, increased enzyme activity, including amylase, dehydrogenase, and protease, as well as the development of biomass levels, has been shown at MF exposure [68]. This is how further improvements in chl-a and both nonphotochemical and photochemical quenching in two plant species of maize can be achieved using MF with different treatments and thus stimulates water shortage stress resistance [69]. Improving the rate of biomass accumulation in soyabean can be achieved by presowing magnetic treatments [46]. Other applications of MF for photosynthetic activity have been documented for several crop plants [70–73]. Decreased photosynthesis rates are often the result of decreased internal CO_2 levels for MF-exposed radish seedlings as well as inhibitory photosynthetic enzymes (e.g., Rubisco) [74–77]. Interruption of energy metabolism due to exposure to UV-B results in increased suppression of biosynthesized enzymes and elimination of PS II in thylakoid membranes [25], including downregulation of Rubisco [78], which could hinder photosynthetic activity and impeded seedling growth with ultimate death [76, 79, 80]. In addition, researchers have recently confirmed that EMR has negatively impacted photosynthesis by disrupting chloroplast membranes and lowering electron-transport channels with concomitant inactivation of photochemical performance of photosystem II potential activity [81]. ATP synthase bioaccumulation as well as photo-synthesizing carbon capture and sequestration are significantly affected by electromagnetic waves, which could have significant effects on *Microcystis aeruginosa* thylakoid cells [82].

11.3.6   EMR: BIOLOGICAL IMPACTS ON ROS AND OXIDATIVE STRESS

The fast-growing environmental damage caused by the widespread use of electromagnetic energy is one of the biggest problems currently identified [83]. Despite the difficulties, the development of interesting data concerning altered morphological characteristics has been successful in recent years, including outcomes of yield and biomass, inner anatomy changes, tissue development, and metabolites production following long-term irradiation [84, 85]. Impact of environmental stress factor on the respiratory and photo-synthetic mechanisms have become the priority of researchers to provide a clearer understanding of the regulatory performances under extreme conditions. In plants, under normal conditions as well, ROS are generated mostly in four cellular compartments: mitochondria, chloroplasts, cytosol, and peroxisomes [75, 86–88]. ROS triggers serious problems that can affect plant growth and development processing. The toxic radicals include hydrogen peroxide, superoxide, and hydroxyl that result in peroxidation of lipids and proteins [89]. Chloroplasts act as major source of ROS in leaves [90]. ROS generated in leaf tissues promote chlorophyll degradation, which is precisely regulated and results in senescence [91]. Stress-induced chlorophyll degradation and the disassembly of photosynthetic machinery are reflected in reduced CO_2 fixation decreases and increased generation of superoxide anions (O_2^-) and singlet oxygen ($^1O_2$) [92].

ROS may be used for dual purposes, as low levels are well known for signaling receptor activation molecules, cell proliferation, and other important metabolic activities [74, 88]. The free radical nonconventional mechanism of cell membranes and ROS accumulation can contribute to the effects of MFs, as well as the antioxidant activity of the enzyme [93]. Oxidant markers, including lipid peroxidation and H_2O_2 overproduction, are correlated with EMR as a potential stress response and have a measurable impact on anti-oxidant defense mechanisms in leaf tissues likely to cause redox disturbance [94]. In addition, these metabolic oxidative stress expressions negatively affect photosynthetic activity as well as growth parameters [74, 80, 95, 96].

Efforts have been made to reduce the destructive impact of ROS in plants, first to minimize additional ROS generated and to protect significant structures. Both are synchronized with the indigenous immune responses. The antioxidants, consisting of both enzymatic and nonenzymatic components, are included within the protective mechanisms. CAT, ascorbate peroxidase (APX), SOD, glutathione reductase, dehydroascorbate reductase, monodehydroascorbate reductase, and glutathione peroxidase are the enzymatic components of this pathway [74, 76, 86, 97]. Whereas ascorbate, tripeptide glutathione (reduced and oxidized), tocopherols, anthocyanins, carotenoids, and secondary metabolites, including polyphenols and flavonoids are included in nonenzymatic antioxidants as well as in detoxifying properties [98]. Overproduction of ROS and lower pigment levels were identifiable factors developed in *Nerium oleander* or *Phlomis fruticoses* with exposure to nonionized radiation [99–101]. Electromagnetic potential applications are commonly defined in recent studies with respect to the protective function of potential antioxidant dopamines, such as ROS scavenger and strong defense molecules [102, 103].

11.3.7 LASER TECHNOLOGY CAN BE A CONVENIENT PHYSICAL STRATEGY TO IMPROVE GERMINATION AND GROWTH

Current developments in global civilization have negatively affected our climate. Rapid urbanization of the earth has contributed to an increase in energy and food use. As a result, the need to protect crop losses and start producing additional feed and food materials to fulfill the standards of the over-growing people has just become extremely relevant [104]. The use of physical factors that may be essential for organic farmers is a current trend toward conjoining the development of plant technology with environmental necessities to track biological change during advancement [105]. Various practical uses of physical factors, including electromagnetic wave treatment, in particular optical, ultrasound, magnetic field, and ionizing radiation, are used to optimize the production of vegetables [105, 106]. As the most effective physical methods in agriculture, the development of new techniques allowing laser radiation to improve crops and plant growth [69]. Germination is the first sensitive stage leading to the emergence of radicles and plums and the reactivation of the metabolic seed metabolites [107, 108]. The germination and growth of the seeds depend primarily on internal and external environmental factors [109, 110]. Laboratory trials with the application of laser technology have already shown that appropriate radiation levels possess a beneficial influence on plant growth and physiological metabolism and can be observed clearly in enhancing seedling tolerance to abiotic stress for example chilling, salinity as well as heavy metal pollution [64, 82, 111]. Light, either visible or IR, has been recognized as a key pivotal factor of the metabolic processes during seedling growth and development. In addition, research has shown that the germination of seed can be improved with milliseconds and 5 s by the weaker exposure to the sun [112].

A significant amount of research is concerned with young plants of different crops that have also shown proliferative activity and cell reproductive capacity in the meristem of plant roots [44]. Multiple metabolic activities are significantly achieved with the potential energy of seeds. This energy supplied as bio-stimulator via light or external resources influences the conversion of cotyledon dry weight into usable food storage and accelerating germination rate and seedling growth [113]. Photon absorption also benefits from preventing tissue breakdown or organ harm and disease infection without risk [55, 106, 114, 115]. In addition, physical treatments provide environmentally sustainable strategies through the reduction of chemical fertilizers and weed growth control [116–118]. The enhancements in germination and growth of tomatoes, cucumbers, and maize have been improved by different MF exposure levels [64, 119, 120]. On the other side, laser stimulation does not induce adverse genetic issues, which are extremely necessary for healthy and safe food production and sustainable agriculture, which cumulatively enhance massive food accumulation with the best quality [113, 121, 122].

11.3.8 EMF EMITTED BY MOBILE PHONES IMPACT GERMINATION AND GROWTH

Mobile telephones are the easiest and fastest way to connect with modern society. However, it is used to add substantially to the environment's EMF-are radiation, and therefore their impacts on biological organisms are to be researched Mobile cell phones as wireless technology and radiation-based devices could threaten human health and safety of all living environments [123]. The wireless EMF interact with the different morphological or physiological processes and are easily representable, depending on the mode of conduct, period exposure to radiation, and tissue thickness [21]. Afzal and Mansoor [124] analyzed the impact of cell phone exposure to both Monocotyledon (wheat) and Dicotyledon (mung) and observed: unchanged germination, inhibited seedling growth, decreased protein content, significantly increased the activity of ROS metabolism [124]. Net photosynthetic incidence and stomatal conductivity in three Aromatic Plant Species had shown negative responses to the Global Mobile Radiation System and a significant reduction was reported [125]. In reverse, species-by-species output and plant growth responses vary, when exposed to low-frequency EMFs (425 MHz, 2 h, 1 mW), root elongation in both mung bean and water convolvulus seedlings is stimulated, whereas duck weed (*Lemna minor,* Araceae) has decreased dramatically with similar exposure levels [67, 126, 127]. In addition, the low-frequency MF (16 Hz) can improve seeds at postharvesting stage for different plant species, in particular temperature-sensitive seeds that germinate at low temperatures [71].

11.3.9 EMR NULLIFIES ABIOTIC ENVIRONMENTAL STRESS-THREATENING AGRICULTURAL PRODUCTIVITY, LASER AS POTENTIAL ENERGY CAN BE A SOLUTION TO MITIGATE SALINITY STRESS

With continuous advances in environmental issues, the need for research on different stresses is expanding. Salinity is one of the main environmental threat and is steadily increasing, despite irrigation as a barrier to the use of brackish and marine water in agriculture regions around the world. The impact of output is almost 70% [128–131]. A global problem, decreased yields of agriculture and food saturated by soil salinity primarily lead to negative effects in the specific growth stage, ranging from plant growth to seedling maturation and yield [79, 129, 132–135]. The early osmotic and subsequent slow ionic phases are two major stresses affecting salinity in plants. In recent decades, the use of laser radiation as physical new technology strategies for measuring plant quality has accelerated in the biology and agriculture fields [115, 123, 136, 137]. Soil salinity can lead to Na+ plant cell toxicity, which affects the absorption and retaining of key ions essential for plant development such as K^+. Diode laser radiation significantly improved germination rate and counteract salinity stress [138]. In addition, Hernandez et al. [111] documented that laser accelerates growing emergence of speed and percentage of seedlings, dry mass, and finally, the final yield will be improved greatly. The yield per plant has increased significantly for seeds exposed to laser radiation of 45 and 30 min and under salt stress [139]. Several of the reasons for the emergence of germination could be due to extra energy from the plant during laser beam emissions. Physically, the increase in plant growth of pretreated laser seeds relative to control plants is due to the breakdown of the kinetic balance of germinated seeds and the increase in the heat content associated with internal seed energy [140, 141]. Qiu et al. [142] documented that plants pretreatment with incident light for a long period would have built up an additional tolerance to abiotic stresses. Moreover, it is confirmed in a study that not only thermal values of germination seed have been hugely enhanced, but also physiological and biochemical pathways are upregulated and seedlings have increased in terms of growth and development as a result of laser radiation manipulation [143, 144]. In addition, the pretreatment with laser light seed resulted in a noticeable rise in phosphorus, protein, and molybdenum content of the plant's dry weight and a decline in crude fiber [145].

11.3.10 ROLE OF LASER AND ELECTROMAGNETIC WAVES AS A USEFUL TOOL TO REDUCE DROUGHT STRESS AND PROMOTING AGRICULTURAL OUTPUT

Drought stress restricts agronomic sustainability and causes a pronounced crop yield damage by interfering biochemical and physiological tasks, which impede the plant growth as well as development [146–150]. Furthermore, oxidative damage occurs under drought condition cause electron leakage to O_2 in photosynthesis and respiration processes, which lead to accumulation of ROS in terms of O_2^-, O_2^1, hydroxyl, or perhydroxyl radicals (OH^-), and hydrogen peroxide [151–153]. A variety of abiotic stress consequences might be diminished with He–Ne laser pretreatment, biological reactions, and phenotypes of wheat plants might be considerably improved under drought condition and along with improved resistance to cold stress [154], UV-B radiation [155], cadmium stress [156], and salt stress [157]. In addition, from molecular modifications, gene expression in He–Ne laser pretreated wheat under water-deficit continues to be mentioned with reaction to transcripts that could feature within the photosynthetic activity upregulation and cellular antioxidant response [155]. Generally, the equilibrium between both formations as well as use of ROS under maximum growing environments is securely managed by a suitable and cost-efficient antioxidant structure [79]. Some other aspects talk about protein modification under EMF exposure, Jangid et al. [20] offered observational proof (RAPD) proposing high-power microwave heating (2450 MHz, 800 W cm^2) alters the transcription of genes within *Vigna aconitifolia*. Various biological parameters such as leaf water content, photosynthetic rate, and chlorophyll builds up an amount and stomatal conductance had been enhanced under drought stress, when plants exposed to MF of various strengths. Totally free radicals as ROS particles may also be inhibited as well as the antioxidant potential is simultaneously higher [158]. Out of the biological element, Nasiri et al. [158] clearly demonstrate

MF raises the seed coating's membrane fluidity by decreasing the osmotic impact plus giving drought tolerance resistance. Multiple experiments have found that the inhibitory impact of heat and drought stress can be decreased with MF-treatment [159], its saline-alkali tolerance improved [160] and an extending life cycle [161].

Drought pressure restricts agronomic sustainability as well as triggered a pronounced crop yield damage by interfering biochemical and physiological tasks, which impede the plant development as well as the development [146–150].

11.3.11 EMR AND FLOODING STRESS

Waterlogging, such as additional environmental stresses, is a crucial area of relevance as well as exploration interest and avital determinant of plant development. Waterlogging became a worldwide issue for agriculture. Climate modification initiated by extra water of rainfall and drivers must have been viewed as the principal explanation for flooding [162–165]. The soil is considered waterlogged if the standing waters are 20% higher than the field capacity [166]. As a direct consequence, groundwater-hypoxia or anoxia constraints due to submergence have been considered to be the key effect impacting the stage of plant development, survival, and furthermore agricultural production [162, 167–169]. Plant growth and development with enhanced yields in anoxic conditions are the most intensive strategy for the monitoring of crop production for sustainable development [170, 171]. Flood stressed plants are adapted phenotypically, biologically, anatomically, and physiologically to minimize the negative impact of low oxygen concentration [172–174].

In addition, development programs are initiated through the gene expression regulation and adaptation mechanisms including expansion of adventitious root [172, 175] or the ion uptake regulation [176, 177] to mitigate the flooding stress [170, 178, 179]. Root modifications play a pivotal role in resisting waterlogging effects. Increased accumulation of both suberin and lignin and emergence of an adventitious root system has been shown in tomato and wheat [180–182]. Hypoxia impairs aerobic metabolism leading to reduced plant growth by influencing photosynthesis, nutrient, and hormonal balance, resulting in sluggish growth and reduced crop yield [183, 184]. In addition, cell energy and metabolic ATP synthesis are significantly inadequate and declined by oxidative phosphorylation pathways [162, 185]. Stimulated stomatal closure restricts CO_2 in plant cells and induces free radical accumulation that induces oxidative damage and eventually programmed cell death due to aerenchyma and ethylene development [186, 187].

With the development of wireless communication technology, further future studies and specialization have been made therefore to contribute to the increase in electromagnetic field exposure (EMF). Despite the large studies for this promotional effect on the plant, not too many studies have investigated the effects of abiotic stress. Waterlogging as abiotic stress impact soybean growth and development has recently been investigated. As for Zhong et al. [188] earlier work, his technique of using low doses of magnetic radiation offers better performance over gene expression related to hormonal levels and metabolic functions and photosynthetic pigment economic output. Besides that, with low irradiation rates and increased tolerance of soybean in the event of a flood disaster, the agronomic traits of the soybean growth criteria, including root and hypocotyl, have improved significantly. Many studies consistently report a close and strong association between Trehalose and Trehalose-6P storage metabolites and improved *Arabidopsis Thaalyan* at mitotic stage and photosynthetic seedlings, as well as excellent results in increased flood tolerance for EMR crop plants [188–191]. Further support is given by Li et al. [192] that more Tre6P synthase upregulated by low-dose UV-B radiation and better carbohydrate utilization. Several published studies examined the influence of lower dose UV-B radiation on the pathways of carbon resources biosynthetic pathway, which include gene expression of Prupe.1G159700 (sugar-phosphate synthase, EC:2.4.1.14) as a vital component of sucrose-6-phosphate production [193, 194]. Balakhnina [195] found that the potential function of plant antioxidants as biochemical parameters most contributes to enhance root hypoxia resistance with MF pretreatment. A recent study has been further supported by typical hormonal balance and UV-B as a link, including photochemical, storage, biosynthesis and/or signaling pathways [196].

11.4   ROLE OF EMR IN ABIOTIC STRESS

11.4.1   EMRS

EMR comprises energy of both wave- and particle-like properties that is formed from continuously varying electric and magnetic fields. Such radiation comprises visible light, radio waves, gamma rays, UV radiation, microwaves, X-rays, and radio waves, etc. Many diverse forms of EMR are crucial for sustaining life on Earth; photosynthesis, thermal energy, photomorphogenesis, and genetic alterations are the four major ways in which EMR interacts with biotic entities. Some specific sources that can emit EMR of very specific wavelengths, such as LEDs and lasers, are used as photostimulators owing to their capability to enhance bioprocesses in microorganisms [197], for instance, improved production of bioactive secondary metabolites in actinomycetes [198]. Regardless, the mechanisms by which monochromatic radiation interacts with biological objects are not established well. This is largely due to challenges in observing the transformation between light and energy within cells and in integrating the complex responses of the numerous levels constituting cellular systems [199].

11.4.2 INFLUENCE OF SALINITY ON PHYSIOLOGICAL FUNCTIONS IN PLANTS

Abiotic stresses, such as salinity, drought, and high or low-temperature influence the health of plants by interacting with their molecular, biochemical, and physiological systems [76, 200]. If a plant species does not have sufficient tolerance, its growth and development can be severely affected [38].

11.4.3 EFFECT OF SALINITY ON PROLINE ACCUMULATION

One of the amino acids that are majorly affected by salinity, extreme temperatures, heavy metals, and UV radiations is proline. Plants can accumulate unusual amounts of proline when they are exposed to drought or salinity [201]. Proline and hydroxyproline are found in specific compounds. Under normal circumstances, proline is involved in proteinogenic functions in plant cells by acting as an osmolyte, radical scavenger, electron sink, macromolecule stabilizer, and even as a structural component of the cell wall [202, 203]. In high amounts, proline induces an increase in the activity of glutamate kinase that, in turn, increases biosynthesis of proline [203]. The increased proline content is used to synthesize proline-rich proteins involved in fighting abiotic stress [204]. As proline accumulates with time, its synthesis is inhibited. Therefore, the accumulation of proline is dependent on many physiological and cellular relationships [203].

11.4.4   GENERATION OF ROS UNDER CONDITIONS OF EXCESS SALINITY

The generation of ROS is triggered under abiotic stress to an extent that their production no longer remains in balance with the quenching activity of antioxidants, which leads to oxidative damage in the cell [75, 76, 205]. Major forms of oxidative damage by ROS are lipid peroxidation, destruction of the cell membrane, protein denaturation, and DNA damage [92]. Naturally, the activity of antioxidant enzymes, such as SOD, CAT, APX, and peroxidases, is increased, when the production of ROS is increased under stress [206]. These enzymes scavenge ROS to prevent cellular damage. Considering that ROS are involved in numerous functions in plant metabolism, it is plausible to assume that protective pathways in plants against different types of stress are not independent of one another. Tolerance mechanisms to varied abiotic stress factors, such as drought, salinity, and exposure to UV radiation, are overlapping and interrelated. Studies on physiological responses in plants have reported levels of antioxidants to be correlated with stress tolerance [206]. For instance, the activity of antioxidant enzymes is proven to be correlated with salt tolerance [75]. The overall response of plants to increased salinity is known to manifest as osmotic stress, toxicity to specific ions, and nutrient deficiency [38, 74, 75, 207]. Other research has similarly reported it to result in a reduction in water and osmotic potential, leaf area, and stomatal density [207]. As a result of compromised stomatal and nonstomatal components of photosynthesis, CO_2 assimilation is reduced as well [208–210]. Therefore, all the plant responses to stress factors negatively influence a range of physiological processes involved in cell metabolism.

11.5   EXPERIMENTAL MEANS TO OBSERVE STRESS RESPONSE IN PLANTS BY IR RADIATION

11.5.1   IR THERMOGRAPHY

IR thermography is a process by which an EMR spectrum can be plotted based on the amount of thermal energy or temperature released from each specific radiation present [211]. This is a promising research technique that allows plant response to be measured in the form of changes in wavelength and other informative parameters of EMR emitted from plants.

The principle of IR thermography is based on the fact that the temperature of a plant is proportional to thermal radiation emitted by it. This technique, therefore, detects low-energy EMR emitted and/or reflected from different parts of a plant using a microbolometer, which is a specialized detector designed to measure changes in temperature. The distribution of temperature all across the surface of the plant is then visualized as a graph.

Drought or salinity can induce several common physiological responses in plants, such as a reduction in water retention and compromised photosynthetic activity [212]. This response is generally a result of cellular dehydration in plants and may particularly occur during the early onset of the stress [74, 210, 213]. Drought and salinity may lead to a partial closure of stomata that not only causes a decrease in CO_2 availability for photosynthesis but also elevates the temperature of leaves [214]. In such a situation, regulation of stomata and water content in plants in response to abiotic stress factors is crucial. Most importantly, these physiological responses help stabilize the temperature of the entire plant or individual leaves. Generally, plants with water deficiency will have a higher temperature, and thus, will display more IR radiation. Under conditions of abiotic stresses, such as drought and salinity, the water content in plants is affected, where plant response may vary from sensitive to tolerant levels. Therefore, IR images captured from plants can indicate their tolerance levels and can be used to perform phenotyping for the selection of drought- or salinity-tolerant genotypes.

11.5.2 APPLICATION OF IR RADIATION TO IMPROVE FOOD QUALITY

IR radiation has been applied in many different ways to achieve improvement in food quality and safety. Some common applications toward this end include quality inspection, drying, blanching, and sterilization [215, 216]. The potential of IR radiation to heat up surfaces, and thus, be used in the peeling process has been investigated well. The earliest report on the use of IR radiation to assist in heating during lye peeling combined dry-caustic peeling with IR irradiation [217]. In this method, heat from IR radiation was used to promote chemical reactions on the surface of vegetables sprayed with lye. The application of IR radiation on plants has to be done very carefully. Extended exposure to IR radiation for heating has been reported to cause deterioration in the quality of peach in the form of nutritional loss, while less heating required may cause reduction in permeability [218].

11.5.3   SCREENING FOR SALINITY TOLERANCE USING IR THERMOGRAPHY

Salinity is a significant limiting factor in crop production that can cause severe damage and harmful physiological changes in crops [76]. Leaf temperature varies with transpiration that indicates that it has a role in stomatal conductance. High temperature causes stomatal closure, while open stomata usually exist at low temperatures [74]. IR thermography is a valuable technique to screen stomatal behavior [214], crop water stress, and water use [218]. It can also be used to assess osmotic stress tolerance in cereal crops. Flexas [214] indicated that variation in stomatal conductance could be detected using a high-resolution imager, and the findings could be applied toward the identification of genotypic variation in stomatal response [219]. The precision of this technique depends on measuring the stomatal conductance of fully expanded leaves: Once before exposure of plants to salt treatment and then once after the desired salt concentration is achieved. Applying these, two steps help to prevent variation in stomatal conductance because of salt-induced changes in leaf morphology and the accumulation of salt in leaves to potentially toxic concentrations. Salt accumulation in plant tissues increases osmotic stress and transpiration rates [76]. As a result, the latent-heat flux in leaves decreases, leading to an increased surface temperature. This manifests has an increased difference in temperatures of plants grown under salt treatment and that of the control plants.

11.5.4   APPLICATION OF IR THERMOGRAPHY TO OBSERVE METHANE EMISSION

It is a well-known fact that methane is a greenhouse gas that has been contributing significantly to global warming [218]. Methane has a high global warming potential, which reach to 36 times more than CO_2 [220]. A high concentration of this gas in environmental systems, other than the atmosphere, may cause much damage [221]. Its concentration can be increased mostly by anthropogenic sources, but natural sinks, such as methane-oxidizing bacteria, can oxidize, and thus, consume methane gas in the atmosphere [17]. In an experiment based on laboratory incubations, the soil was exposed to IR radiation, Harte [222] observed high consumption of methane in soil that was collected at a depth of 12 and 25 cm as compared to soil that was sampled near the surface. Limited microbial activity on the surface was attributed to the lowered moisture in surface soil due to an increase in temperature from exposure to IR radiation [223]. A loss of methane oxidation from reduced microbial activity can lead to positive feedback between methane flux and global warming [224]. It will be reasonable to conclude that the global methane budget can clearly be altered by the extent of methane oxidation in soil. IR imaging can be used to estimate CH_4 by measuring methane leakage via computerized visualization approaches. This technique can effectively detect CH_4 emissions automatically that removes uncertainty associated with operator experience; this technique can locate CH_4 leaks, sizes, and imaging distances [225].

11.6 CONCLUSION

IR radiation is a low-energy portion of the EMR that can readily transfer heat from one body to the other. In the presence of abiotic stress factors, such as salinity, drought, and extreme temperatures, the molecular, biochemical, and physiological systems of plants are compromised. Plants respond in a multitude of ways, for instance higher production of proline and/or the release of ROS. One response to stress is the partial closure of stomata in response to salinity that leads to an increased temperature in plants that is measurable by a technique, IR thermography. This is a promising technique that allows reliable estimation of a plant's response to stress by measuring the amount of thermal energy or temperature released from each specific radiation present from different parts of the plant. Its principle of working further allows its effective use for estimation of N mineralization in soil, as well as emission of CH_4.

KEYWORDS

- abiotic stress
- photobiology
- laser
- ultraviolet
- crop plants
- stress tolerance
- acclimation

REFERENCES

1. Jones, R.L., ed., The molecular life of plants, Wiley-Blackwell, Chichester, West Sussex; Hoboken, NJ, 2011.
2. Björn, L.O., ed., Photobiology, Springer New York, New York, NY, USA, 2015. http:// link.springer.com/10.1007/978-1-4939-1468-5 (accessed July 13, 2015).
3. Rule, D., Cheeptham, N., The effects of UV light on the antimicrobial activities of cave actinomycetes, Int. J. Speleol. 42 (2013) 7.
4. McKenzie, R.L., Aucamp, P.J., Bais, A.F., Björn, L.O., Ilyas, M., Madronich, S., Ozone depletion and climate change: impacts on UV radiation, Photochem. Photobiol. Sci. Off. J. Eur. Photochem. Assoc. Eur. Soc. Photobiol. 10 (2011) 182–198.
5. Bais, A.F., McKenzie, R.L., Bernhard, G., Aucamp, P.J., Ilyas, M., Madronich, S., Tourpali, K., Ozone depletion and climate change: impacts on UV radiation, Photochem. Photobiol. Sci. Off. J. Eur. Photochem. Assoc. Eur. Soc. Photobiol. 14 (2015) 19–52.
6. Williamson, C.E., Zepp, R.G., Lucas, R.M., Madronich, S., Austin, A.T., Ballaré, C.L., Norval, M., Sulzberger, B., Bais, A.F., McKenzie, R.L., Robinson, S.A., Häder, D.-P, Paul, N.D., Bornman, J.F., Solar ultraviolet radiation in a changing climate, Nat. Clim. Change. 4 (2014) 434–441.
7. Damian, D.L., Matthews, Y.J., Phan, T.A., Hallida, G.M.y, An action spectrum for ultraviolet radiation-induced immunosuppression in humans, Br. J. Dermatol. 164 (2011) 657–659.
8. Yuan, Z., Kardynal, B.E., Stevenson, R.M., Shields, A.J., Lobo, C.J., Cooper, K., Beattie, N.S., Ritchie, D.A., Pepper, M., Electrically driven single-photon source, Science. 295 (2002) 102–105.
9. Taniyasu, Y., Kasu, M., Makimoto, T., An aluminium nitride light-emitting diode with a wavelength of 210 nanometres, Nature. 441 (2006) 325–328.
10. Ouf, S.A., Alsarran, A.Q.i, Al-Adly, A.A., Ibrahim, M.K., Evaluation of low-intensity laser radiation on stimulating the cholesterol degrading activity: Part I. Microorganisms isolated from cholesterol-rich materials, Saudi J. Biol. Sci. 19 (2012) 185–193.
11. Abu-Elsaoud, A.M., Abdel-Azeem, A.M., Mousa, S.A., Hassan, S.S., Biosynthesis, optimisation and photostimulation of α-NADPH-Dependent Nitrate Reductase-mediated silver nanoparticles by Egyptian endophytic fungi, Adv. Environ. Biol. 9 (2015) 259–269.
12. Abu-Elsaoud, A.M., Abdel-Azeem, A.M., Elian, M., Loutfy, A., Mycosporine like amino acids and UV-absorbing compounds production in Fungi under enhanced Levels of ultraviolet radiations, 2019. unpublished data.
13. Jiang, Y., Wen, J., Jia, X., Caiyin, Q., Z. Hu, Mutation of Candida tropicalis by irradiation with a He-Ne laser to increase its ability to degrade phenol, Appl. Environ. Microbiol. 73 (2007) 226–231.
14. Zhang, H.-N, Ma, H.-L., Zhou, C.-S, Yan, Y., Yin, X.-L, Yan, J.-K, Enhanced production and antioxidant activity of endo-polysaccharides from Phellinus igniarius mutants screened by low power He-Ne laser and ultraviolet induction, Bioact. Carbohydr. Diet. Fibre. (2016). https://doi.org/10.1016/j.bcdf.2016.11.006.
15. Johnston, S.F., A history of light and colour measurement: science in the shadows, CRC Press, 2015.
16. Kim, H.H., Wheeler, R.M., Sager, J.C., Gains, G.D., Naikane, J.H., Evaluation of lettuce growth using supplemental green light with red and blue light-emitting diodes

in a controlled environment-a review of research at Kennedy Space Center, In: V International Symposium on Light in Horticulture.. 711, 2005: pp. 111–120.

17. Cristiano, L., Use of infrared-based devices in aesthetic medicine and for beauty and wellness treatments, Infrared Phys. Technol. 102 (2019) 102991.

18. Santini, S.J., Cordone, V., Falone, S., Mijit, M., Tatone, C., Amicarelli, F., G. Di Emidio, Role of mitochondria in the oxidative stress induced by electromagnetic fields: focus on reproductive systems, Oxid. Med. Cell. Longev. 2018 (2018) 1–18.

19. Kouzmanova, M., Dimitrova, M., Dragolova, D., Atanasova, G., Atanasov, N., Alterations in enzyme activities in leaves after exposure of *Plectranthus sp.* plants to 900 MHZ electromagnetic field, Biotechnol. Biotechnol. Equip. 23 (2009) 611–615.

20. Jangid, R.K., Sharma, R., Sudarsan, Y., Eapen, S., Singh, G., Purohit, A.K., Microwave treatment induced mutations and altered gene expression in Vigna aconitifolia, Biol. Plant. 54 (2010) 703–706.

21. Vian, A., Davies, E., Gendraud, M., Bonnet, P., Plant responses to high frequency electromagnetic fields, BioMed Res. Int. 2016 (2016).

22. Saberali, S.F., Moradi, M., Effect of salinity on germination and seedling growth of Trigonella foenum-graecum, Dracocephalum moldavica, Satureja hortensis and Anethum graveolens, J. Saudi Soc. Agric. Sci. 18 (2019) 316–323.

23. Walleczek, J., Electromagnetic field effects on cells of the immune system: the role of calcium signaling, FASEB J. 6 (1992) 3177–3185.

24. García-López, J., Lorite, I.J., García-Ruiz, R., Domínguez, J., Evaluation of three simulation approaches for assessing yield of rainfed sunflower in a Mediterranean environment for climate change impact modelling, Clim. Change. 124 (2014) 147–162.

25. Reina, F.G., Pascual, L.A., Influence of a stationary magnetic field on water relations in lettuce seeds. Part I: Theoretical considerations, Bioelectromagnetics. 22 (2001) 589–595.

26. Bolaños, J.A., Longstreth, D.J., Salinity effects on water potential components and Bulk elastic modulus of alternanthera philoxeroides (Mart.) Griseb., Plant Physiol. 75 (1984) 281–284.

27. Astumian, R.D., Adiabatic pumping mechanism for ion motive ATPases, Phys. Rev. Lett. 91 (2003) 118102.

28. Galvanovskis, J., Sandblom, J., Amplification of electromagnetic signals by ion channels, Biophys. J. 73 (1997) 3056–3065.

29. Beaubois, E., Girard, S., Lallechere, S., Davies, E., Paladian, F., Bonnet, P., Ledoigt, G., Vian, A., Intercellular communication in plants: evidence for two rapidly transmitted systemic signals generated in response to electromagnetic field stimulation in tomato, Plant Cell Environ. 30 (2007) 834–844.

30. Davies, E., Stankovic, B., Electrical signals, the cytoskeleton, and gene expression: a hypothesis on the coherence of the cellular responses to environmental insult, In: Communication in Plants, Springer, 2006: pp. 309–320.

31. Goodman, E.M., Greenebaum, B., Marron, M.T., Altered protein synthesis in a cell-free system exposed to a sinusoidal magnetic field, Biochim. Biophys. Acta BBA-Protein Struct. Mol. Enzymol. 1202 (1993) 107–112.

32. Goodman, R., Blank, M., Insights into electromagnetic interaction mechanisms, J. Cell. Physiol. 192 (2002) 16–22.

33. Tuinstra, R., Greenebaum, B., Goodman, E.M., Effects of magnetic fields on cell-free transcription in E. coli and HeLa extracts, Bioelectrochem. Bioenerg. 43 (1997) 7–12.

34. Blank, M., ed., Electromagnetic fields: biological interactions and mechanisms, 1 edition, American Chemical Society, Washington, DC, 1995.
35. Blank, M., Soo, L., Enhancement of cytochrome oxidase activity in 60 Hz magnetic fields, Bioelectrochem. Bioenerg. 45 (1998) 253–259.
36. David, M. Influence of electromagnetic fields on biological signalling: an experimental and theoretical approach, Dissertation thesis, AMS Dottorato (2013). https://doi. org/10.6092/UNIBO/AMSDOTTORATO/5952.
37. Pressley, T.A., Haber, R.S., Loeb, J.N., Edelman, I.S., F. Ismail-Beigi, Stimulation of Na, K-activated adenosine triphosphatase and active transport by low external K+ in a rat liver cell line, J. Gen. Physiol. 87 (1986) 591–606.
38. Munns, R., Comparative physiology of salt and water stress, Plant Cell Environ. 25 (2002) 239–250.
39. C.F. de Lacerda, Cambraia, J., Oliva, M.A., Ruiz, H.A., Osmotic adjustment in roots and leaves of two sorghum genotypes under NaCl stress, Braz. J. Plant Physiol. 15 (2003) 113–118.
40. Wei, W., Bilsborrow, P.E., Hooley, P., Fincham, D.A., Lombi, E., Forster, B.P., Salinity induced differences in growth, ion distribution and partitioning in barley between the cultivar Maythorpe and its derived mutant Golden Promise, Plant Soil. 250 (2003) 183–191.
41. Trewavas, A., Signal perception and transduction, In: Buchanan, B, Biochemistry and Molecular Biology of Plantsm, American Society of Plant Physiologists, 2000: pp 930–988.
42. Trewavas, A.J., Malho, R., Signal perception and transduction: the origin of the phenotype, Plant Cell. 9 (1997) 1181–1195.
43. Aldinucci, C., Garcia, J.B., Palmi, M., Sgaragli, G., Benocci, A., Meini, A., Pessina, F., Rossi, C., Bonechi, C., Pessina, G.P., The effect of exposure to high flux density static and pulsed magnetic fields on lymphocyte function, Bioelectromagn. J. Bioelectromagn. Soc. Soc. Phys. Regul. Biol. Med. Eur. Bioelectromagn. Assoc. 24 (2003) 373–379.
44. Belyavskaya, N.A., Biological effects due to weak magnetic field on plants, Adv. Space Res. 34 (2004) 1566–1574.
45. Nedukha, O., Kordyum, E., Bogatina, N., Sobol, M., Vorobyeva, T., Ovcharenko, Y., The influence of combined magnetic field on the fusion of plant protoplasts, J. Gravitational Physiol. J. Int. Soc. Gravitational Physiol. 14 (2007) P117–P118.
46. Shine, M.B., Guruprasad, K.N., Anand, A., Enhancement of germination, growth, and photosynthesis in soybean by pre-treatment of seeds with magnetic field, Bioelectromagnetics. 32 (2011) 474–484.
47. Szcześ, A., Chibowski, E., Hołysz, L., Rafalski, P., Effects of static magnetic field on electrolyte solutions under kinetic condition, J. Phys. Chem. A. 115 (2011) 5449–5452.
48. Eşitken, A., Turan, M., Alternating magnetic field effects on yield and plant nutrient element composition of strawberry (*Fragaria x ananassa cv. camarosa*), Acta Agric. Scand. Sect. B—Soil Plant Sci. 54 (2004) 135–139.
49. Kondrachuk, A., Belyavskaya, N., The influence of the HGMF on mass-charge transfer in gravisensing cells., J. Gravitational Physiol. J. Int. Soc. Gravitational Physiol. 8 (2001) P37–8.
50. Ferguson, I.B., The plant response: stress in the daily environment, J. Zhejiang Univ.-Sci. A. 5 (2004) 129–132.
51. Galland, P., Pazur, A., Magnetoreception in plants, J. Plant Res. 118 (2005) 371–389.

52. Vian, A., Henry-Vian, C., Schantz, R., Schantz, Davies, M.-L., G'rard Ledoigt, E., Desbiez, M.-O, Effect of calcium and calcium-counteracting drugs on the response of Bidens pilosa L. to wounding, Plant Cell Physiol. 38 (1997) 751–753.

53. Moal, J., Le Coz, J.-R., Samain, J.-F., Daniel, J.-Y., Bodoy, A., Oyster adenylate energy charge: response to levels of food, Aquat. Living Resour. 4 (1991) 133–138.

54. Dobrota, C., Energy dependant plant stress acclimation, Rev. Environ. Sci. Biotechnol. 5 (2006) 243–251.

55. Roux, D., Vian, A., Girard, S., Bonnet, P., Paladian, F., Davies, E., Ledoigt, G., High frequency (900 MHz) low amplitude (5 V m- 1) electromagnetic field: a genuine environmental stimulus that affects transcription, translation, calcium and energy charge in tomato, Planta. 227 (2008) 883–891.

56. Sanmartín, P., Vázquez-Nion, D., Arines, J., Cabo-Domínguez, L., Prieto, B., Controlling growth and colour of phototrophs by using simple and inexpensive coloured lighting: A preliminary study in the Light4Heritage project towards future strategies for outdoor illumination, Int. Biodeterior. Biodegrad. 122 (2017) 107–115.

57. Răcuciu, M., 50 Hz frequency magnetic field effects on mitotic activity in the maize root, Romanian J. Biophys. 21, 2011.

58. Çelik, Ö., Büyükuslu, N., Atak, Ç., Rzakoulieva, A., Effects of magnetic field on activity of superoxide dismutase and catalase in Glycine max (L.) Merr. Roots, Pol. J. Environ. Stud. 18 (2009) 175–182.

59. Han, P., Shen, S., Wang, H.-Y., Sun, Y., Dai, Y., Jia, S., Comparative metabolomic analysis of the effects of light quality on polysaccharide production of cyanobacterium Nostoc flagelliforme, Algal Res. 9 (2015) 143–150.

60. You, T., Barnett, S.M., Effect of light quality on production of extracellular polysaccharides and growth rate of Porphyridium cruentum, Biochem. Eng. J. 19 (2004) 251–258.

61. Arasimowicz, M., Floryszak-Wieczorek, J. Nitric oxide as a bioactive signalling molecule in plant stress responses, Plant Sci. 172 (2007) 876–887.

62. Zlatev, Z., Lidon, F.C., An overview on drought induced changes in plant growth, water relationsand photosynthesis, Emir. J. Food Agric. (2012) 57–72.

63. Jakubowski, T., The impact of microwave radiation at different frequencies on weight of seed potato germs and crop of potato tubers, Inż. Rol. R. 14, nr 6 (2010) 57–64.

64. Aladjadjiyan, A., Study of the influence of magnetic field on some biological characteristics of Zea mais, J. Cent. Eur. Agric. 3 (2002) 89–94.

65. Aladjadjiyan, A., Kakanakova, A., Physical methods in agro-food chain, J. Cent. Eur. Agric. 9 (2008) 789–793.

66. Naz, A., Jamil, Y., Iqbal, M., Ahmad, M.R., Ashraf, M.I., Ahmad, R., Enhancement in the germination, growth and yield of okra (Abelmoschus esculentus) using pre-sowing magnetic treatment of seeds, Indian J. Biochem. Biophys. 49 (2012) 211–214.

67. Ragha, L., Mishra, S., Ramachandran, V., Bhatia, M.S., Effects of low-power microwave fields on seed germination and growth rate, J. Electromagn. Anal. Appl. 2011 (2011).

68. Vashisth, A., Nagarajan, S., Effect on germination and early growth characteristics in sunflower (Helianthus annuus) seeds exposed to static magnetic field, J. Plant Physiol. 167 (2010) 149–156.

69. Javed, N., Ashraf, M., Akram, N.A., Al-Qurainy, F., Alleviation of adverse effects of drought stress on growth and some potential physiological attributes in maize (Zea mays L.) by seed electromagnetic treatment, Photochem. Photobiol. 87 (2011) 1354–1362.

70. Radhakrishnan, R., Kumari, B.D.R., Influence of pulsed magnetic field on soybean (*Glycine max* L.) seed germination, seedling growth and soil microbial population, Indian J Biochem Biophys, 50, 312–317, (2013).

71. Rochalska, M., Orzeszko-Rywka, A., Magnetic field treatment improves seed performance, Seed Sci. Technol. 33 (2005) 669–674.

72. Turker, M., Temirci, C., Battal, P., Erez, M.E., The effects of an artificial and static magnetic field on plant growth, chlorophyll and phytohormone levels in maize and sunflower plants, Phyton Ann Rei Bot. 46 (2007) 271–284.

73. Voznyak, V.M., Ganago, I.B., Moskalenko, A.A., Elfimov, E.I., Magnetic field-induced fluorescence changes in chlorophyll-proteins enriched with P-700, Biochim. Biophys. Acta BBA—Bioenerg. 592 (1980) 364–368.

74. Alhaithloul, H.A., Soliman, M.H., Ameta, K.L., El-Esawi, M.A., Elkelish, A., Changes in ecophysiology, osmolytes, and secondary metabolites of the medicinal plants of mentha piperita and catharanthus roseus subjected to drought and heat stress, Biomolecules. 10 (2020) 43.

75. Soliman, M., Elkelish, A., Souad, T., Alhaithloul, H., Farooq, M., Brassinosteroid seed priming with nitrogen supplementation improves salt tolerance in soybean, Physiol. Mol. Biol. Plants. 26 (2020) 501–511.

76. Soliman, M.H., Abdulmajeed, A.M., Alhaithloul, H., Alharbi, B.M., El-Esawi, M.A., Hasanuzzaman, M., Elkelish, A., Saponin biopriming positively stimulates antioxidants defense, osmolytes metabolism and ionic status to confer salt stress tolerance in soybean, Acta Physiol. Plant. 42 (2020) 114.

77. Yano, A., Ohashi, Y., Hirasaki, T., Fujiwara, K., Effects of a 60 Hz magnetic field on photosynthetic CO_2 uptake and early growth of radish seedlings, Bioelectromagnetics. 25 (2004) 572–581.

78. Sita, K., Sehgal, A., Bhandari, K., Kumar, J., Kumar, S., Singh, S., Siddique, K.H., Nayy, H.ar, Impact of heat stress during seed filling on seed quality and seed yield in lentil (Lens culinaris Medikus) genotypes, J. Sci. Food Agric. 98 (2018) 5134–5141.

79. Elkeilsh, A., Awad, Y.M., Soliman, M.H., Abu-Elsaoud, A., Abdelhamid, M.T., El-Metwally, I.M., Exogenous application of β-sitosterol mediated growth and yield improvement in water-stressed wheat (*Triticum aestivum*) involves up-regulated antioxidant system, J. Plant Res. (2019).

80. Soliman, M., Alhaithloul, H.A., Hakeem, K.R., Alharbi, B.M., El-Esawi, M., Elkelish, A., Exogenous nitric oxide mitigates nickel-induced oxidative damage in eggplant by upregulating antioxidants, osmolyte metabolism, and glyoxalase systems, Plants. 8 (2019) 562.

81. Tufescu, F., Creanga, D.E., Microwave effects upon vegetal cell cultures, In: Recent Advances in Multidisciplinary Applied Physics, Elsevier, 2005: pp. 931–941.

82. Tang, C., Yang, C., H. Yu, Tian, S., Huang, X., Wang, W., Cai, P., Electromagnetic radiation disturbed the photosynthesis of microcystis aeruginosa at the proteomics Level, Sci. Rep. 8 (2018) 479.

83. Sharma, S., Parihar, L., Effect of mobile phone radiation on nodule formation in the leguminous plants, Curr. World Environ. J. 9 (2014) 145–155.

84. Stefi, A.L., Margaritis, L.H., Christodoulakis, N.S., The effect of the non ionizing radiation on exposed, laboratory cultivated upland cotton (*Gossypium hirsutum* L.) plants, Flora. 226 (2017) 55–64.

85. Stefi, A.L., Margaritis, L.H., Christodoulakis, N.S., The effect of the non-ionizing radiation on exposed, laboratory cultivated maize (*Zea mays* L.) plants, Flora. 233 (2017) 22–30.

86. Ahmad, P., Jaleel, C.A., Salem, M.A., Nabi, G., Sharma, S., Roles of enzymatic and nonenzymatic antioxidants in plants during abiotic stress, Crit. Rev. Biotechnol. 30 (2010) 161–175.

87. Mittler, R., Vanderauwera, S., Gollery, M., F. Van Breusegem, Reactive oxygen gene network of plants, Trends Plant Sci. 9 (2004) 490–498.

88. Rogers, H., Munné-Bosch, S. Production and scavenging of reactive oxygen species and redox signaling during leaf and flower senescence: similar but different, Plant Physiol. 171 (2016) 1560–1568.

89. Ahmad, P., Growth and antioxidant responses in mustard (*Brassica juncea* L.) plants subjected to combined effect of gibberellic acid and salinity, Arch. Agron. Soil Sci. 56 (2010) 575–588.

90. Ishida, H., Izumi, M., Wada, S., Makino, A., Roles of autophagy in chloroplast recycling, Biochim. Biophys. Acta BBA - Bioenerg. 1837 (2014) 512–521.

91. Xie, Q., Michaeli, S., Peled-Zehavi, H., Galili, G., Chloroplast degradation: one organelle, multiple degradation pathways, Trends Plant Sci. 20 (2015) 264–265.

92. Pintó-Marijuan, M., Munné-Bosch, S. Photo-oxidative stress markers as a measure of abiotic stress-induced leaf senescence: advantages and limitations, J. Exp. Bot. 65 (2014) 3845–3857.

93. Maffei, M.E., Magnetic field effects on plant growth, development, and evolution, Front. Plant Sci. 5 (2014).

94. Cakmak, T., Cakmak, Z.E., Dumlupinar, R., Tekinay, T., Analysis of apoplastic and symplastic antioxidant system in shallot leaves: Impacts of weak static electric and magnetic field, J. Plant Physiol. 169 (2012) 1066–1073.

95. Rizhsky, L., Liang, H., Shuman, J., Shulaev, V., Davletova, S., Mittler, R., When defense pathways collide. The response of Arabidopsis to a combination of drought and heat stress, Plant Physiol. 134 (2004) 1683–1696.

96. Zhang, H., Sonnewald, U., Differences and commonalities of plant responses to single and combined stresses, Plant J. 90 (2017) 839–855.

97. Ahanger, M.A., Tomar, N.S., Tittal, M., Argal, S., Agarwal, R.M., Plant growth under water/salt stress: ROS production; antioxidants and significance of added potassium under such conditions, Physiol. Mol. Biol. Plants. 23 (2017) 731–744.

98. M.R. McCall, Frei, B., Can antioxidant vitamins materially reduce oxidative damage in humans?, Free Radic. Biol. Med. 26 (1999) 1034–1053.

99. Stefi, A.L., Mitsigiorgi, K., Vassilacopou, D., Christodoulakis, N.S., Response of young Nerium oleander plants to long-term non-ionizing radiation, Planta. 251 (2020) 108

100. Stefi, A.L., Vassilacopou, D., Christodoulakis, N.S., Environmentally stressed summer leaves of the seasonally dimorphic *Phlomis fruticosa* and the relief through the L-Dopa decarboxylase (DDC), Flora. (2019). https://agris.fao.org/agris-search/search.do?recordID=US201900047265 (accessed July 18, 2020).

101. Suzuki, N., Koussevitzky, S., Mittler, R.O.N., Miller, G.A.D., ROS and redox signalling in the response of plants to abiotic stress, Plant Cell Environ. 35 (2012) 259–270.

102. Kanazawa, K., Sakakibara, H., High content of dopamine, a strong antioxidant, in cavendish banana, J. Agric. Food Chem. 48 (2000) 844–848.

103. Kulma, A., Szopa, J., Catecholamines are active compounds in plants, Plant Sci. 172 (2007) 433–440

104. Wani, S.H., Dutta, T., Neelapu, N.R.R., Surekha, C., Transgenic approaches to enhance salt and drought tolerance in plants, Plant Gene. 11 (2017) 219–231.

105. Aladjadjiyan, A., Physical factors for plant growth stimulation improve food quality, In: A. Aladjadjiyan (Ed.), Food Production—Approaches, Challenges and Tasks, InTech, 2012

106. Hasan, M., Hanafiah, M.M., Aeyad Taha, Z., AlHilfy, I.H.H., Said, M.N.M. Laser irradiation effects at different wavelengths on phenology and yield components of pretreated maize seed, Appl. Sci. 10 (2020) 1189.

107. Bahran, A.I, Pourreza, J., Gibberellic acid and salicylic acid effects on seed germination and seedlings growth of wheat (*Triticum aestivum* L.) under salt stress condition, World Appl. Sci. J. 18 (2012) 633–641.

108. De Villiers, A.J., van Rooyen, M.W., Theron, G.K., van de Venter, H.A. Germination of three Namaqualand pioneer species, as influenced by salinity, temperature and light, Proceedings of the International Seed Testing Association. 22 (1994) 427–433.

109. Fuller, M.P., Hamza, J.H., Rihan, H.Z., Al-Issawi, M. Germination of Primed Seed under NaCl Stress in Wheat, ISRN Bot. 2012 (2012). https://doi.org/10.5402/2012/167804.

110. Wahid, A., Farooq, M., Basra, S., Rasul, E., Siddique, K., Germination of seeds and propagules under salt stress, In: M. Pessarakli (Ed.), Handbook of Plant and Crop Stress., CRC Press, 2010: pp. 321–337. https://doi.org/10.1201/b10329-16.

111. Hernandez, A.C., Carballo, C.A., Artola, A., Michtchenko, A., Laser irradiation effects on maize seed field performance, Seed Sci. Technol. 34 (2006) 193–197.

112. Hartmann, K.M., Grundy, A.C., Market, R., Phytochrome-mediated long-term memory of seeds, Protoplasma. 227 (2005) 47–52.

113. Samiya, S. Aftab, Younus, A., Effect of low power laser irradiation on bio-physical properties of wheat seeds, Inf. Process. Agric. (2019) S221431731930099X. https://doi.org/10.1016/j.inpa.2019.12.003.

114. Costilla-Hermosillo, M.G., Ortiz-Morales, M., Loza-Cornejo, S., Frausto-Reyes, C., Ali Metwally, S. Laser biostimulation for improving seeds germinative capacity and seedlings growth of Prosopis laevigata and Jacaranda mimosifolia, Madera Bosques. 25 (2019). https://doi.org/10.21829/myb.2019.2521665.

115. Hernaacute ndez, A.C., Rodriacute guez, P. ez C.L., Domiacute nguez, P.F.A., a Hernaacute ndez, A.A.M., Cruz Orea, A., Carballo, C.A., Laser light on the mycoflora content in maize seeds, Afr. J. Biotechnol. 10 (2011) 9280–9288.

116. Harun, S.N., Hanafiah, M.M., Estimating the country-level water consumption footprint of selected crop production, Appl. Ecol. Environ. Res. 16 (2018) 5381–5403.

117. Hasan, M., Effect of rhizobium inoculation with phosphorus and nitrogen fertilizer on physico-chemical properties of the groundnut soil, Environ. Ecosyst. Sci. 2 (2018) 4–6.

118. Smalley, P.J., Laser safety: risks, hazards, and control measures, Laser Ther. 20 (2011) 95–106.

119. Pietruszewski, S.T., Effect of magnetic seed treatment on yields of wheat, Seed Sci. Technol. 21 (1993) 621–626.

120. Yao, Y., Li, Y., Yang, Y., Li, C. Effect of seed pretreatment by magnetic field on the sensitivity of cucumber (Cucumis sativus) seedlings to ultraviolet-B radiation, Environ. Exp. Bot. 54 (2005) 286–294.

121. Khalifa, N., Ghandoor, H., Investigate the Effect of Nd-Yag Laser Beam on Soybean (Glycin max) Leaves at the Protein Level, Int. J. Biol. 3 (n.d.) p135.

122. Joshi, S., Joshi, G.C., Agrawal, H.M., Study on the effect of laser irradiation on wheat (*Triticum aestivum* L.) variety PBW-373 seeds on zinc uptake by wheat plants, J. Radioanal. Nucl. Chem. 294 (2012) 391–394.

123. Redlarski, G., Lewczuk, B., A. Żak, Koncicki, A., Krawczuk, M., Piechocki, J., Jakubiuk, K., Tojza, P., Jaworski, J., Ambroziak, D., Skarbek, Ł., Gradolewski, D., The influence of electromagnetic pollution on living organisms: historical trends and forecasting changes, BioMed Res. Int. 2015 (2015).

124. Afzal, M., Manso, S., Effect of mobile phone radiations on morphological and biochemical parameters of mung bean (Vigna radiata) and wheat (Triticum aestivum) Seedlings, Asian J. Agric. Sci. (2012). https://agris.fao.org/agris-search/search.do? recordID=DJ2012074362 (accessed July 10, 2020).

125. Soran, M.-L., Stan, M., Niinemets, Ü., Copolovici, L., Influence of microwave frequency electromagnetic radiation on terpene emission and content in aromatic plants, J. Plant Physiol. 171 (2014) 1436–1443.

126. Dymek, K., Dejmek, P., Panarese, V., Vicente, A.A., Wadsö, L., Finnie, C., Galindo, F.G., Effect of pulsed electric field on the germination of barley seeds, LWT—Food Sci. Technol. 47 (2012) 161–166.

127. Stratton, J.A., Electromagnetic theory, McGrow-Hill Book Company, Inc N. Y. Lond. (1941).

128. Kaur, G., Nelson, K., Motavalli, P., Early-season soil waterlogging and N fertilizer sources impacts on corn N uptake and apparent N recovery efficiency, Agronomy. 8 (2018) 102.

129. Munns, R., Day, D.A., Fricke, W., Watt, M., Arsova, B., Barkla, B.J., Bose, J., Byrt, C.S., Chen, Z.-H., Foster, K.J., Gilliham, M., Henderson, S.W., Jenkins, C.L.D., Kronzucker, H.J., Miklavcic, S.J., Plett, D., Roy, S.J., Shabala, S., Shelden, M.C., Soole, K.L., Taylor, N.L., Tester, M., Wege, S., Wegner, L.H., Tyerman, S.D., Energy costs of salt tolerance in crop plants, New Phytol. 225 (2020) 1072–1090.

130. Thakur, M., Sharma, A.D., Salt-stress-induced proline accumulation in germinating embryos: Evidence suggesting a role of proline in seed germination, J. Arid Environ. 62 (2005) 517–523.

131. Vorasoot, N., Songsri, P., Akkasaeng, C., Jogloy, S., Patanothai, A., Effect of water stress on yield and agronomic characters of peanut (*Arachis hypogaea* L.), Songklanakarin J Sci Technol. 25 (2003) 283–288.

132. Acosta-Motos, J., Ortuño, M., Bernal-Vicente, A., Diaz-Vivancos, P., Sanchez-Blanco, M., Hernandez, J., Plant responses to salt stress: adaptive mechanisms, Agronomy. 7 (2017) 18.

133. Elkelish, A.A., Soliman, M.H., Alhaithloul, H.A., El-Esawi, M.A., Selenium protects wheat seedlings against salt stress-mediated oxidative damage by up-regulating antioxidants and osmolytes metabolism, Plant Physiol. Biochem. PPB. 137 (2019) 144–153.

134. Munns, R., Tester, M., Mechanisms of salinity tolerance, Annu. Rev. Plant Biol. 59 (2008) 651–681.

135. Yadav, S.S., Redden, R., Hatfield, J.L., Lotze-Campen, H., Hall, A.J., Crop adaptation to climate change, John Wiley & Sons, 2011.

136. Dinoev, S., Laser: a controlled assistant in agriculture, Probl. Eng Cybern. Robot. 56 (2006) 86–91.

137. Vasilevski, G., Perspectives of the application of biophysical methods in sustainable agriculture, Bulg. J. Plant Physiol. 29 (2003) 179–186.

138. Ferdosizadeh, L., Sadat-Noori, S.A., Zare, N., Saghafi, S., Assessment laser pretreatments on germination and yield of wheat (*Triticum aestivum* L.) under Salinity stress, World J. Agric. Res. 1 (2013) 5–9.

139. Mohammadi, S., Shekari, F., Fotovat, R., Darudi, A., Effect of laser priming on canola yield and its components under salt stress, Int. Agrophysics. 26 (2012) 45–51.

140. Raffa, R.B., Porreca, F., Thermodynamic analysis of the drug-receptor interaction, Life Sci. 44 (1989) 245–258.
141. Wu, J., Gao, X., Zhang, S., Effect of laser pretreatment on germination and membrane lipid peroxidation of Chinese pine seeds under drought stress, Front. Biol. China. 2 (2007) 314–317.
142. Qiu, Z.-B., X. Liu, Tian, X.-J., M. Yue, Effects of CO_2 laser pretreatment on drought stress resistance in wheat, J. Photochem. Photobiol. B. 90 (2008) 17–25.
143. Chen, Y.-P., Yue, M., Wang, X.-L. Influence of He–Ne laser irradiation on seeds thermodynamic parameters and seedlings growth of Isatis indogotica, Plant Sci. 168 (2005) 601–606.
144. Shelden, M.C., Gilbert, S.E., Tyerman, S.D., A laser ablation technique maps differences in elemental composition in roots of two barley cultivars subjected to salinity stress, Plant J. 101 (2020) 1462–1473.
145. Ćwintal, M., Dziwulska-Hunek, A., Wilczek, M., Laser stimulation effect of seeds on quality of alfalfa, Int. Agrophysics. 24 (2010) 15–19.
146. Anjum, S.A., Ashraf, U., Tanveer, M., Khan, I., Hussain, S., Zohaib, A., Abbas, F., Saleem, M.F., Wang, L., Drought tolerance in three maize cultivars is related to differential osmolyte accumulation, antioxidant defense system, and oxidative damage, Front. Plant Sci. 8 (2017) 69.
147. Bano, A., Ullah, F., Nosheen, A., Role of abscisic acid and drought stress on the activities of antioxidant enzymes in wheat, Plant Soil Env. 58 (2012) 181–185.
148. Niu, J., Ahmad Anjum, S., Wang, R., Li, J., Liu, M., Song, J., Zohaib, A., Lv, J., Wang, S., Zong, X., Exogenous application of brassinolide can alter morphological and physiological traits of Leymus chinensis (Trin.) Tzvelev under room and high temperatures, Chil. J. Agric. Res. 76 (2016) 27–33.
149. Shao, H.-B., Chu, L.-Y., Jaleel, C.A., Manivannan, P., Panneerselvam, R., Shao, M.-A. Understanding water deficit stress-induced changes in the basic metabolism of higher plants—biotechnologically and sustainably improving agriculture and the ecoenvironment in arid regions of the globe, Crit. Rev. Biotechnol. 29 (2009) 131–151.
150. Zandalinas, S.I., Mittler, R., Balfagón, D., Arbona, V., Gómez-Cadenas, A. Plant adaptations to the combination of drought and high temperatures, Physiol. Plant. 162 (2018) 2–12.
151. Apel, K., Hirt, H., Reactive oxygen species: metabolism, oxidative stress, and signal transduction, Annu. Rev. Plant Biol. 55 (2004) 373–399.
152. Asada, K., Production and scavenging of reactive oxygen species in chloroplasts and their functions, Plant Physiol. 141 (2006) 391–396.
153. Waraich, E.A., Ahmad, R., Ashraf, M.Y., Saifullah, Ahmad, M., Improving agricultural water use efficiency by nutrient management in crop plants, Acta Agric. Scand. Sect. B-Soil Plant Sci. 61 (2011) 291–304.
154. Chen, Y.-P., Jia, J.-F., Yue, M. Effect of CO_2 laser radiation on physiological tolerance of wheat seedlings exposed to chilling stress, Photochem. Photobiol. 86 (2010) 600–605.
155. Qiu, Z., Yuan, M., He, Y., Li, Y., Zhang, L., Physiological and transcriptome analysis of He-Ne laser pretreated wheat seedlings in response to drought stress, Sci. Rep. 7 (2017) 6108.
156. Chen, H., Han, R. He-Ne laser treatment improves the photosynthetic efficiency of wheat exposed to enhanced UV-B radiation, Laser Phys. 24 (2014) 105602.
157. Gao, L.-M., Li, Y.-F., Han, R. He-Ne laser preillumination improves the resistance of tall fescue (Festuca arundinacea Schreb.) seedlings to high saline conditions, Protoplasma. 252 (2015) 1135–1148.

158. Nasiri, A.A., Mortazaeinezhad, F., Taheri, R., Seed germination of medicinal sage is affected by gibberellic acid, magnetic field and laser irradiation, Electromagn. Biol. Med. 37 (2018) 50–56.

159. Ružič, R., Jerman, I., Weak magnetic field decreases heat stress in cress seedlings, Electromagn. Biol. Med. 21 (2002) 69–80.

160. Xi, G., Fu, Z.D., Ling, J., Change of peroxidase activity in wheat seedlings induced by magnetic field and its response under dehydration condition, Acta Bot Sin. 36 (1994) 113–118.

161. Piacentini, M.P., Fraternale, D., Piatti, E., Ricci, D., Vetrano, F., Dachà, M., Accorsi, A., Senescence delay and change of antioxidant enzyme levels in *Cucumis sativus* L. etiolated seedlings by ELF magnetic fields, Plant Sci. 161 (2001) 45–53.

162. Elkelish, A.A., Alhaithloul, H.A.S., Qari, S.H., Soliman, M.H., Hasanuzzaman, M., Pretreatment with *Trichoderma harzianum* alleviates waterlogging-induced growth alterations in tomato seedlings by modulating physiological, biochemical, and molecular mechanisms, Environ. Exp. Bot. 171 (2020) 103946.

163. Kim, Y.-H, Hwang, S.-J., Waqas, M., Khan, A.L., Lee, J.-H, Lee, J.-D, Nguyen, H.T., Lee, I.-J, Comparative analysis of endogenous hormones level in two soybean (*Glycine max* L.) lines differing in waterlogging tolerance, Front. Plant Sci. 6 (2015).

164. Milly, P.C.D., Wetherald, R.T., Dunne, K.A., Delworth, T.L., Increasing risk of great floods in a changing climate, Nature. 415 (2002) 514.

165. Wright, A.J., Kroon, H., Visser, E.J., Buchmann, T., Ebeling, A., Eisenhauer, N., Fischer, C., Hildebrandt, A., Ravenek, J., Roscher, C., Plants are less negatively affected by flooding when growing in species-rich plant communities, New Phytol. 213 (2017) 645–656.

166. Agarwal, S., Grover, A., Molecular biology, biotechnology and genomics of flooding-associated low O_2 stress response in plants, Crit. Rev. Plant Sci. 25 (2006) 1–21.

167. Jackson, M.B., Colmer, T.D., Response and adaptation by plants to flooding stress, Ann. Bot. 96 (2005) 501–505.

168. Kozdrój, J., J.D. van Elsas, Response of the bacterial community to root exudates in soil polluted with heavy metals assessed by molecular and cultural approaches, Soil Biol. Biochem. 32 (2000) 1405–1417.

169. Paul, M.V., Iyer, S., Amerhauser, C., Lehmann, M., J.T. van Dongen, Geigenberger, P., oxygen sensing via the ethylene response transcription factor RAP2.12 affects plant metabolism and performance under both normoxia and hypoxia, Plant Physiol. 172 (2016) 141–153.

170. Osakabe, Y., Osakabe, K., Shinozaki, K., L.-S.P. Tran, Response of plants to water stress, Front. Plant Sci. 5 (2014) 86.

171. Pedersen, O., Perata, P., Voesenek, L.A.C.J., Flooding and low oxygen responses in plants, Funct. Plant Biol. 44 (2017) iii–vi.

172. Chen, T., Yuan, F., Song, J., Wang, B., Nitric oxide participates in waterlogging tolerance through enhanced adventitious root formation in the euhalophyte Suaeda salsa, Funct. Plant Biol. 43 (2016) 244.

173. Ravanbakhsh, M., Sasidharan, R., Voesenek, L.A., Kowalchuk, G.A., Jousse, A.t, ACC deaminase-producing rhizosphere bacteria modulate plant responses to flooding, J. Ecol. 105 (2017) 979–986.

174. Striker, G.G., Colmer, T.D., Flooding tolerance of forage legumes, J. Exp. Bot. 68 (2017) 1851–1872.

175. Song, J., W. Shi, R. Liu, Y. Xu, N. Sui, Zhou, J., Feng, G., The role of the seed coat in adaptation of dimorphic seeds of the euhalophyte Suaeda salsa to salinity, Plant Species Biol. 32 (2017) 107–114.

176. Carter, J.L., Colmer, T.D., Veneklaas, E.J., Variable tolerance of wetland tree species to combined salinity and waterlogging is related to regulation of ion uptake and production of organic solutes, New Phytol. 169 (2006) 123–134.

177. Colmer, T.D., Flowers, T.J., Flooding tolerance in halophytes, New Phytol. 179 (2008) 964–974.

178. J. Bailey-Serres, Voesenek, L.A.C.J., Flooding stress: acclimations and genetic diversity, Annu. Rev. Plant Biol. 59 (2008) 313–339.

179. Li, X., Wei, J.-P, Scott, E., Liu, J.-W, Guo, S., Li, Y., Zhang, L., Han, W.-Y, Exogenous melatonin alleviates cold stress by promoting antioxidant defense and redox homeostasis in *Camellia sinensis* L., Molecules. 23 (2018) 165.

180. Colmer, T.D., Voesenek, L., Flooding tolerance: suites of plant traits in variable environments, Funct. Plant Biol. 36 (2009) 665–681.

181. Ezin, V., Pena, R.D.L., Ahanchede, A., Flooding tolerance of tomato genotypes during vegetative and reproductive stages, Braz. J. Plant Physiol. 22 (2010) 131–142.

182. Herzog, M., Striker, G.G., Colmer, T.D., Pedersen, O., Mechanisms of waterlogging tolerance in wheat–a review of root and shoot physiology, Plant Cell Environ. 39 (2016) 1068–1086.

183. Terazawa, K., Maruyama, Y., Morikawa, Y., Photosynthetic and stomatal responses of Larix kaempferi seedlings to short-term waterlogging, Ecol. Res. 7 (1992) 193–197.

184. Valliyodan, B., H. Ye, Song, L., Murphy, M., Shannon, J.G., Nguyen, H.T., Genetic diversity and genomic strategies for improving drought and waterlogging tolerance in soybeans, J. Exp. Bot. (2016) erw433.

185. Gibbs, J., Greenway, H., Review: Mechanisms of anoxia tolerance in plants. I. Growth, survival and anaerobic catabolism, Funct. Plant Biol. 30 (2003) 1–47.

186. Lenochová, Z., Soukup, A., Votrubová, O., Aerenchyma formation in maize roots, Biol. Plant. 53 (2009) 263–270.

187. Sairam, R.K., Kumutha, D., Ezhilmathi, K., Chinnusamy, V., Meena, R.C., Waterlogging induced oxidative stress and antioxidant enzyme activities in pigeon pea, Biol. Plant. 53 (2009) 493–504.

188. Zhong, Z., Furuya, T., Ueno, K., Yamaguchi, H., Hitachi, K., Tsuchida, K., Tani, M., Tian, J., Komatsu, S., Proteomic analysis of irradiation with millimeter waves on soybean growth under flooding conditions, Int. J. Mol. Sci. 21 (2020).

189. Gómez, L.D., Gilday, A., Feil, R., Lunn, J.E., Graham, I.A., AtTPS1-mediated trehalose 6-phosphate synthesis is essential for embryogenic and vegetative growth and responsiveness to ABA in germinating seeds and stomatal guard cells, Plant J. 64 (2010) 1–13.

190. Grennan, A.K., The role of trehalose biosynthesis in plants, Plant Physiol. 144 (2007) 3–5.

191. Schluepmann, H., Pellny, T., van Dijken, A., Smeekens, S., Paul, M., Trehalose 6-phosphate is indispensable for carbohydrate utilization and growth in Arabidopsis thaliana, Proc. Natl. Acad. Sci. U. S. A. 100 (2003) 6849–6854.

192. Li, C., Chen, M., Ji, M., Wang, X., Xiao, W., Li, L., Gao, D., Chen, X., Li, D. Transcriptome analysis of ripe peach (Prunus persica) fruit under low-dose UVB radiation, Sci. Hortic. 259 (2020) 108757.

193. Seger, M., Gebril, S., Tabilona, J., Peel, A., C. Sengupta-Gopalan, Impact of concurrent overexpression of cytosolic glutamine synthetase (GS 1) and sucrose phosphate synthase (SPS) on growth and development in transgenic tobacco, Planta. 241 (2015) 69–81.

194. Sun, J., Gu, J., Zeng, J., Han, S., Song, A., Chen, F., Fang, W., Jiang, J., Chen, S., Changes in leaf morphology, antioxidant activity and photosynthesis capacity in two different drought-tolerant cultivars of chrysanthemum during and after water stress, Sci. Hortic. 161 (2013) 249–258.

195. Balakhnina, T.I., Magnetic fields, temperature, and exogenous selenium effect on reactive oxygen species metabolism of plants under flooding and metal toxicity, In: M. Hasanuzzaman (Ed.), Plant Ecophysiology and Adaptation under Climate Change: Mechanisms and Perspectives, Springer, Singapore, 2020: pp. 443–475.

196. Vanhaelewyn, L., Prinsen, E., Van Der Straeten, D., Vandenbussche, F., Hormone-controlled UV-B responses in plants, J. Exp. Bot. 67 (2016) 4469–4482.

197. Abu-Elsaoud, A.M., Abdel-Azeem, A.M., Light, electromagnetic spectrum, and photostimulation of microorganisms with special reference to chaetomium, In: Abdel-Azeem, A.M. (Ed.), Recent Developments on Genus Chaetomium, Springer International Publishing, Cham, 2020: pp. 377–393.

198. Azad, M., Kalam, A., Sarker, M., Li, T., J. Yin, Probiotic species in the modulation of gut microbiota: an overview, BioMed Res. Int. 2018 (2018).

199. Salyaev, R.K., Dudareva, L.V., Lankevich, S.V., Makarenko, S.P., Sumtsova, V.M., Rudikovskaya, E.G., Effect of low-intensity laser irradiation on the chemical composition and structure of lipids in wheat tissue culture, In: Doklady Biological Sciences, Springer Nature BV, 2007: p. 87.

200. Funk, J.L., Larson, J.E., Vose, G., Leaf traits and performance vary with plant age and water availability in *Artemisia californica*, Ann. Bot. 127 (2021) 495–503.

201. Yaish, M.W., Antony, I., Glick, B.R., Isolation and characterization of endophytic plant growth-promoting bacteria from date palm tree (*Phoenix dactylifera* L.) and their potential role in salinity tolerance, Antonie Van Leeuwenhoek. 107 (2015) 1519–1532.

202. Delauney, A.J., Verma, D.P.S., Proline biosynthesis and osmoregulation in plants, Plant J. 4 (1993) 215–223.

203. Hare, P.D., Cress, W.A., Metabolic implications of stress-induced proline accumulation in plants, Plant Growth Regul. 21 (1997) 79–102.

204. B.P. O'regan, Cress, W.A., J. Van Staden, Root growth, water relations, abscisic acid and proline levels of drought-resistant and drought-sensitive maize cultivars in response to water stress, South Afr. J. Bot. 59 (1993) 98–104.

205. Ahanger, M.A., F. Gul, Ahmad, P., Akram, N.A., Environmental stresses and metabolomics—deciphering the role of stress responsive metabolites, In: Plant Metabolites and Regulation under Environmental Stress, 2018 53–67.

206. Kumar, N., Pamidimarri, S., Kaur, M., Boricha, G., Reddy, M., Effects of NaCl on growth, ion accumulation, protein, proline contents and antioxidant enzymes activity in callus cultures of *Jatropha curcas*, Biologia (Bratisl.). 63 (2008) 378–382.

207. Parida, A.K., Das, A.B., Salt tolerance and salinity effects on plants: a review, Ecotoxicol. Environ. Saf. 60 (2005) 324–349.

208. Kao, W.-Y, Tsai, T.-T, Shih, C.-N, Photosynthetic gas exchange and chlorophyll a fluorescence of three wild soybean species in response to NaCl treatments, Photosynthetica. 41 (2003) 415–419.

209. Saleem, M.H., S. Ali, Rehman, M., Rana, M.S., Rizwan, M., Kamran, M., Imran, M., Riaz, M., Soliman, M.H., Elkelish, A., others, Influence of phosphorus on copper phytoextraction via modulating cellular organelles in two jute (*Corchorus capsularis* L.) varieties grown in a copper mining soil of Hubei Province, China, Chemosphere. 248 (2020) 126032.

210. Zaheer, I.E., Ali, S., Saleem, M.H., Imran, M., Alnusairi, G.S.H., Alharbi, B.M., Riaz, M., Abbas, Z., Rizwan, M., Soliman, M.H., Role of iron–lysine on morpho-physiological traits and combating chromium toxicity in rapeseed (*Brassica napus* L.) plants irrigated with different levels of tannery wastewater, Plant Physiol. Biochem. 155 (2020) 70–84.

211. Usamentiaga, R., Venegas, P., Guerediaga, J., Vega, L., Molleda, J., Bulnes, F.G., Infrared thermography for temperature measurement and non-destructive testing, Sensors. 14 (2014) 12305–12348.

212. Chaves, M.M., Flexas, J., Pinheiro, C., Photosynthesis under drought and salt stress: regulation mechanisms from whole plant to cell, Ann. Bot. 103 (2009) 551–560.

213. Kwon, M.Y., Woo, S.Y., Plants' responses to drought and shade environments, Afr. J. Biotechnol. 15 (2016) 29–31.

214. Flexas, J., Barón, M., Bota, J., Ducruet, J.-M, Gallé, A., Galmés, J., Jiménez, M., A. Pou, M. Ribas-Carbó, Sajnani, C., Photosynthesis limitations during water stress acclimation and recovery in the drought-adapted Vitis hybrid Richter-110 (V. berlandieri$\times$ V. rupestris), J. Exp. Bot. 60 (2009) 2361–2377.

215. Lemmens, L., Tibäck, E., Svelander, C., Smout, C., Ahrné, L., Langton, M., Alminger, M., A. Van Loey, Hendrickx, M., Thermal pretreatments of carrot pieces using different heating techniques: Effect on quality related aspects, Innov. Food Sci. Emerg. Technol. 10 (2009) 522–529.

216. Z. Pan, Atungulu, G.G., X. Li, Infrared heating, In: Emerging Technologies for Food Processing, Elsevier, 2014: pp. 461–474.

217. Sproul, J.S., Rau, E. Process for producing sodium carbonate from trona, 1975.

218. Deng, L.-Z, Mujumdar, A.S., Zhang, Q., Yang, X.-H, Wang, J., Zheng, Z.-A, Gao, Z.-J, H.-W. Xiao, Chemical and physical pretreatments of fruits and vegetables: Effects on drying characteristics and quality attributes—a comprehensive review, Crit. Rev. Food Sci. Nutr. 59 (2019) 1408–1432.

219. James, A.T., Lawn, R.J., Cooper, M., Genotypic variation for drought stress response traits in soybean. II. Inter-relations between epidermal conductance, osmotic potential, relative water content, and plant survival, Aust. J. Agric. Res. 59 (2008) 670–678.

220. Alvarez, R.A., Pacala, S.W., Winebrake, J.J., Chameides, W.L., Hamburg, S.P., Greater focus needed on methane leakage from natural gas infrastructure, Proc. Natl. Acad. Sci. 109 (2012) 6435–6440.

221. Abdulmajeed, A.M., Qaderi, M.M., Differential effects of environmental stressors on physiological processes and methane emissions in pea (Pisum sativum) plants at various growth stages, Plant Physiol. Biochem. 139 (2019) 715–723.

222. Harte, J. Ecological feedbacks to global warming: extending results from plot to landscape scale. In Elements of Change: roceedings of the 1997 Aspen Global Change Institute Workshop on Scaling from Site-specific Observations to Global Model Grids. Aspen Global Change Institute, Aspen, Colorado, USA, pp. 69–74.

223. Horn, R., Schlögl, R., Methane activation by heterogeneous catalysis, Catal. Lett. 145 (2015) 23–39.

224. Torn, M.S., Harte, J., Methane consumption by montane soils: implications for positive and negative feedback with climatic change, Biogeochemistry. 32 (1996) 53–67.
225. Wang, G.-L, Ren, X.-Q, Liu, J.-X, Yang, F., Wang, Y.-P, Xiong, A.-S, Transcript profiling reveals an important role of cell wall remodeling and hormone signaling under salt stress in garlic, Plant Physiol. Biochem. 135 (2019) 87–98.

Index

For Product Safety Concerns and Information please contact our EU
representative GPSR@taylorandfrancis.com
Taylor & Francis Verlag GmbH, Kaufingerstraße 24, 80331 München, Germany

www.ingramcontent.com/pod-product-compliance
Lightning Source LLC
Chambersburg PA
CBHW060747220326
41598CB00022B/2355